Parallel Coupled Lines and Directional Couplers

Leo Young

ARTECH HOUSE, INC.
Dedham, Massachusetts
affiliated with
Horizon House-Microwave Inc.

Printed and bound in the United States of America.

Library of Congress catalog card number: 76-168946

PARALLEL COUPLED LINES AND DIRECTIONAL COUPLERS

This book of reprints is one of two companion volumes; the ten papers in the first three sections of the present volume (comprising about one-third of the book) are also basic to the companion volume on MICROWAVE FILTERS USING PARALLEL COUPLED LINES (such as parallel-coupled filters, interdigital filters and combline filters). It is hoped to publish other small volumes of reprints on microwave topics as the opportunity arises.

INTRODUCTION

There is a large bibliography on coupled lines, as well as on directional couplers. In order to keep this book within reasonable bounds, it was decided to reprint only those papers that are frequently referenced, or are of basic importance, or that contain generally useful design information. Some reprints are also of historical interest in the development of our subject.

EARLY HISTORY OF PARALLEL–LINE COUPLERS

The exact beginnings of directional couplers using parallel wires are not clear. Such couplers were used for various applications before any of the modern theories or design data were available. Most of the applications were well below the microwave band. We are indebted to Mr. Harold A. Wheeler for supplying the information in this section.

U.S. Patent 1,615,896, entitled "High Frequency Signaling System" was filed in 1922 and granted in 1927 to Herman A. Affel,[1] and assigned to American Telephone and Telegraph Company. Affel refers to a "loop antenna", which in his Fig. 6 consists of a two-wire transmission line one-quarter-wavelength long, with a resistive termination at one end and a detector at the other end. This quarter-wave line (or "loop antenna," as it was called) is parallel to another longer two-wire transmission line, and could act like a directional coupler at midband if properly designed. However, it is not explicitly referred to as a directional coupler.

A similar quarter-wave backward wave coupler is attributed by Wheeler to Pistolkors and dated 1941.

U.S. Patent 2,606,974, entitled "Directional Coupler" was filed in 1946, and granted in 1952 to Harold A. Wheeler,[2] being assigned to Hazeltine Research, Inc. The cover page is reproduced here (with some notes by Wheeler) as Fig. 1. Wheeler states that it was used in 1944 at the Hazeltine laboratory for the Mark V IFF system. It clearly describes a TEM-mode quarter-wave parallel-line backward coupler, and gives formulas for such a coupler.

On the negative side, it is notable that no mention is made of these couplers in the 44-page long chapter on directional couplers (Chap. 14) written by R.L. Kyhl in Vol. II of the M.I.T. Radiation Laboratory[3] series, which was published in 1947.

Thus, we can state with some certainty that these couplers were not well known at the end of World War II. More historical data in the development of the subject in more recent years will be presented in this Foreword as the occasion arises.

BACKGROUND ON DIRECTIONAL COUPLERS

The parallel-line couplers that are the subject of this book are all **backward** couplers. They perform well over very large bandwidths—multi-octave designs are quite usual. Their simplicity of construction combined with wideband performance makes them very attractive. The main drawbacks are that they are limited to TEM lines (and thus cannot handle the power of waveguide couplers). They are backward couplers, which makes it more difficult to cascade couplers in certain forms of construction, since the connecting lines have to cross. However, for most applications these drawbacks are of little significance, and the TEM-mode parallel-line coupler is by far the most popular coupler in use today.

Waveguide couplers are not TEM-mode couplers and are not included in this volume. One type of directional coupler that can be built equally well in waveguide and in TEM line is the branch-line coupler.[4-6] Branch-line couplers are forward couplers (and can therefore be cascaded without crossing lines).

A coupler that uses parallel lines and is a **forward** coupler has been described by Dalley.[7] This coupler requires an inhomogeneous medium and is relatively narrow-band. Its action depends on **unequal phase velocities** for the even and odd modes, which is achieved by using an inhomogeneous dielectric medium. On the other hand the backward coupler that is treated here depends on equal phase velocities but **unequal characteristic impedances** for the even and odd modes.

When only two in-phase outputs are required, and the fourth port need not be accessible, then another kind of coupler is very useful. Here the fourth (difference) port is internally terminated. Such couplers have been described by Parad and Moynihan,[8] Cohn,[9] and (when many output ports are required) by Wilkinson.[10]

For a brief review of other types of coupler than treated in this book and further references, see two review papers by Young;[11,12] for an extensive bibliography see the reprinted paper by Caswell and Schwartz (p. 101).

An excellent account on all kinds of directional couplers up to 1966 will be found in a chapter by Levy [13] in Volume I of the **ADVANCES IN MICROWAVES.**

A great deal of work has been done in Japan, much of which went unnoticed in Western countries—this situation has now been remedied by two books in English, one edited by Matsumoto,[14] and one written by Uchida.[15]

DISCUSSION OF CONTENTS IN THIS VOLUME

With this brief review of couplers generally, we shall now review the contents of this book, and offer some explanations and perspectives as we go along.

Sections I, II, and III, as already mentioned, contain basic papers in the sense that they provide the link between the electrical design parameters and the dimensions of various physical structures.

Section IV begins the treatment of directional couplers with a bibliography, followed by some general papers in Section V. Stepped parallel-line couplers are treated in Sections VI and VII, the former dealing with symmetrical couplers (which are useful mostly for 3-dB quadrature hybrids) and the latter dealing with asymmetrical couplers (which provide maximum coupling bandwidth for a given number of sections at the expense of phase control, and are thus useful mostly for non-3-dB couplers). Section VIII concerns tapered coupled lines (useful mostly in making printed-circuit magic T's). Section IX is on slot line, and as will be explained below, this data can also be used for slot-line directional couplers (as well as some slot-line filters). Some specific comments appear on the title page of each section. The papers within each section are listed in chronological order, the oldest paper coming first, and the newer papers following (except for related papers).

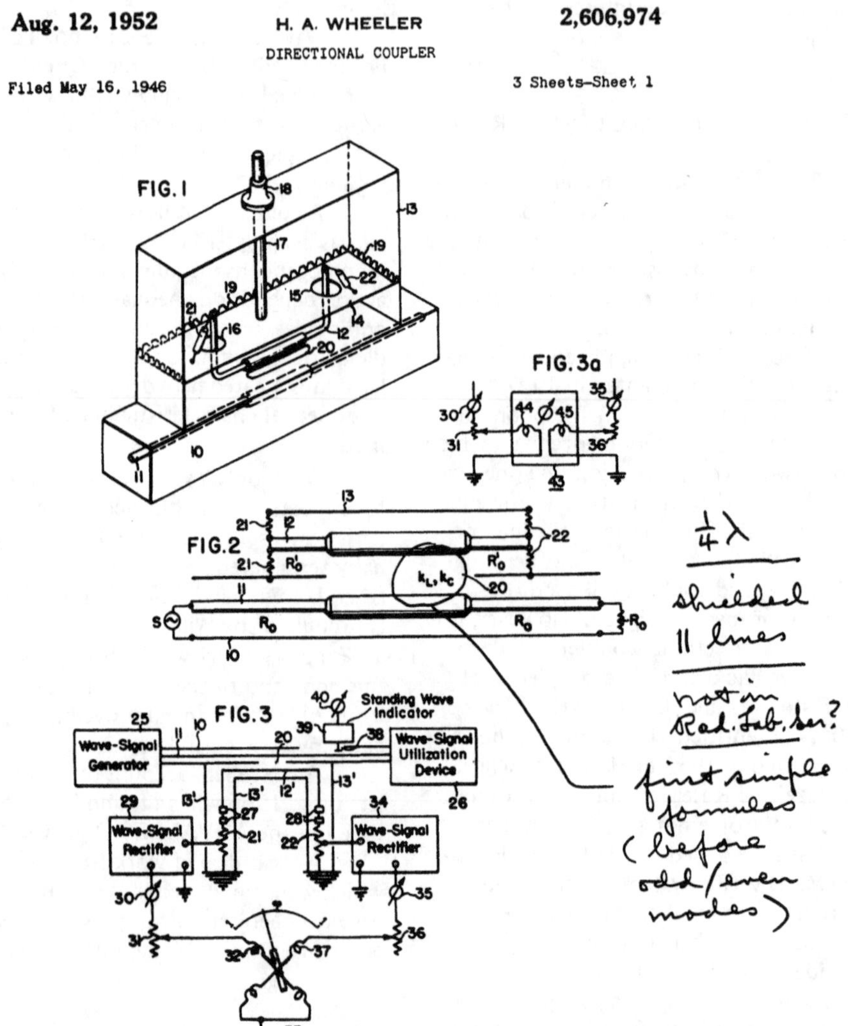

Fig.1 Cover Page from Patent by H. A. Wheeler.

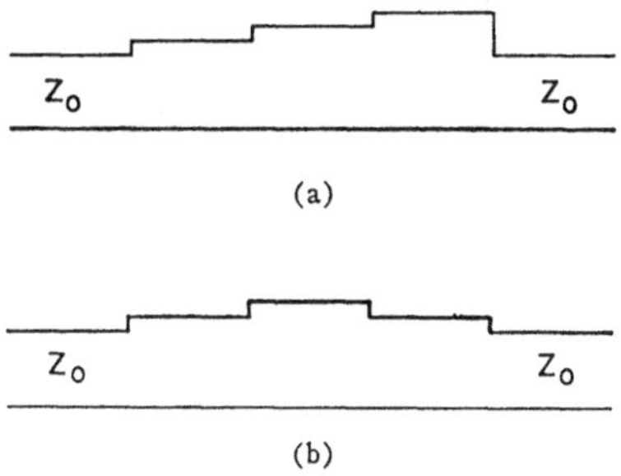

Fig. 38. Stepped-impedance filters that serve as prototypes for TEM-mode backward-wave directional couplers; (a) asymmetrical, and (b) symmetrical.

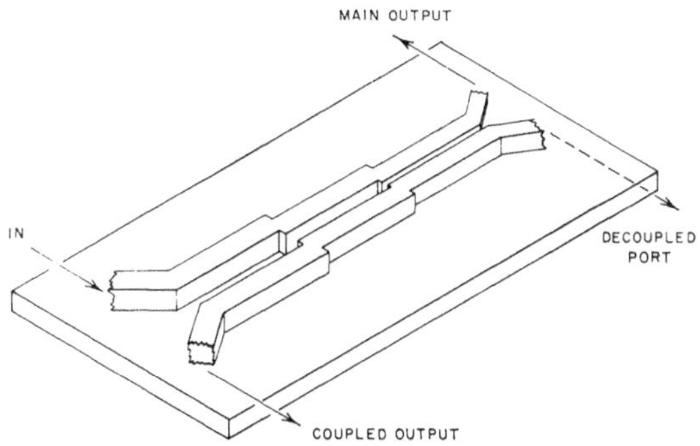

Fig. 39. Typical TEM-mode backward-wave directional coupler in strip line.

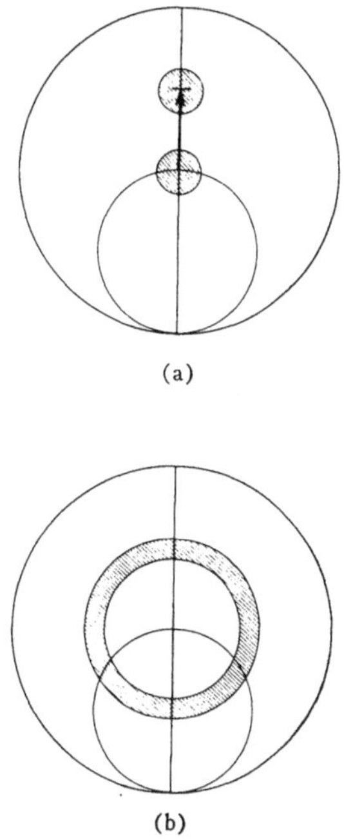

Fig. 40. Diagram showing why the optimum prototype in Fig. 38(a) is not based on a quarter-wave transformer.

Fig. 2 Concerning Prototypes for Optimum Couplers and for Magic-T Couplers.

ACKNOWLEDGMENTS

We wish to express our appreciation to all the authors and publishers for permission to reprint the papers in this volume. The great majority of these papers originated in publications of the IEEE. One frequently quoted paper, by DuHamel and Armstrong, is printed here for the first time in a way that will make it generally accessible. (It was given at a U.S. Air Force Symposium in 1965 and had been available previously only in the proceedings of that Symposium.) Dr. R.H. DuHamel was good enough to rewrite and up-date his paper expecially for this volume, and we are grateful to his for his efforts "beyond the call of duty."

We have also received particularly helpful correspondence or comments from H.J. Carlin, S.B. Cohn, M. Dishal, T. Fjallbrant, W.J. Getsinger, M.C. Horton, J. Ishii, K. Kuroda, R.M. Kurzrok, R. Levy, A. Matsumoto, W.W. Mumford, J. Neirynck, R. Pregla, J.D. Rhodes, H.J. Riblet, B.M. Schiffman, C.P. Tresselt, H. Uchida, R.J. Wenzel and H.A. Wheeler.

It is always a pleasure to acknowledge the assistance made available by Stanford Research Institute. In particular, I am grateful to Mrs. Dianna Tseu for lightening some of my editorial tasks.

Leo Young

TABLE OF CONTENTS

1. H. A. Affel, "High Frequency Signaling System," U.S. Patent 1,615,896, filed 15 December 1922, issued 1 February 1927.
2. H. A. Wheeler, "Directional Coupler," U.S. Patent 2,606,974, filed 16 May 1946, issued 12 August 1952.
17. W. J. Getsinger, "A Coupled Strip-Line Configuration Using Printed-Circuit Construction that Allows Very Close Coupling," **IEEE Trans. on Microwave Theory and Techniques**, Vol. MTT-9, No. 6, pp. 535-544 (November 1961).

REFERENCES

3. R. L. Kyhl, "Directional Couplers," Chapter 14 in **Technique of Microwave Measurements**, edited by C. G. Montgomery, M.I.T. Radiation Laboratory Series, Vol. 11 (McGraw-Hill Book Company, New York, 1947).
4. R. Levy and L. F. Lind, "Synthesis of Symmetrical Branch-Guide Directional Couplers," **IEEE Trans. on Microwave Theory and Techniques**, Vol. MTT-16, No. 2, pp. 80-89 (February 1968).
5. Leo Young, "Synchronous Branch-Guide Directional Couplers for Low and High Power Application," **IEEE Trans. on Microwave Theory and Techniques**, Vol. MTT-10, No. 6, pp. 459-475 (November 1962).
6. G. L. Matthaei, Leo Young, and E. M. T. Jones, **Microwave Filters, Impedance Matching Networks, and Coupling Structures** (McGraw-Hill Book Company, New York, 1964).
7. J. E. Dalley, "A Strip-Line Directional Coupler Utilizing a Non-Homogeneous Dielectric Medium," **IEEE Trans. on Microwave Theory and Techniques**, Vol. MTT-17, No. 9, pp. 706-712 (September 1969).
8. L. I. Parad and R. L. Moynihan, "Split-Tee Power Divider," **IEEE Trans. on Microwave Theory and Techniques**, Vol. MTT-13, No. 1, pp. 91-95 (January 1965).
9. S. B. Cohn, "A Class of Broadband Three-Port TEM-Mode Hybrids," **IEEE Trans. on Microwave Theory and Techniques**, Vol. MTT-16, No. 2, pp. 110-116 (February 1968).
10. E. J. Wilkinson, "An N-Way Hybrid Power Divider," **IEEE Trans. on Microwave Theory and Techniques**, Vol. MTT-8, No. 1, pp. 116-118 (January 1960).
11. Leo Young, "Microwave Filters – 1965," **IEEE Trans. on Microwave Theory and Techniques**, Vol. MTT-13, No. 5, pp. 488-408 (September 1965).
12. Leo Young, "Reflections on Microwave Filters and Couplers," **The Microwave Journal**, Vol. 11, No. 4, pp. 54-63 (April 1968).
13. R. Levy, "Directional Couplers," in **Advances in Microwaves**, Vol. 1, edited by Leo Young (Academic Press, New York, 1966).
14. A. Matsumoto, editor, **Microwave Filters**, Supplement 1 to **Advances in Microwaves** (general editor, L. Young) (Academic Press, New York, 1970).
15. H. Uchida, **Fundamentals of Coupled Lines and Multiwire Antennas**, Sasaki Printing and Publishing Co., Ltd., Sendai, Japan, 1967).
16. R. R. Gupta, "Accurate Impedance Determination of Coupled TEM Conductors," **IEEE Trans. on Microwave Theory and Techniques**, Vol. MTT-17, No. 8, pp. 479-489 (August 1969).
18. W. I. Firestone, "Analysis of Transmission Line Directional Couplers, **Proc. IRE**, Vol. 42, pp. 1529-1538 (October 1954).
19. S. D. Casper and J. E. McFarland, "Multi-Section Asymmetrical Coupler," U.S. Patent 3,113,277, filed 2 May 1960, issued 3 December 1963.
20. T. S. Saad, "Microwave Coupled Line Device Having Insulated Coupled Inner Conductors Within a Common Outer Conductor," U.S. Patent 3,358,248, filed 22 July 1964, issued 12 December 1967.
21. J. P. Shelton and J. A. Mosko, "Synthesis and Design of Wide-Band Equal-Ripple TEM Directional Couplers and Fixed Phase Shifters," **IEEE Trans. on Microwave Theory and Techniques**, Vol. MTT-14, No. 10, pp. 462-473 (October 1966).
22. P. P. Toulios and A. C. Todd, "Synthesis of Symmetrical TEM-Mode Directional Couplers," **IEEE Trans. on Microwave Theory and Techniques**, Vol. MTT-13, No. 5, pp. 536-544 (September 1965).
23. D. I. Kraker, "Asymmetric Coupled Transmission Line Magic T," **IEEE Trans. on Microwave Theory and Techniques**, Vol. MTT-12, No. 6, pp. 595-599 (November 1964).
24. T. E. Rozzi, "Phase-Compensated Asymmetric Directional Couplers for the Design of Decade-Wide Quadrature Hybrids," **Proc. IEE**, Vol. 117, pp. 704-708 (April 1970).
25. G. L. Millican and R. C. Wales, "Practical Strip-Line Microwave Circuit Design, **IEEE Trans. on Microwave Theory and Techniques**, Vol. MTT-17, No. 9, pp. 696-705 (September 1965).
26. D. W. Kammler, "The Design of Discrete N-Section and Continuously Tapered Symmetrical Microwave TEM Directional Couplers," **IEEE Trans. on Microwave Theory and Techniques**, Vol. MTT-17, No. 8, pp. 577-590 (August 1969).
27. F. C. deRonde, "A New Class of Microstrip Directional Couplers," **1970 G-MTT International Microwave Symposium Digest**, IEEE Cat. No. 70 C 10-MTT, pp. 184-189 (May 1970).
28. J. K. Hunton and J. S. Takeuchi, "Recent Developments in Microwave Slot-Line Mixers and Frequency Multipliers," **1970 G-MTT International Microwave Symposium Digest**, IEEE Cat. No. 70 C 10-MTT, pp. 196-199 (May 1970).
29. C. P. Wen, "Coplanar Waveguide Directional Couplers," **IEEE Trans. on Microwave Theory and Techniques**, Vol. MTT-18, No. 6 (June 1970).

COUPLED BARS AND RODS

1

The papers by Getsinger (p. 3) and by Cristal (p. 13) are widely used especially for machined inner conductors in air dielectric. Such transmission lines are used, for example, when printed circuit techniques do not yield high enough Q (low enough loss). The letter by Gupta (p. 11) is an extension to Getsinger's paper. There is another related paper by Gupta[16] in the same issue of the Transactions.

Reprinted from IRE TRANSACTIONS
ON MICROWAVE THEORY AND TECHNIQUES
Volume MTT-10, Number 1, January, 1962

PRINTED IN THE U.S.A.

Coupled Rectangular Bars Between Parallel Plates*

WILLIAM J. GETSINGER†, MEMBER, IRE

Summary—**Curves are presented giving the even-mode fringing capacitance, the odd-mode fringing capacitance, and the difference between odd- and even-mode fringing capacitances for wide ranges of thickness and spacing of rectangular bars centered between parallel plates. Simple formulas are given relating these capacitances to even- and odd-mode characteristic impedances of coupled rectangular bars. Possible applications to strip-line and other circuits are described.**

The Appendix gives the derivation of the fringing capacitances by conformal mapping techniques. The results are exact for bars extending in width infinitely far from the coupling region, and have only small error (less than 1.24 per cent) for bars whose width is greater than about 35 per cent of the difference between plate spacing and bar thickness.

I. General

In working with shielded strip-line, it is sometimes desirable to couple center conductors having appreciable thickness. The cross section of a typical structure of this type is shown in Fig. 1. There are two parallel ground planes spaced a distance b apart, and two rectangular bars located parallel to, and midway between, the ground planes. It is well known[1,2] that TEM propagation along such a structure can be described in terms of two orthogonal modes, usually denoted as the even mode and the odd mode. In the even mode, both center conductors are at the same potential, while in the odd mode, the two center conductors are at opposite potentials with respect to the ground planes. These two TEM modes have different characteristic impedances, which are intimately related to the static capacitances of the bars to ground. These capacitances are designated conventionally as parallel-plate capacitances between bar and ground planes, and fringing capacitances from ends and corners of the bars, as indicated schematically in Fig. 1.

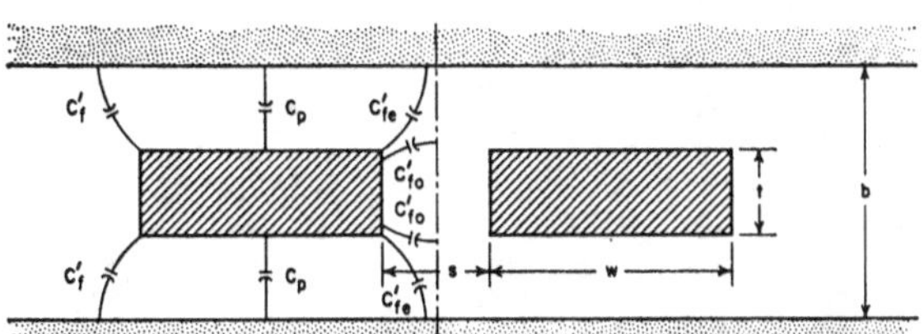

Fig. 1—Coupled rectangular bars centered between parallel plates.

Cohn[3] has previously considered coupled rectangular bars between parallel plates. He gives rigorous results, based on a conformal mapping solution, for the zero- 3
thickness case, and adds correction terms to the zero-thickness case to obtain approximate results for bars of finite thickness. In a later paper,[4] Cohn provides an improved approximation for the odd-mode correction term. Horgan[5] has approached the problem by dividing the cross section of the coupled rectangular bar structure into insulated rectangular boxes, for each of which potential solutions were available. Horgan then corrected these approximations by adding new potential functions such that total potentials were matched across the insulating bounds, and the resulting capacitance was a maximum. Horgan's results, like those of Cohn, were written as exact thin-strip capacitance solutions plus correction terms.

The present paper, however, does not modify a solution for coupled thin strips, but determines the fringing capacitances for bars extending in width infinitely far from the coupling region by an exact conformal mapping method, and presents these fringing capacitances directly on graphs for wide ranges of bar thickness and spacing. The use of accurate graphs, giving the fringing capacitances directly, allows the total capacitance (and thus characteristic impedance, or alternatively, bar width) of any of a variety of structures to be determined quite easily by simply summing the appropriate fringing and parallel plate capacitances, as will be shown subsequently.

II. Technical Description

The characteristic impedance Z_0 of a lossless uniform transmission line operating in the TEM mode is related to its shunt capacitance by

$$Z_0\sqrt{\epsilon_r} = \frac{\eta}{(C/\epsilon)} \text{ ohms} \tag{1}$$

where

$\sqrt{\epsilon_r}$ = the relative dielectric constant of the medium in which the wave travels

η = the impedance of free space = 376.7 ohms per square

* Received by PGMTT, August 23, 1961; revised manuscript received, October 4, 1961. This work was supported by the U. S. Army Signal Res. and Dev. Lab., Fort Monmouth, N. J., as part of Contract DA-36-039 SC-87398.

† Stanford Research Institute, Menlo Park, Calif.

[1] B. M. Oliver, "Directional electromagnetic couplers," Proc. IRE, vol. 42, pp. 1686–1692; November, 1954.

[2] E. M. T. Jones and J. T. Bolljahn, "Coupled-strip-transmission-line filters and directional couplers," IRE Trans. on Microwave Theory and Techniques, vol. MTT-4, pp. 75–81; April, 1956.

[3] S. B. Cohn, "Shielded coupled-strip transmission line," IRE Trans. on Microwave Theory and Techniques, vol. MTT-8, pp. 29–38; October, 1955.

[4] S. B. Cohn, "Thickness corrections for capacitive obstacles and strip conductors," IRE Trans. on Microwave Theory and Techniques, vol. MTT-8, pp. 638–644; November, 1960.

[5] J. D. Horgan, "Coupled strip transmission lines with rectangular inner conductors," IRE Trans. on Microwave Theory and Techniques, vol. MTT-5, pp. 92–99; April, 1957.

C/ϵ = the ratio of the static capacitance per unit length between conductors to the permittivity (in the same units) of the dielectric medium (this ratio is independent of the dielectric constant).

The even- and odd-mode impedances of coupled TEM lines[1,2] can be found by substituting even- and odd-mode capacitances of the lines in (1).

4

A generalized schematic diagram of shielded coupled-strip transmission line is shown in Fig. 2. The circles represent the coupled conductors. The capacitance to ground for a single conductor, when both conductors are at the same potential, is C_{oe}, the even-mode capacitance. The capacitance to ground when the two conductors are oppositely charged with respect to ground is C_{oo}, the odd-mode capacitance.

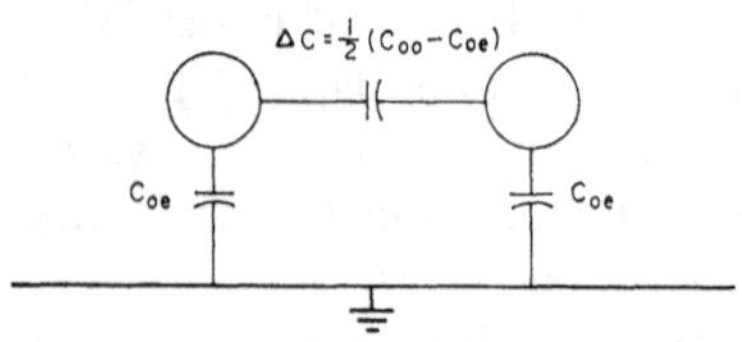

Fig. 2—Generalized schematic diagram.

The structure of Fig. 1 is composed of parallel-planar surfaces. This makes it practical to consider the total capacitance of a given strip to be composed of parallel-plane capacitances plus appropriate fringing capacitances. (Fringing capacitances take into account the distortion of the field lines in the vicinity of the edges of the plane strips.) Fig. 1 relates the various capacitances to the geometry of the structure under consideration. Thus, it can be seen that the total even-mode capacitance C_{oe}/ϵ from one bar to ground is

$$C_{oe}/\epsilon = 2(C_p/\epsilon + C_{fe}'/\epsilon + C_f'/\epsilon) \tag{2}$$

and the total odd-mode capacitance C_{oo}/ϵ from one bar to ground is

$$C_{oo}/\epsilon = 2(C_p/\epsilon + C_{fo}'/\epsilon + C_f'/\epsilon). \tag{3}$$

In (2) and (3), C_p is the parallel-plate capacitance from the top or bottom side of one bar to the nearest ground plane; C_{fe}' is the capacitance to ground from one corner and half the associated vertical wall in the coupling region of a bar for even-mode excitation; C_{fo}' is the capacitance to ground from one corner and half the associated vertical wall in the coupling region of a bar for odd-mode excitation; and C_f' is the capacitance to ground from one corner and half the associated vertical wall away from the coupling region of a bar for any

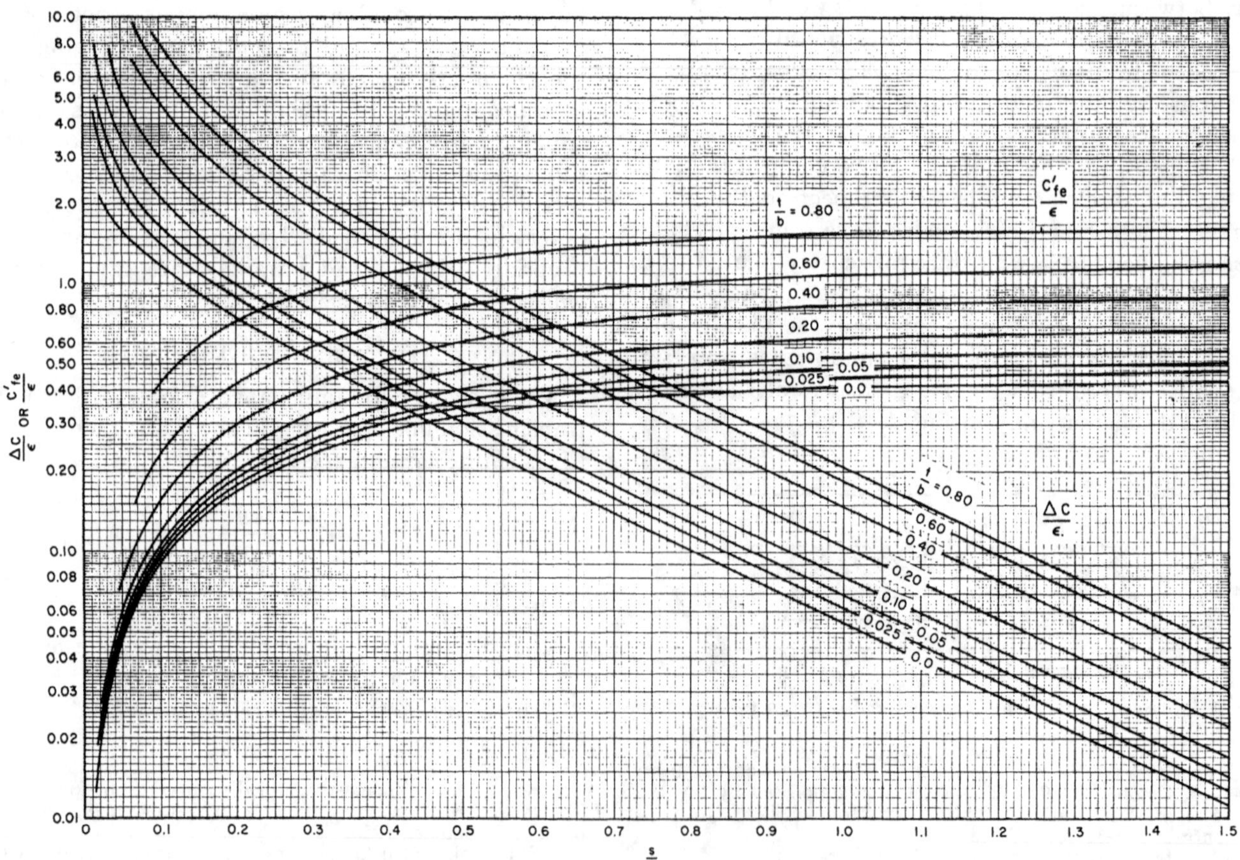

Fig. 3—Fringing capacitances for coupled rectangular bars.

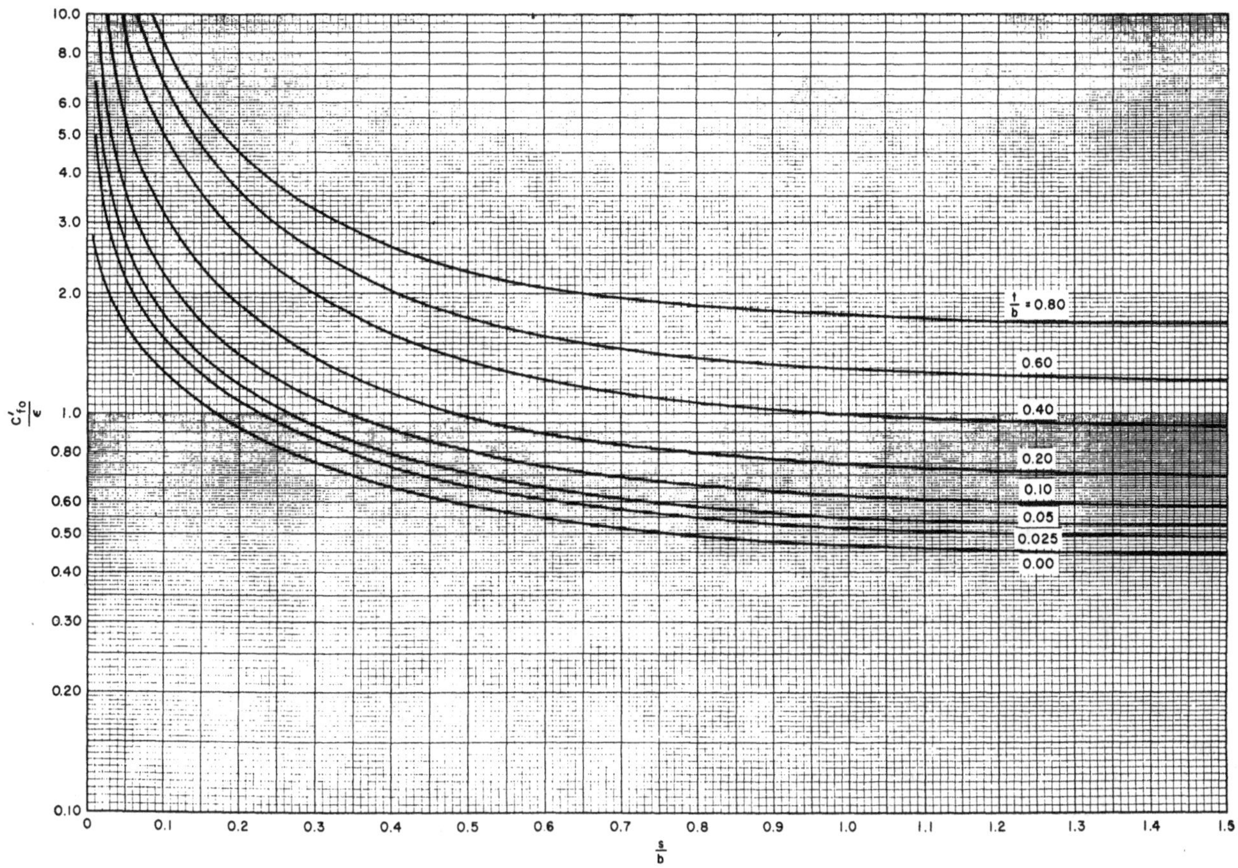

Fig. 4—Odd-mode fringing capacitance for coupled rectangular bars.

excitation. Consideration of Fig. 2 and the definitions of even- and odd-mode capacitances show that the capacitance, $\Delta C/\epsilon$, from one bar to the other is given by

$$\Delta C/\epsilon = \tfrac{1}{2}(C_{oo}/\epsilon - C_{oe}/\epsilon). \tag{4}$$

Subtraction of (2) from (3) shows that $\Delta C/\epsilon$ can be written entirely in terms of the fringing capacitances as

$$\Delta C/\epsilon = C_{fo}'/\epsilon - C_{fe}'/\epsilon. \tag{5}$$

Fig. 3 is a plot of both even-mode fringing capacitances, C_{fe}'/ϵ, and the capacitance, $\Delta C/\epsilon$, between bars as functions of bar thickness and spacing, while Fig. 4 is a similar graph for the odd-mode fringing capacitance, C_{fo}'/ϵ. The derivation of Figs. 3 and 4 is described in the Appendix. Fig. 5 gives the fringing capacitance C_f'/ϵ from the outer edges of the bars as a function of thickness. The parallel plate capacitance C_p/ϵ is given by

$$C_p/\epsilon = 2\,\frac{w/b}{1 - t/b}, \tag{6}$$

where w and t are the width and thickness of the bar. Through the use of the above relations and figures, it is possible to relate physical dimensions of the given configuration to even- and odd-mode capacitances or impedances.

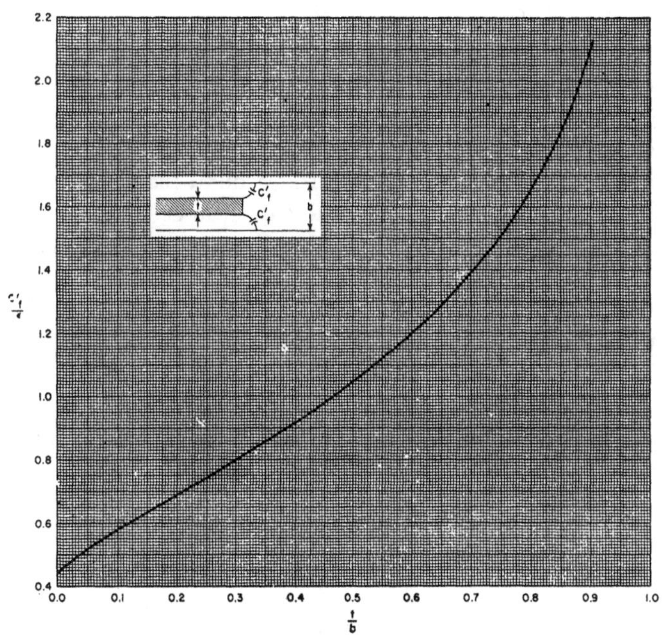

Fig. 5—Fringing capacitance for an isolated rectangular bar.

III. Use of the Graphs

Usually, an engineer designing parallel-coupled lines first determines the values of even- and odd-mode impedances, Z_{oe} and Z_{oo}, or even- and odd-mode capacitances, C_{oe} and C_{oo}, as required by theoretical considerations. He then wishes to determine the corresponding physical line dimensions. A simple procedure accomplishes this. Using (1) in (4) gives

6

$$\Delta C/\epsilon = \frac{\eta}{2\sqrt{\epsilon_r}}\left(\frac{1}{Z_{oo}} - \frac{1}{Z_{oe}}\right). \tag{7}$$

Values of b and t are selected and used with the value of $\Delta C/\epsilon$ found from (7) to determine s/b directly from Fig. 3. Next, C_{oe}/ϵ is determined by using Z_{oe} in (1), and then C_{fe}'/ϵ and C_f'/ϵ are found from t/b and from the graphs of Figs. 3 and 5. These quantities can be substituted into the following equation to give w/b

$$w/b = \frac{1}{2}\left(1 - \frac{t}{b}\right)\left(\frac{C_{oe}/\epsilon}{2} - C_{fe}'/\epsilon - C_f'/\epsilon\right). \tag{8}$$

Eq. (8) results from substitution of (6) in (2), and rearrangement of terms.

Thus, the two unknown dimensions, s/b and q/b, have been determined.

IV. Considerations of Accuracy

If the bar width w is allowed to become too small, there is interaction of the fringing fields from the two edges, and the decomposition of total capacitance into parallel plane capacitance and fringing capacitances (which are based on infinite bar widths) is no longer accurate. Cohn[6] shows that for a single bar centered between parallel planes, the error in total capacitance from interaction of the fringing fields is about 1.24 per cent for $w/(b-t)=0.35$, where w is the width of the bar, t is its thickness, and b is again the ground-plane spacing. If a maximum error in total capacitance of approximately this magnitude is allowed, then it is necessary that $[(w/b)/(1-t/b)]>0.35$.

Should this inequality be too restricting, it is possible to make approximate corrections based on an increase in the parallel-plate capacitance, to compensate for the loss of fringing capacitance due to interaction of fringing fields. If an initial value w_1/b is found to be less than $0.35[1-(t/b)]$, a new value, w_2/b can be used, where

$$w_2/b = \{0.07[1 - (t/b)] + w_1/b\}/1.20 \tag{9}$$

provided that $0.1<(w_2/b)/[1-(t/b)]<0.35$. This formula is based on a linear approximation to the exact fringing capacitance of single thin strip for a $(w/b)/[1-(t/b)]$ ratio between 0.1 and 0.35. As the relative strip width becomes narrower than 0.35, the fringing capacitance, defined as total capacitance less parallel plate capacitance, becomes smaller. The total capacitance is given by substituting into (1) the exact thin-strip formula for Z_o given by Cohn.[6] Eq. (9) adds sufficient parallel-plate capacitance to compensate for the loss of fringing capacitance. The loss of fringing is assumed to vary linearly below a relative width of 0.35. Although the formula is analytically only approximate, it is sufficiently accurate for practical use because it does no more than give a small correction to a quantity that is reasonably close to the exact value. It can be used with both isolated and coupled bars.

The derivations for the fringing capacitances are exact for bars extending in width infinitely far to the right and left, away from the coupling region. The original computed values were accurate to eight places. However, in order to give values of fringing capacitance associated with constant t/b, it was necessary to use graphical interpolation, as pointed out in the Appendix. The plotted points were held to an accuracy of three figures after the decimal point, so that the interpolated results are slightly less accurate. The curves of Figs. 3 and 4 are accurate to within about one or two per cent. However, since fringing capacitances are usually not the predominant part of the total capacitance of a structure, total capacitance can be specified with somewhat greater accuracy.

Fig. 5, for the fringing capacitance C_f'/ϵ of a single bar extending infinitely far in one direction, is based on an exact solution given by Cohn.[6] The same data can be found from Figs. 3 and 4 by reading either C_{fe}'/ϵ or C_{fo}'/ϵ as functions of t/b for large s/b. The accuracy of Fig. 5 is thus limited by the precision to which the graph can be read.

V. Applications

Figs. 3–5, for fringing capacitances, can be used for a variety of structures, as shown in Fig. 6, simply by adding the appropriate fringing capacitances with the parallel plate capacitances to give the even-mode capacitance, the odd-mode capacitance, or the total capacitance. Use of (1) then gives the associated characteristic impedance. Thus, Fig. 6(a) shows ordinary shielded strip-line and the capacitances involved when it is open, closed at one end, or closed at both ends.[7] The structure closed at one end is sometimes called trough-line, the structure closed at both ends is sometimes called rectangular coaxial line. Similarly, the even- and odd-mode capacitances and impedances can be determined for the coupled structures shown in Fig. 6(b) for open or closed ends. This simple technique may also be applicable when the arms of an N-way power divider in shielded strip-line must run parallel for some distance, as shown in plan in Fig. 6(c). The even-mode fringing

[6] S. B. Cohn, "Problems in strip transmission lines," IRE Trans. on Microwave Theory and Techniques, vol. MTT-3, pp. 119–126; March, 1955.

[7] The notation in Fig. 6(a), $C_{fo}'(s_a/b)$ and $C_{fo}'(s_b/b)$, does not indicate multiplication, but merely that C_{fo}' is to be evaluated at s_a/b or s_b/b, as appropriate for the spacing from the nearby wall.

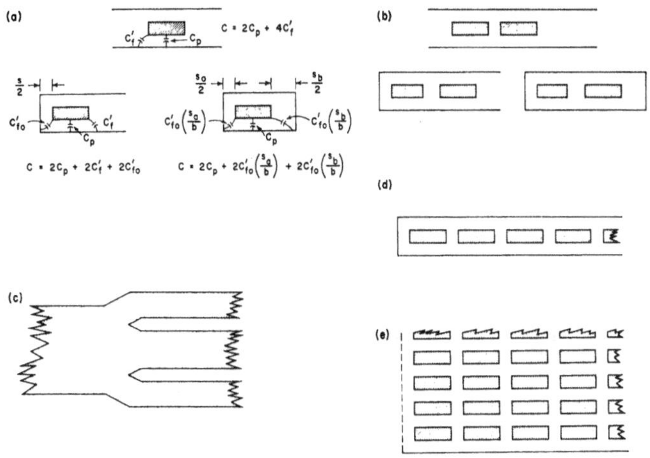

Fig. 6—Possible applications.

capacitance C_{fe}' would then be appropriate for adjacent edges of the arms.

Both even- and odd-mode fringing capacitances would be necessary for multi-element lines, such as are shown in Fig. 6(d) and (e). The cross section shown in (d) could be part of a meander or interdigital line, while that in (e) might be part of a finite or infinite array of elements, which might be used as an artificial dielectric medium.

The curves given herein can be used in the design of wide-band, parallel-coupled, strip-transmission-line filters, such as described by Matthaei.[8] Also, Bolljahn and Matthaei[9] have presented design data for slow-wave structures and filters using parallel-coupled arrays of line elements. Some realizations of those devices use relatively wide rectangular bars to form an interdigital line, comb line, meander line, or similar slow-wave line. In such cases, the curves given herein greatly facilitate the process of precision design.

Coupled rectangular bars may also be applied to strip-line directional couplers, described by Jones and Bolljahn,[2] in which the use of rectangular bars allows closer coupling to be achieved with less critical tolerances.

Appendix

Derivation of Fringing Capacitances

Preliminary

It is desired to determine the static fringing capacitances shown on the structure of Fig. 1 by means of conformal mapping techniques.[10,11] This can be done by subjecting the boundaries of the structure to transformations under which capacitance is invariant, and that lead to a new structure for which capacitance is known. Subtraction of parallel plate capacitances of the original structure from the total capacitance then leaves the fringing capacitances. The analysis will be limited to structures in which the bars are so wide that interaction between fringing fields of the two edges of a single bar is negligible. As discussed in Section IV, this requires that the approximate relation $\{(w/b)/[1-(t/b)]\}>0.35$ be held. Under these conditions it is possible to let the bars extend in width infinitely far to the left and right, without disturbing the fringing fields appreciably in the coupling region where the capacitances interact. The vertical centerline shown on Fig. 1 may be replaced by an electric wall (conductor) for the odd mode, or by a magnetic wall for the even mode, in consideration of the symmetry of the structure. Also, the electric field can lie parallel to the horizontal centerline where no conductor exists, but cannot cross it because of the symmetry. Therefore, a magnetic wall can be placed along the horizontal centerline. These modifications allow analysis of only one-quarter of the total symmetrical structure. The mathematical model is shown on the z-plane in Fig. 7. Conductors are indicated by solid lines and magnetic walls by broken lines. The upper-case letters denote pertinent points of the structure and will serve as references when transformations to different complex planes are made.

The analysis essentially consists in transformation of the contours of the structure on the z-plane into a parallel-plate representation on another complex plane, where capacitance can be computed directly.

The static electric fields of interest lie within the polygon defined by the boundaries of the structure on the z-plane. The interior of this polygon is to be mapped onto the first quadrant of the t-plane shown in Fig. 7. The integral resulting from direct use of the Schwarz-Christoffel transformation is

$$z = \int \frac{(1 - t^2)^{1/2}}{(1 - k^2t^2)^{1/2}(1 - k^2t^2 \operatorname{sn}^2 a)} dt, \tag{10}$$

where, for the present, $1/k$ and $1/(k \operatorname{sn} a)$ are the points on the real t axis to which the corner F and the point $-j_\infty$ map from the z-plane. This integral can be evaluated by further relating the first quadrant of the t-plane to the interior of the fundamental rectangle of Jacobian elliptic functions on the u-plane, also shown in Fig. 7, using the transformation

$$t = \operatorname{sn} u. \tag{11}$$

This substitution gives

$$z = \int \frac{\operatorname{cn}^2 u du}{1 - k^2 \operatorname{sn}^2 u \operatorname{sn}^2 a}. \tag{12}$$

[8] G. L. Matthaei, "Design of wide-band (and narrow-band) band pass microwave filters on the insertion loss basis," IRE Trans. on Microwave Theory and Techniques, vol. MTT-8, pp. 580–593; November, 1960.

[9] J. T. Bolljahn and G. L. Matthaei, "Microwave Filters and Coupling Structures," Stanford Res. Inst., Menlo Park, Calif., Report No. 1, Contract DA 36-039 SC-87398; April, 1961.

[10] W. R. Smythe, "Static and Dynamic Electricity," McGraw-Hill Book Co., Inc., New York, N. Y.; 1939.

[11] E. Weber, "Electromagnetic fields: theory and applications," in "Mapping of Fields," John Wiley and Sons, Inc., New York, N. Y., vol. 1; 1950.

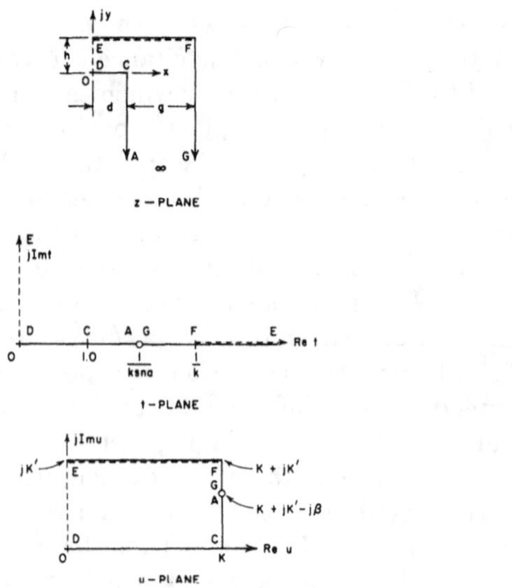

8

Fig. 7—Mathematical models on z-plane, t-plane, and u-plane.

In the above equations, sn u and cn u are Jacobian elliptic functions having a quarter period K determined by k, denoted by convention as the modulus. By virtue of (11), a can be considered to be a point on the perimeter of the fundamental rectangle on the u-plane. It is convenient to let

$$a \triangleq K - j\beta \tag{13}$$

by definition. Ultimately, k and β will be used as independent variables.

The mapping of the z-plane onto the t-plane in this manner has been carried through by Cockroft,[12] whose symbols are retained in Fig. 7 and in the equations given in this section. The distances on the z-plane are given by Cockroft as

$$d = K\left[1 - \frac{\text{dn } a}{k^2 \text{ sn } a \text{ cn } a} Z(a)\right] \tag{14}$$

$$g = -j\frac{\pi}{2} \frac{\text{dn } a}{k^2 \text{ sn } a \text{ cn } a} \tag{15}$$

$$h = jd\frac{K'}{K} - j\frac{\pi}{2} \frac{\text{dn } a}{k^2 \text{ sn } a \text{ cn } a}\left(\frac{a}{K} - 1\right). \tag{16}$$

It should be noted that g is a negative quantity and h an imaginary one. The quantity dn a is also a Jacobian elliptic function and $Z(a)$ is the Jacobian zeta-function. The quantity K' is the same function of the complementary modulus k', as K is of k. The moduli are related by

$$k^2 + k'^2 = 1. \tag{17}$$

[12] J. D. Cockroft, "The effect of curved boundaries on the distributions of electrical stress round conductors," *J. IEE*, vol. 66, pp. 385–409; April, 1926.

Comparison of Fig. 1 with Fig. 7 shows that the conventional normalized dimensions s/b and t/b of the rectangular bar structure are related to Cockroft's dimensions d, g, and h by

$$s/b = \frac{-jh}{d - g}$$

$$t/b = \frac{d}{d - g}. \tag{18}$$

Thus, the physical dimensions of the structure of Fig. 1 have been related to the parameters of the u-plane by (14)–(16) and (18).

Now it is necessary to transform the t-plane to a parallel-plate structure, and determine fringing capacitances as functions of u-plane parameters.

Odd-Mode Capacitance

The two rectangular bars in Fig. 1 are at equal and opposite voltages when energized in the odd mode, so that the plane midway between the bars is at zero potential. Thus, a conductor may be placed in this plane without disturbing the fields. This is indicated by the solid line between E and F on the planes of Fig. 7. For this condition, the t-plane configuration can be transformed to a parallel-plate structure of unit height by the function

$$w = \frac{1}{\pi} \ln\left(\frac{1 + tk \text{ sn } a}{1 - tk \text{ sn } a}\right), \tag{19}$$

which moves the singularity at AG to infinity. The interesting region of the w-plane is shown in Fig. 8.

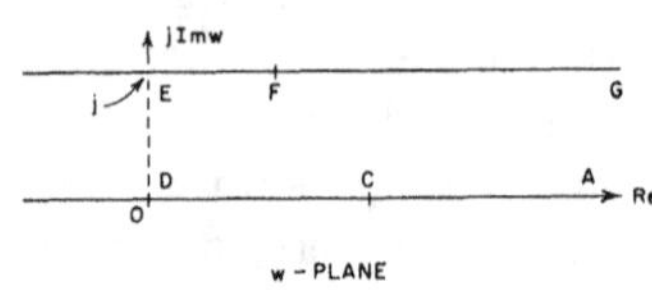

Fig. 8—w-plane for odd-mode capacitance.

The fringing capacitance is the difference between the total capacitance of the structure (total capacitance is the same on both z- and w-planes) and the parallel-plane capacitance of the z-plane structure. A reasonable definition for z-plane parallel-plate capacitance C_{pz} is parallel-plate capacitance existing between the ground plane and the full length of the rectangular bar. Mathematically,

$$\frac{C_{pz}}{\epsilon} = \lim_{z \to -j\infty} \text{Im}(z/g). \tag{20}$$

Then the odd-mode fringing capacitance C_{fo}'/ϵ is

$$C_{fo}'/\epsilon = \lim_{z \to -j\infty} [w(z) - \text{Im}(z/g)] \tag{21}$$

where $w(z)$ is given by (19) related to the z-plane. In order to evaluate C_{fo}'/ϵ, it is necessary that both w and $\text{Im}(z/g)$ be expressed as functions of u. The limit must also be in terms of u. From Fig. 7 it can be seen that as $z \to -j_\infty$ along the path from C to A, then $u \to K + jK' - j\beta$ along the path from C to A. Thus

$$C_{fo}'/\epsilon = \lim_{u \to K+jK'-j\beta} \left[w(u) - \text{Im} \frac{z(u)}{g} \right]. \tag{22}$$

Substitution of (11) in (19) gives

$$w(u) = \frac{1}{\pi} \ln \left(\frac{1 + k \text{ sn } a \text{ sn } u}{1 - k \text{ sn } a \text{ sn } u} \right). \tag{23}$$

The limiting process is simplified by letting

$$u = K + jK' - j\beta - j\delta \tag{24}$$

where $\delta \to 0$ as $u \to K + jK' - j\beta$ along the path from C to A. Assuming very small δ, and using various elliptic function equivalences, such as may be found in Byrd and Friedman,[13] (23) reduces to

$$\underset{\delta \to 0}{w(u)} = \frac{1}{\pi} \ln 2 - \frac{1}{\pi} \ln \left(-j \frac{\text{cn } a \text{ dn } a}{\text{sn } a} \right) - \frac{1}{\pi} \ln \delta. \tag{25}$$

Cockroft's[12] (44) gives $z(\delta)$ as

$$z(\delta) = (K + jK' - j\beta) \left[1 - \frac{\text{dn } a}{k^2 \text{ sn } a \text{ cn } a} Z(a) \right]$$
$$- \frac{1}{2} \frac{\text{dn } a}{k^2 \text{ sn } a \text{ cn } a} \ln \frac{\Theta(jK' - j\delta)}{\Theta(2K + jK' - 2j\beta)} \cdot \tag{26}$$

Using (26) with (15), and passing to the limit of δ approaching zero, gives

$$\underset{\delta \to 0}{\text{Im}} \left[\frac{z(u)}{g} \right] = (K' - \beta) \left[\frac{1}{g} - \frac{2j}{\pi} Z(a) \right]$$
$$- \frac{K'}{4K} - \frac{1}{2\pi} \ln \frac{2kk'K}{\pi} - \frac{1}{\pi} \ln \delta. \tag{27}$$

In (26) and (27) the term Θ is Jacobi's theta function.[14] Now (25) and (27) can be substituted into (22), yielding

$$C_{fo}'/\epsilon = \frac{1}{\pi} \ln \frac{2j \text{ sn } a}{\text{cn } a \text{ dn } a} + \frac{1}{2\pi} \ln \frac{2kk'K}{\pi} + \frac{K'}{4K}$$
$$+ (K' - \beta) \left[\frac{2j}{\pi} Z(a) - \frac{1}{g} \right]$$
$$- \frac{1}{\pi} \ln \Theta(jK' - 2j\beta). \tag{28}$$

This is the final form in which odd-mode fringing capacitance will be presented.

[13] P. F. Byrd and M. D. Friedman, "Handbook of Elliptic Integrals for Physicists and Engineers," Springer-Verlag, Berlin, Germany; 1954.

[14] E. T. Copson, "Theory of Functions of a Complex Variable," Oxford University Press, London; pp. 405–407; 1960.

Even-Mode Capacitance

The two rectangular bars in Fig. 1 are at the same potential when energized in the even mode, so that no electric field crosses the plane midway between them. Thus, a magnetic wall may be placed in this plane without disturbing the fields. This is indicated by the broken line between E and F on the planes of Fig. 7. The upper half of the t-plane is mapped into a strip of unit height on a t_1-plane in such a manner that the singularity at AG is removed to infinity by the transformation 9

$$t_1 = \frac{1}{\pi} \ln \left[\frac{tk \text{ sn } a + 1}{tk \text{ sn } a - 1} \right]. \tag{29}$$

The t_1-plane is shown in Fig. 9. Notice that the upper half of the t-plane maps into the strip directly below the Re t_1 axis. The positive half of this strip is next mapped onto the lower half of a t_2-plane, shown in Fig. 9 by the transformation

$$t_2 = M - 1 + M \cosh \pi t_1, \tag{30}$$

where

$$M \triangleq \frac{2}{1 + \cosh \pi t_1(F)} = - \frac{\text{cn}^2 a}{\text{sn}^2 a} \cdot \tag{31}$$

Finally, the desired parallel-plate configuration is achieved by mapping the lower half of the t_2-plane onto the positive half of a strip of unit height on the w_1-plane, using the transformation

$$w_1 = \frac{1}{\pi} \text{arc cosh } (-t_2). \tag{32}$$

The w_1-plane is also shown on Fig. 9. Combining (11), (29), (30), (31), and (32) gives w_1 as a function of u

$$w_1(u) = \frac{1}{\pi} \text{arc cosh} \left\{ 1 + \frac{cn^2 a}{\text{sn}^2 a} + \frac{\text{cn}^2 a}{\text{sn}^2 a} \cdot \left[\frac{(k \text{ sn } a \text{ sn } u)^2 + 1}{(k \text{ sn } a \text{ sn } u)^2 - 1} \right] \right\}. \tag{33}$$

As with the odd mode, it is convenient to use the variable δ, defined by (24), in passing to the limit. When (24) is substituted in (33) and appropriate approximations made for small δ, manipulation yields

$$\underset{\delta \to 0}{w_1(u)} = \frac{1}{\pi} \ln 2 + \frac{1}{\pi} \ln \left(-j \frac{\text{cn } a}{\text{sn } a \text{ dn } a} \right) - \frac{1}{\pi} \ln \delta. \tag{34}$$

Using the definition of z-plane parallel-plate capacitance given in (20), the even-mode fringing capacitance, C_{fe}'/ϵ is

$$\frac{C_{fe}'}{\epsilon} = \lim_{\delta \to 0} \left[w_1(u) - \text{Im} \frac{z(u)}{g} \right]. \tag{35}$$

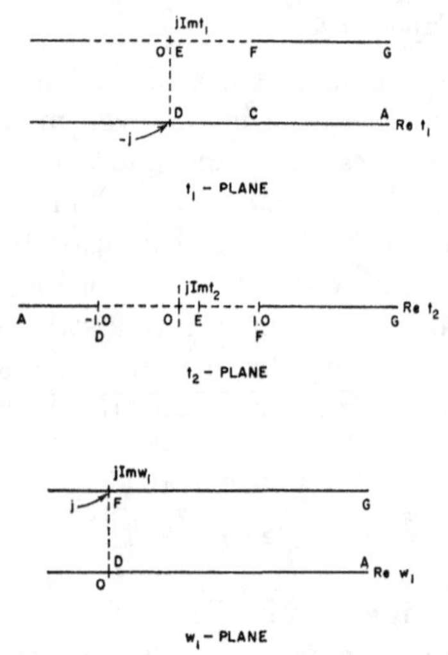

Fig. 9—t_1-plane, t_2-plane, and w_1-plane for even-mode capacitance.

Substitution of (34) and (27) in (35) yields, after simplification,

$$\frac{C_{fe}'}{\epsilon} = \frac{1}{\pi} \ln \left(\frac{-2j \text{ cn } a}{\text{sn } a \text{ dn } a} \right) + \frac{1}{2\pi} \ln \frac{2kk'K}{\pi} + \frac{K'}{4K} + (K' - \beta) \left[\frac{2jZ(a)}{\pi} - \frac{1}{g} \right] - \frac{1}{\pi} \ln \Theta(jK' - 2j\beta). \tag{36}$$

This is the final form in which the even-mode fringing capacitance will be given.

Difference of Fringing Capacitances

The difference between fringing capacitances, $C_{fo}'/\epsilon - C_{fe}'/\epsilon$, is a very useful quantity because in most coupled structures it is also half of the difference between total odd-mode and even-mode capacitances. This difference is found by subtracting (36) from (28), yielding

$$\frac{C_{fo}}{\epsilon} - \frac{C_{fe}'}{\epsilon} = \frac{1}{\pi} \ln \left(- \frac{\text{sn}^2 a}{\text{cn}^2 a} \right). \tag{37}$$

Evaluation of Formulas

The curves of C_{fe}'/ϵ and $C_{fo}'/\epsilon - C_{fe}'/\epsilon$ as functions of s/b and t/b were determined in the following manner: values of $0 < k < 1$ were selected from tables[13,15] that determined K, K', and k'; then for each value of k, a range of values of β/K' was selected from tables[15] that gave sn (β, k'), cn (β, k'), and dn (β, k'). These functions are related to sn a, cn a, and dn a by

$$\text{sn } (a, k) = \frac{1}{\text{dn } (\beta, k')}, \quad \text{cn } (a, k) = jk' \frac{\text{sn } (\beta, k')}{\text{dn } (\beta, k')}, \quad \text{dn } (a, k) = k' \frac{\text{cn } (\beta, k')}{\text{dn } (\beta, k')}. \tag{38}$$

The zeta function can be expressed as

$$Z(a, k) = j \left[Z(\beta, k') + \frac{\pi\beta}{2KK'} - k'^2 \frac{\text{sn } (\beta, k') \text{ cn } (\beta, k')}{\text{dn } (\beta, k')} \right], \tag{39}$$

where

$$Z(\beta, k') = \frac{\Theta'(\beta, k')}{\Theta(\beta, k')}. \tag{40}$$

The theta functions were evaluated using a Fourier series expansion.[14] Values of t/b, s/b, C_{fe}'/ϵ and $C_{fo}'/\epsilon - C_{fe}'/\epsilon$ were then calculated from (18), (26), and (37), and plotted as function of β/K', with k as parameter. Values of t/b were selected to be used as parameters on and final graph, and the related values of k and β/K' were taken from the t/b graph and tabulated. The values of k and β/K' at each point were used to determine related values of s/b, C_{fe}'/ϵ, and $C_{fo}'/\epsilon - C_{fe}'/\epsilon$ from their graphs. In this way it was possible to compile values of s/b, C_{fe}'/ϵ and $C_{fo}'/\epsilon - C_{fe}'/\epsilon$ for constant t/b. This compilation, was used to plot the final sets of curves shown in Figs. 3 and 4.

[15] G. W. Spenceley and R. M. Spenceley, "Smithsonian Elliptic Functions Tables," Smithsonian Misc. Coll., Washington, D. C., vol. 109; 1947.

Reprinted from IEEE TRANSACTIONS
ON *MICROWAVE THEORY AND TECHNIQUES*
Volume MTT-17, Number 8, August, 1969
pp. 637-638

Fringing Capacitance Curves for Coplanar Rectangular Coupled Bars

Abstract—**Even and odd mode fringing capacitance curves are presented, on an expanded scale, for tightly coupled coplanar rectangular bars between parallel ground planes. Accurate values of mode capacitances have been calculated by numerically solving the Laplace equation for the lower and upper bounds on the mode capacitance. The computed capacitance data presented is the average of the two bounds and is correct to within $\pm\frac{1}{4}$ percent of the exact data (before plotting).**

INTRODUCTION

Based on conformal mapping analysis, even and odd mode fringing capacitance curves have been presented by Getsinger [1] for coplanar coupled rectangular bars between parallel ground planes. The transverse cross section of a typical structure of this type is shown in Fig. 1. Although fringing capacitance curves have been provided for wide ranges of bar thickness and coupling, there is insufficient data for very tight coupling. Furthermore, the accuracy with which capacitance data can be read from these curves for tightly coupled bars is inadequate.

The purpose of this correspondence is to present even and odd mode fringing capacitance curves, on an expanded scale, for tightly coupled coplanar rectangular bars between parallel ground planes. Accurate values of mode capacitances have been calculated by numerically solving the Laplace equation for the lower and upper bounds on the mode capacitance. The computed capacitance data presented is the average of the two bounds and is correct to within $\pm\frac{1}{4}$ percent of the exact data (before plotting).

NUMERICAL SOLUTION

The calculation of mode capacitance in a TEM mode can be reduced to the solution of a two-dimensional Laplace equation for electric or magnetic potential, subject to appropriate boundary conditions. To compute the average value of the capacitance, both lower and upper bounds are desired. Therefore, the Laplace equation must be solved, separately, for both the electric and magnetic potentials.

Both electric and magnetic potential Laplace equations are first transformed into two separate sets of finite-difference equations describing discrete potentials at each node of the graded mesh superimposed over the cross sectional domain of the problem. The finite-difference equations are then numerically solved on a computer by successive-over-relaxation. From the independently computed electric and magnetic potentials, respective electrostatic and electromagnetic energies are computed. The electrostatic and magnetostatic energies, respectively, provide stationary upper bounds on the capacitance and inductance for the TEM mode under consideration [2]. The upper bound on inductance results in a lower bound on the capacitance. The average of the lower and upper bounds on capacitance yields a more accurate value than that obtained from the concept of electric charge on a conductor for comparable mesh sizes, since the electric charge formalism does not result in a stationary value of the capacitance.

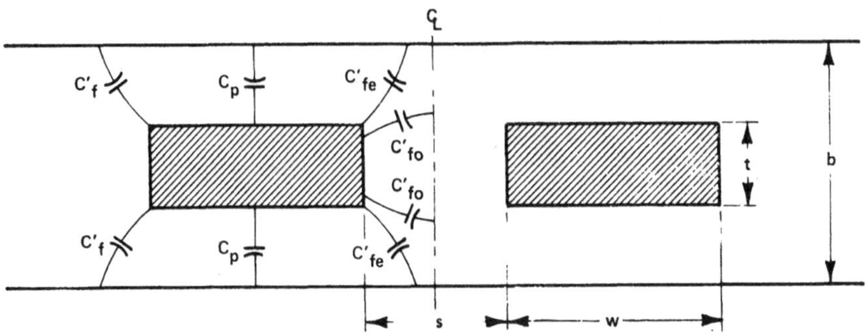

Fig. 1. Coupled rectangular bars centered between parallel plates.

MODE CAPACITANCES

For the configuration shown in Fig. 1, filled with a uniform homogeneous dielectric medium of permittivity ϵ and permeability μ_0, the total even mode capacitance C_{oe} and the total odd mode capacitance C_{oo} from one bar to ground are related by

$$\frac{C_{oe}}{\epsilon} = 2\left(\frac{C_p}{\epsilon} + \frac{C_{fe}'}{\epsilon} + \frac{C_f'}{\epsilon}\right) \quad (1)$$

$$\frac{C_{oo}}{\epsilon} = 2\left(\frac{C_p}{\epsilon} + \frac{C_{fo}'}{\epsilon} + \frac{C_f'}{\epsilon}\right) \quad (2)$$

where all capacitance parameters are as defined by Getsinger [1]. The parallel plate capacitance C_p is given by

$$C_p = 2\frac{w/b}{1-(t/b)} \quad (3)$$

where w and t are the width and thickness of the bar; and b is the spacing between ground planes. The difference between C_{oe} and C_{oo} is usually defined by

$$\frac{\Delta C}{\epsilon} = \frac{1}{2}\left(\frac{C_{oo}}{\epsilon} - \frac{C_{oe}}{\epsilon}\right) = \frac{C_{fo}'}{\epsilon} - \frac{C_{fe}'}{\epsilon}. \quad (4)$$

If $\Delta C/\epsilon$, C_{fe}'/ϵ, and C_f'/ϵ are known for a given configuration even and odd mode ca-

Manuscript received February 13, 1969; revised April 28, 1969.

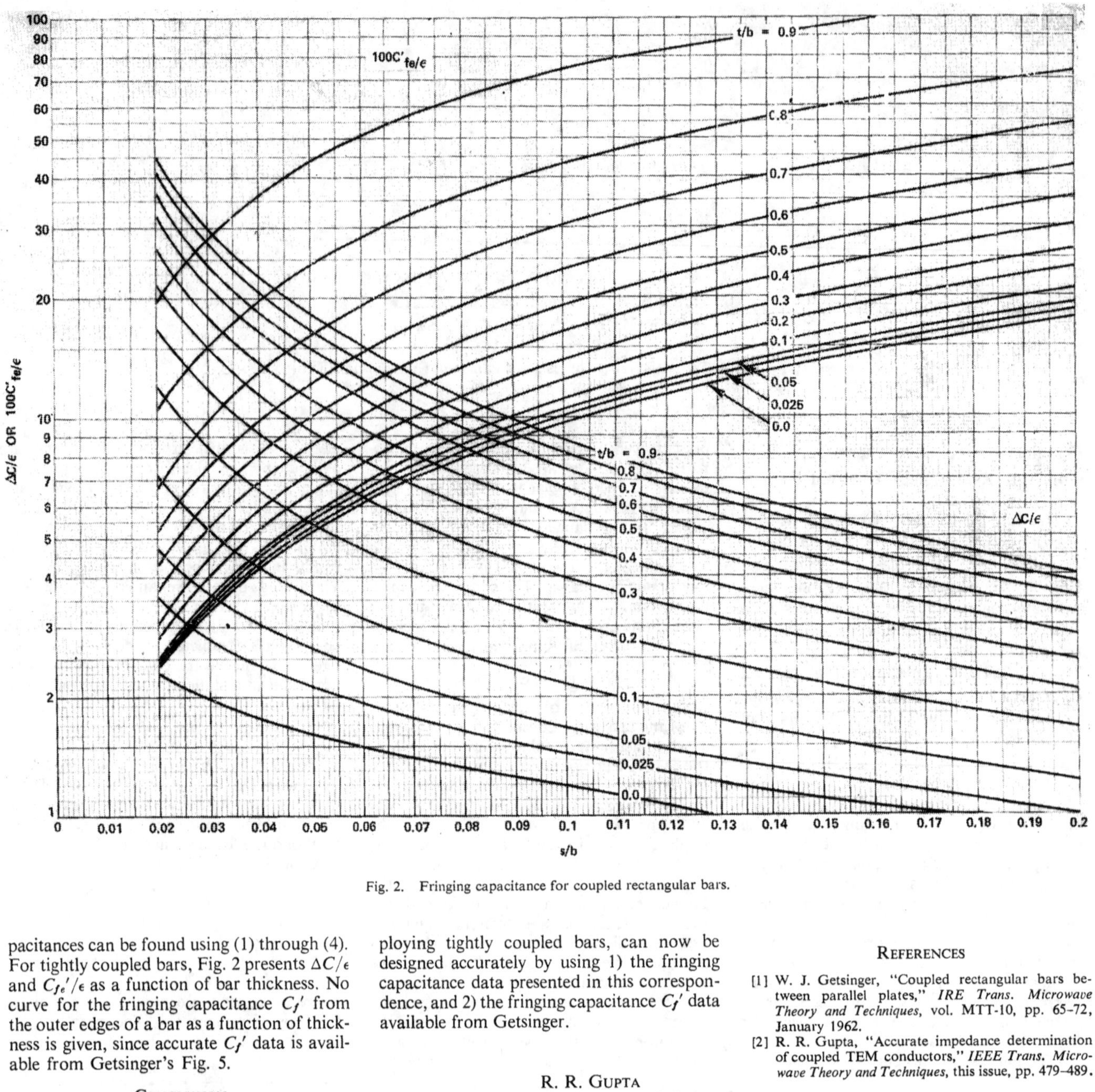

Fig. 2. Fringing capacitance for coupled rectangular bars.

pacitances can be found using (1) through (4). For tightly coupled bars, Fig. 2 presents $\Delta C/\epsilon$ and C_{fe}'/ϵ as a function of bar thickness. No curve for the fringing capacitance C_f' from the outer edges of a bar as a function of thickness is given, since accurate C_f' data is available from Getsinger's Fig. 5.

Conclusion

Precision TEM component designs, employing tightly coupled bars, can now be designed accurately by using 1) the fringing capacitance data presented in this correspondence, and 2) the fringing capacitance C_f' data available from Getsinger.

R. R. Gupta
Bendix Research Labs.
Southfield, Mich. 48075

References

[1] W. J. Getsinger, "Coupled rectangular bars between parallel plates," *IRE Trans. Microwave Theory and Techniques*, vol. MTT-10, pp. 65–72, January 1962.

[2] R. R. Gupta, "Accurate impedance determination of coupled TEM conductors," *IEEE Trans. Microwave Theory and Techniques*, this issue, pp. 479–489.

Coupled Circular Cylindrical Rods Between Parallel Ground Planes

EDWARD G. CRISTAL, MEMBER, IEEE

Summary—The normalized self and mutual capacitances of periodic, circular cylindrical rods located between parallel ground planes are presented graphically. The capacitances were determined by solving the appropriate integral equation by numerical methods. Charts of self and mutual capacitance are given for rod diameter-to-ground plane spacing ratios varying from 0.05 to 0.8 and for very small to very large spacings between rods. Accuracy of the data is believed to be generally better than 2 per cent for the normalized mutual capacitance and generally better than 1 per cent for the normalized self capacitance. An approximate design method is also presented that permits using the data to synthesize filters (such as interdigital and comb-line filters) that require rods of nonequal diameters and spacings. An example of the design method is given, and a filter is constructed from the resulting data. The filter response was measured and found to agree closely with that called for by the theory.

Manuscript received March 17, 1964. This work was sponsored by the U. S. Army Electronics Research and Development Agency, Fort Monmouth, N. J., under Contract No. DA 36-039-AMC-00084(E).
The author is with the Stanford Research Institute, Menlo Park Calif.

I. Introduction

THE DESIGN of many UHF and microwave filters is based on electrically coupling arrays of cylindrical bars located between parallel ground planes of which the comb-line and interdigital filter are two examples [1], [2]. In the past, rectangular bars generally have been used as coupling elements because the necessary data are available [3]. Filter equations and procedures of Matthaei [1], [2], together with design data of Getsinger [3], have been used to design filters which have proved to have excellent electrical characteristics. However, obtaining rectangular bars in practice requires manufacturing processes which are generally costly. On the other hand, the use of circular cylindrical rods as resonators offers several manufacturing advantages and should result in the same excellent electrical filter properties. For these reasons, design data of coupled circular cylindrical rods between parallel ground planes and a design method for using the data to synthesize filters have been prepared and are presented in this paper.

The coupled rod data are presented in Section II. Also in Section II, the filter synthesis method is explained by means of an example. The mathematical derivation of the design data is given in Section III. The physical basis for the design method is presented in Section IV. A discussion of the accuracy of the coupled
14 rod data will be found in Section V. The measured performance of a trial filter constructed from the design data of Section II is described in Section VI.

II. Technical Description

The geometry of the periodic, circular cylindrical rods between parallel ground planes and the dimensional notation are shown in Fig. 1. The circular rods have diameter d and are spaced periodically at a distance c. The ground planes are separated at distance b. The spacing between adjacent rod surfaces is denoted by s and is given by

$$s = c - d. \tag{1}$$

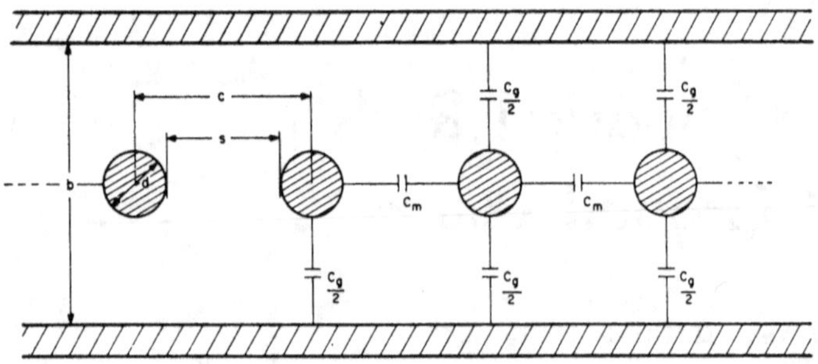

Fig. 1—Coupled circular cylindrical rods between parallel ground planes.

In the derivation of the design data fringing capacitances beyond nearest neighbors were neglected. It is therefore possible to describe TEM propagation along the structure in terms of two orthogonal modes which have been designated as the even mode and the odd mode. In the even mode all center conductors are at the same potential, while in the odd mode successive center conductors are at equal but opposite signed potentials with respect to the ground planes. These two TEM modes have different characteristic impedances which are intimately related to the total static capacitances of the rods to ground when in one or the other mode. The total static capacitances are related to the mutual capacitance between successive rods C_m and the self capacitance C_g of each rod. A consideration of Fig. 1 shows that the total capacitance measured between one rod and ground when the rods are driven in the odd mode is

$$C_0 = C_g + 4C_m, \tag{2}$$

and the total capacitance measured between one rod and ground when the rods are driven in the even mode is

$$C_e = C_g. \tag{3}$$

From (2) and (3) are obtained

$$C_g = C_e \tag{4}$$

and

$$C_m = 1/4(C_0 - C_e). \tag{5}$$

The characteristic impedance Z_0 of a lossless uniform transmission line operating in the TEM mode may be related to its shunt capacitance by

$$Z_0\sqrt{\epsilon_r} = \frac{\eta}{(C/\epsilon)} \text{ ohms} \tag{6}$$

where

ϵ_r is the relative dielectric constant of the medium in which the wave travels
η is the impedance of free space (376.7 ohms)
C/ϵ is the ratio of the static capacitance per unit length between conductors to the permittivity (in the same units) of the dielectric medium. (This ratio is independent of the dielectric constant.)

The even and odd mode impedances of coupled TEM lines can be found by substituting even and odd mode capacitances of the lines into (6).

Graphs of normalized capacitance C/ϵ vs normalized spacing s/b are presented in a form that assists the filter synthesis method. In Fig. 2, C_m/ϵ vs $(1/2)s/b$ is given. In Fig. 3, $(1/2)C_g/\epsilon$ vs $(1/2)s/b$ is given. Although these data are for the periodic structure of Fig. 1, a method has been developed whereby the data can be used for synthesis of filters in which rods of arbitrary diameters and spacings are required. The method is approximate, but is expected to give good results.

The design method will be explained by means of an example. To facilitate the evaluation of the method, a previously constructed interdigital band-pass filter that uses rectangular bars has been chosen as the example.[1] The filter design calls for a 6-resonator filter having a 10-per cent bandwidth and a 0.1-db Chebyshev ripple in the pass band. Because the filter is symmetrical about its center, only one-half the filter is shown in the cross-sectional view in Fig. 4, page 430. The nomenclature used in Fig. 4 is adopted for this discussion, except that the dimensions w_k and t for the rectangular cross section of the bars are replaced by the diameters d_k of the round rods. The required normalized capacitances for the design example are given in Table I [2].

[1] Matthaei, see [2], pp. 483–486.

15

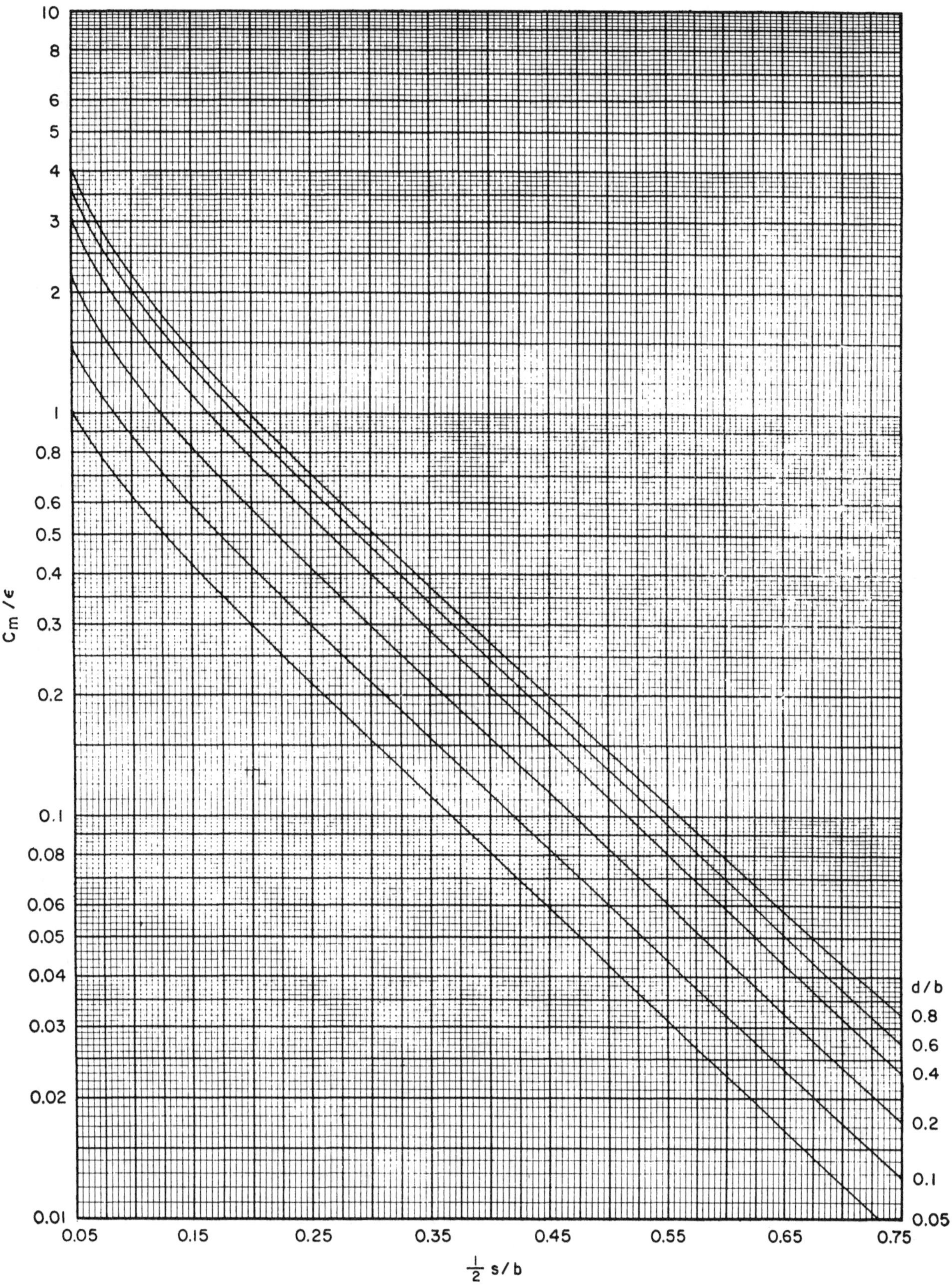

Fig. 2—Graph of C_m/ϵ (normalized mutual capacitance) vs $\frac{1}{2}(s/b)$ (normalized half spacing).

16

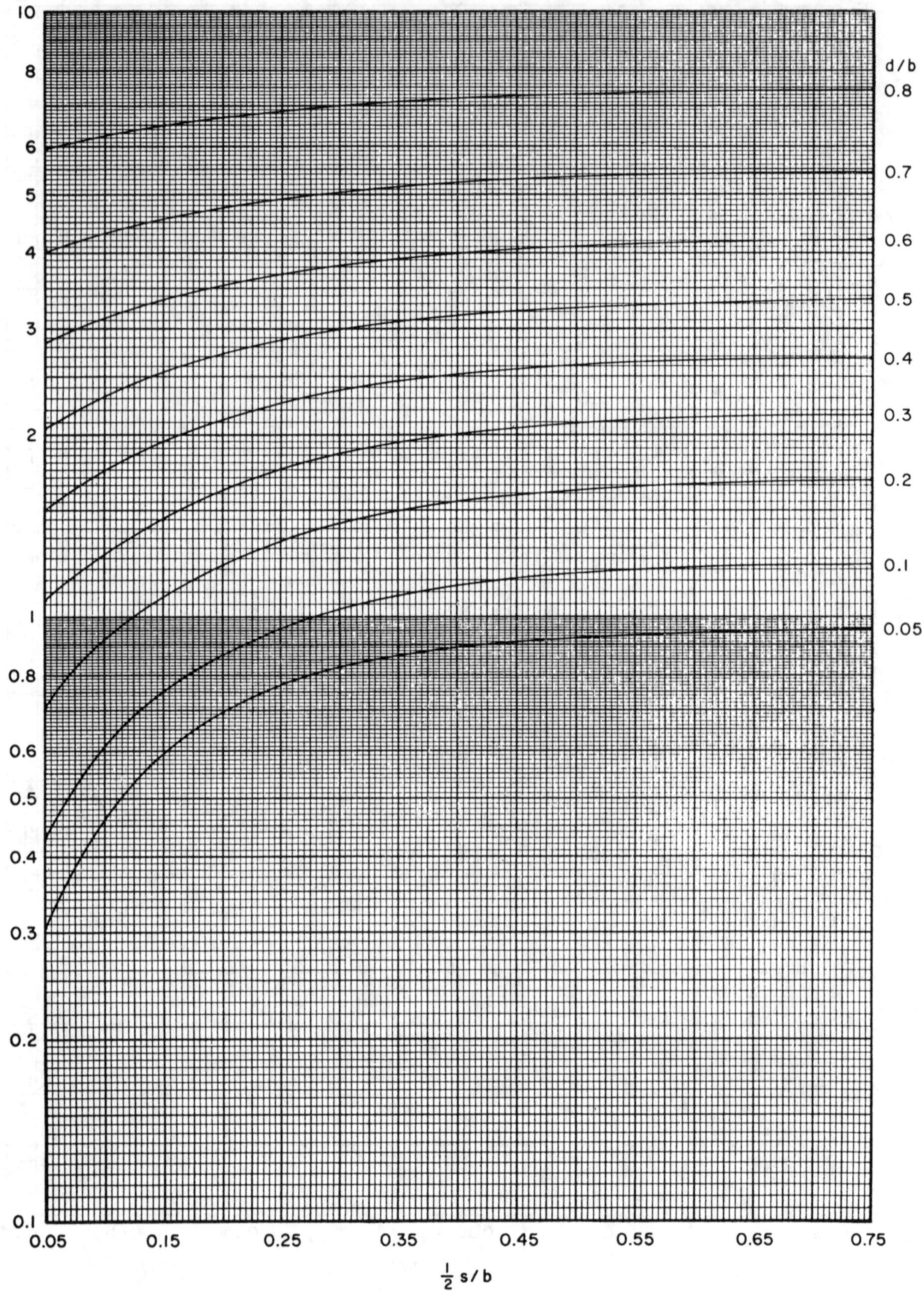

Fig. 3—Graph of $(\frac{1}{2})C_g/\epsilon$ (normalized half self capacitance) vs $(\frac{1}{2})s/b$ (normalized half spacing).

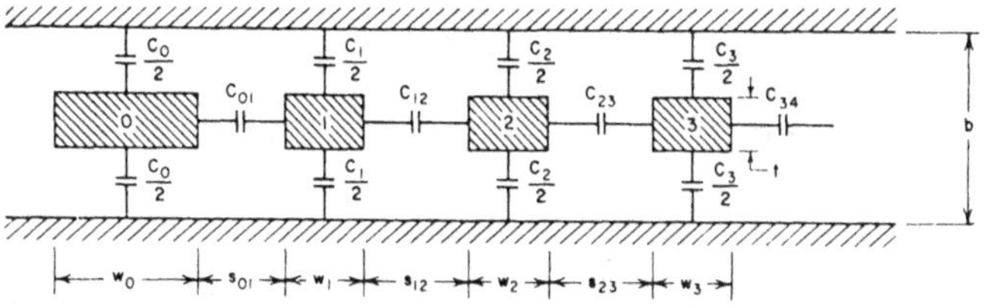

Fig. 4—Cross section of an array of parallel-coupled lines between ground planes.

TABLE I

TABULATION OF QUANTITIES OF A 10-PER CENT BANDWIDTH INTERDIGITAL FILTER DESIGN WITH $n=6$ RESONATORS [2]

k	$C_{k,k+1}/\epsilon$	k	C_k/ϵ
0 and 6	1.582	0 and 7	5.950
1 and 5	0.301	1 and 6	3.390
2 and 4	0.226	2 and 5	4.420
3	0.218	3 and 4	4.496

This filter was designed for 0.10-db Chebyshev ripple.

The design procedure given here assumes that values for the normalized capacitances have been previously obtained [1], [2]. The initial step in obtaining the dimensions of the structure from the normalized capacitances is to mark off on the graph of Fig. 2 horizontal lines corresponding to the values of C_m/ϵ that are called for in the design. In this example four horizontal lines of $C_m/\epsilon = C_{k,k+1}/\epsilon$ (for $k=0$, 1, 2, 3) = 1.58, 0.301, 0.226, and 0.218 are drawn. Next, the coordinates of the intersections of constant $C_m/\epsilon = 0.218$, 0.226, 0.301, and 1.58 with the family of constant d/b curves are noted and plotted on the graph of Fig. 3. Smooth curves are then drawn on the graph of Fig. 3 through points of constant C_m/ϵ, thus obtaining curves of constant C_m/ϵ. This is shown in Fig. 5.

Next, it is useful to partition the filter configuration of Fig. 4 into smaller subsections as shown in Fig. 6. Each subsection consists of a normalized capacitance to ground and the normalized coupling capacitances to the right and to the left. To determine the normalized rod diameters and normalized spacings the designer may choose any of the subsections (a), (b), (c) or (d) of Fig. 6 and proceed in the following manner: Assume that Fig. 6(d) is chosen. On the graph of Fig. 5, assisted by a suitable drawing aid, several auxiliary curves of constant d/b are drawn which intersect the curves of constant $C_m/\epsilon = 0.218$ and 0.226. These auxiliary curves are drawn in the vicinity of one-half the value of the required normalized capacitance to ground, in this case, $(1/2)C_g/\epsilon = (1/2)C_3/\epsilon = 2.25$. *The objective of drawing auxiliary curves is to find the unique intersections with the constant curves of $C_m/\epsilon = 0.218$ and 0.226 such that the sum of the ordinates of the intersections totals $C_g/\epsilon = C_3/\epsilon = 4.5$.* The proper d/b curve is shown in Fig. 7.

It is seen in Fig. 7 that the coordinates of the intercept at $C_m/\epsilon = 0.218$ are $[(1/2)s/b = 0.387,\ (1/2)C_g/\epsilon = 2.25]$, and those at $C_m/\epsilon = 0.226$ are (0.381, 2.25). Note that the sum of the ordinates is 4.5 as required. Next, using linear interpolation, d/b of rod 3 is found to be 0.351.

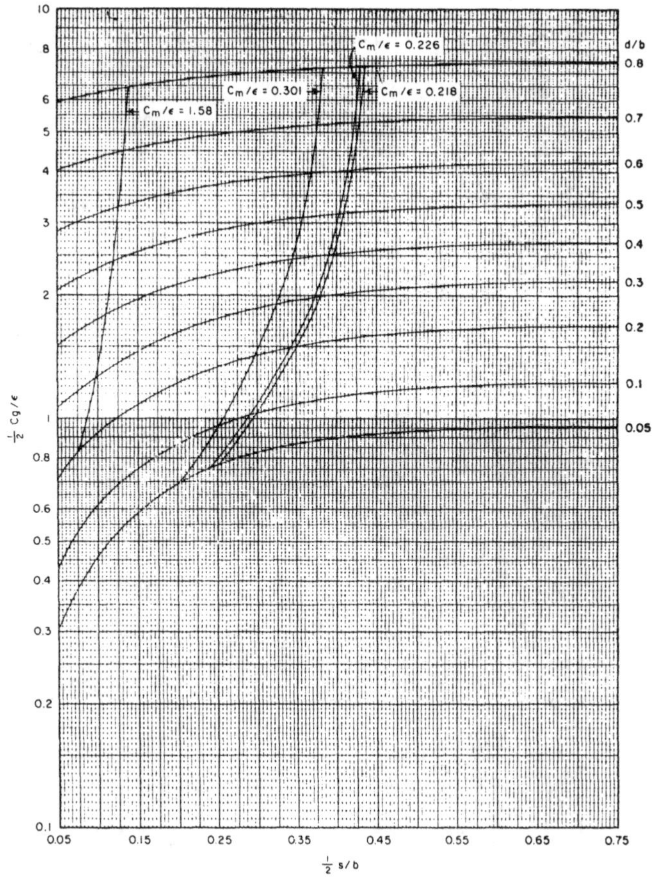

Fig. 5—Curves of constant C_m/ϵ plotted on a graph of $(\frac{1}{2})C_g/\epsilon$ vs $(\frac{1}{2})\ s/b$.

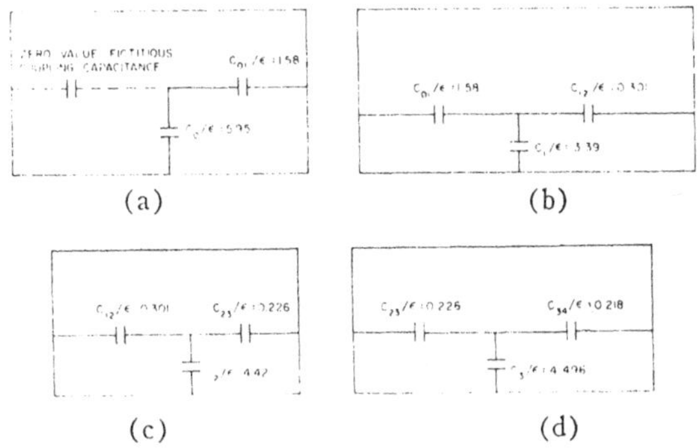

Fig. 6—Groupings of self and mutual capacitances used in the design example.

Next, the same technique is applied to the values of normalized capacitance in Fig. 6(c). The correct auxiliary curve of constant d/b is shown in Fig. 8, in which the coordinates of the intercepts with $C_m/\epsilon = 0.226$ and 0.301 are (0.381, 2.24) and (0.335, 2.18), respectively. Note that the ordinates of the intercepts are unequal; this will usually be the case. However, the sum of the ordinates must equal the normalized capacitance to ground. In this case the sum of the ordinates is 2.24 $+2.18 = 4.42 = C_2/\epsilon$. By linear interpoltion d/b of rod 2 is found to be 0.352.

Proceeding to Fig. 6(b) and again applying the design technique, it is found that the coordinates of the intercepts of the auxiliary constant d/b curve with the $C_{12}/\epsilon = 0.301$ and $C_{01}/\epsilon = 1.582$ curves are (0.330, 2.03) and (0.098, 1.36), respectively. The d/b of rod 1 is found by linear interpolation to be 0.324.

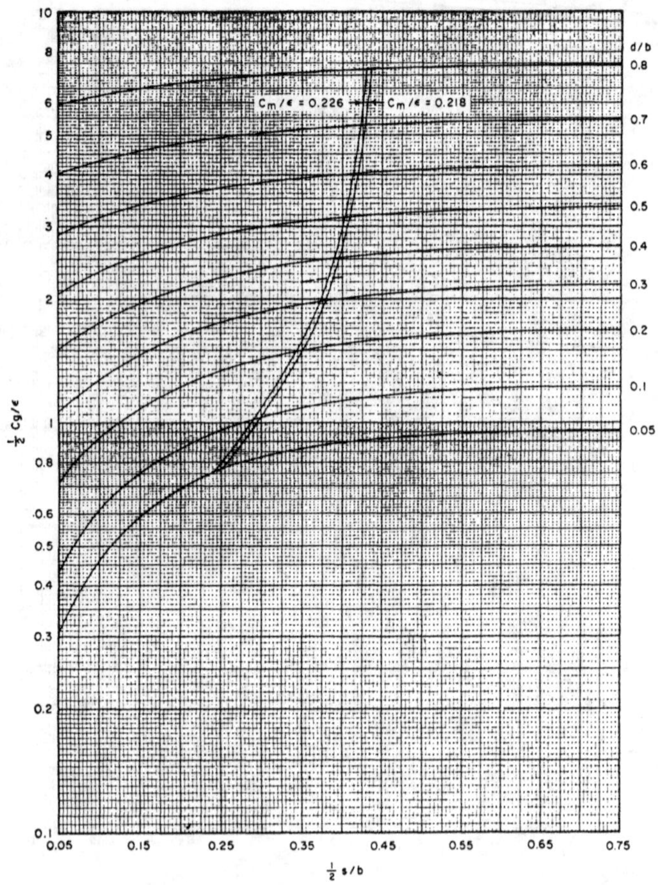

Fig. 7—Determination of the normalized diameter of rod 3 and the partial spacings to rods 2 and 4.

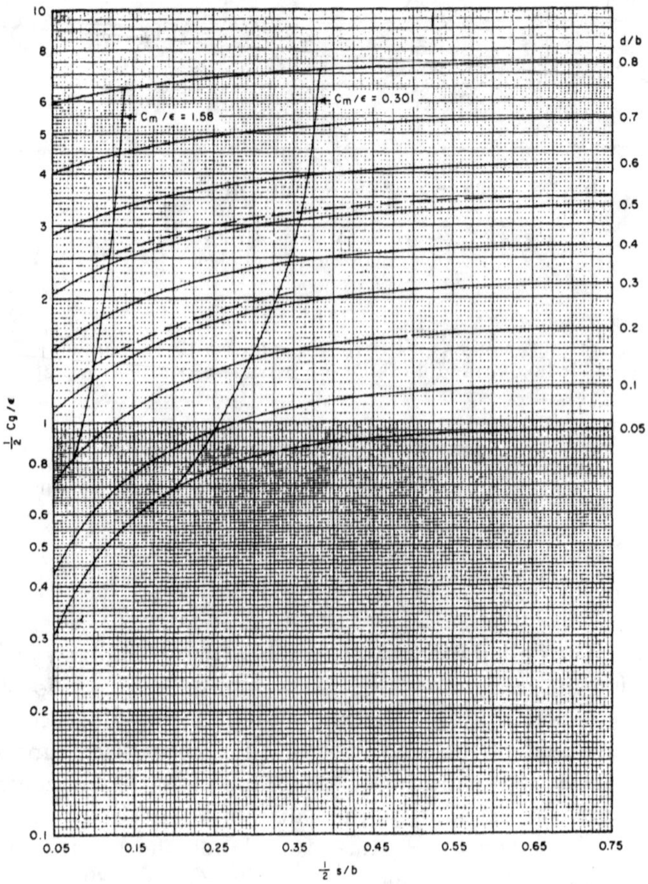

Fig. 9—Determination of the normalized diameter of rods 0 and 1, and the respective partial spacings to adjacent rods.

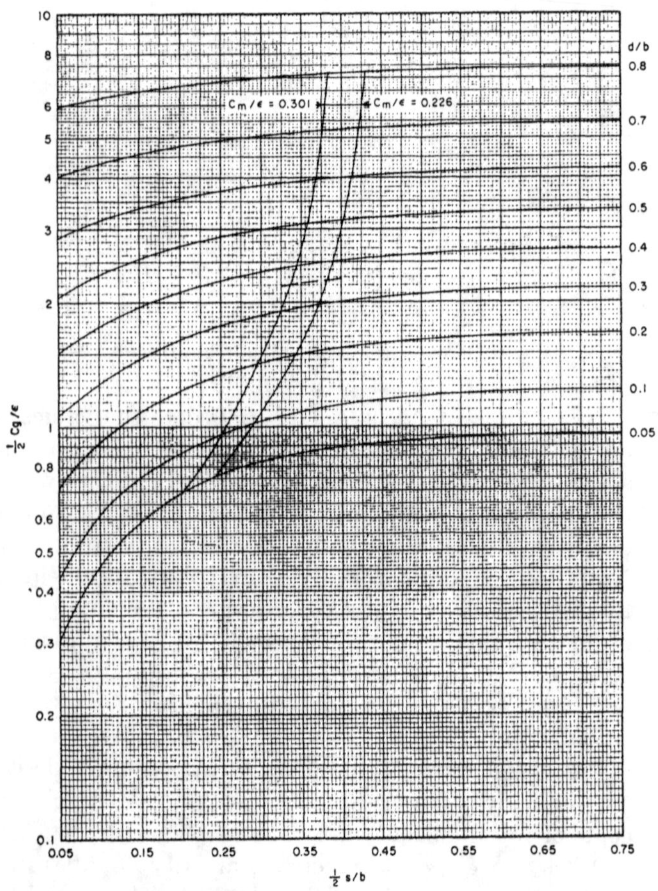

Fig. 8—Determination of the normalized diameter of rod 2 and the partial spacings to rods 1 and 3.

Finally, the design technique is applied to Fig. 6(a), and the coordinates (0.120, 2.5) are determined as the intercept of the auxiliary constant d/b curve and the $C_{01}/\epsilon = 1.582$ curve. The coordinates (∞, 3.45) are found as the intercept of the same constant d/b curve with the fictitious coupling capacitance to the left of C_0 set equal to zero. In the latter instance the right-hand extreme of Fig. 5 is used where the constant d/b curves are virtually flat. By linear interpolation, d/b of rod 0 is found to be 0.516. The auxiliary constant d/b curves used in the last two subsections are shown in Fig. 9.

The respective normalized spacings, $s_{k,k+1}/b$ are obtained by summing abscissa values that correspond to the same mutual capacitance. Bookkeeping is simplified by placing abscissa values along their corresponding mutual capacitances:

Abscissa Value, $(1/2)s/b$	$C_{k,k+1}$	Obtained From
0.120	C_{01}	Fig. 6(a)
0.098 0.330	C_{01} C_{12}	Fig. 6(b)
0.335 0.381	C_{12} C_{23}	Fig. 6(c)
0.381 0.387	C_{23} C_{34}	Fig. 6(d)

Therefore,

$$s_{01}/b = 0.120 + 0.098 = 0.218 = s_{67}/b$$
$$s_{12}/b = 0.330 + 0.335 = 0.665 = s_{56}/b$$
$$s_{23}/b = 0.381 + 0.381 = 0.762 = s_{45}/b$$
$$s_{34}/b = 0.387 + 0.387 = 0.774.$$

Center to center spacings may be obtained by adding to the appropriate s/b value $(1/2)d/b$ of the rod on the left and $(1/2)d/b$ value of the rod on the right.

Thus,

$$c_{k,k+1}/b = s_{k,k+1}/b + (1/2)d_k/b + (1/2)d_{k+1}/b$$
$$c_{01}/b = 0.218 + 0.258 \; + 0.162 \; = 0.638 = c_{67}/b$$
$$c_{12}/b = 0.665 + 0.162 \; + 0.176 \; = 1.003 = c_{56}/b$$
$$c_{23}/b = 0.762 + 0.176 \; + 0.175 \; = 1.114 = c_{45}/b$$
$$c_{34}/b = 0.774 + 0.1755 + 0.1755 = 1.125.$$

The normalized rod diameters obtained as described previously using Figs. 7–9 are

$$d_0/b = 0.516 = d_7/b$$
$$d_1/b = 0.324 = d_6/b$$
$$d_2/b = 0.352 = d_5/b$$
$$d_3/b = 0.351 = d_4/b.$$

The design specifications for the filter using rectangular bars and the same filter using round rods are compared in Table II, where the spacings between successive round rods are closer, in general, than spacings between successive rectangular bars. Also, the round rods are larger in cross-sectional areas than the rectangular bars.

TABLE II
PARAMETERS OF AN INTERDIGITAL FILTER DESIGN USING ROUND RODS COMPARED WITH THE SAME FILTER USING RECTANGULAR BARS

k	Round Rods	Rectangular Bars	k	Round Rods	Rectangular Bars	Rectangular Bars
	$s_{k,k+1}/b$	$s_{k,k+1}/b$		$(d/b)_k$	$(w/b)_k$	$(t/b)_k$
0 and 6	0.218	0.255	0 and 7	0.516	0.648	0.3
1 and 5	0.665	0.670	1 and 6	0.324	0.243	0.3
2 and 4	0.762	0.820	2 and 5	0.352	0.294	0.3
3	0.774	0.830	3 and 4	0.351	0.293	0.3

III. MATHEMATICAL DETERMINATION OF C_m AND C_g

The boundary-value problems associated with the periodic cylindrical rods between parallel ground planes depend on the mode of excitation. When the structure is excited in the odd mode the mathematical problem is to satisfy Laplace's equation in the region shown in Fig. 10(a), and the boundary conditions as given in the figure. This is a Dirichlet problem. When the structure is excited in the even mode the mathematical problem is to satisfy Laplace's equation in the region shown in Fig. 10(b) and the boundary conditions as shown in this figure. This is a boundary-value problem with mixed boundary conditions.

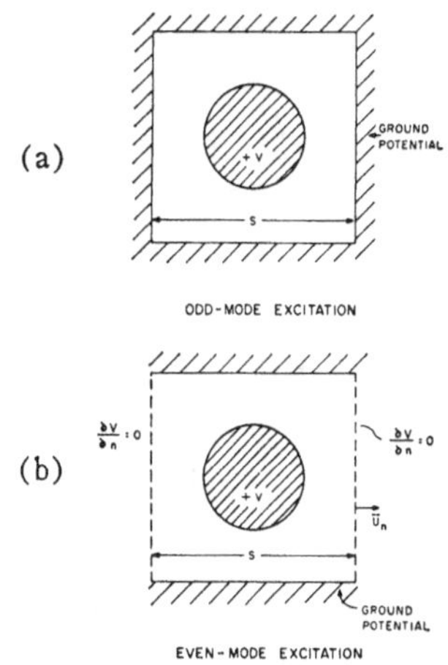

Fig. 10—Equivalent boundary value problems associated with the even and odd mode excitation of the coupled circular rods.

As stated in Section II, the capacitance associated with the solution of the Dirichlet problem is given by (2), and the capacitance associated with the solution of the mixed boundary problem is given by (3). From these are obtained C_m and C_g.

The method of determining the even and odd mode capacitances was to solve the integral equation associated with the boundary-value problem and thus obtain the normalized surface charge density; then the total normalized charge was obtained by integration. The normalized charge was divided by the potential difference to give the normalized capacitance. The generalized integral equation for both boundary-value problems is

$$V(\bar{r}_0) = -\frac{1}{2\pi}\oint_P \left[\frac{\eta(\bar{r})}{\epsilon}\right] \ln |\bar{r} - \bar{r}_0| \, ds(\bar{r}) + \frac{1}{2\pi}\oint_P V(\bar{r}) \frac{\bar{u}_r \cdot \bar{u}_n}{|\bar{r} - \bar{r}_0|} ds(\bar{r}) + \frac{1}{2} V(\bar{r}_0) \quad (7)$$

wherever the boundary curve has a unique tangent.[2] P denotes the principal value of the integral.[3]

[2] In Fig. 11, it is shown that the path of integration for (7) has an interior right angle. If the position vector $\bar{r}_0$ lies at this point, the term $(1/2)V(\bar{r}_0)$ on the right-hand side of (7) should be replaced by $(3/4)V(\bar{r}_0)$.

[3] Equation (7) may be derived by taking the two-dimensional integral formulation for the potential at a point within a closed boundary and permitting the observation point to approach the boundary (see for example, J.A. Stratton, "Electromagnetic Theory," McGraw-Hill Book Co., Inc., New York, N.Y., pp. 166-170 and problem 8, p. 219; 1941). The limiting operation may be performed as in Mai and Van Bladel [11], or alternatively, the observation point may be placed on the boundary and the boundary curve in the vicinity of the observation point deformed into an appropriate segment of a circle of infinitesimally small radius. The deformation is such as to leave the observation point within the boundary. The segment of circle is π radius where the boundary curve has a unique tangent and is $(3/2)\pi$ radians at the right angle.

For the Dirichlet problem (7) reduces to

$$0 = \oint_P \left[\frac{\eta(\bar{r})}{\epsilon}\right] \ln |\bar{r} - \bar{r}_0| \, ds(\bar{r})$$

on the outer boundary

$$-2\pi(V_i - V_o) = \oint_P \left[\frac{\eta(\bar{r})}{\epsilon}\right] \ln |\bar{r} - \bar{r}_0| \, ds(\bar{r})$$

20

on the inner boundary (8)

where V_i and V_o are the respective potentials on the inner and the outer boundary. The notation used in (7) and (8) is clarified by Fig. 11. Vector $\bar{r}$ is a position vector from an arbitrary origin to a point on the boundary at which the integrand is being evaluated. Vector $\bar{r}_0$ is a position vector to a point on the boundary where the voltage is being evaluated. Voltages $V(r)$ and $V(\bar{r}_0)$ are the voltages at the vector positions $\bar{r}$ and $\bar{r}_0$, respectively. Vector $\bar{u}_r$ is a unit vector in the direction of $\bar{r}-\bar{r}_0$. Vector $\bar{u}_n$ is the outward pointing unit normal to the boundary surface. Finally, $ds(\bar{r})$ is the differential length of the boundary curve at the vector position $\bar{r}$, and $\eta(\bar{r})/\epsilon$ is the normalized surface charge density at the vector position $\bar{r}$.

The method of solution was to divide the boundary curve into n subintervals, and assign to each subinterval an unknown constant normalized surface charge density or an unknown constant potential. These unknown constants were then removed from under the integral sign and the integrations performed analytically or numerically in each subinterval. In the general case, (7) takes the form

$$V(\bar{r}_i) = \frac{1}{2\pi}\sum_{j=1}^{n}\left\{-\frac{\eta(\bar{r}_j)}{\epsilon}\int_{\text{subinterval } j} \ln |\bar{r}_j - \bar{r}_i| \, ds(\bar{r}_j) + V(\bar{r}_j)\int_{\text{subinterval } j} \frac{\bar{u}_r \cdot \bar{u}_n}{|\bar{r}_j - \bar{r}_i|} \, ds(\bar{r}_j)\right\} + \left(\frac{1}{2}\right)V(\bar{r}_i) \quad (9)$$

for $i=1, 2, \cdots, m$. Because of the symmetry of the boundary curve and boundary conditions, (9) may be reduced by a factor of 4 or, alternatively, the problem may be reformulated using one quadrant of the structure. (The former procedure was used in this work.) After the integrations in (9) had been carried out, the equations were arranged into a matrix. For (8) the matrix takes the form

$$\begin{pmatrix} 0 \\ 0 \\ \cdot \\ \cdot \\ V \\ V \\ \cdot \\ \cdot \\ V \end{pmatrix} = \begin{pmatrix} g_{11} & g_{12} & \cdots & g_{1n} \\ g_{21} & & \cdots & g_{2n} \\ \cdot & & & \cdot \\ \cdot & & & \cdot \\ \cdot & & & \cdot \\ g_{m1} & & & g_{mn} \end{pmatrix} \begin{pmatrix} \eta_1/\epsilon \\ \eta_2/\epsilon \\ \cdot \\ \cdot \\ \cdot \\ \eta_n/\epsilon \end{pmatrix} \quad (10)$$

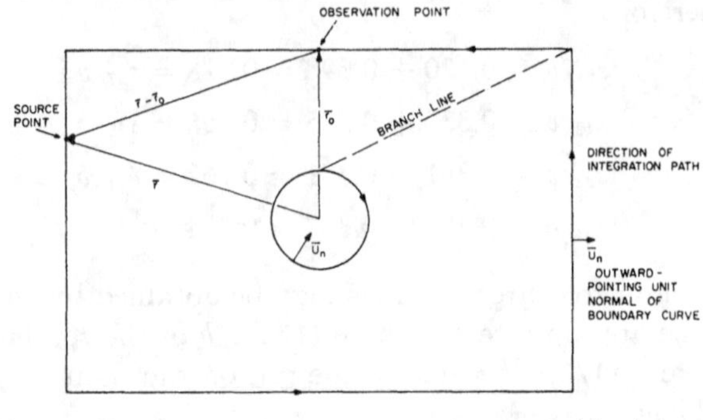

Fig. 11—Geometry and vector nomenclature used in the even and odd mode boundary value integral equation.

where V is the potential difference between the boundaries. For (7), the matrix takes the form

$$\begin{pmatrix} \tilde{V}_1 \\ \tilde{V}_2 \\ \cdot \\ \cdot \\ \cdot \\ \tilde{V}_m \end{pmatrix} = \begin{pmatrix} G_{11} & G_{12} & \cdots & G_{1n} \\ G_{21} & & \cdots & G_{2n} \\ \cdot & & & \cdot \\ \cdot & & & \cdot \\ \cdot & & & \cdot \\ G_{m1} & & & G_{mn} \end{pmatrix} \begin{pmatrix} \eta_1/\epsilon \\ \cdot \\ \cdot \\ \eta_s/\epsilon \\ V_1 \\ \cdot \\ \cdot \\ V_r \end{pmatrix} \quad (11)$$

where V_i $(i=1, 2, \cdots, r)$ represents the unknown potential on the part of the boundary where $\partial V/\partial n = 0$; and $\tilde{V}_i$ $(i=1, 2, \cdots, m)$ are constants which take into account contributions from the second term on the right in (7). The entries g_{ij} and G_{ij} are constants obtained from the indicated integrations in (9).

In solving (10) and (11) it was found that convergence was quicker in many cases when an overdetermined set of equations was compiled (*i.e.*, $m > n$) and a least squares method used to solve them [4]. This was done in all cases. In order to guarantee accuracy of the inverted matrix, which was generally a 40 by 40 matrix, perturbation methods were used which required the inverse matrix to be accurate to 0.01 per cent [5]. This was found necessary because of the loss of significant digits in inverting the very large matrices in (10) and (11).

In order to determine whether or not a solution to (10) or (11) gave a sufficiently close approximation to the even or odd mode capacitance, the number of subintervals of the boundary curve was increased, thereby increasing the number of simultaneous equations. The capacitance obtained from the solution of the new set of equations was next compared with that of the former set. When the capacitance remained sufficiently invariant to successive enlargements of the matrix, convergence was assumed.[4] In order to avoid having to test every solution by this method, several geometries were picked that appeared to offer the greatest difficulty of obtaining convergence. The largest order that was required to obtain convergence for these geometries was used for all subsequent cases.

[4] The problem of convergence and of the accuracy of the solutions will be discussed in Section V.

One additional refinement was explored in attempting to obtain fast convergence. The side condition that the total charge be zero was attached, although the condition is redundant in that it is satisfied by the proper solution. This technique was applied only to (11). In the few cases examined, the additional side condition did appear to result in a better approximation to the correct answer with smaller matrices than the former method without side conditions. However, for large matrices no general trend was clear. Also for rods of very small diameter (*i.e.*, $d/b<0.3$), the inverse matrix of (11) could not be obtained in the time allowed. This inability to obtain an inverse matrix was probably due to the side condition making the matrix of (11) extremely skew [6] and thereby causing too great a loss of significant digits when the matrix was inverted.

IV. Physical Basis for the Method of Designing Filters that Use Rods of Unequal Diameters and Spacings

Fig. 12 shows three rods of a hypothetical multirod filter configuration with the self capacitances divided into left and right components parts. The following assumptions are made concerning the coupled-rod configuration.

1) Coupling beyond nearest neighbors is negligible.
2) A moderate change in the size or spacing of a given rod has only a second-order effect on the charge distribution on the far side of adjacent rods.

Therefore, by assumptions 1) and 2), the mutual capacitance $C_{i,i+1}$ is dependent only on the right half of rod i and the left half of $i+1$, and (from Fig. 12) C_i^L is independent of the spacing and diameter of the adjacent rod on the right; C_i^R is independent of the spacing and diameter of the adjacent rod on the left. The self capacitance of a rod is given by

$$C_i = C_i^L + C_i^R. \quad (12)$$

Consider the rods in Fig. 12 to be excited in the odd mode. This is shown in Fig. 13 where the electric flux lines are sketched and a ground value equipotential line is approximated. The equipotential line will be approximately straight for rods having moderate and far spacings and should not be strongly dependent on the rod diameters with these spacings. (In most filter designs the rods in the center of the filter are nearly the same diameter and only vary greatly at the ends of the filter.) If the rods were excited in the even mode, then the equipotential line in Fig. 13 would approximately be the line where, also, $\partial V/\partial n=0$.[5]

Again from Fig. 13, it may be stated on the basis of the previous approximations that the left half of rod 2 sees boundary conditions which are equivalent to a system of equispaced, equidiameter rods of diameter d_2 spaced at $s=s_2^L$ and having a mutual capacitance

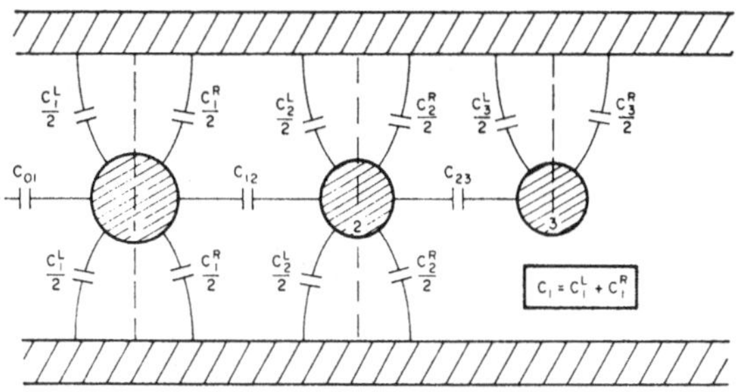

Fig. 12—Hypothetical multirod filter configuration showing division of self capacitance into left and right component parts.

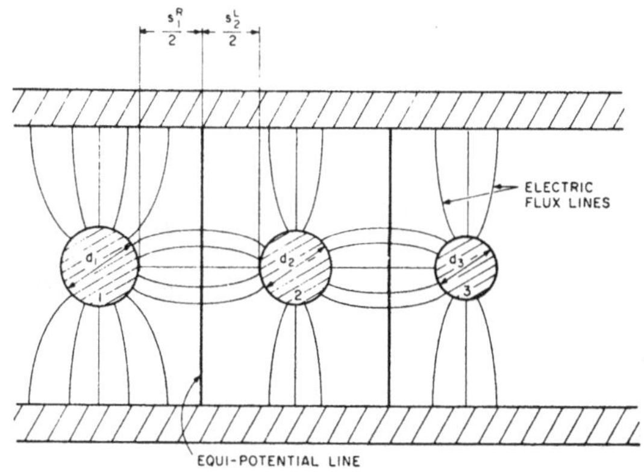

Fig. 13—Hypothetical multirod filter configuration excited in odd mode.

$C_m=C_{12}$. (Note that $s_2^L/2$ is indicated in Fig. 13.) At the same time, the right half of rod 1 sees boundary conditions equivalent to a system of equispaced, equidiameter rods of diameter d_1 spaced at $s=s_1^R$ but also with a mutual capacitance of $C_m=C_{12}$.[6] Through the use of this artifice (*i.e.*, approximating left and right boundary conditions independently) the designer is able to use the data given in Figs. 2 and 3 to obtain arbitrary values of mutual and self capacitance needed in a filter design. The design method described in Section II is one method of utilizing the above artifice. The method described was found easy to apply and also efficient in terms of design time.

V. Accuracy of the Solutions

A difficulty that arises from solving (7) and (8) by the method discussed in Section III is that the accuracy of the solution cannot be precisely stated. Although other methods of solutions are possible which in theory give upper and lower bounds for the answer [7] (at least in the Dirichlet problem), practical considerations make these methods undesirable in this particular problem. However, in order to be sure of the results obtained by the method of Section III, several comparisons are made in Table III with data taken from other sources.

By replacing the round rods with infinitesimal line charges located at the center of each cylinder and using the method of images, a solution has been obtained

[5] This approximation is perhaps in error more than previous approximations. However, for moderately spaced rods the even mode capacitance is not strongly dependent on spacing. Therefore, a larger error in the position of the $\partial V/\partial n=0$ line can be tolerated.

[6] This concept (*i.e.*, that the proper left and right half spacings are determined when successive rods have the same mutual capacitance) is credited to Dr. G. L. Matthaei.

21

22

TABLE III

Comparisons of Characteristic Impedances Obtained from Solutions of (7) and (8) With Those Obtained From Published Sources

Case d/b	Case s/b	Z_{even} ohms This Work	ITT-Handbook [8]* $Z_{even}=138\log_{10}[(4/\pi)/(d/b)]$ Accuracy not specified	Per cent Difference
0.6	1.5	44.1	45.1	1.5 per cent
0.4	1.5	68.7	69.4	1.0 per cent
0.2	1.5	109.8	110.9	0.99 per cent
Case d/b	**Case s/b**	**Z_{odd} ohms This Work**	**ITT-Handbook [9]** Z_{odd} ohms Accuracy not specified**	**Per cent Difference**
0.5	0.5	46.1	45.0	2.5 per cent
0.1	0.9	142.6	141.5	0.78 per cent
Case d/b	**Case s/b**	**Z_{odd} ohms This Work**	**Chisholm [10]*** Z_{odd} ohms Accuracy about 0.6 per cent, when the impedance is of the order of 50 ohms.**	**Worst Per cent Difference**
			Min — Max	
0.8	1.5	25.2	25.26 — 25.86	2.31 per cent
0.6	1.5	44.1	44.25 — 44.40	0.68 per cent

* The case here is for a single wire between parallel ground planes which is similar to the case of widely spaced coupled rods excited in even mode.
** The case here is for a single wire centered in a square waveguide which is equivalent to the odd mode case wherein $d/b+s/b=1$.
*** The case here is the trough line; wide spacing is used to make the comparison practical.

which gives the correct asymptotic form to the solutions of the actual round-rod problems in the even and odd mode. A second-order correction to the solution was made which gives the following equations for the even and odd mode normalized capacitance:

$$\frac{\epsilon}{C_{odd}}=\frac{1}{2\pi}\left|\ln\frac{\frac{\pi}{4}\frac{d}{b}}{\sqrt{1-\left(\frac{d}{2b}\right)^4}}-\frac{1}{2}\ln\left[1-\left(\frac{d/b}{2(c/b)}\right)^4\right]+2\sum_{m=1}^{\infty}(-1)^m\ln\tanh m\frac{\pi}{2}\frac{c}{b}\right| \tag{13}$$

$$\frac{\epsilon}{C_{even}}=\frac{1}{2\pi}\left|\ln\frac{\frac{\pi}{4}\frac{d}{b}}{\sqrt{1-\left(\frac{d}{2b}\right)^4}}+\frac{1}{2}\ln\left[1-\left(\frac{d/b}{2(c/b)}\right)^4\right]+2\sum_{m=1}^{\infty}\ln\tanh m\frac{\pi}{2}\frac{c}{b}\right|. \tag{14}$$

These equations should be quite accurate for small d/b, say $d/b<0.3$, and $c/b>3(d/b)$. Table IV shows a comparison of C_o/ϵ and C_m/ϵ obtained from (13) and (14) with those obtained from the method in Section III.

In conclusion, the number and consistency of very good checks indicates that the accuracy of C_m/ϵ and G_o/ϵ obtained from the numerical solutions of (7) and (8) is generally better than 2 and 1 per cent, respectively.

TABLE IV

Comparison of Normalized Capacitances Obtained from (13) and (14) with those Obtained from Solutions of (7) and (8)

d/b	s/b	Per Cent Difference Mutual Capacitances	Per Cent Difference Self Capacitances
0.4	0.40	2.43	1.60
	0.50	2.68	1.08
	0.70	2.83	0.53
	0.90	2.77	0.30
	1.20	2.46	0.16
	1.50	3.82	0.16
0.2	0.20	0.06	1.057
	0.30	0.17	0.01
	0.40	0.72	0.64
	0.50	0.39	0.21
	0.70	1.50	0.40
	0.90	1.28	0.16
	1.20	1.21	0.04
	1.50	1.22	0.01
0.1	0.20	0.07	0.12
	0.30	0.06	0.08
	0.40	0.26	0.30
	0.50	1.01	0.64
	0.70	2.06	0.58
	0.90	1.22	0.14
	1.20	1.06	0.05
	1.50	1.09	0.01
0.05	0.10	0.52	0.54
	0.15	0.12	0.204
	0.20	0.02	0.13
	0.30	0.03	0.08
	0.40	0.16	0.14
	0.50	0.59	0.33
	0.70	3.55	0.87
	0.90	1.33	0.17
	1.20	1.09	0.01
	1.50	1.03	0.02

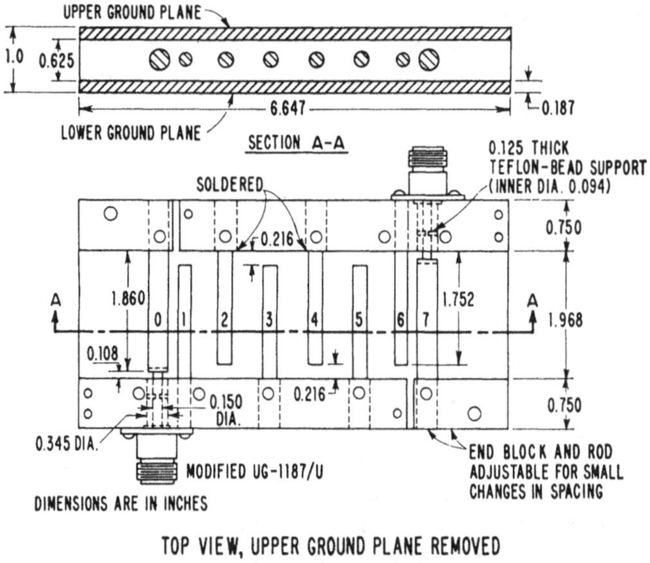

ROD DIAMETERS AND CENTER TO CENTER SPACING IN INCHES

K	DIAMETER OF ROD K	K	CENTER TO CENTER SPACING ($C_{K, K+1}$) OF RODS K AND K+1
0 AND 7	0.322	0 AND 6	0.399*
1 AND 6	0.200	1 AND 5	0.627
2 AND 5	0.220	2 AND 4	0.696
3 AND 4	0.219	3	0.703

* ADJUSTED TO 0.391 IN FINAL STRUCTURE

Fig. 14—Drawing of interdigital filter using round rods.

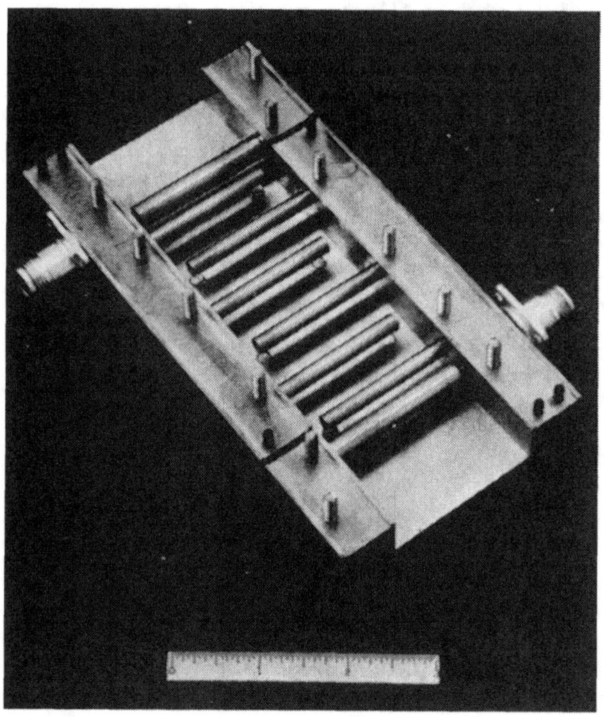

Fig. 15—Photograph of a trial interdigital filter using round rods.

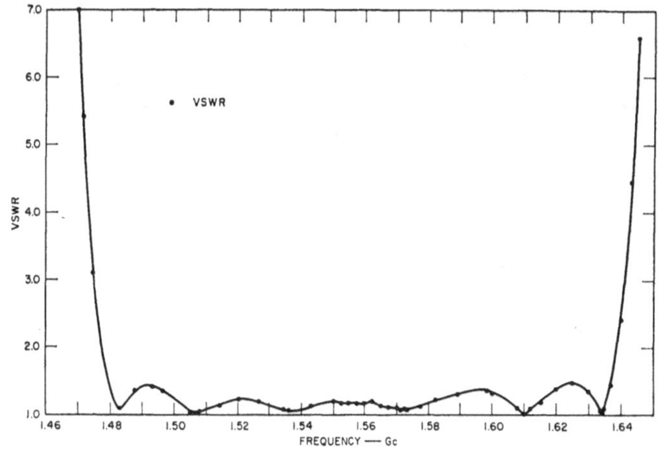

Fig. 16—Measured VSWR of interdigital filter using round rods.

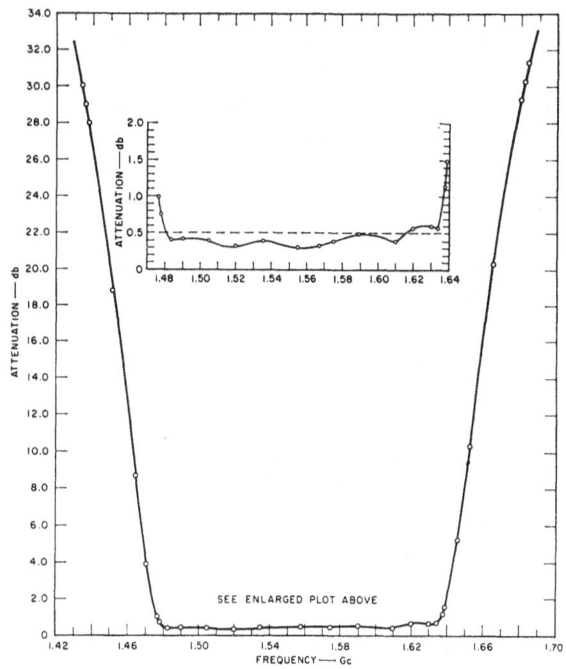

Fig. 17—Measured attenuation of interdigital filter using round rods.

VI. Trial Interdigital Filter

The interdigital filter whose design was described in Section II was constructed with $\frac{5}{8}$-inch ground-plane spacing and design center frequency of 1.5 Gc. Previous work with interdigital filters using rectangular bars showed that the spacing between the end resonators and the impedance-transforming section (*i.e.*, rods 0 and 1 and rods 6 and 7 in the present case) should be adjustable for small changes in spacing [2]. Therefore, the filter was constructed accordingly. A drawing of the filter is given in Fig. 14, and a photograph of the constructed filter is given in Fig. 15.

When the filter was initially tested, the pass band VSWR was slightly high. This condition was corrected by reducing the spacing between the terminations and

23

the first resonator on each end (*i.e.*, s_{01} and s_{67}) by about 0.008 inch—or about 6 per cent. The resulting VSWR is shown in Fig. 16. It is seen to be very nearly Chebyshev although the ripple peaks near the band edge are slightly high. (A 0.1-db Chebyshev ripple calls for VSWR peaks of 1.36.)

Fig. 17 shows the measured attenuation characteristic of the filter. The measured fractional bandwidth is 0.0996 which is very near the design value of 0.10. 24 (With rectangular bars, the same filter design has a measured fractional bandwidth of 0.0935, a shrinkage of 7 per cent in bandwidth [2].) The shift of the center frequency of the filter to 1.557 results from the resonators being slightly short, as noted by Matthaei [2].

VII. Conclusions

The normalized self and mutual capacitances of periodic, circular cylindrical rods located between parallel ground planes were given graphically. The capacitances were determined by solving the appropriate integral equation by numerical methods. Data were presented for rod diameter-to-ground-plane spacing ratios varying from 0.05 to 0.8 for very near to very far rod spacings. Accuracy of the data is believed to be generally better than 2 per cent for the normalized mutual capacitance, and generally better than 1 per cent for the normalized self capacitance.

An approximate design method was also presented which permits using the data to synthesize filters that require rods of nonequal diameters and spacings. The design method should be most reliable for moderately and far spaced rods, but it should also be suitable for more closely spaced rods. (Closely spaced rods would probably require some experimental adjustments.)

An example of the design method was given. The relative rod diameters and spacings were determined for a 10-per cent 6-resonator 0.1-db Chebyshev ripple, interdigital filter. This design was constructed and tested, and its performance was found to conform closely to that called for by theory, thus tending to confirm the accuracy of the design data and the validity of the design method. Using round rods rather than rectangular bars in filters where these types of resonators are required (such as in comb-line and interdigital filters) should reduce manufacturing cost while retaining the electrical characteristics of the filter.

Acknowledgment

The computer programming of Stanford University's IBM 7090 computer for the computation of the data for the design of round rods between parallel ground planes was done by P. H. Omlor.

Robert Pierce fabricated the interdigital filter described in this paper, and York Sato made the laboratory tests.

Dr. G. L. Matthaei proposed a way of adapting the periodic, coupled-rod data to synthesis of filters that require arbitrary rod diameters and spacings. The author is indebted to him for his contribution to this paper and also for other helpful suggestions made during the course of this work.

References

[1] G. L. Matthaei, "Comb-line band-pass filters of narrow or moderate bandwidth," *Microwave J.*, vol. 6, pp. 82–96; August, 1963.

[2] G. L. Matthaei, "Interdigital band-pass filters," IRE Trans. on Microwave Theory and Techniques, vol. MTT-10, pp. 479–491; November, 1962.

[3] W. J. Getsinger, "Coupled rectangular bars between parallel plates," IRE Trans. on Microwave Theory and Techniques," vol. MTT-10, pp. 65–72; January, 1962.

[4] C. Lanczos, "Applied Analysis," Prentice-Hall, Inc., Englewood Cliffs, N. J., pp. 156–161; 1956.

[5] *Ibid.*, p. 148.

[6] *Ibid.*, pp. 161–163.

[7] H. J. Greenberg, "The determination of upper and lower bounds for the solution of the Dirichlet problem," *J. Math. and Phys.*, vol. 27, pp. 161–181; July, 1948.

[8] "Reference Data for Radio Engineers," International Telephone and Telegraph Corp., American Book-Stratford Press, Inc., New York, N. Y., 4th Edition, ch. 20, Part P, p. 592; 1956.

[9] *Ibid.*, ch. 20, p. 590, Part H.

[10] R. M. Chisholm, "The characteristic impedance of trough and slab lines," IRE Trans. on Microwave Theory and Techniques, vol. MTT-4, pp. 166–172; July, 1956.

[11] K. K. Mei, J. G. Van Bladel, "Scattering by perfectly-conducting rectangular cylinders," IEEE Trans. on Antennas and Propagation, vol. AP-11, pp. 191–192; March, 1963.

Reprinted from IEEE TRANSACTIONS
ON MICROWAVE THEORY AND TECHNIQUES
Volume MTT-12, Number 4, July, 1964

PRINTED IN THE U.S.A.

The two papers by Cohn (p. 27), (p. 37) are very useful for strip-line couplers in which the two strips lie in the same plane. The paper by Shelton (p. 42) treats coupled strips that lie in different planes. The principal advantage of this form of coupling is that the strips can then cross without touching. This remedies very neatly the problem of the crossing connecting lines in cascaded couplers, mentioned earlier. As far as the mechanical layout is concerned, this technique converts a **backward** coupler into a **forward** coupler. The spacing between the two planes containing the strip lines is usually so chosen that the two strips are exactly on top of each other where the coupling is required to be most tight. The main disadvantage of the offset strips is that it upsets the symmetry and excites other modes (notably the parallel-plate mode) which have to be suppressed by rows of screws through the printed boards, lining both sides of the strips. Another problem is that three dielectric sheets (instead of two) are now required; the center sheet may have to be quite thin. To avoid the asymmetry due to the two offset strips, a third strip may be introduced as was shown by Getsinger.[6,17] However this is mechanically more awkward, and Shelton's construction (p. 42) is nearly always preferred wherever coupled lines have to cross.

We have reproduced four design charts computed by J.A. Mosko and kindly supplied by him, immediately following the paper by Shelton, whose design formulas were used in the computation. These four charts were also reproduced (but much reduced in size) in J.A. Mosko's paper in the Microwave Journal, Vol. 10, No.5, pp. 35-37, April 1967, entitled "Coupling Curves for Offset Parallel-Coupled Strip Transmission Lines." The charts are self-explanatory. Note that the coupling coefficient **k** is defined as usual by

$$k = \frac{Z_{0e} - Z_{0o}}{Z_{0e} + Z_{0o}}$$

Where Z_{0e} and Z_{0o} are the even- and odd-mode impedances, respectively, of the coupled lines. The two dielectric constants chosen were 2.32 for polyolefin and 2.65 for Rexolite 2200. The tightest coupling obtains for $w_c = w$, when the two strips are not offset. This is the best place to make the strips cross over to obtain a forward coupler, as for example when two or more weaker couplers are cascaded to make a single stronger coupler. As w_c (the overlap) decreases, the coupling decreases. Negative w_c values are used for loose couplings—in this condition a projection of the two strips would reveal a gap between them instead of an overlap.

Reprinted from IRE TRANSACTIONS
ON MICROWAVE THEORY AND TECHNIQUES
Volume MTT-3, Number 5, October, 1955

PRINTED IN THE U.S.A.

Shielded Coupled-Strip Transmission Line

S. B. COHN†

Summary—**An analysis is made of the odd and even TEM modes on a pair of parallel co-planar strips midway between ground planes. Rigorous formulas are presented for the case of zero-thickness strips, while approximate formulas are given for strips of finite thickness and for strips printed on opposite sides of a thin dielectric sheet supported in air between ground planes (AIL construction). The characteristic impedances and the phase velocities of the two modes are necessary and sufficient information for the design of directional couplers, coupled-line filters, and other components utilizing the coupling between parallel-strip lines. In order to facilitate design work, nomograms are included in the paper which give the dimensions of the coupled-strip cross section in terms of the odd- and even-mode characteristic impedances. The characteristic-impedance scales of these nomograms may be read to an accuracy of better than one per cent over a wide range of values that is sufficient for most purposes.**

Introduction

NUMEROUS strip-line components utilize the coupling between parallel strips as a basic parameter in their design. Several examples of such components are shown in Fig. 1 (next page), where coupled lines are used to achieve a particular effect in each case. In order to design these circuits to meet prescribed performance specifications, it is necessary to have accurate data on the coupling effects of parallel strips. Solutions for the most important parameters have been obtained, and are presented in this paper.

Fig. 2 (next page) shows transverse field distributions for two fundamental TEM modes that can exist on a pair of parallel conducting strips between parallel ground planes. In Fig. 2(a), strips are at same potential and carry equal currents in the same direction. Because of the even symmetry of the electric field about the vertical axis, this mode will be called the *even* coupled-strip mode. In Fig. 2(b), strips are at equal but opposite potentials and carry equal currents in opposite directions. Due to the odd symmetry of the electric field, this mode will be called the *odd* coupled-strip mode. In the case of the odd mode, the vertical plane of symmetry is at ground potential, and may be replaced by a thin conducting wall joined electrically to the horizontal ground plates. It is clear from the field plots that the capacitance per strip to ground is less for the even case

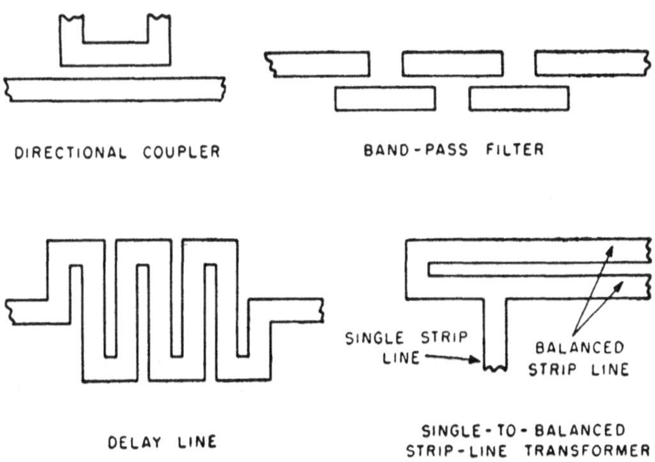

Fig. 1—Several applications of coupled-strip line construction.

and more for the odd case than for a single isolated strip of the same width. Consequently, the characteristic impedances of the two modes are unequal, being greater for the even than for the odd. In this paper, solutions for the two characteristic impedances will be given.[1] These quantities (plus the mode phase velocities, which are also treated) provide sufficient information for the

† Stanford Res. Inst., Menlo Park, Calif.

[1] (After this paper was prepared, a paper by D. Park appeared with a solution for Z_0 of the odd mode. The use of elliptic-integral identities shows Park's formula to be the same as mine.) D, Park. "Planar transmission lines," TRANS. IRE, vol. MTT-3, pp. 8–12; April, 1955.

design of directional couplers, filters, and other coupled-strip-line components. The graphs and nomograms given in this paper will be of sufficient accuracy for most of such design work.

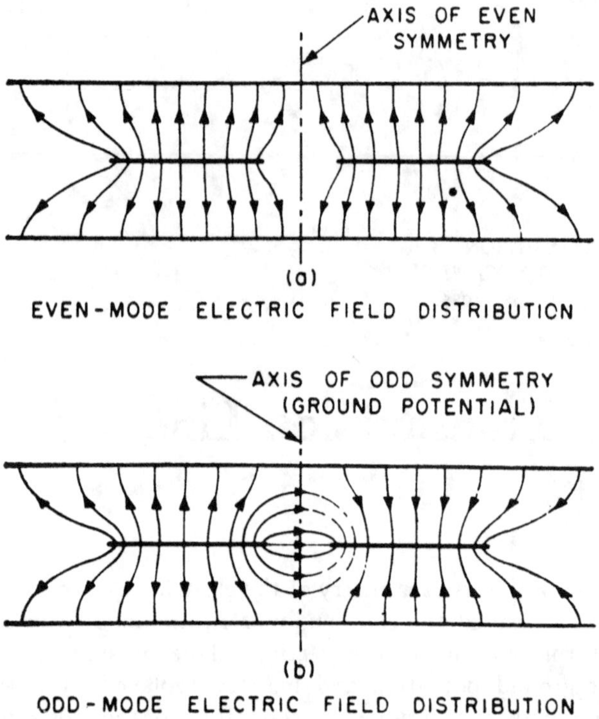

28

Fig. 2—Field distribution of the even and odd modes in coupled-strip line.

The particular cross sections considered here are shown in Fig. 3. The first, containing zero-thickness strips, is solved rigorously by the conformal-mapping technique. This construction may be closely realized in practice by copper-foil strips sandwiched between dielectric plates that are clad with copper on their outer surfaces. Since the cross section is virtually completely filled with the dielectric material, the phase velocity for both modes is

$$v = c/\sqrt{\epsilon_r}, \tag{1}$$

where c is the velocity of light in free space ($3.00(10)^8$ m/s) and ϵ_r is the relative dielectric constant of the medium in the cross section.

The second cross section in Fig. 3 is of importance in air-dielectric lines, where the thick strips may be supported by spaced dielectric beads, by metallic members at zero-voltage points, or by connectors at the strip ends. The third cross section utilizes foil strips printed on both sides of a thin dielectric sheet supported in air midway between ground plates (AIL construction). Approximate formulas for the second and third cross sections are given in this paper. The phase velocity for Fig. 3(b) is equal to that of light, or, if periodic bead support is used, it is given by (1) where an equivalent ϵ_r is selected to represent the effect of the spaced beads. However, the case of Fig. 3(c) is more complex, since the dielectric constant is not uniform in the cross section. The electric field energy in the dielectric sheet is greater for the odd mode than for the even mode, and therefore the phase velocity for the former will be less than for the latter. An approximate formula is given in this paper for the ratio of these velocities.

Rigorous Formulas for Zero-Thickness Strips

The following exact formulas are derived in Appendix A for the cross section of Fig. 3(a):

$$Z_{0e} = \frac{30\pi}{\sqrt{\epsilon_r}} \cdot \frac{K(k_e')}{K(k_e)} \text{ ohms} \tag{2}$$

where Z_{0e} is the even-mode characteristic impedance measured from *one* strip to *ground*, K is the complete elliptic integral of the first kind, ϵ_r is the relative dielectric constant of the material filling the cross section, and

$$k_e = \tanh\left(\frac{\pi}{2}\cdot\frac{w}{b}\right)\cdot\tanh\left(\frac{\pi}{2}\cdot\frac{w+s}{b}\right) \tag{3}$$

$$k_e' = \sqrt{1 - k_e^2}. \tag{4}$$

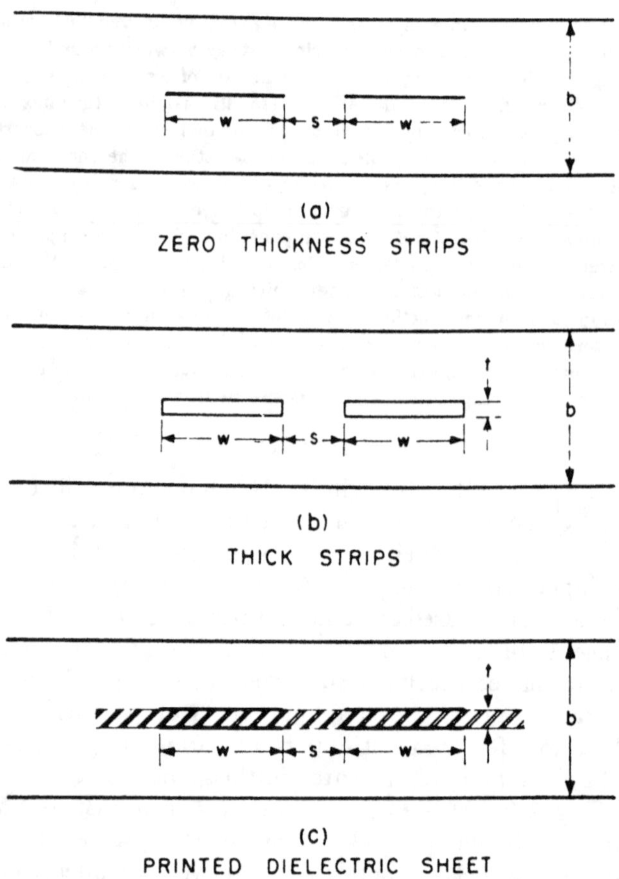

Fig. 3—Coupled-strip-line cross sections considered in this paper.

The variables w, b, and s are dimensions defined in Fig. 3(a). For the odd mode, the characteristic impedance from *one* strip to *ground* is

$$Z_{0o} = \frac{30\pi}{\sqrt{\epsilon_r}} \cdot \frac{K(k_o')}{K(k_o)} \text{ ohms}, \tag{5}$$

where

$$k_o = \tanh\left(\frac{\pi}{2}\cdot\frac{w}{b}\right)\cdot\coth\left(\frac{\pi}{2}\cdot\frac{w+s}{b}\right), \tag{6}$$

$$k_o' = \sqrt{1 - k_o^2}, \tag{7}$$

and all symbols are defined as above. If one allows s to become infinite, both Z_{0e} and Z_{0o} reduce to the characteristic impedance of an isolated strip between ground planes. The nature of the dependence of Z_{0e} and Z_{0o} on w/b and s/b is illustrated graphically by Fig. 4.

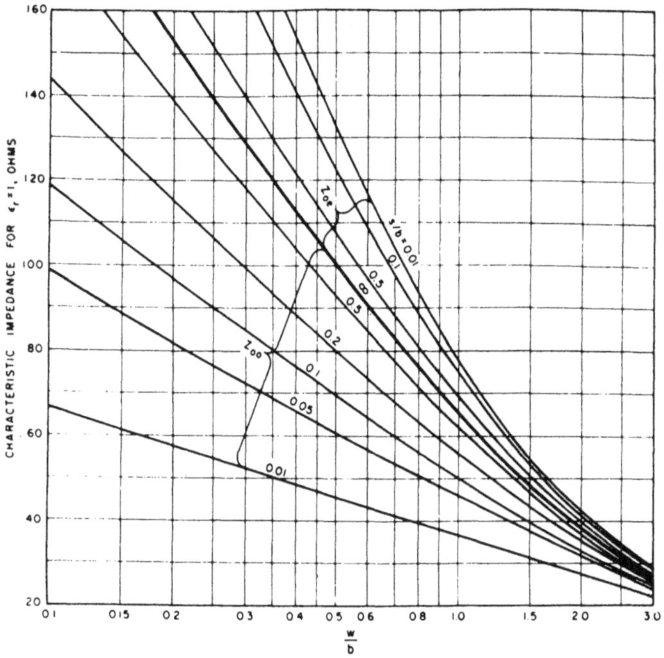

Fig. 4—Graph of Z_{0e} and Z_{0o} vs w/b and s/b.

When a pair of strips are used as a balanced transmission line, the characteristic impedance Z_b measured *between* the strips for the odd-mode distribution is the most useful quantity. This is related to Z_{0o} by

$$Z_b = 2Z_{0o}. \tag{8}$$

In the design of a directional coupler, filter, or other component, Z_{0o} and Z_{0e} are first computed to provide the desired circuit performance. Then the dimensions w, s, and b that will yield these characteristic impedances are determined. The following equations are convenient for calculating w/b and s/b,

$$\frac{w}{b} = \frac{2}{\pi}\tanh^{-1}\sqrt{k_e k_o} \tag{9}$$

and

$$\frac{s}{b} = \frac{2}{\pi}\tanh^{-1}\left(\frac{1-k_o}{1-k_e}\cdot\sqrt{\frac{k_e}{k_o}}\right). \tag{10}$$

The parameters k_e and k_o may be obtained as functions of Z_{0e} and Z_{0o} from a single-curve graph of either (2) or (5).

Nomograms for Zero-Thickness Strips

Eqs. (9) and (10) are expressed in nomogram form in Figs. 5 and 6 (pages 32 and 33). In Fig. 5, a straight edge aligned with the specified values of Z_{0e} and Z_{0o} on the outer scales will intersect the proper value of w/b, while in Fig. 6 a similar procedure will yield s/b. These nomograms are designed to be read accurately over a range of Z_{0e} and Z_{0o} large enough for most purposes. In this range, w/b and s/b can be determined with sufficient precision to ensure that Z_{0e} and Z_{0o} will be within one per cent of the desired values.

Fringing Capacitances for Zero Thickness Strips

A quantity of frequent interest is the equivalent fringing capacitance for zero-thickness coupled strips.[2] It will be assumed for simplicity that the strips are wide enough to allow the fringing capacitances at the two edges of each strip to be independent of w/b. This will hold with good accuracy for $w/b \geq 0.35$, which covers most practical applications. Subject to this assumption, (1) of a previous paper[3] may be written as follows:

$$Z_{0e} = \frac{94.15}{\sqrt{\epsilon_r}\left[\frac{w}{b} + \frac{1}{2\epsilon}\left(C_f'(0) + C_{fe}'\left(0, \frac{s}{b}\right)\right)\right]} \text{ ohms} \tag{11}$$

and

$$Z_{0o} = \frac{94.15}{\sqrt{\epsilon_r}\left[\frac{w}{b} + \frac{1}{2\epsilon}\left(C_f'(0) + C_{fo}'\left(0, \frac{s}{b}\right)\right)\right]} \text{ ohms} \tag{12}$$

where $\epsilon = 0.0885\,\epsilon_r$ in mmf per cm, $C_f'(0)$ is the fringing capacitance from one edge to one ground plane in mmf per cm for a single strip of zero thickness (i.e., for $s/b \to \infty$), and $C_{fe}'(0, s/b)$ and $C_{fo}'(0, s/b)$ are the corresponding fringing capacitances in mmf per cm at the adjacent edges of a pair of strips for the even and odd electric-field distributions. (See the sketches in Fig. 7 on page 34). The following rigorous formulas for the fringing capacitances were obtained by letting $w/b \to \infty$ in the analysis for zero-thickness coupled strips:

$$\frac{C_f'(0)}{\epsilon} = \frac{2}{\pi}\ln 2 = 0.4407 \tag{13}$$

$$\frac{C_{fe}'\left(0, \frac{s}{b}\right)}{\epsilon} \quad \frac{2}{\pi}\ln\left(1 + \tanh\frac{\pi s}{2b}\right)$$

$$= \frac{s}{b} - \frac{2}{\pi}\ln\left(\cosh\frac{\pi s}{2b}\right) \tag{14}$$

[2] (I have been informed in a private communication that C. A. Hachemeister has previously derived general formulas for these capacitances. His results are in a different form than mine, and their equivalence was not checked.)

[3] S. B. Cohn, "Characteristic impedance of the shielded-strip transmission line," Trans. IRE, vol. MTT-2, pp. 52–57; July, 1954.

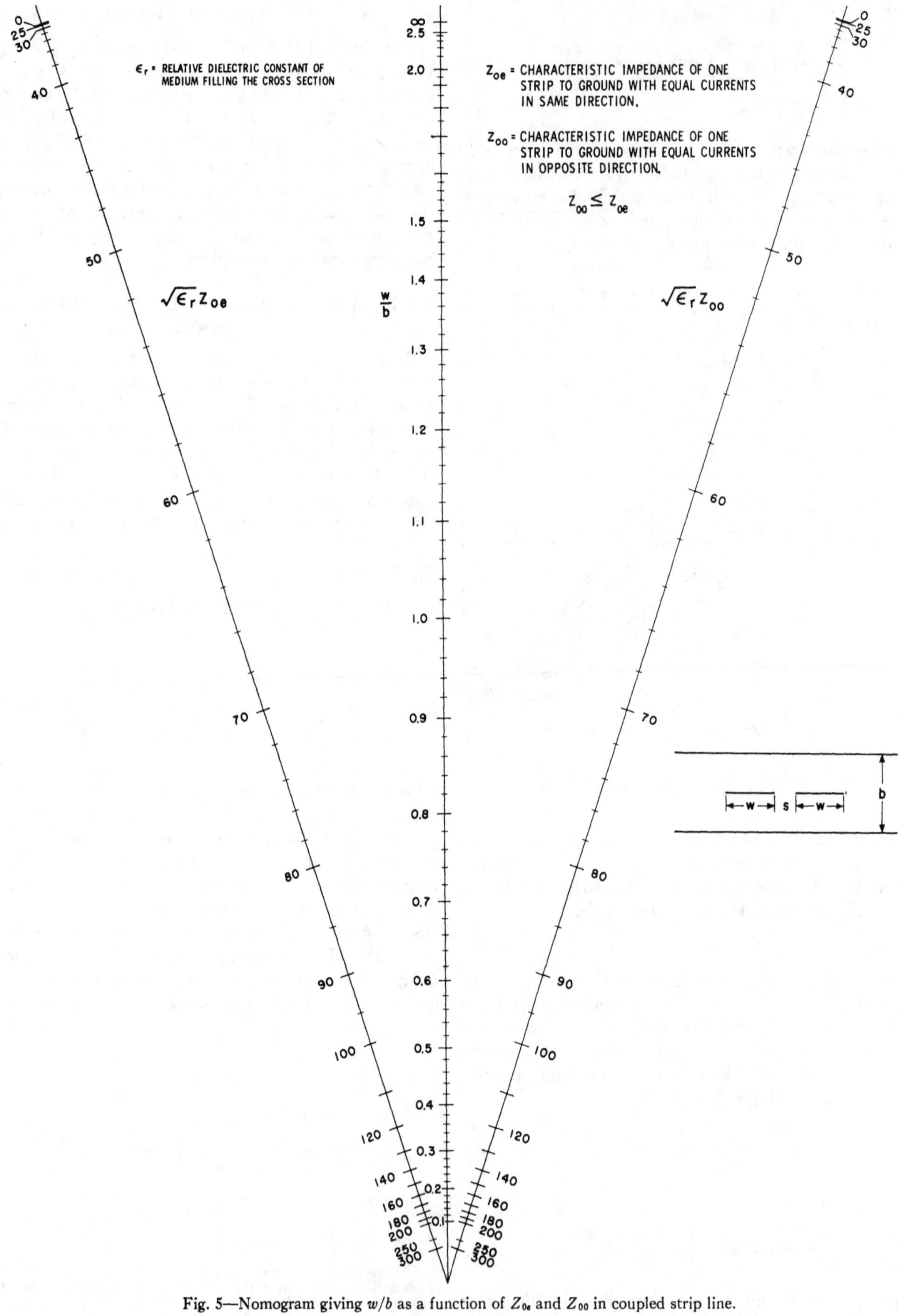

Fig. 5—Nomogram giving w/b as a function of Z_{oe} and Z_{oo} in coupled strip line.

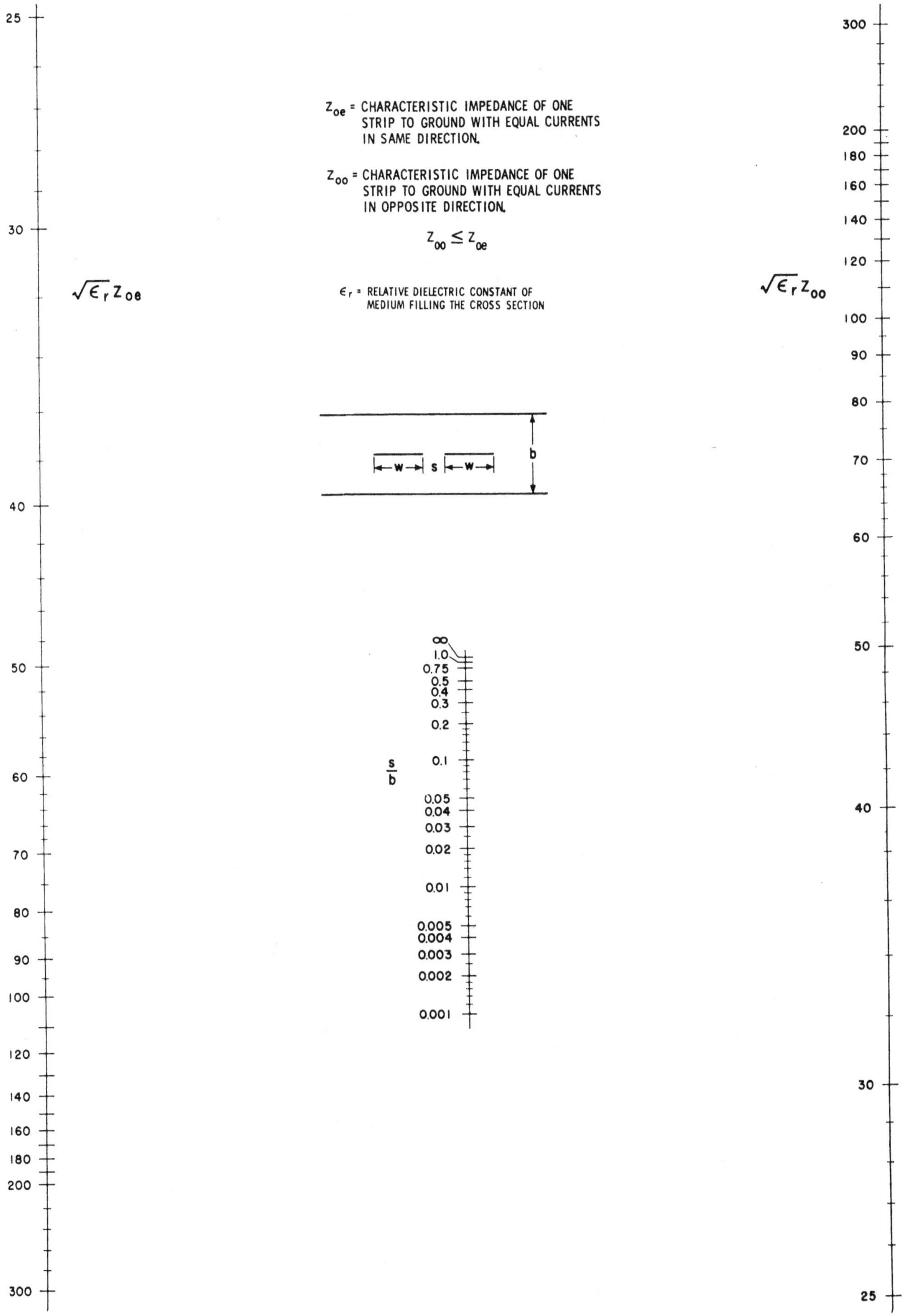

Fig. 6—Nomogram giving w/b as function of Z_{0e} and Z_{00} in coupled strip line.

$$\frac{C_{fo}'\left(0, \frac{s}{b}\right)}{\epsilon} = \frac{2}{\pi} \ln\left(1 + \coth \frac{\pi s}{2b}\right)$$

$$= \frac{s}{b} - \frac{2}{\pi} \ln\left(\sinh \frac{\pi s}{2b}\right). \tag{15}$$

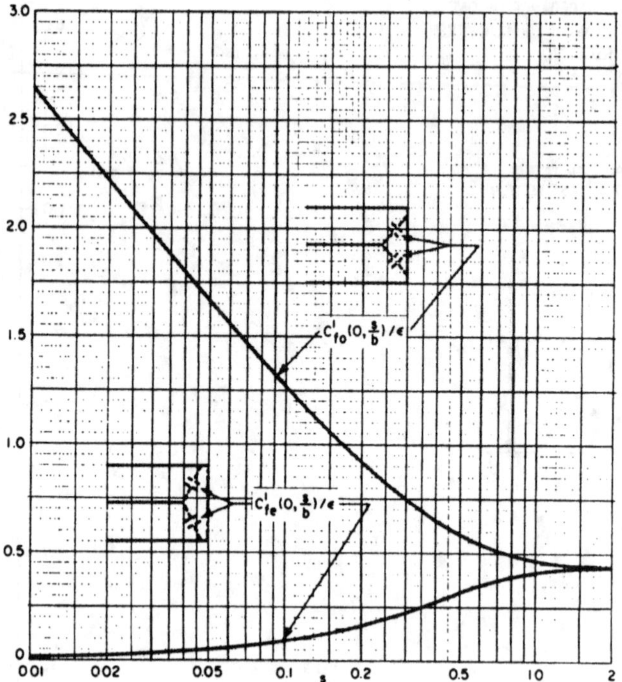

Fig. 7—Fringing capacitances for coupled strips of zero thickness.

Eqs. (14) and (15) are plotted in Fig. 7, where it is seen that they merge into (13) for $s/b \geqq 2$. As $s/b \to 0$, $C_{fe}' \to 0$ and $C_{fo}' \to \infty$. Eqs. (14) and (15) agree, as they should, with Marcuvitz's equivalent circuit for a transverse slit in the common broad wall between two wave-guides.[4] Also, they agree with Oliner's[5] equivalent circuit for a transverse gap in a single strip, which he deduced from Marcuvitz's expressions.

Coupled Thick Strips in Uniform Dielectric

A rigorous solution for coupled thick strips has not been obtained, but the approximate formulas given below should be sufficiently accurate for most purposes. These formulas are based on the solution for zero-thickness strips, and on the fringing-capacitance formula for a thick semi-infinite plate between ground planes. Consequently, the resulting formulas are expected to have their best accuracy for t_b less than about 0.1, and $w/b \geqq 0.35$. However, they may be used with diminished accuracy considerably outside these ranges. In all cases, the phase velocity is given by (1).

[4] N. Maurcuvitz, "Waveguide Handbook," McGraw-Hill Co., Inc., New York, N. Y., pp. 373–375; 1951.

[5] A. A. Oliner, "Equivalent circuits for discontinuities in balanced strip transmission line," Trans. IRE, vol. MTT-3, pp. 134–143; March, 1955.

Even-Mode Case

For the even mode, the characteristic impedance is given by

$$Z_{0e}\left(\frac{w}{b}, \frac{t}{b}, \frac{s}{b}\right) = \frac{94.15 \text{ ohms}}{\sqrt{\epsilon_r}\left\{\frac{w}{b-t} + \frac{1}{2\epsilon}\left(C_f'\left(\frac{t}{b}\right) + C_{fe}'\left(\frac{t}{b}, \frac{s}{b}\right)\right)\right\}}, \tag{16}$$

where $C_f'(t/b)$ is the fringing capacitance for a single thick strip of thickness t.[1] $C_{fe}'(t/b, s/b)$ is the fringing capacitance at the coupled edge of a thick strip in the even mode. The following approximation for $C_{fe}'(t/b, s/b)$ in terms of known quantities is assumed:

$$C_{fe}'\left(\frac{t}{b}, \frac{s}{b}\right) = C_f'\left(\frac{t}{b}\right) \cdot \frac{C_{fe}'\left(0, \frac{s}{b}\right)}{C_f'(0)}. \tag{17}$$

This relation gives the correct result for $t/b \to 0$, $s/b \to 0$, and $s/b \to \infty$. It is expected to give good results throughout the range of interest. Substitution of (17) in (16) (with a certain amount of manipulation) yields the following approximation for Z_{0e}:

$$Z_{0e}\left(\frac{w}{b}, \frac{t}{b}, \frac{s}{b}\right) = \left\{\frac{1}{Z_0\left(\frac{w}{b}, \frac{t}{b}\right)} - \frac{C_f'\left(\frac{t}{b}\right)}{C_f'(0)} \cdot \left[\frac{1}{Z_0\left(\frac{w}{b}, 0\right)} - \frac{1}{Z_{0e}\left(\frac{w}{b}, 0, \frac{s}{b}\right)}\right]\right\}^{-1} \text{ohms}. \tag{18}$$

The values $Z_0(w/b, t/b)$ and $Z_0(w/b, 0)$ are characteristic impedances of single strips, which may be read from a graph in reference [1]. $Z_{0e}(w/b, 0, s/b)$ is the even-mode zero-thickness-strip characteristic impedance given above and $C_f'(t/b)$ and $C_f'(0)$ are fringing capacitances that are also given graphically in reference [2].

Odd-Mode Case

For the odd mode, the following definition for C_{fo}' is assumed when s is large compared to t.

$$C_{fo}'\left(\frac{t}{b}, \frac{s}{b}\right) = C_f'\left(\frac{t}{b}\right) \cdot \frac{C_{fo}'\left(0, \frac{s}{b}\right)}{C_f'(0)}. \tag{19}$$

As in the even mode case, this leads to

$$Z_{0o}\left(\frac{w}{b}, \frac{t}{b}, \frac{s}{b}\right) = \left\{\frac{1}{Z_0\left(\frac{w}{b}, \frac{t}{b}\right)} + \frac{C_f'\left(\frac{t}{b}\right)}{C_f'(0)}\right.$$

$$\cdot\left[\frac{1}{Z_{0o}\left(\frac{w}{b},0,\frac{s}{b}\right)}-\frac{1}{Z_0\left(\frac{w}{b},0\right)}\right]\Bigg]^{-1} \text{ohms.} \qquad (20)$$

The definition of terms in (20) is similar to that given for (18).

For s small, (19) can not be expected to be valid, and a better approximation for C_{fo}' is

$$C_{fo}'\left(\frac{t}{b},\frac{s}{b}\right)=C_f'\left(0,\frac{s}{b}\right)+\epsilon\frac{t}{s}, \qquad (21)$$

where $\epsilon(t/s)$ is the parallel-plate capacitance across the gap per unit length. Manipulation of (21) gives

$$Z_{0o}\left(\frac{w}{b},\frac{t}{b},\frac{s}{b}\right)$$
$$=\left[\frac{1}{Z_{0o}\left(\frac{w}{b},0,\frac{s}{b}\right)}+\left\{\frac{1}{Z_0\left(\frac{w}{b},\frac{t}{b}\right)}-\frac{1}{Z_0\left(\frac{w}{b},0\right)}\right\}\right.$$
$$\left.-\frac{2}{377}\left\{\frac{C_f'\left(\frac{t}{b}\right)}{\epsilon}-\frac{C_f'(0)}{\epsilon}\right\}+\frac{2t}{377s}\right]^{-1} \text{ohms.} \qquad (22)$$

A judicious estimate of the valid ranges for (20) and (22) is $s/t \geqq 5$ for the former and $s/t \leqq 5$ for the latter.

Strips Printed on Dielectric Sheet

For the cross section of Fig. 3(c), (18) should be used for the even-mode case for all s/t, while (20) should yield fair results for the odd mode when $s/t \geqq 10$. However, for $s/t \leqq 10$, the following approximation for C_{fo}' is proposed:

$$C_f'\left(\frac{t}{b},\frac{s}{b}\right)=C_f'\left(0,\frac{s}{b}\right)+\frac{\epsilon_s Z_0\left(\frac{s}{t},0\right)}{377}. \qquad (23)$$

The last term in this expression (derived in Appendix B) is the capacitance across the gap due to the field flux inside the dielectric sheet, subject to the assumption that only a tangential component of electric field exists on the dielectric sheet surfaces in the gap region. This assumption gives good results for s/t small, but is poor for s/t large. The quantity $\epsilon_s=\epsilon_0\epsilon_{rs}$ is the absolute dielectric constant of the dielectric sheet, while $Z_0(s/t,\ 0)$ is the characteristic impedance of a shielded single-strip air-dielectric line whose ground plate spacing is t, strip width is s, and strip thickness is zero. $Z_0(s/t,\ 0)$ may therefore be obtained from the characteristic-impedance graph.[1]

Based on (23), the capacitance per strip, and per unit length, for the odd mode is

$$C_{0o}\left(\frac{w}{b},\frac{t}{b},\frac{s}{b}\right)$$
$$=\frac{377\epsilon_0}{Z_{0o}\left(\frac{w}{b},0,\frac{s}{b}\right)}+377\epsilon_0\left[\frac{1}{Z_0\left(\frac{w}{b},\frac{t}{b}\right)}-\frac{1}{Z_0\left(\frac{w}{b},0\right)}\right]$$
$$-2\left[C_f'\left(\frac{t}{b}\right)-C_f'(0)\right]+\frac{2\epsilon_0\epsilon_{rs}Z_0\left(\frac{s}{t},0\right)}{377}. \qquad (24)$$

The corresponding inductance is not affected by ϵ_{rs}. It can be deduced from C_{0o} by applying elementary transmission-line theory as follows:

$$L_{0o}=\frac{1}{c^2C_{0o}\big|_{\epsilon_r=1}} \qquad (25)$$

and

$$Z_{0o}=\sqrt{\frac{L_{0o}}{C_{0o}}}=\frac{1}{c}\sqrt{\frac{1}{C_{0o}\big|_{\epsilon_{rs}>1}}\cdot\frac{1}{C_{0o}\big|_{\epsilon_{rs}=1}}}. \qquad (26)$$

Therefore

$$Z_{0o}\left(\frac{w}{b},\frac{t}{b},\frac{s}{b}\right)$$
$$=\left\{\frac{1}{Z_{0o}\left(\frac{w}{b},0,\frac{s}{b}\right)}+\left\{\frac{1}{Z_0\left(\frac{w}{b},\frac{t}{b}\right)}-\frac{1}{Z_0\left(\frac{w}{b},0\right)}\right\}\right.$$
$$\left.-2\left\{\frac{C_f'\left(\frac{t}{b}\right)-C_f'(0)}{377\epsilon_0}\right\}+\frac{2\epsilon_{rs}Z_0\left(\frac{s}{t},0\right)}{377^2}\right\}^{-1/2}$$
$$\cdot\left\{\frac{1}{Z_{0o}\left(\frac{w}{b},0,\frac{s}{b}\right)}+\left\{\frac{1}{Z_0\left(\frac{w}{b},\frac{t}{b}\right)}-\frac{1}{Z_0\left(\frac{w}{b},0\right)}\right\}\right.$$
$$-2\left\{\frac{C_f'\left(\frac{t}{b}\right)-C_f'(0)}{377\epsilon_0}\right\}$$
$$\left.+\frac{2Z_0\left(\frac{s}{t},0\right)}{377^2}\right\}^{-1/2} \text{ohms.} \qquad (27)$$

This should give good results for $s/t \leqq 10$.

The phase velocity of the even mode cannot be computed accurately, but will usually be only a per cent or so less than the velocity of light in free space. However, the phase velocity of the odd mode will be affected considerably by the field in the dielectric sheet in the gap region. The following approximate formula gives the ratio of the phase velocities of the two modes.

$$\frac{v_{0o}}{v_{0e}}=\sqrt{\frac{1+2Z_{0o}\left(\frac{w}{b},0,\frac{s}{b}\right)\cdot Z_0\left(\frac{s}{t},0\right)/377^2}{1+2\epsilon_{rs}Z_{0o}\left(\frac{w}{b},0,\frac{s}{b}\right)\cdot Z_0\left(\frac{s}{t},0\right)/377^2}}. \qquad (28)$$

Appendix A

Derivation of Z_{0e}

Fig. 8 shows the successive transformations used in the conformal-mapping solution for the odd mode in coupled strip line. The solid boundaries denote "electric" (or perfectly conducting) walls, for which $E \times n = 0$ and $H \cdot n = 0$, while the dotted boundaries denote "magnetic" walls, for which $E \cdot n = 0$ and $H \times n = 0$. Therefore, inside the rectangle in Fig. 8(a) the electric and magnetic field lines form a uniform rectangular grid, and in terms of the dimensions of the rectangle shown in the figure, the characteristic impedance of the transmission line whose cross section is the rectangle is given by

$$Z_0 = 120\pi \cdot \frac{K'}{2K} \text{ ohms.} \tag{29}$$

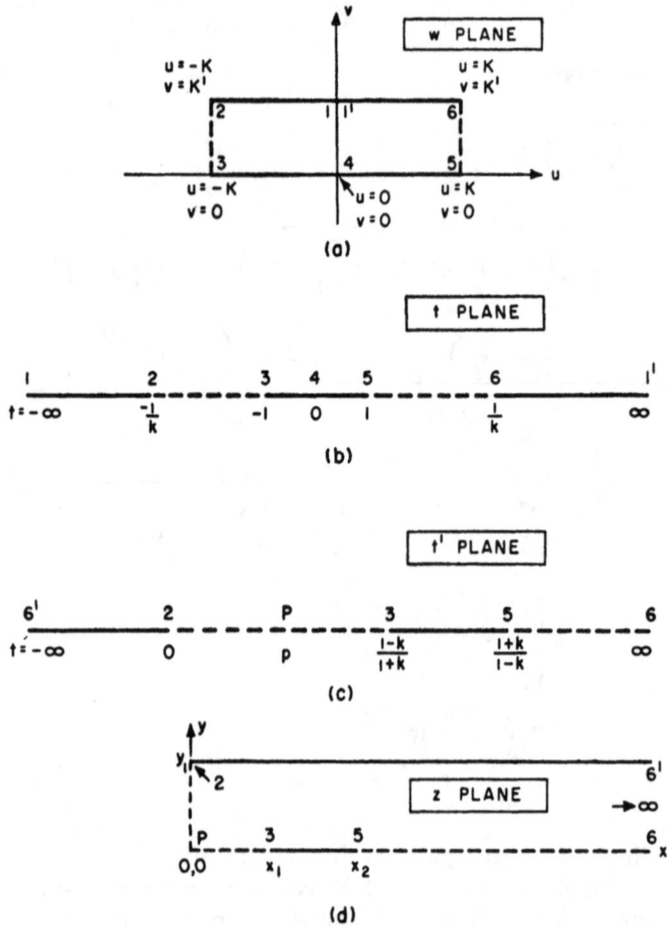

Fig. 8—Transformations used in the derivation of Z_{00}.

Due to the properties of conformal mapping, the characteristic impedance of a given cross section is not changed by successive analytic transformations. In this derivation, three successive transformations are used: transformation of the region inside the rectangle in the w plane into the entire upper half of the t plane; the upper-half t plane into the upper-half t' plane; and the upper-half t' plane into the bounded region in the z plane shown in Fig. 8(d). Hence the characteristic impedance of the z plane cross section is given by (29). Examination of this cross section shows it to be equivalent to one-quarter of the coupled-strip cross section, and to have the field distribution shown in Fig. 2(a) for the even mode. Thus, Z_{0e} is related to Z_0 by

$$Z_{0e} = \frac{Z_0}{2} = 30\pi \cdot \frac{K'}{K}. \tag{30}$$

It now remains to relate K and K' to the dimensions of the coupled strip line.

The transformation from the interior of the rectangle in the w plane to the upper-half t plane is given by Oberhettinger and Magnus,[6] and the information essential to our purpose is shown in Figs. 8(a) and 8(b). Corresponding points on the boundaries in the w and t plane are indicated by numerals 1 to 6. The coordinates of these points in the two planes are related by

$$K = K(k) \tag{31}$$

$$K' = K(k') \tag{32}$$

where $K(k)$ and $K(k')$ are complete elliptic integrals of the first kind with parameter k and k' respectively. k' is related to k by

$$k' = \sqrt{1 - k^2}. \tag{33}$$

The transformation from the t plane to the t' plane is made purely for convenience, since it greatly simplifies the evaluation of the various constants in the transformation to the z plane. The function used to connect t and t' is

$$t' = \frac{1 + kt}{1 - kt}. \tag{34}$$

This "bilinear" function transforms the upper-half t plane into the upper-half t' plane, and the real t axis into the real t' axis. However, the points on the real t' axis are distributed in a manner better suited for the next step. The coordinates of corresponding points are shown in Figs. 8(b) and (c).

The final transformation to the z plane is carried out by the Schwartz-Christoffel method. This technique is described in various books, such as "Static and Dynamic Electricity," by W. R. Smythe,[7] and therefore the reader will be assumed to be familiar with its principles. The differential equation relating the t' and z planes is

$$\frac{dz}{dt'} = C(t' - t_2')^{\beta_2}(t' - t_6')^{\beta_6}(t' - t_P')^{\beta_P}, \tag{35}$$

where t_n' is the value of t' at the point corresponding to the nth corner in the z-plane boundary, and

[6] F. Oberhettinger and W. Magnus, "Anwendung der Elliptischen Funktionen in Physik und Technik," Springer-Verlag, Berlin, p. 32; 1949.

[7] W. R. Smythe, "Static and Dynamic Electricity," McGraw-Hill Book Co., Inc., New York, N. Y., p. 80 ff.; 1939.

$$\beta_n = \frac{\alpha_n}{\pi} - 1 \tag{36}$$

where α_n is the angle in radians inside the z plane boundary at the nth corner. The values of t_n', z_n, α_n and β_n are summarized below:

Point	t_n'	z_n	α_n	β_n
2	0	jy_1	$\pi/2$	$-\frac{1}{2}$
6	∞	∞	0	-1
P	p	0	$\pi/2$	$-\frac{1}{2}$

thus

$$\frac{dz}{dt'} = \frac{C}{\sqrt{t'-p}\sqrt{t'}(t'-\infty)} = \frac{C'}{\sqrt{t'(t'-p)}} \tag{37}$$

where $C/(t'-\infty)$ is treated as a new constant C'. (This fails only at $t'=\infty$, where it does not matter.) Eq. (37) may be integrated with the aid of a table of inetgrals (e.g., B. O. Peirce[8]):

$$z = C' \cdot 2\tanh^{-1}\sqrt{\frac{t'-p}{t'}} + C''. \tag{38}$$

We see at once that $C''=0$, since z is required to be zero when $t'=p$. Also, insertion of the corresponding points $z=jy_1$ and $t'=0$ shows that $C'=y_1/\pi$. Hence (38) reduces to

$$z = \frac{2y_1}{\pi} \cdot \tanh^{-1}\sqrt{1-\frac{p}{t'}} \,. \tag{39}$$

Next, we will substitute corresponding values of t' and z at the edges of the strip.

$$x_1 = \frac{2y_1}{\pi}\tanh^{-1}\left(1 - \frac{1+k}{1-k}\cdot p\right)^{1/2} \tag{40}$$

$$x_2 = \frac{2y_1}{\pi}\tanh^{-1}\left(1 - \frac{1-k}{1+k}\cdot p\right)^{1/2}. \tag{41}$$

But the dimensions b, s, and w of the coupled strip line are related to the z plane dimensions by

$$b = 2y_1, \qquad s = 2x_1, \qquad 2w + s = 2x_2 \tag{42}$$

and therefore

$$\frac{s}{b} = \frac{2}{\pi}\tanh^{-1}\left(1 - \frac{1+k}{1-k}\cdot p\right)^{1/2} \tag{43}$$

$$\frac{2w+s}{b} = \frac{2}{\pi}\tanh^{-1}\left(1 - \frac{1-k}{1+k}\cdot p\right)^{1/2}. \tag{44}$$

It now remains to eliminate p from (43) and (44) and to solve for k. The use of several hyperbolic-function identities leads to the following result:

[8] B. O. Pierce, "A Short Table of Integrals," 3rd Edition, Ginn and Co.; p. 18, Eq. 113; 1929.

$$k = \tanh\left(\frac{\pi}{2}\cdot\frac{w}{b}\right)\cdot\tanh\left(\frac{\pi}{2}\cdot\frac{w+s}{b}\right). \tag{45}$$

When the subscript "*e*" is added to k and k' to denote the even mode, (45), (30), (31), (32) and (33) are equivalent to (5), (6) and (7).

Derivation of Z_{0o}

The procedure and transformation functions are exactly the same for Z_{0o} as for Z_{0e}, except that the point P is placed between points 6′ and 2 instead of between 2 and 3 in Figs. 8(c) and (d). Also, point 2 is made to occur at $z=0$, and point P at $z=jy_1$. These changes cause the vertical section of boundary between $z=0$ and $z=jy_1$ to be an electric wall instead of a magnetic wall. As a result, the field inside the z-plane boundary is equivalent to that in the coupled strip line for the odd mode, and

$$Z_{0o} = \frac{Z_0}{2} = 30\pi\frac{K(k')}{K(k)}.$$

After the same steps are carried out as above, the formula for k is found to be

$$k = \tanh\left(\frac{\pi}{2}\cdot\frac{w}{b}\right)\cdot\coth\left(\frac{\pi}{2}\cdot\frac{w+s}{b}\right).$$

When the subscript "*0*" is added, (2), (3) and (4) result.

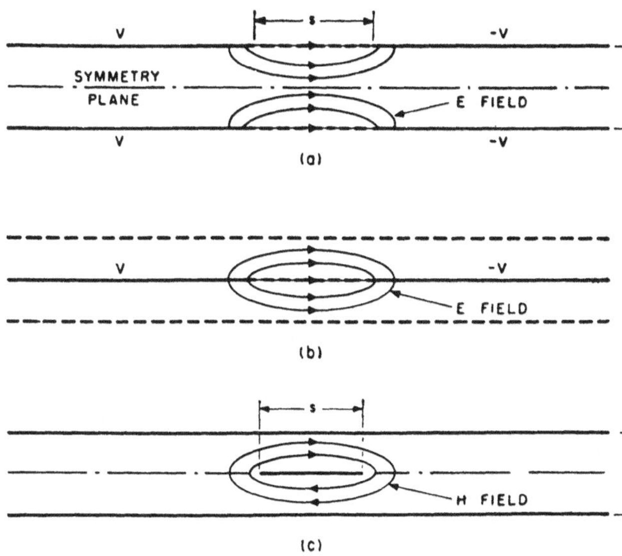

Fig. 9—Equivalence between the region insided a printed dielectric sheet and a shield strip line.

Appendix B

Gap Capacitance in the Printed-Sheet Case

Fig. 9(a) shows the region that is assumed to be equivalent to the printed dielectric sheet in the vicinity of the gap between parallel strip pairs. A magnetic wall may be inserted in the horizontal plane of symmetry without disturbing the field, and then the separated halves of the region can be arranged as in Fig. 9(b) without disturbing the field or the capacitance. Finally, Babinet's transformation is applied to Fig. 9(b), causing all electric and magnetic walls and E and H fields to be interchanged. Thus the E field pattern in (b) is equivalent to the H field pattern in (c). By reversing the direction of the H field lines in the lower half of (c), the magnetic walls on each side of the electric-wall strip may be removed. By Babinet's principle, capacitance per unit length $C_0 = C_{gap}$ in (b) is related to inductance per unit length L_0 in (c) by $C_0/\epsilon = L_0/\mu$. Then since

$$L_0 = Z_0/c,$$

where c is the velocity of light and Z_0 is the characteristic impedance of the cross section in Fig. 9(c), we may write

$$C_{gap} = \frac{Z_0}{377^2 c}.$$

However, if the dielectric sheet has a relative dielectric constant ϵ_{rs}, this should be rewritten as

$$C_{gap} = \frac{\epsilon_{rs} Z_0}{377^2 c} = \frac{\epsilon_0 \epsilon_{rs} Z_0}{377}$$

where Z_0 and c are evaluated for $\epsilon_r = 1$.

Reprinted from IRE TRANSACTIONS
ON *MICROWAVE THEORY AND TECHNIQUES*
Volume MTT-8, Number 6, November, 1960

PRINTED IN THE U.S.A.

Characteristic Impedances of Broadside-Coupled Strip Transmission Lines*

SEYMOUR B. COHN†, FELLOW, IRE

Summary—Formulas are given for the even- and odd-mode characteristic impedances of shielded coupled strip-transmission-line configurations that are especially useful when close coupling is desired. Applications may be made to wideband coupled-strip-line filters, 3-db directional couplers, and many other components. The cross sections considered are thin broadside-coupled strips either parallel or perpendicular to the ground planes. Modification of the formulas for thick strips is discussed. The derivations are outlined, with particular attention given to the underlying assumption that restricts the use of the formulas to cases of close coupling.

I. Introduction

COUPLING effects between parallel transmission lines have applications in the design of many components, such as filters,[1,2] directional couplers,[1,3,4] baluns,[5] and differential-phase-shift networks Three useful coupled-strip-line configurations are shown in Fig. 1. The coplanar-strip cross section of Fig. 1(a) was analyzed previously and design data are conveniently available,[7] while the broadside-coupled strip-line cross sections of Fig. 1(b) and 1(c) are treated in this paper. In all three cases, the theory applies to strips of zero thickness, which may be approximated by metal-foil conductors sandwiched between dielectric plates filling the cross section. If air dielectric is desired, the strips must be given a moderate thickness to provide

* Received by the PGMTT, June 23, 1960. The work described in this paper was performed at Stanford Res. Inst. with the support of the USASRDL.

† Rantec Corp., Calabasas, Calif. Formerly at Stanford Res. Inst., Menlo Park, Calif.

[1] E. M. T. Jones and J. T. Bolljahn, "Coupled-strip-transmission-line filters and directional couplers," IRE Trans. on Microwave Theory and Techniques, vol. MTT-4, pp. 75–81; April, 1956.

[2] S. B. Cohn, "Parallel-coupled transmission-line-resonator filters," IRE Trans. on Microwave Theory and Techniques, vol. MTT-6, pp. 224–231; April, 1958.

[3] J. K. Shimizu, "Strip-line 3-db directional couplers," 1957 IRE WESCON Convention Record, pt. 1, pp. 4–15.

[4] J. K. Shimizu and E. M. T. Jones, "Coupled-transmission-line directional couplers," IRE Trans. on Microwave Theory and Techniques, vol. MTT-6, pp. 403–410; October, 1958.

[5] E. M. T. Jones and J. K. Shimizu, "A wide-band strip-line balun," IRE Trans. on Microwave Theory and Techniques, vol. MTT-7, pp. 128–134; January, 1959.

[6] B. M. Schiffman, "A new class of broad-band microwave 90 degree phase shifters," IRE Trans. on Microwave Theory and Techniques, vol. MTT-6, pp. 232–237; April, 1958.

[7] S. B. Cohn, "Shielded coupled-strip transmission lines," IRE Trans. on Microwave Theory and Techniques, vol. MTT-3, pp. 29–38; October, 1955.

mechanical strength. In this case, adequate computational accuracy may be obtained by suitably modifying the zero-thickness formulas.

The coplanar configuration of Fig. 1(a) is particularly well suited to photo-etched strip-line circuits, but cannot be used where very close coupling is required. For example, in directional couplers having couplings greater than about −8 db, the spacing between the strips becomes prohibitively small. However, closer coupling may be obtained with a reasonable spacing if thick strips are used in the arrangement of Fig. 1(a). This is clear if one considers that a sizable parallel-plate capacitance is added to the fringing capacitance between the two strips. To achieve close coupling with thin
38 strips, a broadside coupling arrangement as in Fig. 1(b) and 1(c) is necessary. Both of these configurations have been used successfully in directional couplers having midband couplings as great as −2.7 db.[3]

The details of the analysis of the two broadside-coupled strip configurations are not given here, since they involve a fairly routine application of the well-known Schwartz-Christoffel transformation method. However, the specific transformation geometries are described in Section IV, where they reveal a simplifying approximation that was made in each derivation. It is believed that the error in computed values of characteristic impedance due to these approximations will not be greater than about 0.1 per cent in close-coupling applications. The error increases as the coupling is decreased, but is not likely to exceed 1 per cent in any practical application for which a broadside-coupled strip arrangement would be preferred over a coplanar arrangement. The advantage gained by these approximations is that the elliptic functions necessary in exact solutions are avoided, thus simplifying the use of the formulas, as well as their derivation. However, the reader will be interested to know that rigorous solutions valid for all values of the parameters have recently been carried out by Hachemeister.[8]

II. Formulas for Broadside-Coupled Strips Parallel to Ground Planes

Two orthogonal TEM modes can propagate on a pair of parallel-coupled transmission lines. These are the *even* mode, for which the respective voltages and currents on the two conductors are equal and of the same sign, and the *odd* mode, for which the respective voltages and currents are equal but of opposite sign. All other voltage, current, and field conditions on the conductors can be expressed as a linear combination of these two modes and, therefore, the complete performance of the coupled conductors in a circuit may be computed in terms of the characteristic impedances and propagation constants of these modes. It will be assumed that losses are small and that the cross sections are uniformly filled with a medium of relative dielectric constant ϵ_r, so that the characteristic impedances are essentially real, the attenuation is small, and the phase velocities are equal to $c/\sqrt{\epsilon_r}$, where c is the velocity of light in free space.

The following formulas give the even-mode characteristic impedance, Z_{oe}, and the odd-mode characteristic impedance, Z_{oo}, of each conductor with respect to ground for the cross section of Fig. 2. These formulas hold for any ratio of w/b and s/b, as long as w/s is greater than about 0.35:

$$Z_{oe} = \frac{188.3}{\sqrt{\epsilon_r}} \frac{K(k')}{K(k)} \tag{1}$$

$$Z_{oo} = \frac{296.1}{\sqrt{\epsilon_r}\,\dfrac{b}{s}\tanh^{-1} k} \tag{2}$$

where

k = a parameter,
$k' = \sqrt{1-k^2}$,
$K(k)$ and $K(k')$ = complete elliptic integrals of the first kind.

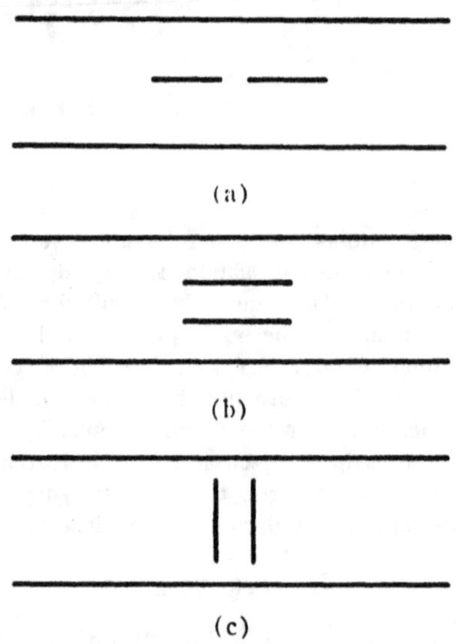

Fig. 1—Three useful configurations of very thin coupled strips between parallel ground planes.

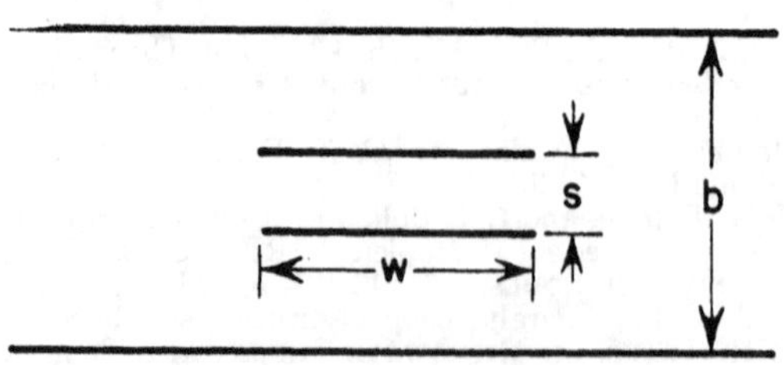

Fig. 2—Cross-section dimensions of broadside-coupled strips parallel to the ground planes.

[8] C. A. Hachemeister, "The Impedances and Fields of Some TEM-Mode Transmission Lines," Microwave Res. Inst., Polytechnic Inst. of Brooklyn, N. Y., Res. Rept. No. R-623-57, PIB-551, Contract No. AF 19(604)-2031, ASTIA Doc. No. AD-160802; April 16, 1958.

The ratio w/b is given by

$$\frac{w}{b} = \frac{2}{\pi}\left\{\tanh^{-1}\sqrt{\frac{k\frac{b}{s}-1}{\frac{1}{k}\frac{b}{s}-1}} - \frac{s}{b}\tanh^{-1}\left[\frac{1}{k}\sqrt{\frac{k\frac{b}{s}-1}{\frac{1}{k}\frac{b}{s}-1}}\right]\right\}. \quad (3)$$

The quantity $K(k')/K(k)$ has been tabulated vs k by Oberhettinger and Magnus.[9] With the aid of their table, a cross section may be designed in a straightforward process to have the desired characteristic impedances Z_{oe} and Z_{oo}, as follows. First, from (1), solve for $K(k')/K(k)$. Next determine k from the above-mentioned table.[10] Then, obtain b/s from (2). Finally, calculate w/b from (3).

A. Simplification for $(w/b)/(1-s/b)\geq 0.35$

If, in addition to the condition $w/s\geq 0.35$, one imposes the condition

$$\frac{w/b}{1-s/b}\geq 0.35,$$

the fringing fields at opposite edges of the strips will be sufficiently isolated that fringing capacitances can be specified that are essentially independent of the strip width. Then, the characteristic impedances are given by

$$Z_{oe} = \frac{188.3/\sqrt{\epsilon_r}}{\frac{w/b}{1-s/b}+\frac{C_{fe}'}{\epsilon}} \quad (4)$$

and

$$Z_{oo} = \frac{188.3/\sqrt{\epsilon_r}}{\frac{w/b}{1-s/b}+\frac{w}{s}+\frac{C_{fo}'}{\epsilon}}, \quad (5)$$

where the fringing capacitance, C_{fe}', is the capacitance per unit length that must be added at each edge of each strip to the parallel-plate capacitance so that the total capacitance to ground for the even-mode field distribution will be correct. C_{fo}' is the corresponding quantity for the odd mode, and ϵ is the absolute dielectric constant, equal to 8.85 ϵ_r mmf per meter. The fringing capacitances are functions only of s/b, and are given by

$$\frac{C_{fe}'}{\epsilon} = 0.4413 + \frac{1}{\pi}\left[\log_e\left(\frac{1}{1-s/b}\right) + \frac{s/b}{1-s/b}\log_e\frac{b}{s}\right] \quad (6)$$

$$\frac{C_{fo}'}{\epsilon} = \frac{b/s}{\pi}\left[\log_e\left(\frac{1}{1-s/b}\right) + \frac{s/b}{1-s/b}\log_e\frac{b}{s}\right]. \quad (7)$$

Eqs. (6) and (7) are plotted vs s/b in Fig. 3.

B. Effect of a Small Thickness of the Strips

When foil strips are used in the cross section of Fig. 2, the strip thickness of about 0.0015 inch is usually not great enough to affect seriously the zero-thickness values of C_{fe}'/ϵ and C_{fo}'/ϵ, but it can have an appreciable effect on the parallel-plate capacitance between the strips in the odd-mode case, depending on how the dimension s is defined. In the usual case of $s/b<0.5$, s should be taken to be the spacing between the strips, as in Fig. 4; then (1)–(3) or (4)–(7) may be used with good accuracy. For greater strip thickness, the fringing capacitances will be larger than the values given by (6) and (7) and Fig. 3. The parallel-plate capacitance between each strip and its adjacent ground plane will also be increased. A paper on thickness corrections that

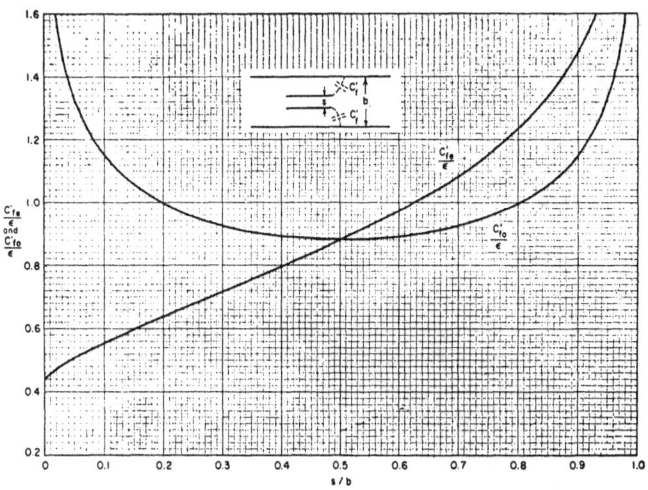

Fig. 3—Even- and odd-mode fringing capacitances for broadside-coupled very thin strips parallel to the ground planes.

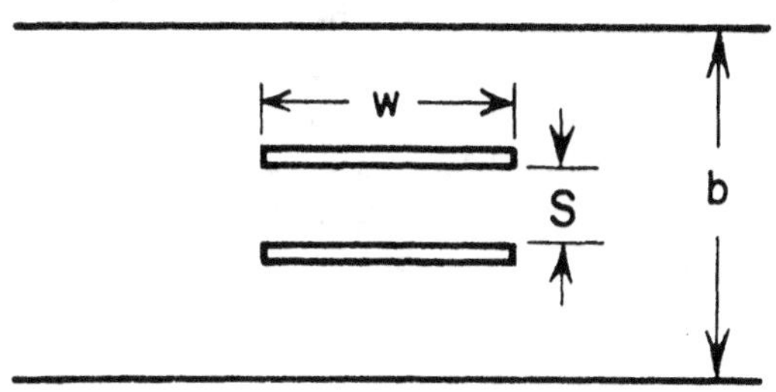

Fig. 4—Thick strips in the broadside-coupled parallel arrangement.

[9] F. Oberhettinger and W. Magnus, "Anwendung der Elliptischen Functionen in Physik und Technik," Springer-Verlag, Berlin, Germany; 1949.

[10] For k small, one may use

$$\frac{K(k')}{K(k)} = \frac{2}{\pi}\log_e\frac{4}{k}.$$

The error in this formula increases with k, being about $\frac{1}{4}$ per cent for $k=0.2$, and 1 per cent for $k=0.3$.

applies to this and numerous other practical configurations is published in this issue.[11]

III. Formulas for Broadside-Coupled Strips Perpendicular to Ground Planes

The even- and odd-mode characteristic impedances of the cross section of Fig. 5 are given by these formulas:

$$Z_{oe} = \frac{188.3}{\sqrt{\epsilon_r}} \frac{K(k)}{K(k')} \tag{8}$$

$$Z_{oo} = \frac{295.9/\sqrt{\epsilon_r}}{\dfrac{b}{s}\cos^{-1} k + \log_e \dfrac{1}{k}} \tag{9}$$

40

where k is a parameter and k', $K(k)$, and $K(k')$ are as defined in Section II. The ratio w/b is given by

$$\frac{w}{b} = \frac{2}{\pi}\left\{ \tan^{-1}\left[\frac{k'}{k} \sqrt{\frac{1 - \dfrac{k}{k'}\dfrac{s}{b}}{1 + \dfrac{k'}{k}\dfrac{s}{b}}} \right] - \frac{s}{b}\tanh^{-1}\left[\sqrt{\frac{1 - \dfrac{k}{k'}\dfrac{s}{b}}{1 + \dfrac{k'}{k}\dfrac{s}{b}}} \right] \right\}. \tag{10}$$

The inverse cosine and tangent functions in (9) and (10) are evaluated in radians between 0 and $\pi/2$. As in the case of broadside strips parallel to the ground planes, the dimension ratios may be determined in a straightforward manner for given values of Z_{oe} and Z_{oo}. With the aid of the table of $K(k)/K(k')$ vs k in Oberhettinger and Magnus,[9] the parameter k is determined from (8). Then b/s is obtained from (9), and finally w/b from (10).

Eqs. (8)–(10) are accurate for all values of w/b and s/b, as long as w/s is greater than about 1.0. As s/b is made small compared to 1.0, the formulas remain accurate for smaller values of w/s. This range of validity covers all practical cases except that of very loose coupling.

A. Effect of a Small Thickness of the Strips

In the case of foil strips, the effect of strip thickness can be neglected if s is defined to be the spacing between the strips, as shown in Fig. 6. For thicker strips the formulas should be used with discretion. A companion paper on thickness corrections, referred to in Section II, will enable an improvement in accuracy in the case of thick strips.[11]

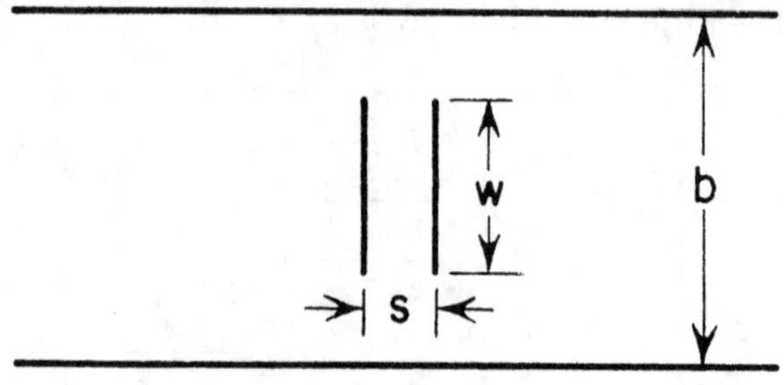

Fig. 5—Cross-section dimensions of broadside-coupled strips perpendicular to the ground planes.

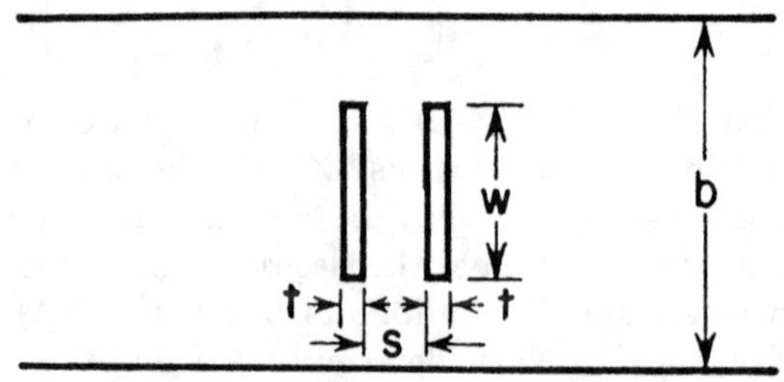

Fig. 6—Thick strips in the broadside-coupled, perpendicular arrangement.

IV. Comments on the Derivations

The formulas in this paper were derived by the Schwartz-Christoffel conformal-transformation method.[12,13] This method enables one to evaluate the capacitance and characteristic impedance between straight-sided conductors when the problem can be reduced to two dimensions, as in the cross-section plane of a transmission line. By means of one or more transformations in the complex plane, the boundary of the cross section is transformed into a simpler boundary for which the solution is known. Then because of the special properties of the conformal transformation, the capacitance and characteristic impedance of the original boundary are equal to the respective quantities of the transformed boundary.

Fig. 7(a) shows the cross section of broadside-coupled strips parallel to the ground planes. Vertical and horizontal planes of symmetry are indicated, and because of this symmetry the solution can be obtained from the geometry of Fig. 7(b), which is the upper right-hand quarter of the complete cross section. The boundary in Fig. 7(b) consists of half of the upper ground plane, the two sides of half of the upper strip, two vertical magnetic-wall segments, and a horizontal plane which is a magnetic wall for the even-mode case and an electric wall for the odd-mode case. In addition, the open end of the boundary is assumed to be closed at infinity. An exact solution of Fig. 7(b) requires elliptic functions. In order to avoid these functions, it is necessary to reduce the number of right-angle corners in the boundary. One way of doing this is shown in Fig. 7(c), where the lower

[11] S. B. Cohn, "Thickness corrections for capacitive obstacles and strip conductors," this issue, p. 638.

[12] W. R. Smythe, "Static and Dynamic Electricity," McGraw-Hill Book Co., Inc., New York, N. Y., 1st ed., p. 80 ff.; 1939.

[13] E. Weber, "Electromagnetic Fields," vol. 1, John Wiley and Sons, Inc., New York, N. Y., p. 325 ff.; 1950.

vertical segment of magnetic wall is shifted to minus infinity. The new boundary is not exactly equivalent to Fig. 7(a) but the approximation is excellent in the usual case of $w>s$. In that case it is clear that for the even mode very little electric-field energy penetrates into the region on the left of the $x=0$ plane in the lower left-hand part of Fig. 7(c), while for the odd mode, the electric field is virtually uniform in that region. Thus very good accuracy is obtained in the even-mode case by evaluating the total capacitance between the conductors of Fig. 7(c) and assuming it to be equal to the capacitance of Fig. 7(b), and in the odd-mode case by calculating the total capacitance minus the parallel-plate capacitance between $x=0$ and $x=-\infty$. This procedure has been followed in the derivation of (1)–(3), and no further approximations were made. Experience with other geometries of strip conductors indicates that an accuracy of the order of 1 per cent should ensue for w/s as low as 0.35, and of the order of 0.1 per cent for w/s greater than 1.0.

A similar approximation was made in the analysis of broadside-coupled strips perpendicular to the ground planes. The cross section is shown with its symmetry planes in Fig. 8(a), and the exact quarter model is shown in Fig. 8(b). As in the previous case, elliptic functions are needed in the solution of Fig. 8(b), but are avoided in the solution of Fig. 8(c), which is the geometry that was used in the derivation of (8)–(10). The total even-mode capacitance in Fig. 8(c) is assumed equal to the even-mode capacitance in Fig. 8(b), while the total odd-mode capacitance minus the parallel plate capacitance from $y=0$ to $y=-\infty$ is assumed equal to the odd-mode capacitance in Fig. 8(b). The considerations about accuracy are the same as in the previous case, but it is believed that w/s should not be made smaller than 1.0 unless further evidence justifies a lower limit.

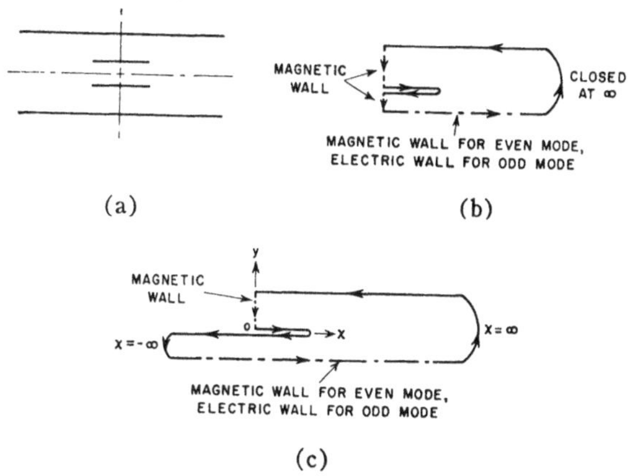

Fig. 7—Boundaries considered for the broadside-coupled parallel case.

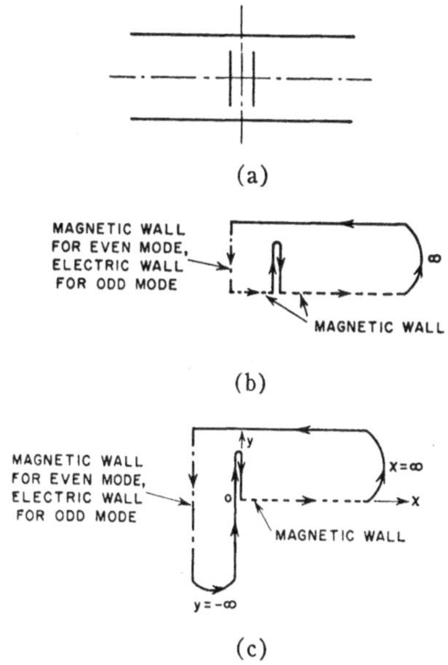

Fig. 8—Boundaries considered for the broadside-coupled perpendicular case.

41

Impedances of Offset Parallel-Coupled Strip Transmission Lines

J. PAUL SHELTON, JR., MEMBER, IEEE

Abstract—**An offset parallel-coupled strip configuration is described, in which the mechanical parameters are strip width, strip offset, and ratio of strip spacing to ground-plane spacing. The electrical parameters are dielectric constant, characteristic impedance, even and odd mode impedance. The configuration is analyzed by conformal mapping techniques. Explicit design equations are derived in which the mechanical parameters are expressed in terms of the electrical parameters. Illustrative results are presented, and the limitations on coupling strength, characteristic impedance, and strip configuration are discussed.**

INTRODUCTION

MANY MICROWAVE components are based upon parallel-coupled transmission-line sections. Examples are found among directional couplers, baluns, hybrid junctions, phase shifters, and filters [1]–[3]. In some cases, several coupled regions are used to obtain increased control over the theoretical characteristics of the component. In general, multisection components require different coupling values for the various sections. The subject of this paper is the analysis of a parallel-coupled strip transmission-line configuration that permits smooth variation of coupling from some designed maximum level to any lower value.

Manuscript received March 25, 1965; revised August 25, 1965.
The author is with the Institute for Defense Analyses, Arlington, Va. He was formerly with Radiation Systems, Inc., Alexandria, Va.

The first strip transmission-line technique for realizing variable coupling was edge-coupled coplanar strips [4]. The drawback to this method is the limitation of ρ, the ratio of even to odd mode impedance, to a maximum of two or three.

Getsinger proposed a technique for achieving very tight coupling in which a line with single center strip is sandwiched between the two strips of a second line [5]. This method requires the use of four layers of dielectric, and the transmission lines are unlike, one having single strip and one having double strips. Furthermore, it is somewhat difficult to determine the maximum coupling that is available for a given strip spacing.

Impedance relations for parallel strips, one above the other between ground planes, were derived by Cohn [6]. This configuration requires three dielectric layers and provides maximum coupling for given layer thicknesses. For this configuration, variation in coupling can be easily achieved in practice by offsetting the strips without changing the thicknesses of the dielectric layers. In general, both strip width and overlap are functions of coupling. This offset configuration has not previously been available to the microwave engineer because it has not been analyzed. An analysis is presented here.

In the following paragraphs, the even and odd mode impedances are derived for the offset configuration shown in Fig. 1, in which zero-thickness strips are located in the dielectric-filled region between infinite ground planes. First, expressions for the necessary fringing capacitances are obtained. Then the equations for the various impedances are derived. Finally, curves of coupling vs. strip width and offset are presented for sample configurations. The derivation of fringing capacitance is described in the Appendix.

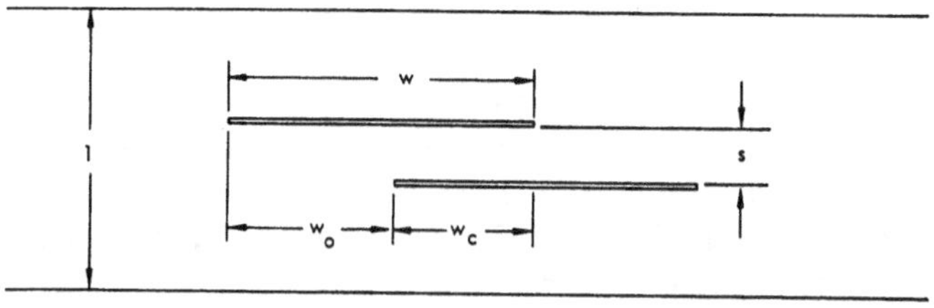

Fig. 1. Offset parallel-coupled strip transmission lines.

It will be noted that the uncoupled single strip, which is obtained when the separation is made very large, is not centered between the ground planes. This configuration has been avoided by some workers in the field because of concern over parallel-plate radiation. This concern is unfounded. The same care must be taken with any unbounded transmission line to avoid radiation, whether it uses a centered strip, an off-center strip, or an open strip over single ground plane. The configuration described in this paper has been used successfully over three years by the author and his associates.

Fringing Capacitances

The technique used to determine even and odd mode impedance is an approximate one. That is, the capacitance between strip and ground is separated into parallel-plate and fringing components. As a result, there are two cases to be considered, depending on the relative positions of the strip edges. In Fig. 2(a), the case for tight coupling is shown, and the approximate limitations on strip dimensions are

$$\frac{w}{1-s} \geqq 0.35$$

$$\frac{w_c}{s} \geqq 0.7.$$

For loose coupling, shown in Fig. 2(b), the limitations are

$$\frac{w}{1-s} \geqq 0.35$$

$$\frac{2w_o}{1+s} \geqq 0.85.$$

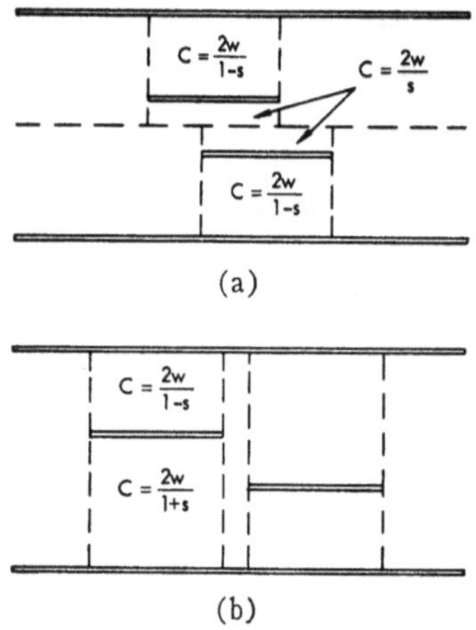

Fig. 2. Coupling configurations. (a) Configuration for tight coupling. (b) Configuration for loose coupling.

The effects of these limitations on the allowable range of strip configuration, maximum coupling, and characteristic impedance are discussed in a later section.

The derivation of the fringing capacitances is discussed in the Appendix. The Schwartz-Cristofel transformation is used to achieve the following results:

Tight coupling:

$$\pi C_{fo} = -\frac{2}{1-s}\log s + \frac{1}{s}\log\left[\frac{pr}{(p+s)(p+1)(r-s)(1-r)}\right] \quad (1)$$

$$\pi C_{fe} = -\frac{2s}{1-s}\log s - 2\log s + 4\log(s+pr) - \log pr(p+s)(p+1)(r-s)(1-r) \quad (2)$$

where

$$r = \frac{p+\dfrac{1+s}{2}}{1+p\left(\dfrac{1+s}{2s}\right)} \qquad (o < p < \infty).$$

Loose coupling:

$$\pi C_{fo} = \frac{2}{1+s}\log\left(\frac{1+a}{a(1-q)}\right) - \frac{2}{1-s}\log q \quad (3)$$

$$\pi C_{fe} = \frac{2}{1+s}\log\left[\frac{1+a}{a(1-q)}\right] - \frac{2}{1-s}\log q - 2\log\left(\frac{1+aq}{aq}\right) \quad (4)$$

where

$$q = \left(\frac{s+1}{2}\right)\left|\frac{a+\dfrac{2s}{s+1}}{a+\dfrac{s+1}{2}}\right|.$$

Furthermore, the conformal mapping procedure yields the following expressions for offset dimensions:

For tight coupling,

$$\pi w_o = \frac{1+s}{2}\log\frac{p}{r} + \frac{1-s}{2}\log\left(\frac{1+p}{s+p}\right)\left(\frac{r-s}{1-r}\right). \tag{5}$$

For loose coupling,

$$\pi w_c = s\log\frac{q}{a} + (1-s)\log\left(\frac{1-q}{1+a}\right). \tag{6}$$

44 IMPEDANCE EQUATIONS

Although it is possible to plot families of curves relating C_{fo}, C_{fe}, w, w_c, and s, it is very useful to relate the fringing and parallel-plate capacitances and the associated strip dimensions to the even and odd mode impedances, which are the basic design parameters of any component in which the coupled lines are used. One listing of parameters, which can be considered the independent variables, is

Z_o, the characteristic impedance

ϵ_r, the relative dielectric constant of the medium

ρ, the ratio of Z_{oe} to Z_{oo}.

The basic relationships are

$$\begin{aligned} Z_{oe} &= \sqrt{\rho}\,Z_o \\ Z_{oo} &= Z_o/\sqrt{\rho} \\ C_e &= 120\pi/\sqrt{\epsilon_r}\,Z_{oe} \\ C_o &= 120\pi/\sqrt{\epsilon_r}\,Z_{oo}. \end{aligned} \tag{7}$$

For tight coupling, the mode capacitances may be expressed

$$\begin{aligned} C_o &= 2w\left(\frac{1}{1-s}+\frac{1}{s}\right)+C_{fo} \\ &= \frac{2w}{s(1-s)}+C_{fo} \end{aligned}$$

$$C_e = \frac{2w}{1-s}+C_{fe} \tag{8}$$

where C_{fo} and C_{fe} are given by (1) and (2), and the parallel-plate capacitances are those indicated in Fig. 2(a). Note that the capacitance between strips is approximated by $2w/s$ rather than $2w_c/s$, an artifice which simplifies the expressions for fringing capacitances and allows explicit solution.

For loose coupling, the capacitances are

$$\begin{aligned} C_o &= 2w\left(\frac{1}{1-s}+\frac{1}{1+s}\right)+C_{fo}-C_f(a=\infty) \\ &= \frac{4w}{1-s^2}+C_{fo}+C_f(a=\infty) \end{aligned}$$

$$C_e = \frac{4w}{1-s^2}+C_{fe}+C_f(a=\infty) \tag{9}$$

where C_{fo} and C_{fe} are given by (3) and (4), and $C_f(a=\infty)$ is the fringing capacitance at the edge of a single, uncoupled strip. The parallel-plate capacitances are those indicated in Fig. 2(b). It is easily found that

$$\pi C_f(a=\infty) = -\frac{2}{1+s}\log\left(\frac{1-s}{2}\right) - \frac{2}{1-s}\log\left(\frac{1+s}{2}\right). \tag{10}$$

At this point, the continuity of the development is strengthened if the cases are treated separately. It is noted that the equations so far obtained do not even present an implicit solution to the problem. It is necessary to find some connection between the results of the conformal mapping and the basic relationships of (7)–(9).

Derivation for Tight Coupling

For tight coupling, the required connection results from

$$\begin{aligned} C_e - s\,C_o &= C_{fe} - s\,C_{fo} && \text{[from (8)]} \\ &= \frac{2}{\pi}\log\left[\frac{\left(1+\frac{pr}{s}\right)^2}{\left(\frac{pr}{s}\right)}\right] && \text{[from (1) and (2)]} \\ &= \left(\frac{1-\rho s}{\sqrt{\rho}}\right)\frac{120\pi}{\sqrt{\epsilon_r}\,Z_o} && \text{[from (7)]} \end{aligned}$$

Therefore, the implicit solution for tight coupling is

$$\frac{1-\rho s}{\sqrt{\rho}} = \frac{\sqrt{\epsilon_r}\,Z_o}{60\pi^2}\log\left[\frac{\left(1+\frac{pr}{s}\right)^2}{\left(\frac{pr}{s}\right)}\right].$$

It is possible to eliminate p and r as follows:

$$\text{Let } A = \frac{(1+pr/s)^2}{(pr/s)} = \exp\left[\frac{60\pi^2}{\sqrt{\epsilon_r}\,Z_o}\left(\frac{1-\rho s}{\sqrt{\rho}}\right)\right] \tag{11}$$

$$\text{Let } B = \frac{pr}{s} = \frac{A-2\pm\sqrt{A^2-4A}}{2}. \tag{12}$$

The two signs of the square root represent the plus or minus values of w_o which produce the same value of $C_e - sC_o$. Using the equation for r in terms of p, p is found to be

$$p = \frac{(B-1)\left(\frac{1+s}{2}\right) + \sqrt{\left(\frac{1+s}{2}\right)^2 (B-1)^2 + 4sB}}{2} \quad (13)$$

where the plus sign on the root gives $p > o$.

The steps in the explicit solution, given Z_o, ϵ_r, ρ, and s, are now complete. The procedure is to solve (11) for A, (12) for B, (13) for p and r, then use (1) or (2) for C_{fo} or C_{fe}, (7) and (8) for w, and, finally, (5) for w_o.

It is worthwhile making a comparison with the results of Cohn for parallel-coupled strips. Examination of (5) reveals that $w_o = 0$ for $p = r = \sqrt{s}$. For this condition, (1) and (2) reduce to

$$C_{fo} = -\frac{2}{\pi}\left[\frac{1}{1-s}\log s + \frac{1}{s}\log(1-s)\right]$$

$$C_{fe} = -\frac{2}{\pi}\left[\frac{s}{1-s}\log s + \log(1-s) - \log 4\right],$$

which are equivalent to Cohn's approximate results, as given in his equations (4), (5), (6), and (7) [6]. Furthermore, the implicit solution reduces to

$$\frac{1 - \rho_{max} s}{\sqrt{\rho_{max}}} = \frac{\sqrt{\epsilon_r} Z_o}{60\pi^2} \log 4. \quad (14)$$

Equation (14) is extremely useful. It relates ρ_{max}, s, ϵ_r, and Z_o in a single expression and allows direct solution of the zero-offset configuration. In many practical cases, one has a maximum value of ρ to be realized; (14) specifies the required corresponding value of s. It is emphasized, however, that (14) is limited by the approximate analysis used. For very large ρ_{max} or small s, one must use Cohn's virtually exact equations, (1), (2), and (3) [6]. In general, (14) holds for

$$\frac{\sqrt{\rho_{max} \epsilon_r} Z_o}{60\pi^2} \log 4 < 0.5,$$

provided $\rho_{max} > 2$.

Derivation for Loose Coupling

For loose coupling, the required combination of expressions is

$$\Delta C = C_o - C_e = \frac{120\pi}{\sqrt{\epsilon_r} Z_o}\left(\frac{1+\rho}{\sqrt{\rho}}\right) \quad \text{[from (8)]}$$

$$= \frac{2}{\pi}\log\left(\frac{1+aq}{aq}\right). \quad \text{[from (3) and (4)]}$$

Letting

$$k = aq,$$

$$\Delta C = \frac{2}{\pi}\log\left(\frac{1+k}{k}\right)$$

or

$$k = \frac{1}{\exp\left(\frac{\pi \Delta C}{2}\right) - 1}. \quad (15)$$

Using the equation for q in terms of a, then solving for a in terms of k,

$$a = +\sqrt{\left(\frac{s-k}{s+1}\right)^2 + k} - \frac{s-k}{s+1}.$$

The positive root guarantees $a > 0$.

The explicit solution for loose coupling, given Z_o, ϵ_r, ρ, and s, is now accomplished by solving (8) for ΔC, (15) for k, (16) for a and q, (3) or (4) for C_{fo} and C_{fe}, (9) for w, and (6) for w_C. 45

Collected Equations

An orderly arrangement of the design equations resulting from the previous section is presented in order to facilitate the computation of parameters.

Equations for tight coupling:

$$A = \exp\left[\frac{60\pi^2}{\sqrt{\epsilon_r} Z_o}\left(\frac{1-\rho s}{\sqrt{\rho}}\right)\right]$$

$$B = \frac{pr}{s} = \frac{A - 2 + \sqrt{A^2 - 4A}}{2}$$

$$p = \frac{(B-1)\left(\frac{1+s}{2}\right) + \sqrt{\left(\frac{1+s}{2}\right)^2 (B-1)^2 + 4sB}}{2}$$

$$r = \frac{sB}{p} \quad \left[\text{also, } r = \frac{p + \frac{1+s}{2}}{1 + p\left(\frac{1+s}{2s}\right)}, \quad o < p < \infty\right]$$

$$C_{fo} = \frac{1}{\pi}\left\{-\frac{2}{1-s}\log s + \frac{1}{s}\log\left[\frac{pr}{(p+s)(1+p)(r-s)(1-r)}\right]\right\}$$

$$C_o = \frac{120\pi\sqrt{\rho}}{\sqrt{\epsilon_r} Z_o}$$

$$w = \frac{s(1-s)}{2}(C_o - C_{fo})$$

$$w_o = \frac{1}{2\pi}\left[(1+s)\log\frac{p}{r} + (1-s)\log\left(\frac{1+p}{s+p}\right)\left(\frac{r-s}{1-r}\right)\right].$$

The condition for tightest coupling $w_o = 0$ is given by

$$\frac{1 - \rho_{max} s}{\sqrt{\rho_{max}}} = \frac{\sqrt{\epsilon_r} Z_o}{60\pi^2} \log 4.$$

Equations for loose coupling:

$$C_o = \frac{120\pi\sqrt{\rho}}{\sqrt{\epsilon_r}\,Z_o}$$

$$\Delta C = \frac{120\pi}{\sqrt{\epsilon_r}\,Z_o}\,\frac{\rho - 1}{\sqrt{\rho}}$$

$$k = \frac{1}{\exp\dfrac{\pi\Delta C}{2} - 1}$$

$$a = +\sqrt{\left(\frac{s-k}{s+1}\right)^2 + k} - \frac{s-k}{s+1}$$

$$q = \frac{k}{a}\left[\text{or } q = \left(\frac{s+1}{2}\right)\left(\frac{a + \dfrac{2s}{s+1}}{a + \dfrac{s+1}{2}}\right)\right]$$

$$C_{fo} = \frac{2}{\pi}\left[\frac{1}{1+s}\log\frac{1+a}{a(1-q)} - \frac{1}{1-s}\log q\right]$$

$$w_c = \frac{1}{\pi}\left[s\log\frac{q}{a} + (1-s)\log\left(\frac{1-q}{1+a}\right)\right]$$

$$C_f(a=\infty) = -\frac{2}{\pi}\left[\frac{1}{1+s}\log\left(\frac{1-s}{2}\right) + \frac{1}{1-s}\log\left(\frac{1+s}{2}\right)\right]$$

$$w = \frac{1-s^2}{4}\left[C_o - C_{fo} - C_f(a=\infty)\right].$$

Illustrative Results and Limitations on Parameters

In the particular application for which the coupled impedances were evaluated, two of the parameters of (14) were held fixed. The dielectric constant ϵ_r was 2.32, corresponding to polyolefin materials, and the characteristic impedance was 50 ohms. It is conceivable that many similar situations, in which the operating impedance and the dielectric material are predetermined, might be encountered. In such a case, the remaining parameters s and $\rho_{\max}$ can be related by a curve, as shown in Fig. 3. The value of $\rho_{\max}$ varies from two to six, and the approximate range of s is 0.1 to 0.4, accounting for the probable range of coupling values amenable to this strip configuration.

Also shown on the curve are the values of s which correspond to layer thickness ratios involving small integral numbers. These values are more easily realized in terms of readily available sheet thicknesses, although it must be noted that material manufacturers are developing considerable flexibility in this respect.

Since the fringing capacitances were determined by an approximate procedure and the accuracy of the results was based on maintaining certain limitations on the strip dimensions, it is of interest to estimate the effects of these limitations on the allowable physical and electrical parameters of the coupled strip configuration. The following limitations were imposed by the fringing capacitance calculation:

In general:

$$\frac{w}{1-s} \geqq 0.35.$$

For loose coupling:

$$\frac{2w_o}{1+s} \geqq 0.85.$$

For tight coupling:

$$\frac{w_c}{s} \geqq 0.7.$$

The first inequality concerns the parallel-plate region between the outer surface of the strip and the ground plane. The second expression deals with the region between the "exposed" inner strip surface and the opposite ground plane; the constant chosen for this condition is 0.85 rather than 0.7 because the fringing fields from the edge of the intervening strip extend into this region. The third relation defines the parallel-plate region between strips.

The effects on strip spacing s and on $\sqrt{\epsilon_r}Z_o$ are considered in Fig. 4. Curve A is the limit imposed by the relationship $2w/(1+s) \geqq 0.7$ for a single uncoupled strip. Curve B is obtained when the strips are located one over the other ($w_o = 0$), and the limitation $w/s \geqq 0.7$ is applied. Thus, curves A and B define the allowable parameters for parallel coupled strips when offsetting is not used. Curve C is obtained by setting $w_c/s = 0.7$ and $2w_o/(1+s) = 0.85$, then assuming $w = w_o + w_c$ for the uncoupled strip and calculating the corresponding value of $\sqrt{\epsilon_r}Z_o$. Since the width of coupled strips is always smaller than that of uncoupled strips, curve C is too high. Two calculated points are available which satisfy the criteria of curve C, and they are shown on the dashed curve. The dashed curve is a rough estimate of the correct location of curve C. The shaded region represents the range of parameters for which the strip dimensions for all values of coupling are very accurately obtained. That is, the curves based on tight- and loose-coupling approximation can be joined by a negligible discontinuity.

The results of Fig. 4 should not be interpreted to forbid the use of parameters falling outside of the shaded region. In some cases, the coupling values which correspond to inaccurate strip dimensions are not used. In other cases, the inaccurate portions of the curves can be corrected by interpolation. However, it is seen that the most versatile application of this design procedure is obtained for values of s in the range of approximately 0.14 to 0.20, corresponding to values of $\rho_{\max}$ in the neighborhood of 3.5 to 4.5.

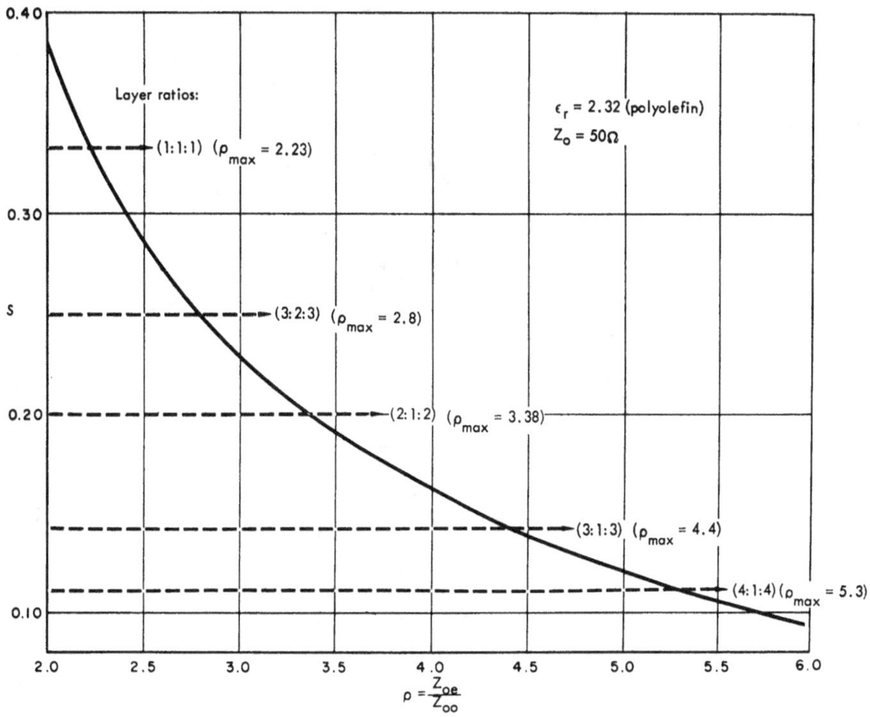

Fig. 3. Relationship between s and ρ.

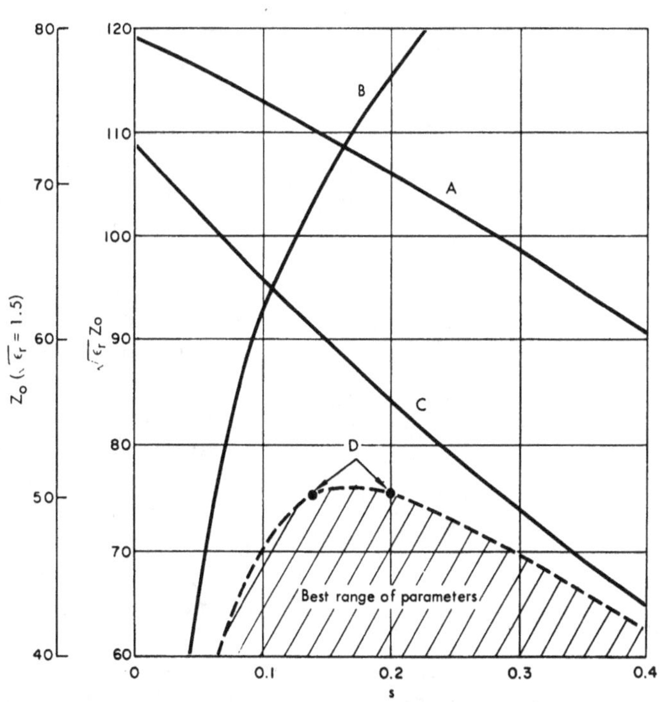

A $\frac{2w}{1+s} > .7$ for single uncoupled strip

B $\frac{w}{s} > .7$ for parallel strips, maximum coupling

C $\frac{w_c}{s} > .7$, $\frac{2w_o}{1+s} > .85$ for offset strips; approximate

D Computed points for $\frac{w_c}{s} > .7$, $\frac{2w_o}{1+s} > .85$

Fig. 4. Limitation on parameters, s and $\sqrt{\epsilon_r}Z_0$.

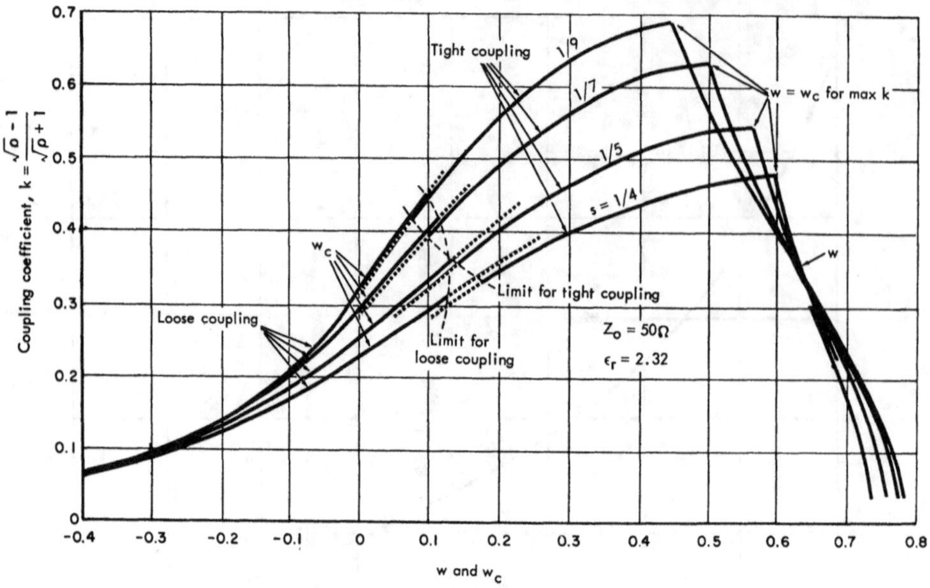

Fig. 5. Representative coupling curves.

Figure 5 contains curves relating coupling coefficient

$$k = \frac{\sqrt{\rho} - 1}{\sqrt{\rho} + 1}$$

to strip dimensions for representative values of s, for $Z_o = 50\Omega$ and $\epsilon_r = 2.32$. An indication of the approximation involved in the analysis is seen in the imperfect match of the strong- and weak-coupling curves.

Conclusions

A design procedure for offset parallel-coupled strips has been derived. The equations are arranged in terms of the electrical parameters, coupling strength (ρ) and characteristic impedance (Z_o). Because of the approximations involved in the conformal mapping analysis, limitations are imposed on the range of characteristic impedance that can be used and on the ratio of strip separation to ground-plane spacing(s). However, the ranges in which good accuracy is achieved are those most likely to be used in practical strip transmission-line components and circuits.

Appendix

Derivation of Fringing Capacitances

The procedure for determining the fringing capacitances for a strip transmission-line configuration is straightforward and is outlined for review purposes. The term "fringing" is generally used when the fields at one edge of the strip do not significantly influence those at the opposite edge. Thus, the strips can be assumed to be infinitely wide, as indicated in Fig. 6. The fringing capacitance is defined as the difference between the ideal parallel-plate capacitance and the actual value. For infinite strips, of course, the capacitances become infinite, but the difference is finite.

The steps in the procedure are as follows:

1) The conductor configuration of Fig. 6 is mapped onto the real axis of the complex plane by means of the Schwartz-Cristofel transformation. In the cases under consideration, the transformation is particularly simple. For both configurations considered, the polygon in the w-plane being mapped onto a real axis of the z-plane has six sides, and the vertex angles are all ± 180 degrees. One result of this condition is that the differential equation of the transformation is readily soluble in terms of logarithmic functions.
2) The parallel-plate capacitances for even and odd modes are selected, and the appropriate expressions are written.
3) The actual capacitances are determined by calculating charge density for unit voltages on the strips. The result is in the form of capacitance distribution.
4) The fringing capacitance is obtained by subtracting the expression of Step 2 from that of Step 3 and integrating over the infinite strip cross section.

If the reader is unfamiliar with this procedure, an excellent description and illustrative example has recently been presented by Wheeler [7].

Case for Large ρ: Considering first the case for strong coupling or large ρ, the conformal transformation is shown in Fig. 7. The equation of the transformation is

$$w = \frac{1}{\pi}\left[s \log z + \frac{1-s}{2} \log (z-1)\left(z + \frac{\rho r}{s}\right)\right]$$

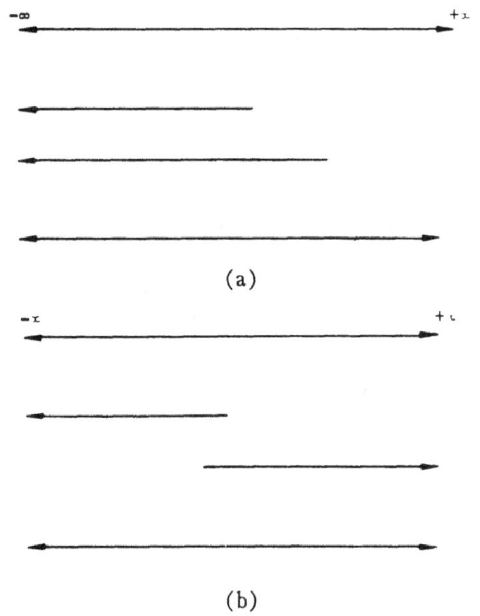

Fig. 6. Infinite strip approximation. (a) Strong coupling. (b) Loose coupling.

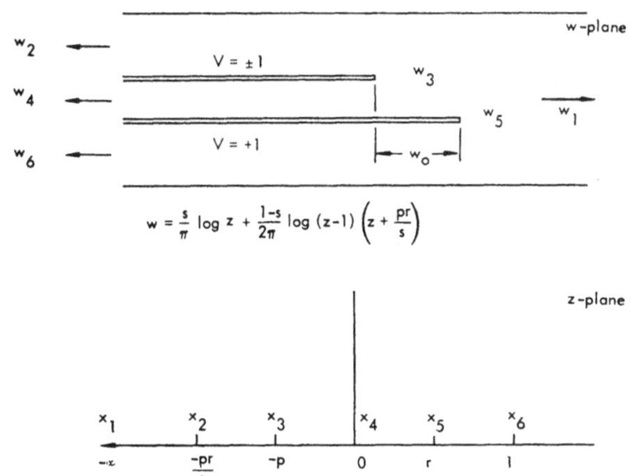

Fig. 7. Conformal transformation—large ρ.

where

$$r = \frac{p + \dfrac{1+s}{2}}{1 + p\left(\dfrac{1+s}{2s}\right)} \qquad 0 < p < \infty \qquad \frac{2s}{1+s} < r < \frac{1+s}{2}.$$

It is immediately possible to express the offset width w_o, in terms of s, p, and r,

$$w_o = \frac{1}{\pi}\left[\frac{1+s}{2}\log\frac{p}{r} + \frac{1-s}{2}\log\left(\frac{1+p}{s+p}\right)\left(\frac{r-s}{1-r}\right)\right].$$

The basis for the parallel-plate capacitances is indicated in Fig. 8. For the even mode, $C_{pe} = C_1 + C_2$; for the odd mode, $C_{po} = C_1 + C_2 + C_3 + C_4$. The capacitance of both strips to ground is computed for this case because the strip arrangement is not symmetrical. The resulting

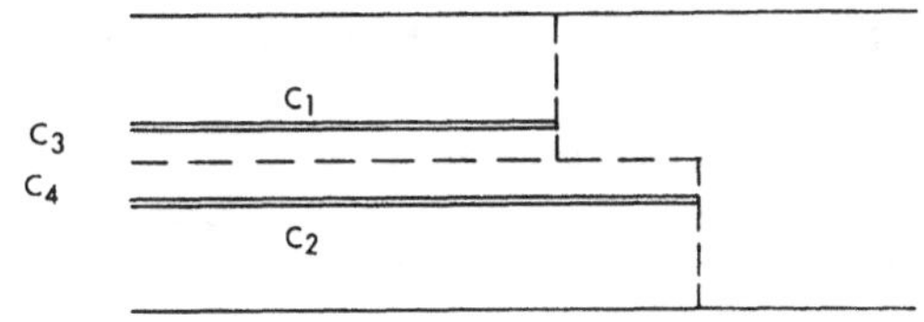

Fig. 8. Parallel-plate capacitances—large ρ.

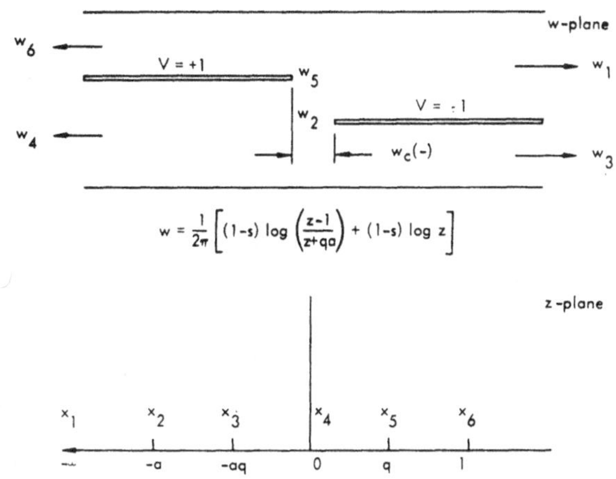

Fig. 9. Conformal transformation—small ρ.

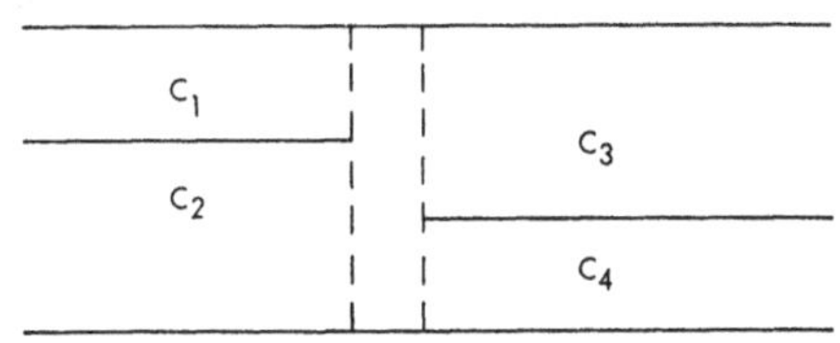

Fig. 10. Parallel-plate capacitances—small ρ.

fringe capacitance is equal to the sum of the fringe capacitances of both edges of a single strip of finite width. Application of the previously mentioned conventional techniques yields the required results for the fringing capacitances.

Case for Small ρ: The conformal transformation for the loose coupling configuration is shown in Fig. 9. The equation of the transformation is

$$w = \frac{1}{2\pi}\left[(1-s)\log\left(\frac{z-1}{z+qa}\right) + (1+s)\log z\right]$$

where

$$q = \frac{s+1}{2}\left(\frac{a + \dfrac{2s}{s+1}}{a + \dfrac{s+1}{2}}\right) \qquad 0 < a < \infty \qquad \frac{2s}{s+1} < q < \frac{s+1}{2}.$$

As for the strong coupling case, it is possible to determine w_c directly:

$$w_c = \frac{1}{\pi}\left[s\log\left(\frac{q}{a}\right) + (1-s)\log\left(\frac{1-q}{1+a}\right)\right].$$

A negative value of w_c represents a gap rather than overlap between strip edges.

The basis for parallel-plate capacitance is shown in Fig. 10. For both even and odd modes, $C_p=C_1+C_2$. Since the strip arrangement is symmetrical in this case, the fringe capacitance is calculated for only one strip. Again, the required expressions for fringing capacitance are found by conventional methods.

ACKNOWLEDGMENT

J. Mosko of Naval Ordnance Test Station, China Lake, Calif., has programmed the design equations for electronic computation. He supplied the data from which Fig. 5 was prepared. He can provide a copy of the Fortran program to anyone who wishes to use it.

REFERENCES

[1] E. M. T. Jones and J. T. Bolljahn, "Coupled-strip-line filters and directional couplers," *IRE Trans. on Microwave Theory and Techniques*, vol. MTT-4, pp. 75–81, April 1956.
[2] J. K. Shimizu and E. M. T. Jones, "Coupled-transmission-line directional couplers," *IRE Trans. on Microwave Theory and Techniques*, vol. MTT-6, pp. 403–410, October 1958.
[3] S. B. Cohn, "Direct-coupled-resonator filters," *Proc. IRE*, vol. 45, pp. 187–196, February 1957.
[4] S. B. Cohn, "Characteristic impedances of broadside-coupled strip transmission lines," *IRE Trans. on Microwave Theory and Techniques*, vol. MTT-8, pp. 633–637, November 1960.
[5] W. J. Getsinger, "A coupled strip-line configuration using printed-circuit construction that allows very close coupling," *IRE Trans. on Microwave Theory and Techniques*, vol. MTT-9, pp. 535–544, November 1961.
[6] S. B. Cohn, "Shielded coupled-strip transmission line," *IRE Trans. on Microwave Theory and Techniques*, vol. MTT-3, pp. 29–38, October 1955.
[7] H. A. Wheeler, "Transmission-line properties of parallel wide strips by a conformal-mapping approximation," *IEEE Trans. on Microwave Theory and Techniques*, vol. MTT-12, pp. 280–289, May 1964.

Reprinted from IEEE TRANSACTIONS
ON *MICROWAVE THEORY AND TECHNIQUES*
Volume MTT-14, Number 1, January, 1966
Pp. 7-15

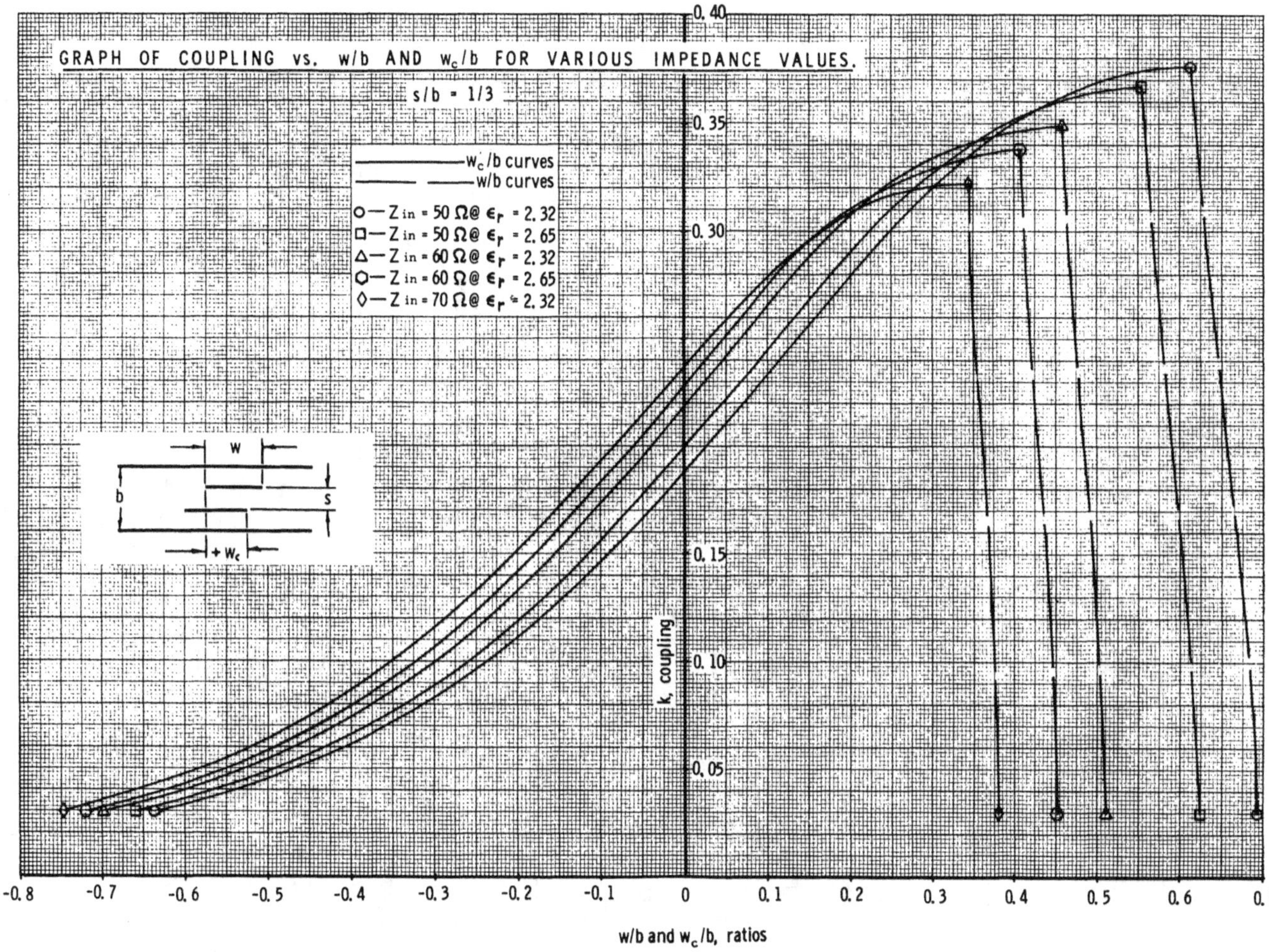

Courtesy of J.A. Mosko, see
Microwave Journal, Vol. 10, No. 5,
pp. 35-37, April 1967.

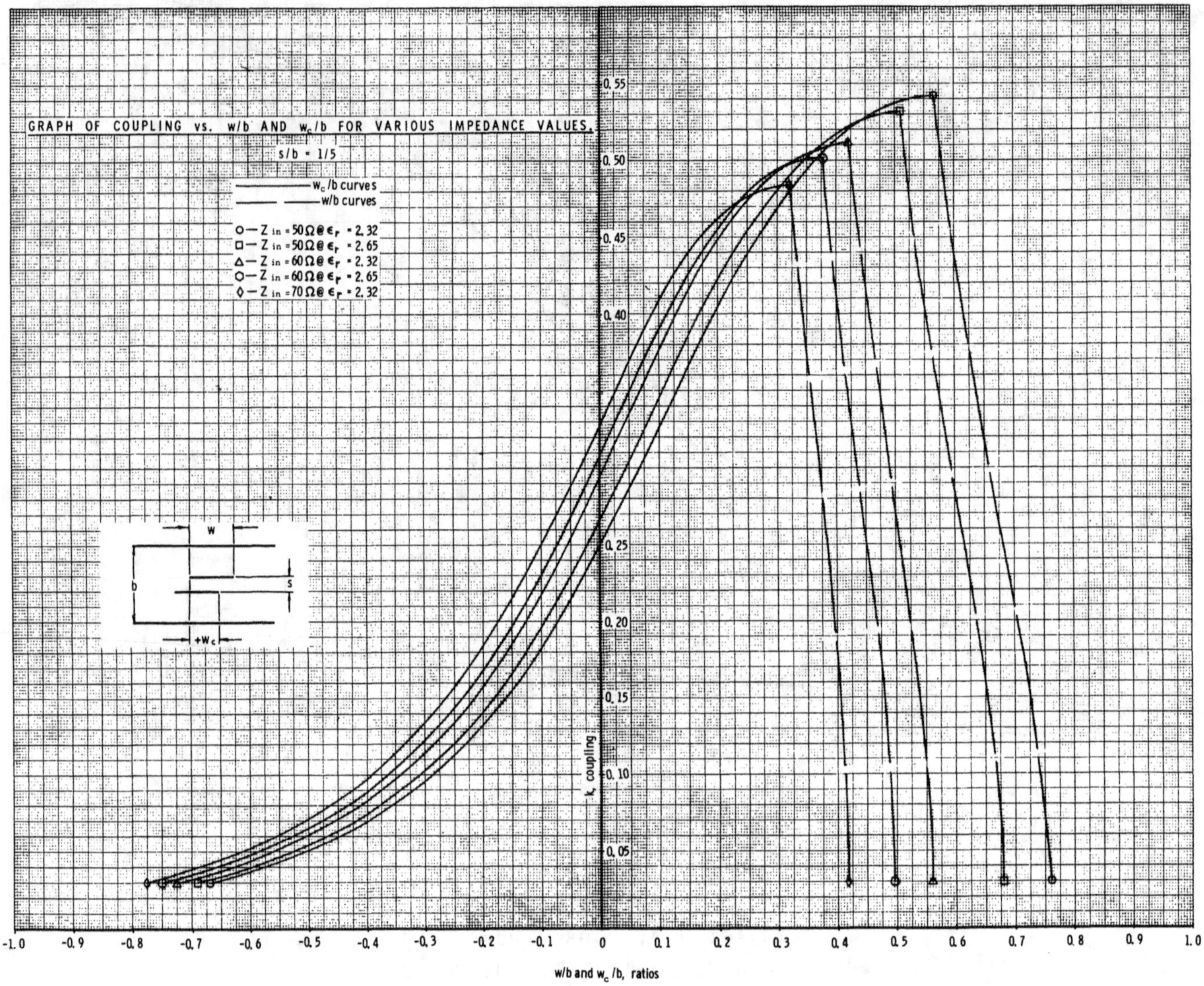

Courtesy of J.A. Mosko, see
Microwave Journal, Vol. 10, No. 5
pp. 35-37, April 1967

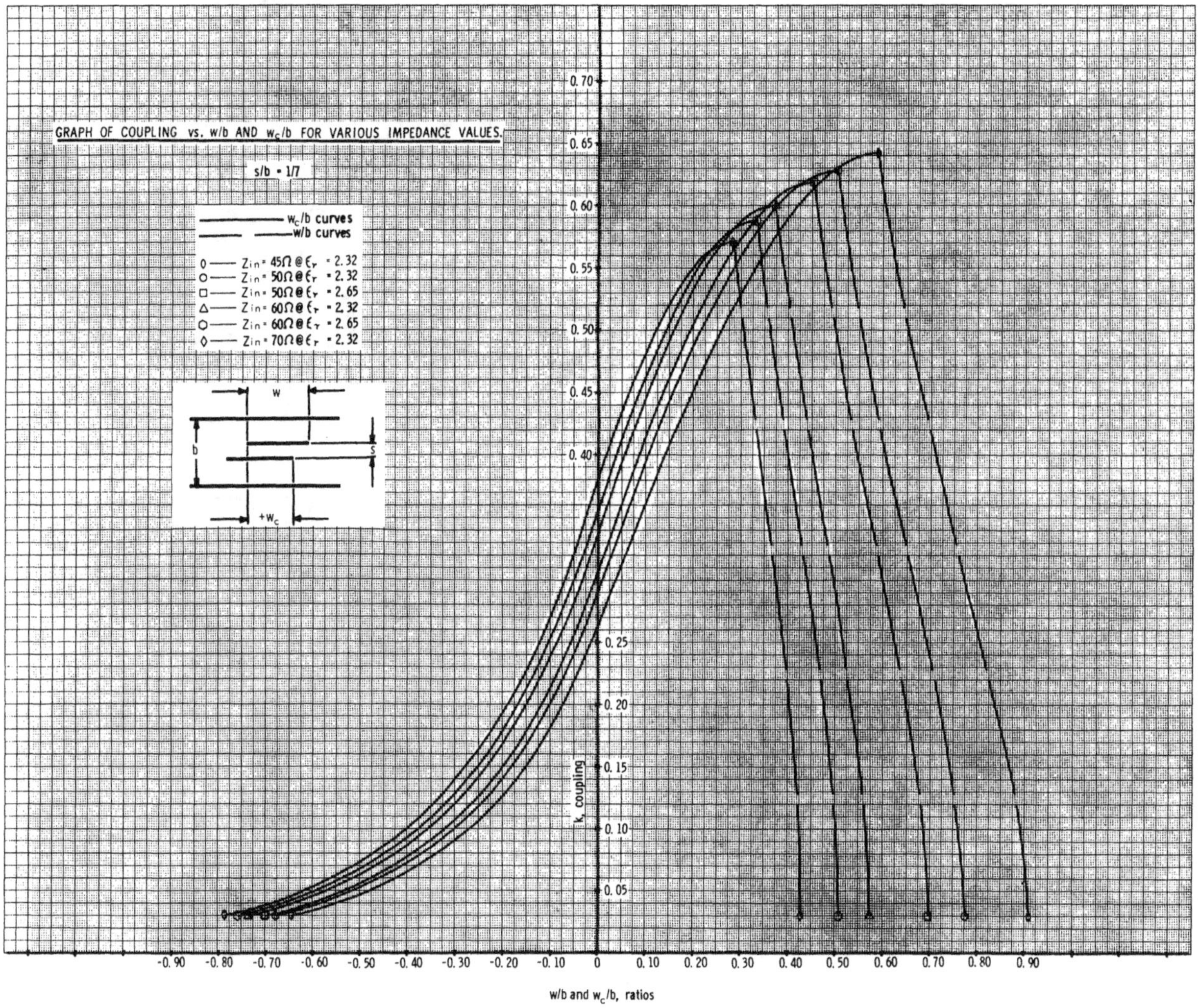

Courtesy of J.A. Mosko, see
Microwave Journal, Vol. 10, No. 5
pp. 35-37, April 1967

54

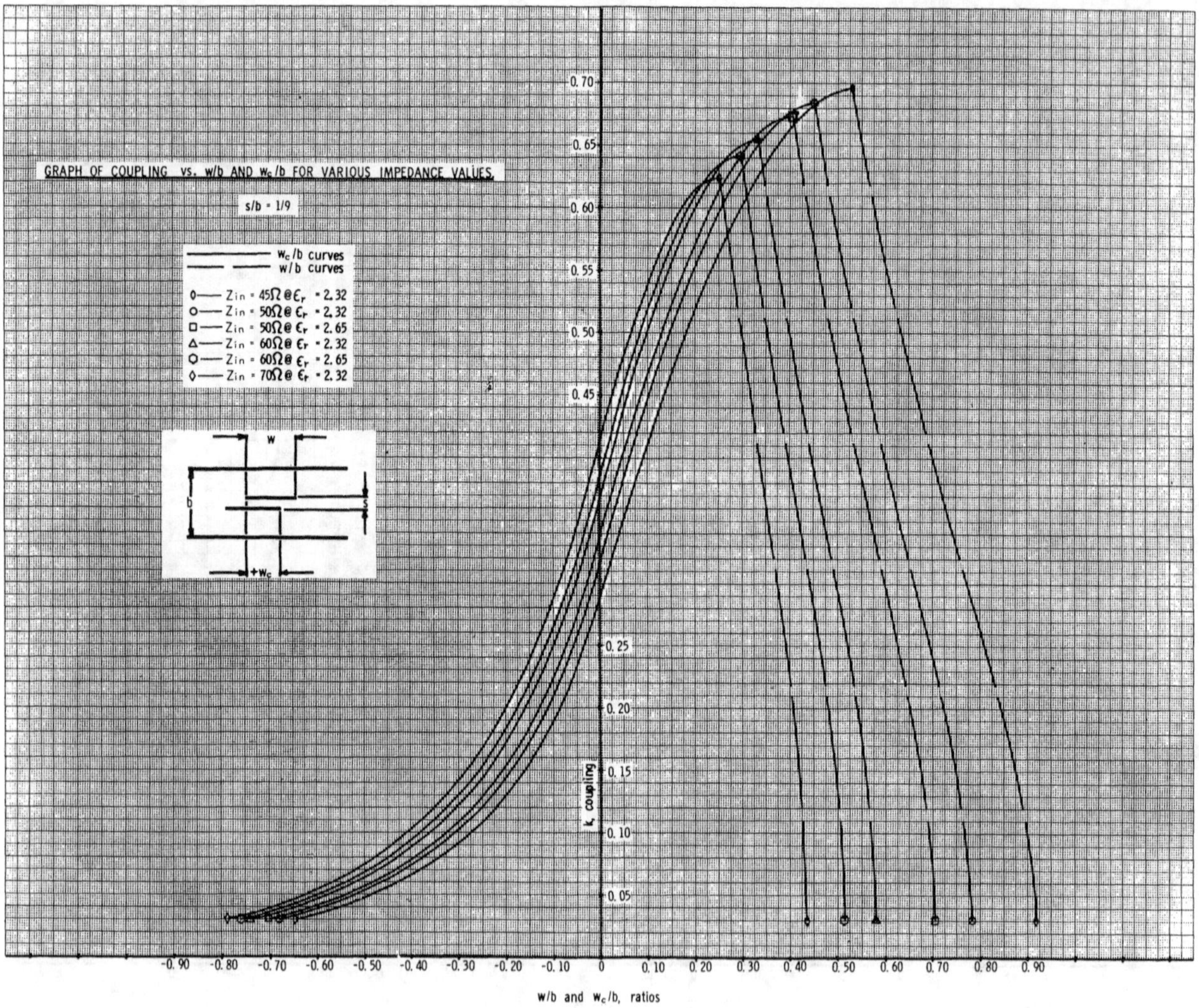

Courtesy of J.A. Mosko, see
Microwave Journal, Vol. 10, No. 5,
pp. 35-37, April 1967.

Calculation of Characteristic Admittances and Coupling Coefficients for Strip Transmission Lines

DAVID W. KAMMLER

Abstract—**An integral equation technique is presented which may be used to efficiently compute the Maxwellian capacitance matrix, i.e., the coefficients of capacitance and inductance, for any system of zero-thickness strip conductors located parallel to and between two ground planes. The TEM characteristic admittances for various operating modes and the coupling coefficients can then be obtained from the elements of this matrix. A single computer program based upon this technique can be used to compute the capacitance matrix for any particular strip line configuration desired and would thus be especially valuable to the design engineer who would like to quickly obtain accurate design curves for a previously unstudied configuration of strip conductors. This procedure gives much more accurate results in but a fraction of the computer time required when the more common finite difference equation approach is used, and it avoids the necessity for a separate mathematical analysis for each new strip line configuration, as would be required when using a conformal mapping technique. Illustrative results are given for several different strip-line configurations.**

Introduction

THE computation of the characteristic admittances of TEM mode transmission lines is one of considerable importance in the design of microwave circuits. For TEM mode propagation, this problem is essentially one of two-dimensional electrostatics with the essential parameters being the elements of the Maxwellian capacitance matrix $\{C_{ij}\}$. This matrix may be defined as follows. Let Q_1, $Q_2, \ldots, Q_N$ be the charge per meter on the strips of an N-conductor (plus ground) transmission line with $V_1, V_2, \cdots, V_N$ the corresponding potentials (see Fig. 1). Then $Q_1, \cdots, Q_N$ and $V_1, V_2, \cdots, V_N$ are related by $\{C_{ij}\}$ such that

$$\begin{bmatrix} Q_1 \\ Q_2 \\ \cdot \\ \cdot \\ Q_N \end{bmatrix} = \begin{bmatrix} C_{11} & C_{12} & \cdots & C_{1N} \\ C_{21} & C_{22} & \cdots & C_{2N} \\ \cdot & \cdot & & \cdot \\ \cdot & \cdot & & \cdot \\ C_{N1} & C_{N2} & \cdots & C_{NN} \end{bmatrix} \cdot \begin{bmatrix} V_1 \\ V_2 \\ \cdot \\ \cdot \\ V_N \end{bmatrix}. \tag{1}$$

The diagonal elements are called coefficients of capacitance and the off-diagonal elements are called the coefficients of induction. The following properties of this matrix are well known [1].

Manuscript received November 9, 1967; revised March 11, 1968. This work was supported by the U. S. Naval Weapons Center, China Lake, Calif., under Contract N123-(60530)52059A.

The author is with the Equipment Research and Development Laboratory, Texas Instruments Inc., Dallas, Tex. 75222

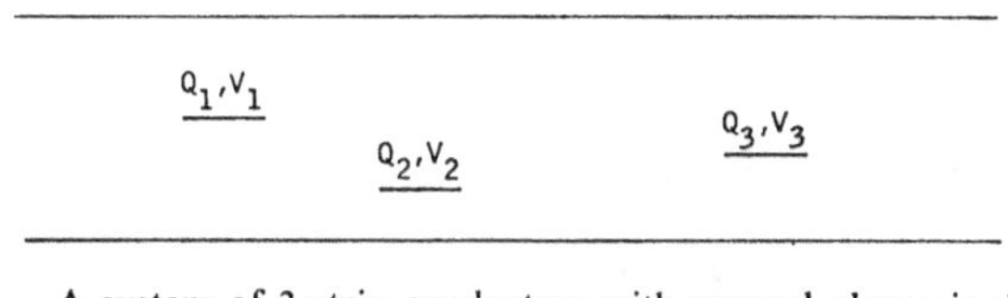

Fig. 1. A system of 3 strip conductors with ground planes is shown with the charges and potentials on the strips related by (1).

$$\begin{aligned} C_{ij} &= C_{ji}, & i, j &= 1, 2, \cdots, N \\ C_{ii} &> 0, & i &= 1, 2, \cdots, N \\ C_{ij} &< 0, & i \neq j &= 1, 2, \cdots, N \\ \sum_{j=1}^{N} C_{ij} &> 0, & i &= 1, 2, \cdots, N. \end{aligned} \tag{2}$$

From this matrix the characteristic admittances may be readily obtained. If but a single line is involved ($N=1$), the admittance is given by

$$Y = vC \tag{3}$$

where C is the capacitance (the only element of $\{C_{ij}\}$ in this case) and v is the velocity of TEM-mode wave propagation in the medium surrounding the line.

When two lines are involved ($N=2$), there are two basic modes of operation of the line, $(++)$ and $(+-)$. The even mode $(++)$ corresponds to $V_1 = V_2 = 1$, and from (1) we have

$$Q_i(++) = (C_{i1} + C_{i2}), \qquad i = 1, 2 \tag{4}$$

with the corresponding even-mode line admittances

$$Y_i(++) = v(C_{i1} + C_{i2}), \qquad i = 1, 2. \tag{5}$$

Analogously, the odd mode $(+-)$ corresponds to $V_1 = 1$, $V_2 = -1$, with

$$Q_i(+-) = (C_{i1} - C_{i2}), \qquad i = 1, 2 \tag{6}$$

and thus

$$Y_i(+-) = v\,|\,C_{i1} - C_{i2}\,|, \qquad i = 1, 2. \tag{7}$$

For $N>2$, the various line admittances for a particular operational mode may be obtained in an analogous fashion from (1) and (3). At the present time, the single-strip transmission line ($N=1$) and the microwave couplers (using $N=2$) are of primary importance, but more complex coupling circuitry now under consideration will require the use of multiconductor strip systems (with $N>2$).

The Mathematical Problem

Following Collin [2], we may approach the problem of the multistrip line in the following manner. Let $V^i(x, y)$ be the potential function of two variables satisfying the following boundary value problem:

$$\begin{aligned} \nabla^2 V^i(x, y) &= 0 \\ V^i(x, y) &= 1 \ \text{on conductor } j \\ V^i(x, y) &= 0 \ \text{on conductor } i \neq j = 1, 2, \cdots, N \\ V^i(x, y) &= 0 \ \text{on the ground planes.} \end{aligned} \tag{8}$$

56 Then since $V_1 = V_2 = \cdots = V_{j-1} = V_{j+1} = \cdots = V_N = 0$ with $V_j = 1$, we have from (1) and Gauss's theorem

$$C_{ij} = Q_i = -\epsilon \oint_{P_i} \frac{\partial V^i}{\partial n} dl, \quad i, j = 1, 2, \cdots, N, \tag{9}$$

where P_i is a closed path encircling the ith conductor about which the line integral of the outward normal component of $-\nabla V^i$ is to be taken, and ϵ is the permittivity constant.

There have been four primary methods used to solve this boundary value problem for systems of practical importance to microwave engineers.

1) The conformal mapping approach [3]–[8].
2) The finite difference equation approach [9], [10].
3) The expansion of V^i in a series of suitable "coordinate functions" [11], [12].
4) The Green's function–integral equation approach [13]–[15].

For a single-strip conductor or for a system of two-strip conductors where a high degree of symmetry is present, the Schwarz–Christoffel conformal mapping may be used to obtain mathematically exact results in terms of elliptic integrals. When the system is not highly symmetric or when more than two strip conductors are involved, however, the Schwarz–Christoffel method becomes hopelessly cumbersome unless a number of simplifying approximations are made, in which case the results are of unknown accuracy and must be checked by some other method. Besides this, the amount of actual computation required often exceeds that of methods 2)–4), so that the "mathematically exact" formulas which explicitly display the capacitances, impedances, coupling coefficients, etc., in terms of the geometric parameters of the system are of much less value to the design engineer than a good computational algorithm which allows him to calculate the desired solution to any desired degree of accuracy.

Consequently, a great deal of interest has been generated in the finite difference equation approach for calculating admittances, attenuation coefficients, etc., due to the fact that it is fairly straightforward, and a general computer program can be written to calculate the capacitances, admittances, etc., for a system of conductors and dielectric slabs which may conveniently be specified on a set of input cards [9]. This technique requires the determination of the potential field at each of the "grid points" used, and if, say, 10 000 grid points are used, one must solve a system of 10 000 linear equations in as many unknowns. This is usually done by the method of successive over relaxation (SOR) [16]. For an N-conductor system (plus ground), this must be done for each of the N-boundary value problems ($j=1, 2, \cdots, N$) of (8) with the line integrals of (9), then computed numerically using the approximate potential field. When combined with a proper "back extrapolation" technique, relatively accurate results can be obtained although the amount of computer time and storage requirements are limiting factors. For example, when using the finite difference technique described by H. Green [9], the capacitance matrix for a two-conductor system (such as that shown in Fig. 7) can be computed to an accuracy of about 0.1 percent in about 8 to 9 minutes on the IBM 7044 computer, using a uniform grid of about 10 000 points.

A number of modifications of the finite difference technique have been used, with varying degrees of success, to improve accuracy and reduce the number of grid points (and thereby the amount of computer time) required to solve the boundary value problem. When a single conductor is involved, the approximate potential field obtained can be used to obtain a variational estimate of the capacitance which is usually more accurate than the corresponding line integral estimates [17], [18]. However, this method is ill-adapted to systems with $N \geq 2$, where the nondiagonal elements of the capacitance matrix must be obtained. In a region where a singularity in the charge density naturally occurs (e.g., at the edge of a zero-thickness strip conductor), the potential function may be approximated analytically and then "matched" to the finite difference grid. This process also results in improved accuracy, [19].

The expansion of the potential function as a series of suitable coordinate functions, e.g., sines, cosines, Bessel functions, exponentials, etc., is a fairly standard method for solving "textbook" type problems, but for most practical problems a good deal of difficulty is encountered in matching coefficients over boundaries, etc. However, for some problems, such as the rectangular coaxial line [12], excellent results have been obtained with this procedure.

In contrast to the other techniques discussed, the Green's function–integral equation technique does not attempt to determine the potential field (a function of two variables) which is usually of little or no interest, but rather attempts to determine the charge densities (functions of a single variable) on the various conductors. For example, by obtaining numerical estimates of the charge density at 20 to 50 grid points, one can often obtain more accurate estimates of the elements of the capacitance matrix than could be obtained from ordinary finite difference numerical estimates of the potential function at 5000 to 10 000 grid points. The method has been successfully used for a number of problems [14], [15], and the greater part of the remainder of this paper will describe the application of this technique to systems of zero-thickness strip conductors which are located between and parallel to two ground planes as in Fig. 1.

The Necessary Green's Function

To avoid the necessity of calculating the charge densities on the two ground planes, we will make use of the Green's function $G(x, y; x', y')$ for the region between the ground planes at $y=0$ and $y=1$. $G(x, y; x', y')$ is the potential at the point (x, y) which results when a unit line charge of 1 C/m is placed at the point (x', y'). When no ground planes are present, the potential at distance r from a unit line charge may be taken (in MKS units) to be

$$V(r) = -\frac{1}{4\pi\epsilon} \ln r^2. \qquad (10)$$

Using this result, we have by the method of images (see Fig. 2)

$$G(x, y; x', y') = -\frac{1}{4\pi\epsilon} \sum_{n=-\infty}^{\infty} \ln \left(\frac{(x-x')^2 + (y-y'-2n)^2}{(x-x')^2 + (y+y'-2n)^2} \right). \qquad (11)$$

This series may be summed (e.g., see [20]) to the closed form

$$G(x, y; x', y') = -\frac{1}{4\pi\epsilon} \ln \left\{ \frac{\left[\sinh \frac{\pi}{2}(x-x')\right]^2 + \left[\sin \frac{\pi}{2}(y-y')\right]^2}{\left[\sinh \frac{\pi}{2}(x-x')\right]^2 + \left[\sin \frac{\pi}{2}(y+y')\right]^2} \right\} \qquad (12)$$

but neither of these forms is suitable for computation, as the series of (11) converges slowly, and integration with respect to x' (which is needed later) in the closed form of (12) is somewhat unappealing. With some manipulation, however, we may combine the $\pm n$ terms of the series of (11) to obtain

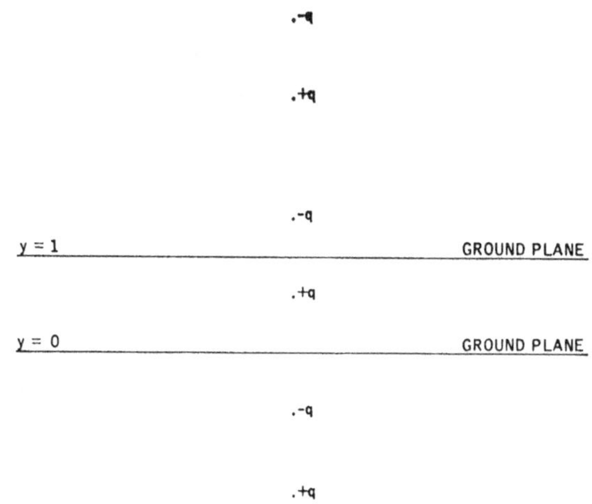

Fig. 2. A single charge placed between two conducting planes produces the same potential in the strip between $y=0$ and $y=1$ as does the system of image charges when no conducting planes are present.

$$-\frac{1}{4\pi\epsilon}\left\{\alpha_2\left[\binom{1}{1}P_2\right] + \alpha_4\left[\binom{3}{3}P_4 - \binom{3}{1}P_2(x-x')^2\right] + \alpha_6\left[\binom{5}{5}P_6 - \binom{5}{3}P_4(x-x')^2 + \binom{5}{1}P_2(x-x')^4\right] + \alpha_8\left[\binom{7}{7}P_8 - \binom{7}{5}P_6(x-x')^2 + \binom{7}{3}P_4(x-x')^4 - \binom{7}{1}P_2(x-x')^6\right] + \cdots\right\} \qquad (15)$$

where

$$\alpha_{2m} = 2 \cdot 4^{1-m} \sum_{k=6}^{\infty} \frac{1}{k^{2m}} \qquad (16)$$

$$\frac{1}{4} \ln \left(\frac{[(x-x')^2 + (y-y'-2n)^2] \cdot [(x-x')^2 + (y-y'+2n)^2]}{[(x-x')^2 + (y+y'-2n)^2] \cdot [(x-x')^2 + (y+y'+2n)^2]} \right)$$
$$= \frac{\binom{1}{1}P_2}{(2n)^2} + \frac{\binom{3}{3}P_4 - \binom{3}{1}P_2(x-x')^2}{(2n)^4} + \frac{\binom{5}{5}P_6 - \binom{5}{3}P_4(x-x')^2 + \binom{5}{1}P_2(x-x')^4}{(2n)_6} + \frac{\binom{7}{7}P_8 - \binom{7}{5}P_6(x-x')^2 + \binom{7}{3}P_4(x-x')^4 - \binom{7}{1}P_2(x-x')^6}{(2n)^8} + \cdots \qquad (13)$$

where

$$P_{2m} = P_{2m}(y, y') = \frac{1}{2m}\left[(y+y')^{2m} - (y-y')^{2m}\right]. \qquad (14)$$

By using (13) we may now perform the summation over the $n = \pm 6, \pm 7, \pm 8, \cdots,$ terms of (11) to obtain the form

$$G(x, y; x', y') = -\frac{1}{4\pi\epsilon} \sum_{n=-5}^{5} \left[\ln\left[(x-x')^2 + (y-y'-2n)^2\right] - \ln\left[(x-x')^2 + (y+y'-2n)^2\right]\right]$$

can easily be computed from the Riemann–Zeta function [21]. When $|x-x'| < 3$, the use of terms through α_{20} will give G to 12-place accuracy.

Approximation of the Charge Densities

Suppose we wish to solve the boundary value problem of (8) for the case where N-strip conductors are located parallel to and between two conducting planes at $y=0$ and $y=1$. Let the ith strip be located by $y=y_i$ and $a_i \leq x \leq b_i$ for $i=1, 2, 3 \cdots, N$. Then if $V^i(x, y)$ is the solution to (8), i.e.,

57

the potential at the point (x, y) when the jth strip is at unit potential and the other strips are at zero potential, then $V^j(x, y)$ satisfies the integral equation

$$V^j(x, y) = \sum_{i=1}^{N} \int_{x=a_i}^{b_i} \sigma_i^j(x') G(x, y; x', y_i) dx' \tag{17}$$

where

$$\sigma_i^j(x) = -\epsilon \frac{\partial V^j}{\partial y}(x, y)\bigg|_{(x, y_i)}, \qquad a_i < x < b_i \tag{18}$$

is the charge density on the ith strip. The elements of the capacitance matrix are then given by

58

$$C_{ij} = \int_{x=a_i}^{b_i} \sigma_i^j(x) dx, \quad i, j = 1, 2, \cdots, N. \tag{19}$$

Although $\sigma_i^j(x)$ becomes infinite as $x \to a_i$ and $x \to b_i$, it is possible to obtain excellent approximate solutions to the boundary value problem of (8) by approximating each $\sigma_i^j(x)$ by a suitable piecewise linear function. (The use of piecewise linear functions in the solution of inhomogeneous integral equations such as (17) is a common one [22].) We shall now proceed as follows. Let the ith strip be divided into M subintervals by the points

$$a_i = x_{i0} < x_{i1} < x_{i2} < x_{i3} < \cdots < x_{i,M-1} < x_{iM} = b_i. \tag{20}$$

Let the approximate charge density functions $\tau_i^j(x)$ be defined by

$$\tau_i^j(x) = \tau_{ik}^j + (\tau_{i,k+1}^j - \tau_{i,k}^j) \cdot \frac{x - x_{ik}}{x_{i,k+1} - x_{ik}} \tag{21}$$

for $x_{ik} \leq x \leq {}_{i,k+1}$ and $k = 0, 1, 2, \cdots, M-1$

so that $\tau_i^j(x)$ is piecewise linear with $M+1$ free parameters τ_{ik}^j, $k=0, 1, \cdots, M$, as shown in Fig. 3. The approximate potential $V_\tau^j(x, y)$ which results from these approximate charge densities is given by

$$V_\tau^j(x, y) = \sum_{i=1}^{N} \int_{x'=a_i}^{b_i} \tau_i^j(x') G(x, y; x', y_i) dx'. \tag{22}$$

An examination of (21) and (22) will show that for a fixed choice of the grid points x_{ik} on the strips, the potential $V_\tau^j(x, y)$ is a linear combination of the $N(M+1)$ parameters τ_{ik}^j, $i=1, 2, \cdots, N$, $k=0, 1, \cdots, M$. Thus we have

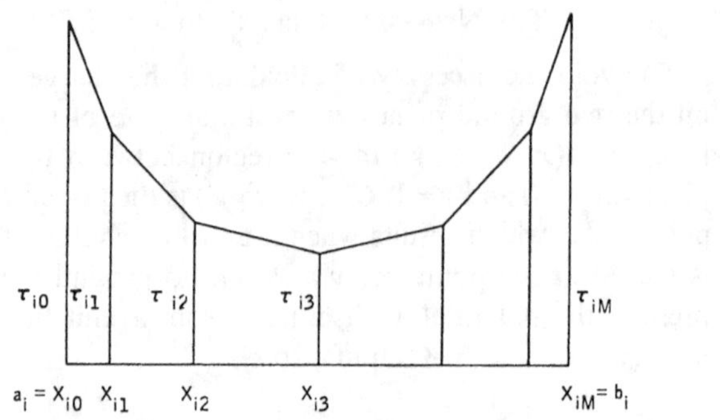

Fig. 3. The approximate charge density on the ith strip is piecewise linear with $M+1$ free parameters: τ_{ik}.

$$V_\tau^j(x, y) = \sum_{i=1}^{N} \sum_{k=0}^{M} \tau_{ik}^j \cdot \Phi_{ik}(x, y) \tag{23}$$

where from (21) and (22) we have

$$\Phi_{ik}(x, y) = \begin{cases} \displaystyle\int_{x_{i0}}^{x_{i1}} \left[1 - \frac{x' - x_{i0}}{x_{i1} - x_{i0}}\right] G(x, y; x', y_i) dx' & \text{for } k = 0 \\ \displaystyle\int_{x_{i,k-1}}^{x_{ik}} \left[\frac{x' - x_{i,k-1}}{x_{ik} - x_{i,k-1}}\right] G(x, y; x', y_i) dx' + \int_{x_{ik}}^{x_{i,k+1}} \left[1 - \frac{x' - x_{ik}}{x_{i,k+1} - x_{ik}}\right] G(x, y; x', y_i) dx' & \\ & \text{for } k = 1, 2, \cdots, M-1 \\ \displaystyle\int_{x_{i,M-1}}^{x_{iM}} \left[\frac{x' - x_{i,M-1}}{x_{iM} - x_{i,M-1}}\right] G(x, y; x', y_i) dx' & \text{for } k = M. \end{cases} \tag{24}$$

A number M' of "test points" are now chosen on each of the N strips, and the $N(M+1)$ parameters τ_{ik}^j, $i=1, 2, \cdots, N$, $k=0, 1, 2, \cdots, M$ are chosen in such a manner as to make $V^j(x, y)$ the best (least squares) fit to the exact potential on the test points, i.e., 1 on the jth strip and 0 on the others.

We may formulate this analysis as follows. Let the k'th test point on the i'th strip be given by $y = y_{i'}$, $x = \xi_{i'k'}$. Then for each i, k, i', and k' combination we must calculate

$$\Phi_{iki'k'} = \Phi_{ik}(\xi_{i'k'}, y_{i'}), \tag{25}$$

i.e., the contribution to the potential at the k'th test point on the i'th strip per unit value of the parameter τ_{ik}, which is the approximate charge density at the kth grid point on the ith strip. It is convenient to organize the calculation in the following manner. Let us define

$$S_{iki'k'} = \int_{x_{ik}}^{x_{i,k+1}} G(\xi_{i'k'}, y_{i'}; x, y_i) dx \tag{26}$$

$$T_{iki'k'} = \int_{x_{ik}}^{x_{i,k+1}} x G(\xi_{i'k'}, y_{i'}; x, y_i) dx, \tag{27}$$

where in both cases $i=1, 2, \cdots, N$, $k=0, 1, \cdots, M-1$, $i'=1, 2, \cdots, N$, and $k'=0, 1, \cdots, M'$. The form of the Green's function in (15) has been chosen so that the integrals may be easily evaluated in a readily computable form [23]. Thus from (24) through (27) we obtain

$$\Phi_{iki'k'} = \begin{cases} \dfrac{x_{i1}S_{i0i'k'} - T_{i0i'k}}{x_{i1} - x_{i0}} & \text{for } k = 0 \\ \dfrac{x_{i,k+1}S_{iki'k'} - T_{iki'k'}}{x_{i,k+1} - x_{ik}} - \dfrac{x_{i,k-1}S_{i,k-1,i'k'} - T_{i,k-1,i'k'}}{x_{ik} - x_{i,k-1}} & \text{for } k = 1, 2, \cdots, M-1 \\ -\dfrac{x_{i,M-1}S_{i,M-1,i'k'} - T_{i,M-1,i'k'}}{x_{iM} - x_{i,M-1}} & \text{for } k = M, \end{cases} \tag{28}$$

with $i=1, 2, \cdots, N$, $i'=1, 2, \cdots, N$, and $k'=0, 1, 2, \cdots, M'$. Now $V_\tau^j(x, y)$ will be the best least squares fit to $V^j(x, y)$ over the $N(M'+1)$ test points $(\xi_{i'k'}, y_{i'})$ when the $N(M+1)$ parameters τ_{ik}^j are chosen such that

$$\sum_{i',k'} \left[V^j(\xi_{i'k'}, y_{i'}) - \sum_{i,k} \tau_{ik}^j \Phi_{iki'k'} \right] = \text{minimum}. \tag{29}$$

This is a standard least squares problem which must be solved for $j=1, 2, \cdots, N$. Thus one obtains the Lth order (where $L=N\cdot(M+1)$) system of linear equations

$$\begin{bmatrix} a_{11} & a_{12} & \cdots & a_{1L} \\ a_{21} & a_{22} & \cdots & a_{2L} \\ \vdots & \vdots & & \vdots \\ a_{L1} & a_{L2} & \cdots & a_{LL} \end{bmatrix} \begin{bmatrix} \tau_{10}^1 & \tau_{10}^2 & \cdots & \tau_{10}^N \\ \tau_{11}^1 & \tau_{11}^2 & \cdots & \tau_{11}^N \\ \vdots & \vdots & & \vdots \\ \tau_{NM}^1 & \tau_{NM}^2 & \cdots & \tau_{NM}^N \end{bmatrix} = \begin{bmatrix} b_{11} & b_{12} & \cdots & b_{1N} \\ b_{21} & b_{22} & \cdots & b_{2N} \\ \vdots & \vdots & & \vdots \\ b_{L1} & b_{L2} & \cdots & b_{LN} \end{bmatrix} \tag{30}$$

where the coefficients a_{nm} and b_{nm} are determined from (29) in the usual manner [24]. It will be noted that coefficients a_{nm} depend only on the elements $\Phi_{iki'k'}$, whereas the coefficients b_{nm} depend upon the elements $\Phi_{iki'k'}$ and the boundary conditions, i.e.,

$$V^j(\xi_{ik}, y_i) = \delta_{ij}. \tag{31}$$

From the solution of (30) we then obtain

$$\begin{aligned} C_{ij} &\approx \int_{a_i}^{b_i} \tau_i^j(x)dx \\ &= \sum_{k=1}^{M} \tfrac{1}{2}[\tau_{ik}^j + \tau_{i,k-1}^j]\cdot[x_{ik} - x_{i,k-1}]. \end{aligned} \tag{32}$$

Location of the Grid Points

The location of the grid points x_{ik} and the test points $\xi_{i'k'}$ considerably affects the results of the calculational procedure which has been outlined above. The following two particular placement procedures have been considered.

1) Equally spaced with $M'=2M$:

$$\begin{aligned} x_{ik} &= a_i + \frac{k}{M}(b_i - a_i), \qquad k = 0, 1, 2, \cdots, M \\ \xi_{i'k'} &= a_i' + \frac{k'}{2M}(b_{i'} - a_{i'}), \quad k' = 0, 1, 2, \cdots, 2M. \end{aligned} \tag{33}$$

2) Sinusoidally spaced with $M'=2M$:

$$\begin{aligned} x_{ik} &= \frac{a_i + b_i}{2} + \frac{b_i - a_i}{2} \sin\left(\frac{2k - M}{2M}\pi\right), \\ &\qquad k = 0, 1, 2, \cdots, M \\ \xi_{i'k'} &= \frac{a_{i'} + b_{i'}}{2} + \frac{b_{i'} - a_{i'}}{2} \sin\left(\frac{k' - M}{2M}\pi\right) \\ &\qquad k' = 0, 1, 2, \cdots, 2M. \end{aligned} \tag{34}$$

The choice of $M'=2M$ results in about twice as many test points as parameters so that a near optimum use is made of the "smoothing" effect of the least squares process. The sinusoidal spacing results from the following considerations. Maxwell, by means of the conformal mapping $w=\cos^{-1} z$, showed that a charge of 1 C/m, when placed on the isolated conducting strip at $y=0$, $-1\leq x\leq 1$, will distribute itself according to the density [25]

$$\sigma(x) = \frac{1}{\pi}\frac{1}{\sqrt{1 - x^2}}, \qquad -1 < x < 1. \tag{35}$$

From this it follows that if $x_0, x_1, \cdots, x_M$ are specified by

$$x_k = \sin\frac{2k - M}{2M}\pi, \qquad k = 0, 1, 2, \cdots, M \tag{36}$$

then

$$\int_{x_{k-1}}^{x_k} \sigma(x)dx = \frac{1}{M} \quad \text{for } k = 1, 2, \cdots, M, \tag{37}$$

i.e., the same amount of charge is located between any two consecutive sinusoidally spaced points. Consequently, there are more points located near the ends of the strip where the singularities in $\sigma(x)$ occur. The general form, i.e., overall shape, singular behavior at the endpoints, etc., of the charge density for a conducting strip between parallel conducting planes is quite similar to that for an isolated strip. This suggests that the sinusoidal spacing might be a useful one to use in the computational procedure and this has, in fact, been found to be the case.

As an example, we shall present the results of the calculation of the capacitance and charge distribution on a single conducting strip of width $w=0.6$ centered between conducting planes at $y=0$ and $y=1$. The capacitance of such a centered strip of width w may be derived from a Schwarz–Christoffel transformation [26] and is given by

TABLE I

CALCULATION OF THE CAPACITANCE OF A SINGLE-STRIP CONDUCTOR OF WIDTH $w=0.6$ CENTERED BETWEEN CONDUCTING PLANES WITH SEPARATION $b=1$. THE EXACT VALUE IS GIVEN BY $C/\epsilon=4.157676 \cdots$

.6

Using Linear Spacing (33) M	C/ϵ	Using Sinusoidal Spacing (34) M	C/ϵ
1	4.41902	1	4.41902
2	4.34798	2	4.27441
3	4.25548	3	4.21911
4	4.23777	4	4.19480
6	4.20968	6	4.17560
8	4.19646	8	4.16816
10	4.18865	10	4.16454
12	4.18347	12	4.16251
16	4.17700	16	4.16045
18	4.17485	18	4.15988
24	4.17055	24	4.15893
30	4.16797	30	4.15849
∞	4.15768	∞	4.15768

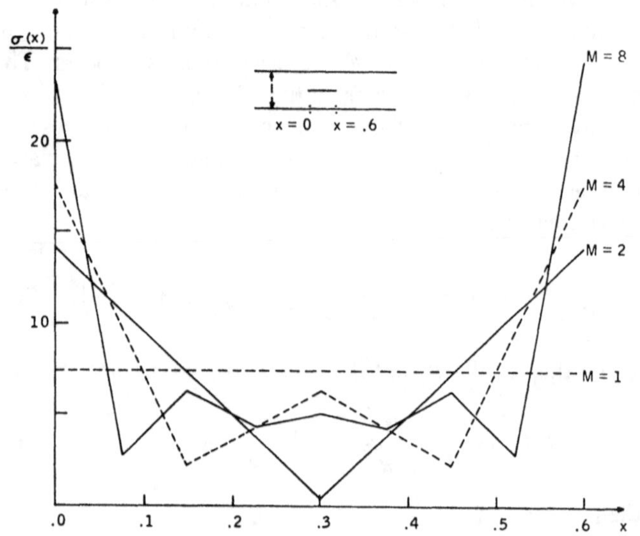

Fig. 4. The approximate charge density function is given for selected values of M for the case where the grid points and test points are linearly spaced as in (33).

$$\frac{C}{\epsilon}=4\,\frac{K\left(\tanh\frac{\pi}{2}w\right)}{K\left(\operatorname{sech}\frac{\pi}{2}w\right)} \tag{38}$$

where $K(u)$ is the complete elliptic integral

$$K(u)=\int_0^{\pi/2}\frac{dx}{\sqrt{1-u^2\sin^2 x}}\cdot \tag{39}$$

For $w=0.6$ we obtain $C/\epsilon=4.157676 \cdots$. Successive estimates, $C_l(M)$, based on $M+1$ equally spaced grid points (33), and $C_s(M)$, based on $M+1$ sinusoidally spaced grid points (34), are given in Table I. The corresponding approximate charge density functions are shown in Figs. 4 and 5. It will be noted that when equally spaced grid points are used the approximate charge density functions oscillate, whereas when the sinusoidally spaced grid points are used the charge density functions smoothly converge to their limiting form. The same behavior has been observed for a variety of two-strip ($N=2$) transmission lines which have been studied.

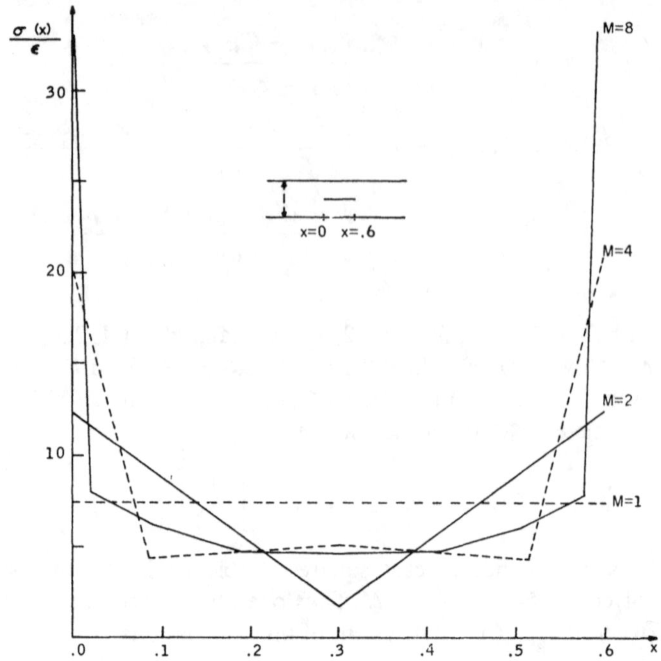

Fig. 5. The approximate charge density function is given for selected values of M for the case where the grid points and test points are sinusoidally spaced as in (34).

EXTRAPOLATION PROCEDURE

If the estimates of $C_l(M)$ and $C_s(M)$ which are given in Table I are plotted as functions of M^{-1}, it is easily seen that $C_l(M)$ and $C_s(M)$ have errors of the form

$$\epsilon_l(M)=C-C_l(M)\approx\frac{\alpha}{M} \tag{40}$$

$$\epsilon_s(M)=C-C_s(M)\approx\frac{\beta}{M^2} \tag{41}$$

where α and β are constants. Consequently, from two computations using $M=M_1$ and $M=M_2$ we may use the first-order extrapolations

$$C\approx\frac{M_2C_l(M_2)-M_1C_l(M_1)}{M_2-M_1} \tag{42}$$

$$C\approx\frac{M_2^2C_s(M_2)-M_1^2C_s(M_1)}{M_2^2-M_1^2} \tag{43}$$

to eliminate errors of the type assumed. Further consideration of the error in $C_s(M)$ (e.g., by plotting $M^2(C-C_s(M))$ as a function of M^{-1}) shows that

$$\epsilon_s(M)=C-C_s(M)\approx\frac{\beta_1}{M^2}+\frac{\beta_1}{M^3} \tag{44}$$

TABLE II

RESULTS OF THE CALCULATED CAPACITANCES OF TABLE I WHEN EXTRAPOLATED BY MEANS OF (42), (43), AND (45) TO OBTAIN INCREASED ACCURACY

Extrapolation (42)			Extrapolation (43)			Extrapolation (45)			
M_1	M_2	C/ϵ	M_1	M_2	C/ϵ	M_1	M_2	M_3	C/ϵ
1	2	4.27694	1	2	4.22621				
2	4	4.12755	2	4	4.16826	1	2	4	4.15998
3	6	4.16387	3	6	4.16110	2	3	6	4.15797
4	8	4.15516	4	8	4.15928	2	4	8	4.15800
6	12	4.15726	6	12	4.15815	4	6	12	4.15768
8	16	4.15754	8	16	4.15788	4	8	16	4.15768
12	24	4.15762	12	24	4.15774	8	12	24	4.15768

TABLE III

THE ESTIMATED CAPACITANCE OF A CENTERED STRIP OF WIDTH w WHEN USING 1-2-4, 2-3-6, AND 4-6-12 EXTRAPOLATIONS, WHERE THE TIME REQUIRED IS 0.17, 0.27, AND 0.75 MIN, RESPECTIVELY, ON THE IBM 7044. THE EXACT RESULT IS GIVEN IN TERMS OF AN ELLIPTIC INTEGRAL BY (38)

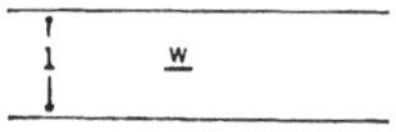

w	C/ϵ, 1-2-4 Extrapolation	C/ϵ, 2-3-6 Extrapolation	C/ϵ, 4-6-12 Extrapolation	C/ϵ, Exact
1.	5.76745	5.76639	5.76451	5.76449
0.3	2.91483	2.91334	2.91347	2.91347
0.1	1.94017	1.93948	1.93964	1.93965
0.03	1.41489	1.41454	1.41466	1.41467
0.01	1.13429	1.13408	1.13416	1.13417
0.003	0.93176	0.93162	0.93169	0.93169
0.001	0.80122	0.80112	0.80117	0.80117
0.0003	0.69458	0.69451	0.69456	0.69455
0.0001	0.61936	0.61930	0.61935	0.61933

is a better approximation where β_1 and β_2 are constants. Thus from computations made using $M=M_1$, $M=M_2$, and $M=M_3$ we may effect the second-order extrapolation

$$C = \frac{M_1{}^3(M_3 - M_2)C_s(M_1) - M_2{}^3(M_3 - M_1)C_s(M_2) + M_3{}^3(M_2 - M_1)C_s(M_3)}{M_1{}^3(M_3 - M_2) - M_2{}^3(M_3 - M_1) + M_3{}^3(M_2 - M_1)} \tag{45}$$

which eliminates any error of the type assumed by (44). This extrapolation gives vastly improved results as seen from Table II. An analogous second-order extrapolation cannot be used when the grid points are linearly spaced due to the fact that the second-order error in $C_l(M)$ is an oscillatory function of M^{-1}.

The extrapolation procedure of (45) was deduced from a series of calculations for the single-strip conductor, but from numerical experimentation it has been found that this extrapolation procedure gives excellent results for a wide variety of strip-line problems where $N\geq 2$. When $N\geq 2$, one repeats the calculation of $\{C_{ij}\}$ for $M=M_1$, M_2, M_3, and from these extrapolates each element C_{ij} by means of (45). It is convenient to choose M_1, M_2, M_3 with M_1 and M_2 being divisors of M_3, e.g., $(M_1, M_2, M_3)=(1, 2, 4)$, $(2, 3, 6)$, $(2, 4, 8)$, $(4, 6, 12)$, etc., because when this is done the S and T arrays of (26) and (27) need be computed only once for $M=M_3$. The Φ array of (28) can then be determined in the obvious manner from these for all three cases $M=M_1$, M_2, M_3, and hence three estimates of each C_{ij} are obtained from (29) and (32). As the bulk of the calculation time is spent in computing the S and T arrays, this procedure enables one to obtain the second-order extrapolation based on $M=M_1$, M_2, M_3 intervals per strip for about the same cost as the single calculation based on M_3 intervals per strip.

ACCURACY OF RESULTS

To give some idea of the accuracy attainable from this computational scheme, it will be noted that for a single strip ($N=1$) the estimates of C are good to about 3 places for a 1–2–4 extrapolation, to about 4 places for a 2–3–6 extrapolation, and to about 6 places for a 4–6–12 extrapolation. This is true for the complete range of strip widths which are of practical importance, e.g., see Table III. Thus it seems unlikely that the use of different numbers of subintervals on the conductors of a multistrip problem would improve the accuracy of the calculated capacitance matrix except, of course, in cases where the strips are very wide or are extremely close to one another. For a two-conductor system the estimates of C_{11}, $C_{12}=C_{21}$, and C_{22} are good to about 3 places for a 2–3–6 extrapolation, to about 4 places for a 2–4–8 extrapolation, and to about 5 places for a 4–6–12 extrapolation. When $N\geq 2$ it is possible to obtain a fairly good estimate of the accuracy by noting how closely the extrapolated values of $\{C_{ij}\}$ satisfy the symmetry requirement of (2). One can also repeat the calculation for several increasing values of M_3 until the desired degree of convergence has obviously been achieved.

Fig. 6 has been included to give some indication of the computer time required for the calculations. The time (in minutes on the IBM 7044) required to calculate the extrapolated value of $\{C_{ij}\}$ is given as a function of $N(M+1)$, where N is the number of conducting strips and $M=M_3$ is the maximum number of subintervals per strip used in approximating the charge density functions. Rough estimates of computation times on other computers can then be obtained from Fig. 6 by using the data in [27].

A SUMMARY OF THE METHOD

We may summarize the suggested calculational procedure of this paper as follows.

1) Compute the sinusoidally spaced grid points x_{ik} and test points ξ_{ik} for each of the N strips using (34). This may be done in single precision.
2) Compute the $S_{iki'k'}$ and $T_{iki'k'}$ arrays of (26) and (27)

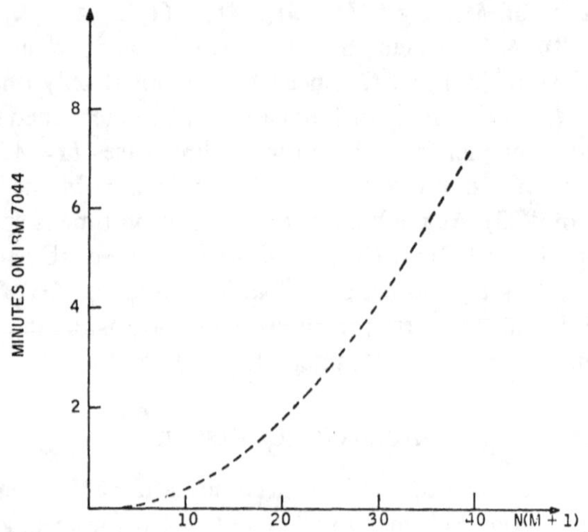

Fig. 6. The time to calculate (C_{ij}) for N conducting strips when a maximum of M subintervals per strip is used in approximating the charge densities on the strips.

using $M=M_3$. The computations should be done in double precision but may then be truncated and stored in single-precision arrays.

3) Using the $S_{iki'k'}$, $T_{iki'k'}$, x_{ik}, and ξ_{ik} arrays computed for $M=M_3$, set up the $\Phi_{iki'k'}$ array using (28) (with obvious modifications) for $M=M_i$ ($i=1, 2, 3$). Then set up the A and B arrays of (30) corresponding to $M=M_i$. This should be done in double precision.
4) Solve the $A\tau=B$ linear system (e.g., by a Crout reduction) to obtain the approximate charge densities on each of the strips. This should be done in double precision.
5) Using the approximate charge densities corresponding to $M=M_i$, compute the approximate Maxwellian capacitance matrix $\{C_{ij}\}$ using (32). Steps 3) through 5) should be carried out for $M=M_1$, M_2, M_3 where M_1 and M_2 are divisors of M_3.
6) Using the three estimates of $\{C_{ij}\}$, extrapolate using (45) for each of the N^2 elements of the matrix.
7) The various admittances, coupling coefficients, etc., for the transmission line can then be calculated in the usual manner from $\{C_{ij}\}$.

Sample Results

Table IV presents a series of capacitances of an off-center single-strip transmission line. (In principle, these results could be computed by means of hyperelliptic functions.) Tables V and VI present the three essential parameters C_{11}, C_{12}, and C_{22} for an unusual configuration of a two-strip transmission line to be used as a coupler. Note that computations for this configuration may be performed just as simply as those for the more familiar symmetric couplers which have previously been considered. Table VII presents the capacitance matrices for systems of 1, 2, 3, 4, and 5 esually spaced zero-thickness strip conductors centered between the ground planes with each strip having width $w=0.2$ and with a strip-to-strip spacing distance of 0.1.

It will be noted, of course, that these results could be computed by means of the finite difference technique. In order to give some comparison of the two methods, Table VIII and the accompanying Fig. 7 have been included. It will be noted that the Green's function–integral equation technique gives much more accurate results in but a fraction of the computer time required when the finite difference method is used.

TABLE IV

Selected Values of C/ϵ for a Strip of Width w at Distance t from the Lower Ground Plane. Exact Results May Be Computed in Terms of Hyperelliptic Functions

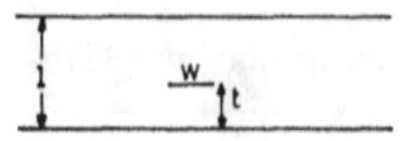

w	C/ϵ, $t=0.1$	C/ϵ, $t=0.2$	C/ϵ, $t=0.3$	C/ϵ, $t=0.4$	C/ϵ, $t=0.5$
1.	13.3996	8.2355	6.6112	5.9509	5.7645
0.3	5.4757	3.7654	3.2095	2.9791	2.9135
0.1	3.0034	2.3154	2.0742	1.9700	1.9397
0.03	1.9216	1.6067	1.4855	1.4308	1.4147
0.01	1.4392	1.2545	1.1793	1.1445	1.1342
0.003	1.1281	1.0114	0.9619	0.9387	0.9317
0.001	0.9423	0.8594	0.8234	0.8063	0.8012

Extension of the Technique

We consider now an extension of the Green's function–integral equation technique to include higher-order representations of the charge density function than the piecewise linear function $\tau(x)$ which has been used up to this point. (For convenience, we shall drop the subscripts on $\tau(x)$ in this discussion.) From (21) we see that the continuous piecewise linear function $\tau(x)$ can be written in the form

$$\tau(x) = c_0\phi_0(x) + c_1\phi_1(x) + c_2\phi_2(x) + \cdots + c_M\phi_M(x) \tag{46}$$

where the functions $\phi_i(x)$ are defined (with respect to the grid points $a=x_0<x_1<x_2<\cdots<x_M=b$) by

$$\phi_0(x) = \begin{cases} \dfrac{x_1 - x}{x_1 - x_0} & \text{for } x_0 \le x \le x_1 \\ 0 & \text{otherwise} \end{cases}$$

$$\phi_i(x) = \begin{cases} \dfrac{x - x_{i-1}}{x_i - x_{i-1}} & \text{for } x_{i-1} \le x \le x_i \\ \dfrac{x_{i+1} - x}{x_{i+1} - x_i} & \text{for } x_i \le x \le x_{i+1}, \quad i = 1, 2, \cdots, M-1 \\ 0 & \text{otherwise} \end{cases} \tag{47}$$

$$\phi_M(x) = \begin{cases} \dfrac{x_M - x}{x_M - x_{M-1}} & \text{for } x_{M-1} \le x \le x_M \\ 0 & \text{otherwise,} \end{cases}$$

e.g., see Fig. 8. Thus $\tau(x)$ is a linear combination of the functions $\phi_i(x)$, and hence the resulting approximate potential may be written in the form

$$V_\tau(x, y) = \sum_{i=0}^{M} c_i\Phi_i(x, y) \tag{48}$$

TABLE V

VALUES OF (C_{ij}) FOR THE TWO-CONDUCTOR SYSTEM SHOWN IN THE DIAGRAM (SEE ALSO TABLE VI)

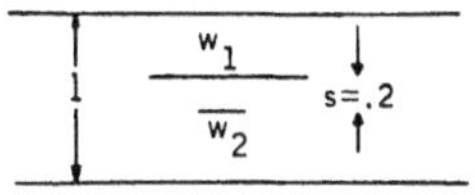

w_1	w_2	C_{11}/ϵ	$C_{12}/\epsilon=C_{21}/\epsilon$	C_{22}/ϵ
1.0	1.0	9.136	−5.355	9.136
1.0	0.8	8.862	−4.743	7.760
1.0	0.6	8.434	−3.953	6.301
1.0	0.5	8.179	−3.518	5.558
1.0	0.4	7.904	−3.064	4.807
1.0	0.3	7.611	−2.592	4.048
1.0	0.2	7.300	−2.100	3.268
1.0	0.1	6.956	−1.561	2.424
1.0	0.05	6.746	−1.234	1.916
1.0	0.01	6.485	−0.828	1.286
1.0	0.001	6.314	−0.563	0.874
0.6	0.6	6.134	−3.354	6.134
0.6	0.5	6.018	−3.075	5.459
0.6	0.4	5.859	−2.740	4.751
0.6	0.3	5.660	−2.357	4.017
0.6	0.2	5.422	−1.931	3.251
0.6	0.1	5.140	−1.445	2.416
0.6	0.05	4.962	−1.145	1.911
0.6	0.01	4.737	−0.769	1.284

Note: These figures are exact to within ±0.001. The time required was 2.80 minutes per line on the IBM 7044.

TABLE VI

VALUES OF (C_{ij}) FOR THE TWO-CONDUCTOR SYSTEM SHOWN IN THE DIAGRAM. IT WILL BE NOTED THAT FOR $w>0.2$ THE CAPACITANCES C_{11}, C_{12}, AND C_{22} ARE LINEAR FUNCTIONS OF w

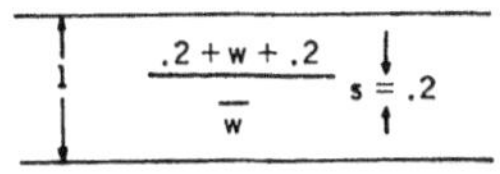

w	C_{11}/ϵ	$-C_{12}/\epsilon$	C_{22}/ϵ
1.0	11.434	5.954	9.303
0.8	9.934	4.954	7.802
0.6	8.434	3.953	6.301
0.4	6.933	2.951	4.794
0.2	5.422	1.931	3.251
0.1	4.649	1.386	2.409
0.05	4.241	1.072	1.904
0.02	3.965	0.825	1.489
0.01	3.853	0.704	1.279
0.00	3.422		

TABLE VII

$C_{11}/\epsilon=2.4618$

$C_{11}/\epsilon=C_{22}/\epsilon=2.8888$
$C_{12}/\epsilon=C_{21}/\epsilon=-1.0379$

$C_{11}/\epsilon=C_{33}/\epsilon=2.8914$
$C_{22}/\epsilon=3.2915$
$C_{12}/\epsilon=C_{21}/\epsilon=C_{23}/\epsilon=C_{32}/\epsilon=-1.0064$
$C_{13}/\epsilon=C_{31}/\epsilon=-0.0841$

$C_{11}/\epsilon=C_{44}/\epsilon=2.8914$
$C_{22}/\epsilon=C_{33}/\epsilon=3.2938$
$C_{12}/\epsilon=C_{21}/\epsilon=C_{34}/\epsilon=C_{43}/\epsilon=-1.0061$
$C_{23}/\epsilon=C_{32}/\epsilon=-0.9767$
$C_{13}/\epsilon=C_{31}/\epsilon=C_{24}/\epsilon=C_{42}/\epsilon=-0.0795$
$C_{14}/\epsilon=C_{41}/\epsilon=-0.0125$

$C_{11}/\epsilon=C_{55}/\epsilon=2.8914$
$C_{22}/\epsilon=C_{44}/\epsilon=3.2939$
$C_{33}/\epsilon=3.2961$
$C_{12}/\epsilon=C_{21}/\epsilon=C_{45}/\epsilon=C_{54}/\epsilon=-1.0061$
$C_{23}/\epsilon=C_{32}/\epsilon=C_{34}/\epsilon=C_{43}/\epsilon=-0.9764$
$C_{13}/\epsilon=C_{31}/\epsilon=C_{35}/\epsilon=C_{53}/\epsilon=-0.0794$
$C_{24}/\epsilon=C_{42}/\epsilon=-0.0751$
$C_{14}/\epsilon=C_{41}/\epsilon=C_{25}/\epsilon=C_{52}/\epsilon=-0.0117$
$C_{15}/\epsilon=C_{51}/\epsilon=-0.0020$

Note: The separation of the ground planes is 1, the width of each strip is 0.2, and the spacing between adjacent strips is 0.1. All of the strips are centered between the ground planes.

TABLE VIII

COMPARISON OF THE FINITE DIFFERENCE TECHNIQUE AND THE GREEN'S FUNCTION–INTEGRAL EQUATION TECHNIQUE

	C_{11}/ϵ	$-C_{12}/\epsilon$	$-C_{21}/\epsilon$	C_{22}/ϵ	Time
Finite Difference					
$h=1/10$	8.6532	3.4125	3.4402	5.2765	
$h=1/20$	8.1688	3.2416	3.2398	5.0721	
Extrapolated	7.6842	3.0709	3.0394	4.8677	0.82 min
$h=1/20$	8.1688	3.2415	3.2398	5.0721	
$h=1/40$	8.0417	3.1550	3.1556	4.9440	
Extrapolated	7.9145	3.0685	3.0714	4.8159	3.48 min
$h=1/30$	8.0936	3.1854	3.1876	4.9891	
$h=1/60$	7.9943	3.1244	3.1244	4.8981	
Extrapolated	7.8950	3.0634	3.0612	4.8071	8.99 min
Green's Function–Integral Equation					
(1,2,4) Extrapolation	7.9414	3.0979	3.0651	4.8119	0.48 min
(2,3,6) Extrapolation	7.9391	3.0841	3.1066	4.8310	0.84 min
(2,4,8) Extrapolation	7.9078	3.0653	3.0682	4.8094	1.36 min
(4,6,12) Extrapolation	7.9035	3.0643	3.0631	4.8076	2.80 min
Exact Value	7.9039	3.0640	3.0640	4.8074	

64

The finite difference approach in [9] may be used to calculate the capacitance matrix for the two-conductor system of Fig. 7(a). The extrapolation ($C=2C_h-C_{2h}$) assumes an error proportional to the mesh size h. One-half of the field for each of the boundary value problems ($V_1=1$, $V_2=0$; $V_1=0$, $V_2=1$) is computed using the SOR technique with a near-optimum relaxation constant, and the total time (minutes on the IBM 7044) is given. For comparison purposes, the Green's function–integral equation approach discussed in this paper has also been used to calculate the matrix for the same system with the end planes removed [Fig. 7(b)], and several extrapolated values [using (45)] are given along with the required computer time.

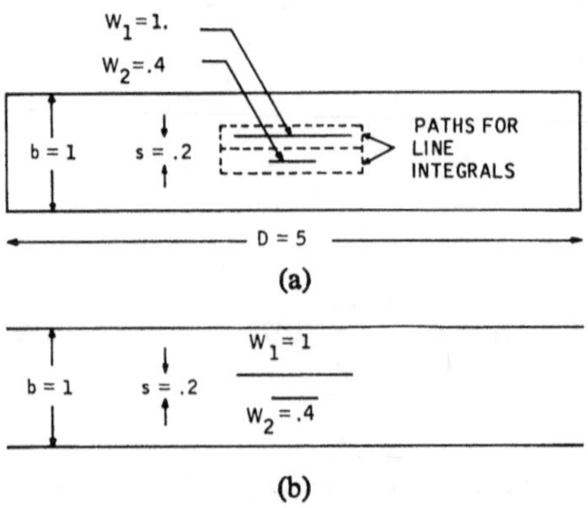

Fig. 7. Computation of the capacitance matrix for the two-conductor system. (a) By the finite difference technique. (b) By the Green's function–integral equation technique. The results are given in Table VIII.

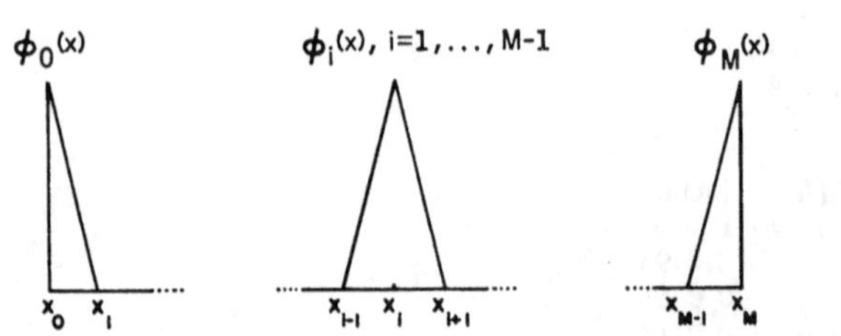

Fig. 8. The continuous piecewise linear charge density $\tau(x)$ which passes through the points (x_i, τ_i) can be written in the form $\tau(x)=\sum c_i\phi_i(x)$ where the functions ϕ_i are as shown here.

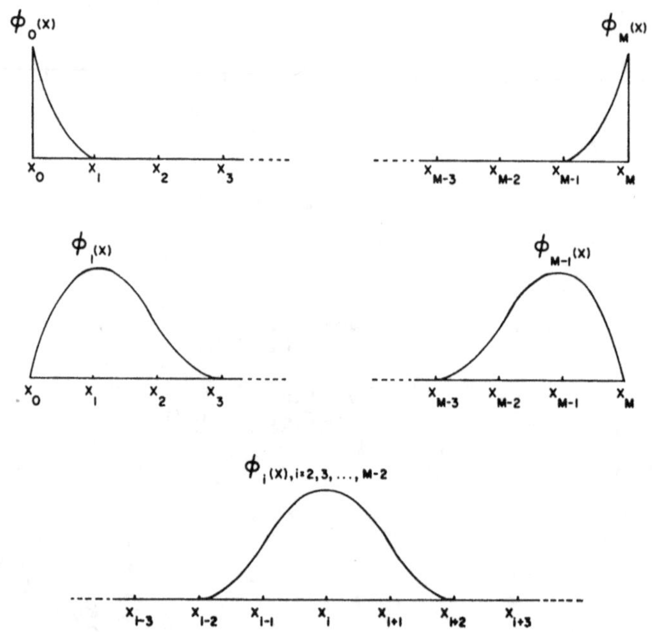

Fig. 9. A continuous piecewise quadratic charge density $\tau(x)$ which has a continuous first derivative for $a \le x \le b$ and which passes through the points (x_i, τ_i) can be written in the form $\tau(x)=\sum c_i\phi_i(x)$ where the functions $\phi_i(x)$ are as shown here.

where

$$\Phi_i(x, y) = \int_a^b \phi_i(x')G(x, y; x', y')dx'. \tag{49}$$

By construction, each of the functions $\Phi_i(x, y)$ satisfies Laplace's equation and vanishes on the ground planes so that V_τ also has these properties. Since the parameters c_i enter linearly, the least squares process can conveniently be used to force V_τ to approximately satisfy the remaining boundary conditions on the strip conductors. The numerical efficiency of the method depends, of course, on the fact that the integrals in (49) can be analytically evaluated in a simple-to-compute form.

From (46) and (47) and Fig. 9 we see that a continuous piecewise quadratic charge density function $\tau(x)$, which has a continuous first derivative and passes through the selected grid points (x_i, τ_i), may be written in the form of (46) if now the functions $\phi_i(x)$ are defined by

The analysis would then be performed exactly as was done previously except that now, in addition to the $S_{iki'k'}$ and $T_{iki'k'}$ integrals of (26) and (27), the integral

$$U_{iki'k'} = \int_{x_{ik}}^{x_{i,k+1}} x^2 G(\xi_{i'k'}, y_{i'}; x, y_i)dx \tag{51}$$

(which is readily computable from (15) and [23]) would be needed in order to compute $\Phi_{iki'k'}$. The use of the piecewise quadratic charge density could be expected to give considerably more accuracy than that which is obtainable by the piecewise linear process, but additional computer time and storage (for the U array) would be required. This process could obviously be extended to include piecewise cubic, piecewise quartic, etc., charge densities, but it is doubtful that the use of these higher-order processes would be as 65
effective as the alternative procedure of using more grid points in the analysis.

$$\phi_0(x) = \begin{cases} \dfrac{(x_1 - x)^2}{(x_1 - x_0)^2} & \text{for } x_0 \le x \le x_1 \\ 0 & \text{otherwise} \end{cases}$$

$$\phi_1(x) = \begin{cases} \dfrac{(x_1 - x_0)^2 - (x_1 - x)^2}{(x_1 - x_0)^2} & \text{for } x_0 \le x \le x_1 \\ \dfrac{(x_2 - x_1)\cdot(x_3 - x_1) - (x - x_1)^2}{(x_2 - x_1)\cdot(x_3 - x_1)} & \text{for } x_1 \le x \le x_2 \\ \dfrac{(x_3 - x)^2}{(x_3 - x_1)\cdot(x_3 - x_2)} & \text{for } x_2 \le x \le x_3 \\ 0 & \text{otherwise} \end{cases}$$

$$\phi_i(x) = \begin{cases} \dfrac{(x - x_{i-2})^2}{(x_{i-1} - x_{i-2})\cdot(x_i - x_{i-2})} & \text{for } x_{i-2} \le x \le x_{i-1} \\ \dfrac{(x_i - x_{i-1})\cdot(x_i - x_{i-2}) - (x_i - x)^2}{(x_i - x_{i-1})\cdot(x_i - x_{i-2})} & \text{for } x_{i-1} \le x \le x_i \\ \dfrac{(x_{i+1} - x_i)\cdot(x_{i+2} - x_i) - (x - x_i)^2}{(x_{i+1} - x_i)\cdot(x_{i+2} - x_i)} & \text{for } x_i \le x \le x_{i+1}, \quad i = 2, \cdots, M-2 \\ \dfrac{(x_{i+2} - x)^2}{(x_{i+2} - x_i)\cdot(x_{i+2} - x_{i+1})} & \text{for } x_{i+1} \le x \le x_{i+2} \\ 0 & \text{otherwise} \end{cases} \tag{50}$$

$$\phi_{M-1}(x) = \begin{cases} \dfrac{(x - x_{M-3})^2}{(x_{M-2} - x_{M-3})\cdot(x_{M-1} - x_{M-3})} & \text{for } x_{M-3} \le x \le x_{M-2} \\ \dfrac{(x_{M-1} - x_{M-2})\cdot(x_{M-1} - x_{M-3}) - (x_{M-1} - x)^2}{(x_{M-1} - x_{M-2})\cdot(x_{M-1} - x_{M-3})} & \text{for } x_{M-2} \le x \le x_{M-1} \\ \dfrac{(x_M - x_{M-1})^2 - (x - x_{M-1})^2}{(x_M - x_{M-1})^2} & \text{for } x_{M-1} \le x \le x_M \\ 0 & \text{otherwise} \end{cases}$$

$$\phi_M(x) = \begin{cases} \dfrac{(x - x_{M-1})^2}{(x_M - x_{M-1})^2} & \text{for } x_{M-1} \le x \le x_M \\ 0 & \text{otherwise.} \end{cases}$$

Application to Other Problems

The Green's function–integral equation technique which has been applied to the specific boundary value problem for systems of zero-thickness strip conductors located between and parallel to two ground planes can obviously be extended to a number of other problems of current interest. The charge density must be expressed in a linear form analogous to that of (46), and the corresponding integrals analogous to (49) must be easily evaluated numerically. It is clear that the problem of zero-thickness strip conductors located at arbitrary angles with respect to parallel ground planes can be handled in this manner. For example, suppose that the conducting strip is parametrically defined by

$$\begin{aligned} x(t) &= a_1 + (a_2 - a_1)t \\ y(t) &= b_1 + (b_2 - b_1)t \end{aligned} \qquad 0 \leq t \leq 1, \qquad (52)$$

e.g., see Fig. 10. Let $\tau(t)dt$ denote the charge at position t on the conducting strip. Since the integrals

$$S = \int G(x, y; a_1 + [a_2 - a_1]t, b_1 + [b_2 - b_1]t)dt$$

$$T = \int tG(x, y; a_1 + [a_2 - a_1]t, b_1 + [b_2 - b_1]t)dt \qquad (53)$$

$$U = \int t^2G(x, y; a_1 + [a_2 - a_1]t, b_1 + [b_2 - b_1]t)dt$$

may be analytically done using the form of the Green's function given in (15) [23], we may choose to work with either a piecewise linear or piecewise quadratic charge density. The remainder of the analysis is then carrier out as before. Moreover, since the boundary of an arbitrarily shaped conductor can be approximated by a series of line segments, it follows that the Green's function–integral equation technique can be extended to handle problems where the conductors are not zero-thickness strips.

The use of piecewise linear or piecewise quadratic charge densities is certainly not necessary for the use of the Green's function–integral equation technique, however. For example, the boundary value problem involving a system of circular cylindrical rods placed between parallel ground planes (see Fig. 11) has been considered by Cristal [14]. This problem is particularly well-suited for solution by the Green's function–integral equation technique. In this case, the charge density on the surface of a circular conductor can be well-approximated by a trigonometric series

$$\tau(\theta) = a_0 + \sum_{m=1}^{M} [a_m \cos m\theta + b_m \sin m\theta] \qquad (54)$$

where the angular coordinate θ locates the position of the increment of charge, $\tau(\theta)\, d\theta$, on the surface of the conductor. Again the parameters a_i, b_i enter linearly so that the approximate potential V can be written in the form

$$V_\tau(x, y) = a_0A_0(x, y) + \sum_{m=1}^{M} [a_mA_m(x, y) + b_mB_m(x, y)] \qquad (55)$$

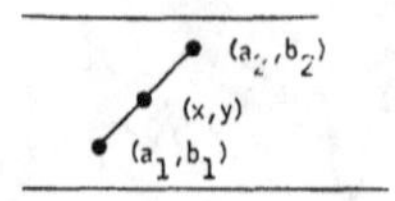

Fig. 10. We may use $(x, y) = (a_1 + [a_2 - a_1]t,\ b_1 + [b_2 - b_1]t)$, $0 \leq t \leq 1$, to locate a point on a strip conductor which is arbitrarily placed between the two parallel ground planes.

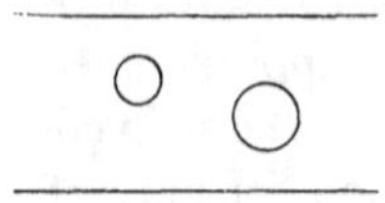

Fig. 11. A system of circular conducting rods is placed between two parallel ground planes.

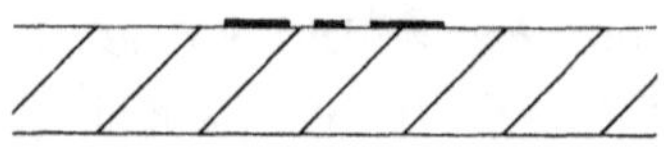

Fig. 12. For the microstrip problem we have a system of zero-thickness strip conductors located on a dielectric slab which rests on a ground plane.

where

$$A_m(x, y) = \int_0^{2\pi} \cos m\theta G(x, y; x_0 + R\cos\theta, y_0 + R\sin\theta)d\theta$$
$$(56)$$
$$B_m(x, y) = \int_0^{2\pi} \sin m\theta G(x, y; x_0 + R\cos\theta, y_0 + R\sin\theta)d\theta$$

and the conductor is assumed to have center (x_0, y_0) and radius R. Using the form of the Green's function given in (15), these integrals can be evaluated analytically [28]. In this case (since no singularities occur in the exact charge density) the test points are chosen to be equally spaced on the surface of the conductors. It will be noted that the use of $M = 6$ in (54) will allow the capacitance matrices to be computed to about 6-place accuracy (unless the conductors are very close together), and the amount of computer time required is about the same as that for the 2–3–6 extrapolation values for the strip conductors. Needless to say, the circular boundary conditions are difficult to incorporate into a finite difference scheme.

We next consider the application of the technique to the microstrip problem where we have a system of strip conductors on top of a dielectric slab which rests on a ground plane (see Fig. 12). In this case we have a troublesome normal derivative condition which must be met along the dielectric interface. For this particular problem we might use the simple Green's function

$$G(x, y; x', y') = \frac{1}{4\pi\epsilon} \cdot \ln\left[\frac{(x - x')^2 + (y - y')^2}{(x - x')^2 + (y + y')^2}\right] \qquad (57)$$

to take care of the boundary conditions on the ground plane. The charge density on the conducting strips and the surface charge density at the dielectric interface can then be approximated with piecewise linear or piecewise quadratic functions with the parameters entering linearly. A number of "test points" on the strips and along the dielectric interface would be selected and a least squares condition used to select the parameters to insure that the remaining boundary conditions (for V on the strips and its normal derivative at the dielectric interfaces) are met. This method could be expected to give considerably more accuracy per computer time invested than the finite difference method, and a detailed development would be well worthwhile if extensive calculations of this type are required.

References

[1] J. C. Maxwell, *A Treatise on Electricity and Magnetism*, 3rd ed., vol. 1. New York: Dover, 1954, pp. 103–118.

[2] R. E. Collin, *Field Theory of Guided Waves*. New York: McGraw-Hill, 1960, p. 124.

[3] D. Park, "Planar transmission lines," *IRE Trans. Microwave Theory and Techniques*, vol. MTT-3, pp. 8–12, April 1955.

[4] ——, "Planar transmission lines: II, *IRE Trans. Microwave Theory and Techniques*, vol. MTT-3, pp. 7–11, October 1955.

[5] S. B. Cohn, "Shielded coupled-strip transmission lines," *IRE Trans. Microwave Theory and Techniques*, vol. MTT-3, pp. 29–38, October 1955.

[6] ——, "Characteristic impedance of broadside-coupled strip transmission lines," *IRE Trans. Microwave Theory and Techniques*, vol. MTT-8, pp. 633–737, November 1960.

[7] H. A. Wheeler, "Transmission-line properties of parallel wide strips by a conformal-mapping approximation," *IEEE Trans. Microwave Theory and Techniques*, vol. MTT-12, pp. 280–289, May 1964.

[8] J. P. Shelton, Jr., "Impedances of offset parallel-coupled strip transmission lines," *IEEE Trans. Microwave Theory and Techniques*, vol. MTT-14, pp. 7–15, January 1966.

[9] H. E. Green, "The numerical solution of some important transmission-line problems," *IEEE Trans. Microwave Theory and Techniques*, vol. MTT-13, pp. 676–692, September 1965.

[10] M. V. Schneider, "Computation of impedance and attenuation of TEM-lines by finite difference methods," *IEEE Trans. Microwave Theory and Techniques*, vol. MTT-13, pp. 793–800, November 1965.

[11] N. A. Begovich, "Capacity and characteristic impedance of strip transmission lines with rectangular inner conductors," *IRE Trans. Microwave Theory and Techniques*, vol. MTT-3, pp. 127–133, March 1955.

[12] O. R. Cruzan and R. V. Garver, "Characteristic impedance of rectangular coaxial transmission lines," *IEEE Trans. Microwave Theory and Techniques*, vol. MTT-12, pp. 488–495, September 1964.

[13] K. G. Black and T. J. Higgins, "Rigorous determination of the parameters of microstrip transmission lines," *IRE Trans. Microwave Theory and Techniques*, vol. MTT-3, pp. 93–113, March 1955.

[14] E. G. Cristal, "Coupled circular cylindrical rods between parallel grounded planes," *IEEE Trans. Microwave Theory and Techniques*, vol. MTT-12, pp. 428–439, July 1964.

[15] J. W. Duncan, "Characteristic impedance of multiconductor strip transmission lines," *IEEE Trans. Microwave Theory and Techniques*, vol. MTT-13, pp. 107–118, January 1965.

[16] R. S. Varga, *Matrix Iterative Analysis*. Englewood Cliffs, N. J.: Prentice-Hall, 1962.

[17] R. G. de Buda, "A method of calculating the characteristic impedance of a strip transmission line to a given degree of accuracy," *IRE Trans. Microwave Theory and Techniques*, vol. MTT-6, pp. 440–446, October 1958.

[18] C. T. Carson and G. K. Cambrell, "Upper and lower bounds on the characteristic impedance of TEM mode transmission lines," *IEEE Trans. Microwave Theory and Techniques (Correspondence)*, vol. MTT-14, pp. 497–498, October 1966.

[19] J. W. Duncan, "The accuracy of finite-difference solutions of Laplace's equation," *IEEE Trans. Microwave Theory and Techniques*, vol. MTT-15, pp. 575–582, October 1967.

[20] L. V. Bewley, *Two Dimensional Fields in Electrical Engineering*. New York: Dover, 1963, pp. 157–160.

[21] National Bureau of Standards, *Handbook of Mathematical Functions*. Washington, D.C.: U. S. Government Printing Office, 1964, p. 811.

[22] E G. Cristal and J. van Bladel, "Fields in cavity-excited accelerators," *J. Appl. Phys.*, pp. 1715–1724, September 1961.

[23] H. B. Dwight, *Tables of Integrals and Other Mathematical Data*, 4th ed. New York: Macmillan, 1961, pp. 144–145, #623, #623.1, #623.2.

[24] F. B. Hildebrand, *Introduction to Numerical Analysis*. New York: McGraw-Hill, 1956, pp. 261–263.

[25] J. C. Maxwell, *A Treatise on Electricity and Magnetism*, 3rd ed., vol. 1. New York: Dover, 1954, pp. 296–297.

[26] R. E. Collin, *Field Theory of Guided Waves*. New York: McGraw-Hill, 1960, pp. 139–143.

[27] E. C. Berkeley, "Characteristics of general purpose digital computers," *Computers and Automation*, vol. 16, pp. 91–121, June 1967.

[28] D. W. Kammler, "Calculation of characteristic admittances, coupling coefficients, and breakdown fields for systems of cylindrical transmission lines between parallel ground planes," Texas Instruments Inc., Dallas, Tex., Rept. 921-02/47-178, May 1967.

"Microstrip"* has become very popular with the advent of integrated circuits, since it allows one free and accessible surface (for example, for laying solid-state or ferrimagnetic devices on that surface).

The solution of a single strip on an infinite dielectric substrate over an infinite ground plane has been given in a paper by Wheeler (p. 71), which is basic to most microstrip circuits. Although it does not deal with coupled lines **per se**, we have not only reprinted the paper, but we reproduce the original four design charts enlarged (courtesy of Mr. H.A. Wheeler) for the convenience of the reader and to enable the designer to read numerical values more accurately.

COUPLED MICROSTRIP LINES III

The numerical solution of microstrip coupled lines requires large digital computers (whereas strip-line problems can generally be handled by conformal transformations and much less numerical calculation). Bryant and Weiss (p. 89) have presented design data for single and coupled microstrip lines.

Waves or modes on microstrip are only quasi-TEM, therefore the theory of TEM coupled lines applies only approximately. One problem on microstrip is radiation, due to discontinuities like short-circuit posts, corners, etc., which should therefore be avoided or made as gentle as possible. From the point of view of coupled-line theory, perhaps the biggest problem is the fact that the even and odd mode velocities are no longer equal. The effect of this inequality on coupler and filter performance had been analyzed by several workers, but Podell (p. 96) has come up with a neat solution, which restores equality between the two velocities in a rather simple way. He serrates or undulates ("wiggles") the facing edges of the two strips, thereby slowing down the mode that normally travels the faster, while having only a minor effect on the other (slower) mode. By a judicious amount of wiggling the two velocities can be equalized over a large bandwidth (several octaves), thus making possible high directivities for couplers and predictable performance for couplers and filters in microstrip line.

* Wheeler prefers the term **strip line**, and distinguishes between shielded and unshielded strip lines. He would suggest "open strip line", but the term microstrip is well entrenched, and we shall also use it here.

Transmission-Line Properties of Parallel Strips Separated by a Dielectric Sheet

HAROLD A. WHEELER, FELLOW, IEEE

Abstract—**A transmission line is made of a symmetrical pair of strip conductors, or a single strip and a ground plane, on opposite faces of a sheet of dielectric material. There is computed, to a close approximation, the relations among the dielectric constant of the sheet, the effective dielectric constant of the sheet and the empty space, the shape ratio, and the wave resistance, for the entire range of possible values. These relations are summarized in a graphical chart covering the range of practical interest.**

The computation is based on conformal mapping of the dielectric boundary on coordinates such that its effect can be most closely evaluated by simple principles. All relations are approximated in terms of ordinary functions (exponential and hyperbolic). Of particular interest is the effective filling fraction of the dielectric material, which depends mainly on the shape ratio and only slightly on the dielectric constant. Explicit formulas are given for analysis or synthesis.

I. Introduction

IN THE DECADE of 1945 to 1955, there was much interest in the subject of electromagnetic-wave transmission lines formed of strips of sheet conductor in various configurations. This culminated in a special issue of the IRE Transactions on Microwave Theory and Techniques in March 1955.

In a previous paper by the author [1], it was pointed out that one simple configuration had been resistant to solution in useful form. This is the symmetrical case of parallel flat strips, face-to-face, or the corresponding asymmetrical case of one flat strip parallel to a ground plane. That paper has presented a solution of this case, in useful form, for strip width comparable with the separation, or greater. That solution is limited to a wave medium of uniform properties, such as free space.

There remains the need for a useful solution for such parallel strips, separated by a sheet of dielectric material serving to support the strips (or one strip relative to a ground plane). This problem of mixed dielectric (air and sheet) has received little attention in the literature, probably because it is difficult to solve by the usual methods. The writer has developed an approach to this problem, which is found to yield close approximations in terms of simple mathematical functions (slide-rule functions, no more advanced than exponential and hyperbolic).

It is the purpose of this paper to present a solution to this problem, leading to graphical charts covering a wide range of shape ratio of the strips and of dielectric constant of the sheet material. The solution employs conformal mapping, based largely on the preceding article [1]. It serves as an introduction to a particular concept for the solution of any problem of mixed dielectric, using the present problem as an example.

It is required to express a relation between the transmission-line properties, such as wave resistance and the shape ratio (strip-width/separation), this relation depending on the dielectric constant of the sheet material. The solution will be developed separately for two ranges of shape ratio, characterized by strips "wide" or "narrow" relative to their separation.

The present approach is based on several steps involving conformal mapping. First, assuming the strips to be surrounded by only one kind of dielectric (free space), the actual field configuration is mapped on a rectangular area of uniform field. Secondly, the straight boundary of the dielectric sheet is mapped as a curve on this rectangle, so the presence of this dielectric distorts the field. Thirdly, some rules are perceived for approximating the effective filling fraction of the dielectric material, from which the effective dielectric constant can be simply computed.

This solution in its complete form does not yield an explicit formula for either synthesis or analysis, but is well adapted for computing a graphical chart for interpolation. However, it is found possible to devise explicit formulas for synthesis or analysis, based on the extremes of "wide" and "narrow" strips, which are useful approximations.

The conformal mapping will be introduced with reference to the two planes here involved, the plane of space coordinates and the plane of flux-potential coordinates (based on free space). Then there will be presented the concepts of approximation for mixed dielectric. The solution of the problem will be developed separately for "wide" strips and for "narrow" strips. The former will be based on the preceding monograph [1]. The latter will rely on some unpublished studies by the author. After some general discussion of significant relations, some procedures for computation, will be given followed by graphical charts in a form convenient for practical applications.

The derivation is based essentially on the separate evaluation of inductance and capacitance along the line. This is valid if substantially all of the energy is in the TEM mode of propagation, the operating frequency being far below cutoff in all higher modes. This condi-

Manuscript received August 11, 1964; revised November 13, 1964.

The author is with Wheeler Laboratories, Great Neck, N. Y.

tion is met if the transverse dimensions are much less than ½ wavelength in the dielectric material.

II. Conformal Mapping of Dielectric Boundary

Figure 1 shows the cross section of two parallel strips on opposite faces of a dielectric sheet. The essential dimensions are the strip width $2a$, the separation equal to the thickness of the dielectric sheet $2b$, and the dielectric constant of the sheet k. The strips form a balanced transmission line having certain values of wave resistance R and effective dielectric constant k'. The strips are assumed to be made of thin sheets of perfect conductor. The dielectric material is assumed to be homogeneous, isotropic and free of dissipation, so it is completely described by one constant k.

The same parallel strips, but without the dielectric sheet, can be formulated more simply, since the space is filled with only one kind of dielectric. For the case of "wide" strips, this is the subject of the writer's preceding monograph [1], which provides much background
72 for the present treatment; equations therein are referenced by their number with this prefix (A- · · ·).

One quadrant of the cross section of the line can be represented on one quadrant of space coordinates (the plane of $x+jy=z$). This is shown in Fig. 2, which is similar to the corresponding Fig. 2 in [1]. Here we add the intervening sheet of dielectric k. Attention is directed to the free-space flux line between points ③ and ⑦. This is the line which would terminate on the edge of the strip if the dielectric sheet were not present. The dielectric outside of this line, in the shaded area, causes a distortion of the flux lines, which complicates the effect of the dielectric sheet. The principal objective in this article is the evaluation of this effect.

By the procedure described in [1], the conductor boundaries are mapped on the z' plane of flux and potential coordinates, as shown in Fig. 3(a). The same symbols are retained. Furthermore, the boundary of the dielectric sheet is mapped on this plane as a curve between points ③ and ⑧. This operation reduces the problem to one of mixed dielectric in the space between parallel-plane conductors of unlimited extent. The area mapped in Fig. 3(a) is one quadrant of the cross section as shown in Fig. 2. The related diagram in Fig. 3(b) will be explained further on.

In Fig. 3(a), the area inside of the curve represents the region outside of the dielectric sheet, bounded by points ①, ②, ③, and ⑧. The area outside of the curve represents the region occupied by the dielectric sheet, bounded by points ③, ④, ⑥, and ⑧. It is required to evaluate the effective dielectric constant of the entire region, which will be a weighted mean between the dielectric constant of the material and unity for free space.

It is noted that the conformal mapping of a boundary between two dielectrics is valid because it retains the angles of "refraction" of the electric field at the boundary.

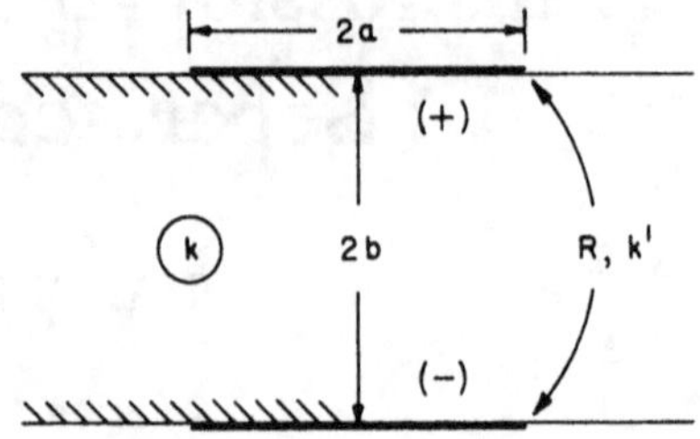

Fig. 1. Transmission line formed of parallel strips on opposite faces of dielectric sheet.

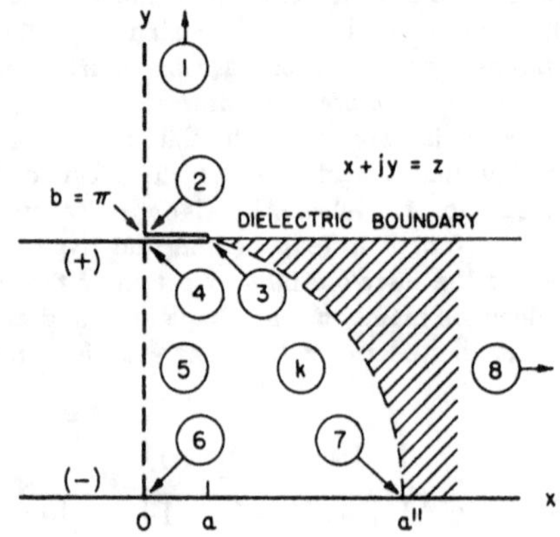

Fig. 2. Space coordinates showing dielectric sheet.

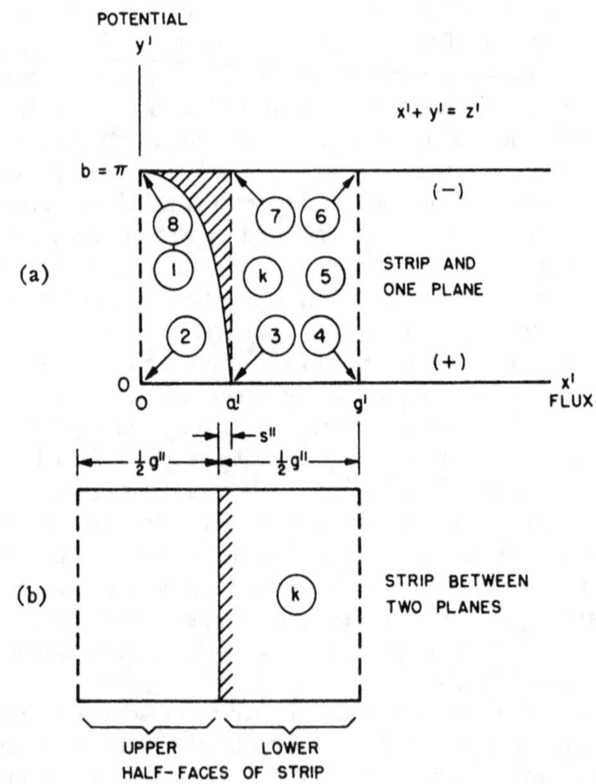

Fig. 3. Flux-potential coordinates, showing dielectric boundary.

III. Principles of Approximation for Mixed Dielectrics

Two parameters are required to determine any example in the present study; these may be taken as the shape ratio a/b and the sheet dielectric constant k. In terms of these parameters, all examples may be charted on a square region as in the diagram of Fig. 4. Every point in this region has a value of effective dielectric constant k' between the limits indicated. For the center of this diagram, it is found that $k'=1.70$, which is between the limits 1.5 and 2.

On this square diagram, definite coordinate scales could be assigned and then contours of k' could be plotted. In Fig. 4, the scales are so chosen that the contour through the center is nearly a straight line. Since the top and bottom contours are horizontal straight lines, it is conjectured that all the intermediate contours would be nearly straight lines of intermediate slopes.

The principal purpose of Fig. 4 is to indicate the limits of the principal parameters and the ranges of particular interest. To this end, the shape ratio is designated as follows for various ranges, in terms of strip width relative to separation.

Very narrow:	$a/b \ll 1$
Narrow:	$a/b < 1$
Square:	$a/b = 1$
Wide:	$a/b > 1$
Very wide:	$a/b \gg 1$

Then there are simple rules for the extreme values of effective dielectric constant k' corresponding to the extremes of shape.

Very wide: $k' = k$

Very narrow: $k' = k_a = \dfrac{k+1}{2}; \quad k = 2k' - 1$

The former is based on the principle that nearly all of the flux is in the dielectric. The latter is based on the principle that the electric field is symmetrically concentrated near the strip, with free space on one side and dielectric on the other side.

Another parameter is introduced, for describing the partial filling of dielectric. This is the effective filling fraction q which is to be evaluated in this presentation. This fraction is mainly dependent on the shape. For any particular shape, its value lies between an upper bound q' and a lower bound q'', respectively, corresponding to the extremes of lo-k and hi-k. These bounds are so close that one suffices for fairly close approximation, preferably the latter q'' because the resulting effect of the dielectric is greater for hi-k.

Referring still to Fig. 4, the present problem may be regarded as one of determining the extreme conditions on the square diagram, and then interpolating for intermediate conditions, in order to evaluate the effective dielectric constant for all conditions. This is to be accomplished for formulating the bounds q', q'' of the effective filling fraction q and then interpolating for intermediate values of dielectric constant k.

There are only two cases of mixed dielectric that can be directly evaluated by conformal mapping. In one case, the dielectric boundaries are all located on flux lines; in the other case, on potential contours. On the coordinates of flux and potential, as in Fig. 3(a), every such boundary would be parallel to one axis or the other. The curve representing the actual boundary meets neither of these conditions, but it can be interpreted as a mixture of both conditions. To develop this idea, the same region is expanded in Fig. 5(a), showing the curve representing the dielectric boundary.

First we note that a certain rectangle of space $a' < x' < g'$ is filled with dielectric. It happens that this is always more than half of the space as mapped on the 73
z' plane. Then we note that the remaining rectangle $0 < x' < a'$ is mostly empty of dielectric, the exception being the shaded area in Fig. 3(a) or 5(a). It is this extra area of dielectric which defies exact analysis but will be found susceptible of close approximation in its effect.

The shaded area, just outside of the curved boundary in Fig. 5(a) can be divided into two parts, as shown in Fig. 5(b). The entire area is expressed as $\pi s'$, in which s' is its effective width on the x' axis. One part of the area is expressed as $\pi s''$, which effectively adds s'' to the width of the dielectric region on the right; this is termed the "parallel" component. The remainder of the area is $\pi(s'-s'')$, which is effectively in "series" with the free-space region outside of the dielectric. This analysis gives a close approximation of the effect of the shaded area of dielectric, which is a small part of the total effect of the dielectric.

From the viewpoint of Fig. 5, the effective width s of the shaded area is between two bounds s', s''. A simple rule for interpolation is to be formulated here. In Fig. 5(b), the effect of the areas can be formulated in terms of dielectric and air capacitors in series and parallel connections. As a first approximation for small shaded area, the resultant effective width is found to be

$$s = s'' + \frac{s' - s''}{k} = \frac{1}{k} s' + \frac{k-1}{k} s'' \tag{1}$$

In words, the parallel component of shaded area is fully effective while the series component is $1/k$ as effective. While this rule is based on simplifying assumptions, it is probably valid as a close approximation for the curved boundary.

In Figs. 3(a) and 5(a), the filling fraction is defined as the ratio of dielectric area over total area in the rectangle of field mapping. It is most easily formulated in

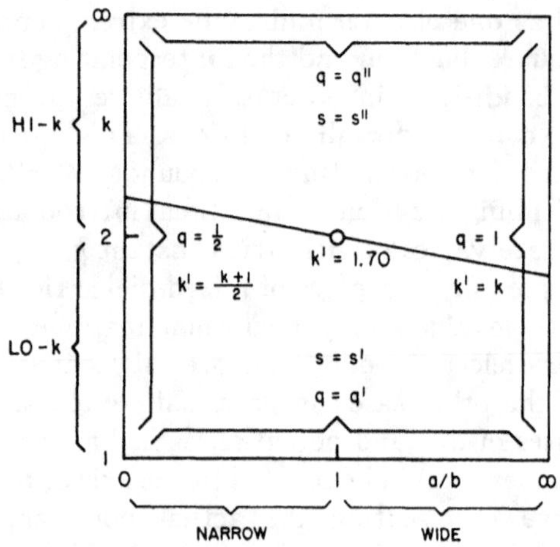

Fig. 4. Diagram of entire range of shape ratio and dielectric constant.

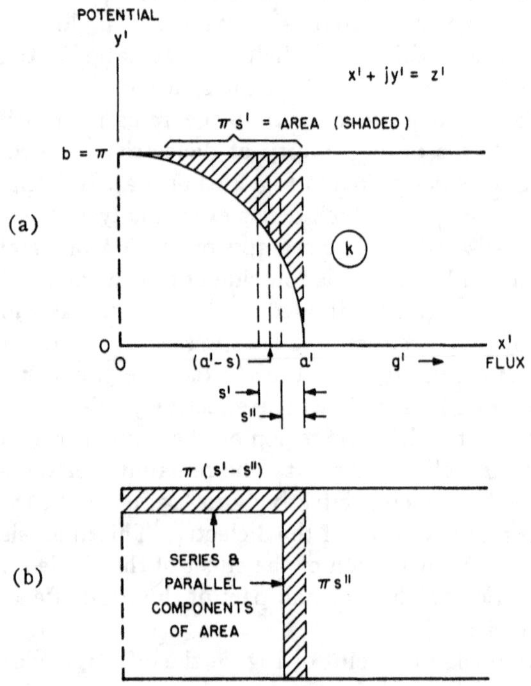

Fig. 5. Extra effective width analyzed in terms of parallel and series components.

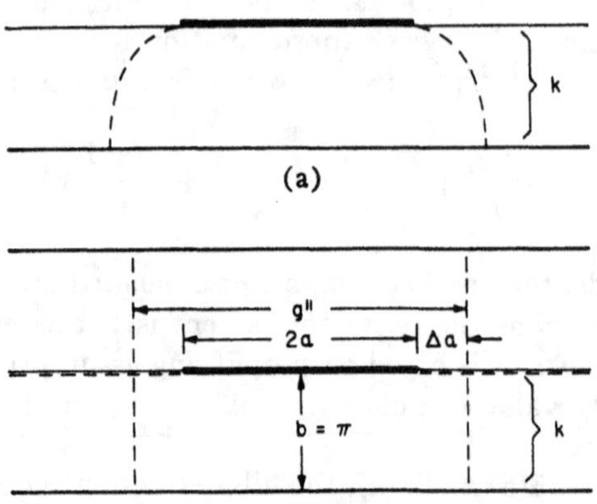

Fig. 6. Comparison with strip between parallel planes.

terms of effective width. The actual filling fraction is

$$q' = \frac{g' - a' + s'}{g} = 1 - \frac{a' - s'}{g'} = 1 - \frac{a'}{g'}\left(1 - \frac{s'}{a'}\right) \tag{2}$$

This includes all of the area outside of the curve, which is the area filled with dielectric. The "parallel" part of the filling fraction is only slightly less.

$$q'' = \frac{g' - a' + s''}{g'} = 1 - \frac{a' - s''}{g'} = 1 - \frac{a'}{g'}\left(1 - \frac{s''}{a'}\right)$$

$$= q' - \frac{s' - s''}{g'} \tag{3}$$

The effective filling fraction is found by (1).

$$q = \frac{g' - a' + s}{g'} = 1 - \frac{a' - s}{g'} = q'' + \frac{q' - q''}{k}$$

$$= q'' + \frac{s' - s''}{kg'} \tag{4}$$

The required components (s', s'' or q', q'') remain to be evaluated by various methods to be described further on.

The relation between the effective filling fraction q and the effective dielectric constant k' can be stated from the concept of parallel capacitors.

$$q = \frac{k' - 1}{k - 1} \tag{5}$$

$$k' = (1 - q) + qk = 1 + q(k - 1)$$

$$= k - (k + 1)\frac{a'}{g'}\left(1 - \frac{s''}{a'} - \frac{s' - s''}{ka'}\right) \tag{6}$$

The last form is in terms of the free-space flux ratio a'/g' from [1].

Referring to Figs. 3(a) and 5(a), there are simple concepts that are valid for the extremes of dielectric constant k. In the "lo-k" extreme, the dielectric does not cause appreciable distortion of the field. Therefore the dielectric area is entirely effective, but its effect is small, being proportional to $k-1$. In the "hi-k" extreme, there is no appreciable energy outside of the dielectric area. Therefore only the "parallel" part of the area is effective. The entire area is defined by conformal mapping of the dielectric boundary, and can be computed to some degree of approximation. The parallel part can be defined and computed by another case of conformal mapping to be described here.

Figure 6(a) shows the difference between the dielectric-sheet boundary and the free-space flux line from the edge of the strip. Figure 6(b) shows another configuration in which the boundary and the flux line coincide [3]. This is accomplished by adding above the strip an image of the neutral plane below the strip. In the resulting configuration of a strip between parallel planes, the effect of the dielectric sheet can be evaluated simply and exactly. In the limiting case of hi-k, these two cases become identical, because all of the flux is in the dielectric. In one quadrant of Fig. 6(a), as represented in Fig. 3(a), the effective width of the dielectric filling is between $\frac{1}{2}g'$ and g'. In the corresponding quadrant of Fig. 6(b), as represented in Fig. 3(b), the entire effective width is just half filled with dielectric, so the effective width of the dielectric filling is $\frac{1}{2}g''$. The latter can be exactly described and evaluated, thereby providing an evaluation of the former. Formulas for the latter are given in the last section herein for reference. The lower bound of the effective filling fraction becomes

$$q'' = \tfrac{1}{2}g''/g' > \tfrac{1}{2} \tag{7}$$

As previously stated, the principal problem in this derivation is the evaluation of the shaded area outside of the curved contour in Figs. 3(a) and 5(a). This will be computed for various cases, as a fraction of the rectangle in which the curve is inscribed. This "area fraction," s'/a', is graphed in Fig. 7, showing its variation with shape ratio. The curve is found to be always outside of an inscribed elliptic quadrant, so the area fraction is less than that for an ellipse, $1-\pi/4$, as seen from the graph of s' in Fig. 7. (This will be discussed further with reference to Fig. 8.) The area fraction has a maximum value near the "square" shape, and decreases toward zero for both extreme shapes. It is plotted on a special scale of abscissas, so chosen that this graph has a simple shape close to a sine wave. This scale and the properties of the graph will be explained in the course of the derivations and computations. The lower graph s'' is somewhat similar in shape, and represents the parallel component of the area fraction s''/a'. The effective area fraction s/a' is a weighted mean between these two graphs, the weighting depending on the dielectric constant k as indicated.

In developing mathematical approximations for the entire range of the parameters, it is helpful to work from the extremes of shape to some more elaborate formulas for intermediate shapes. Therefore this approach will be exploited in the derivations to follow. For wide strips, the derivation is a continuation of [1]. For narrow strips, the required background will be given without proof.

In the literature, the closest reference found by the writer is the 1956 and 1958 papers by Dukes [3], [4]. He does approach the problem of mixed dielectric in the configuration of a strip and ground plane separated by a dielectric sheet. His principal contribution is the concept of Fig. 6(b) herein, which gives the asymptotic solution for high dielectric constant. This concept should have led to fairly close approximations, which he expected. However, he seems to have made an error in application of this principle, since his main formula contains a fallacy in relations. His graphs, published in the later paper, contain errors as great as 15 per cent of the wave resistance. The form of his graphs is followed here for wave resistance in Fig. 9.

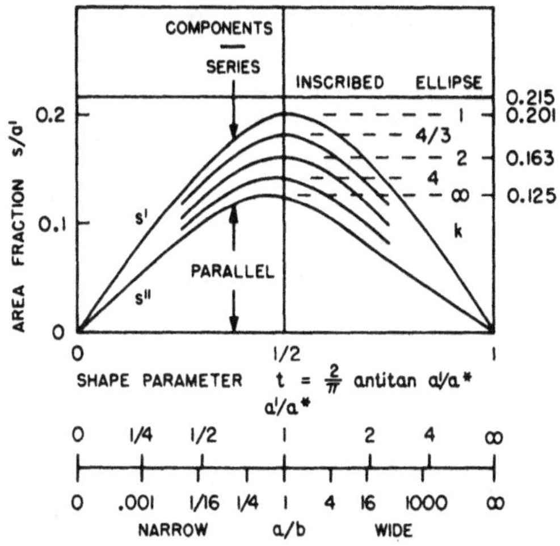

Fig. 7. Variation of extra area with shape.

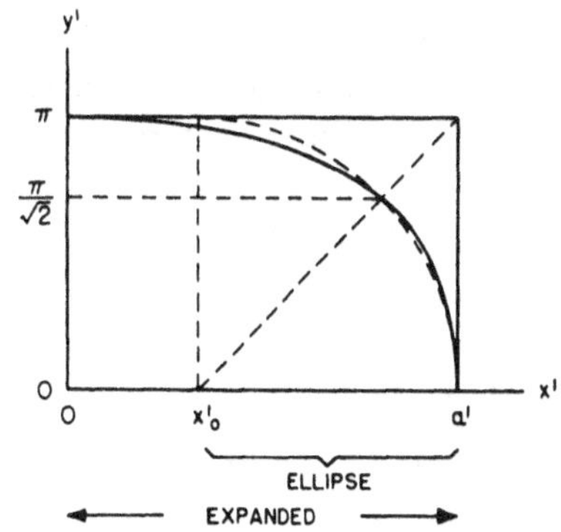

Fig. 8. Approximation of extra area by ellipse.

75

IV. Derivation for Wide Strips

This derivation for wide strips is based on the preceding monograph [1], which will be duplicated here only to the extent desired for continuity. As mentioned before, equations in that paper will be identified by the prefix A followed by the number (A- · · ·).

As in [1], the effective half-width g' on the x' plane is related to the free-space wave resistance R_1, the inductance L and the free-space capacitance C_1.

$$\begin{aligned} R_1 &= R_c\pi/g'; \qquad & g' &= \pi R_c/R_1 \\ L &= \mu_0 l\pi/g'; \qquad & g' &= \pi\mu_0 l/L \\ C_1 &= \epsilon_0 l g'/\pi; \qquad & g' &= \pi C_1/\epsilon_0 l \end{aligned} \tag{8}$$

(The symbols are listed in a later section.) In the present problem of mixed dielectric, a significant parameter is the effective dielectric constant k' of the entire space which is partially filled with the dielectric material k. The resulting values of wave resistance and capacitance are

$$R = \frac{1}{\sqrt{k'}}R_1; \qquad C = k'C_1 \tag{9}$$

Also the wave velocity is decreased in the same ratio as the resistance.

The marginal condition for "wide" strips (the "square" condition, $a/b=1$) corresponds roughly to $R=\frac{1}{2}R_c/\sqrt{k'}$, for which $g'=2\pi$. Therefore this value of g' may be taken as representative of this condition. It happens that the "wide" approximations are useful for even narrower strips, so the case of $g'=\pi$ may be taken as the extreme condition for the "wide" formulas.

The conformal-mapping formulas are given in [1], for the relation between the space coordinates on the z plane of Fig. 2 and the flux-potential coordinates on the z' plane of Fig. 3(a). Some of the formulas to be-presented here are more simply expressed directly in terms of the parameter d, which is nearly equal to the effective width g' and therefore may be used interchangeably in all except the closest approximations. Complete computations, especially in the marginal region, may include the small difference between these two parameters $d-g'$.

The mapping approximation in [1] is simple enough to enable explicit formulation of the curve in Figs. 3(a)
76 and 5(a), representing the dielectric boundary on the flux-potential coordinates. Here several parameters d, d', g' are taken as equal. From (A-52) and (A-53),

$$z = x + jy = j\pi + d \tanh \tfrac{1}{2}z' - z' \tag{10}$$

At the dielectric boundary,

$$z = j\pi; \qquad x = 0; \qquad y = \pi \tag{11}$$

Separating the real and imaginary parts of (10), the imaginary part determines a curve on the z' plane, representing the dielectric boundary.

$$y = \pi = \pi + d\frac{\tan \frac{1}{2}y'(1 - \tanh^2 \frac{1}{2}x')}{1 + \tan^2 \frac{1}{2}y' \tanh^2 \frac{1}{2}x'} - y' \tag{12}$$

Here it is possible to express x' of y' explicitly in closed form. By routine transformations, this is reduced to the following simple form:

$$\cosh x' = d\frac{\sin y'}{y'} - \cos y' \tag{13}$$

The shaded area is found to approach an upper limit for the "very wide" condition of increasing d and a', where the mapping approximation approaches perfection. This is based on the following integration carried to the limit of increasing d; the integrand is based on (13) for the entire curve.

$$a' - x' = \ln \frac{\dfrac{\sin y'}{y'} - \dfrac{\cos y'}{d}}{1 - 1/d} = \ln \frac{\sin y'}{y'} \tag{14}$$

$$\max s' = \frac{1}{\pi}\int_0^\pi (a' - x')dy' = 2 \ln \frac{\pi}{2}$$
$$= 0.903 \tag{15}$$

In the marginal "wide" condition, near the "square" shape, it is found that the dielectric boundary curve (13) comes closest to the shape of an elliptic quadrant inscribed in the same rectangle. This fact is utilized to obtain a close approximation for the area fraction, as will be developed with reference to Fig. 8. Outside of the actual curve, or of an ellipse, most of the area is near the corner of the rectangle. Therefore an ellipse is specified which crosses the actual curve near the corner. This ellipse is one inscribed in a rectangle which is smaller than the entire rectangle, having a lesser dimension along the x' axis. This dimension is chosen so that the ellipse crosses the actual curve at the point on the diagonal of the ellipse, as shown. Then the area outside of this ellipse is taken as an approximation for the area outside of the actual curve.

On this basis, the point of intersection is located as indicated in Fig. 8. The area outside of the curve is taken to be that outside of the ellipse. The resulting effective width can be expressed in terms of a', as follows:

$$s' = 0.732[a' - \text{anticosh}\,(0.358 \cosh a' + 0.953)] \tag{16}$$

This formula for the effective width might not give a close approximation for the "very wide" case, so it may be compared with (15) for this case.

$$(16){:}\quad \max s' = 0.732 \ln \frac{1}{0.358} = 0.752 \tag{17}$$

$$(15){:}\quad \max s' = 2 \ln \frac{\pi}{2} = 0.903 \tag{18}$$

The fact that the ellipse approximation comes so close ($\frac{5}{6}$ of the correct value) in the "very wide" limit is an indication of its probable very close approximation in the intermediate "wide" range for which it is intended (including the "square" shape).

The area fraction is found to have a maximum value near the "square" condition, which has the following properties (see [1], Table I). If

$$a/b = 1{:} \qquad g' = d = 6.65; \qquad a' = 2.416 \tag{19}$$

A slightly wider condition gives the maximum area fraction, which may be compared with that outside of an inscribed elliptic quadrant.

$$(16){:}\quad \max s'/a' = 0.201$$
$$\text{Ellipse:}\ (1 - \pi/4) = 0.215 \tag{20}$$

The area fraction is seen to be always somewhat less than that outside of an elliptic quadrant inscribed in the entire rectangle. (This is not the same as the elliptic quadrant used to approximate the curve, which is inscribed in a smaller rectangle.)

Referring to Fig. 7, formula (16) approximates the area fraction in the intermediate range, while (15) evaluates the limiting slope in the "very wide" range. The full development of this graph is still deferred for more background.

With reference to Figs. 5 and 6, there has been introduced the concept of series and parallel components of the shaded area just outside of the dielectric-boundary curve on the z' plane. The parallel component has an effective area $\pi s''$ and effective width s'' somewhat less than the corresponding values $\pi s'$ and s' for the entire shaded area. The parallel component is to be formulated from the configuration in Fig. 6(b).

As previously explained, the lower half of Fig. 6(b) is a representation of the space occupied by the dielectric sheet k. This set of boundaries can be evaluated exactly, as described in the last section herein. Here, it seems preferable to use the simple close approximation for the "wide" case. The fringing field at each edge causes the effective half-width $\frac{1}{2}g''$ to exceed the actual half-width a by a small amount Δa. This excess approaches a limit very rapidly as the strip width approaches and exceeds the spacing, the limiting value for "very wide" strips being

$$\Delta a = \tfrac{1}{2}g'' - a = \ln 4 = \frac{b}{\pi} \ln 4 \tag{21}$$

Referring to Fig. 3, the dielectric space mapped from Fig. 2 is the same as that in Fig. 6(b). The region indicoted includes both faces of the strip half-width a. Here we are concerned with the lower face, next to the dielectric. The actual shape in Fig. 2 or Fig. 6(a) has a certain effective half-width $g'-a'$ of the lower face as seen in Fig. 3(a). The special shape in Fig. 6(b) has a slightly greater effective half-width $\frac{1}{2}g''$ of the lower face as seen in Fig. 3(b). The excess s'' of the latter over the former is effective in the extreme of hi-k, so it is the parallel component of the shaded area in Fig. 3(a), as described with reference to Fig. 5.

$$s'' = \tfrac{1}{2}g'' - (g' - a') = (\tfrac{1}{2}g'' - a) - (g' - a' - a) \tag{22}$$

From (A-13), (A-14), (A-54), (A-57) [1], and still ignoring small differences (such as $d-g'$),

$$\begin{aligned} a &= (d-1) - \text{exp-}a' - a' \\ &= g' - a' - 1 - \text{exp-}a' \\ &= g' - a' - 1 - 1/2(d-1) \end{aligned} \tag{23}$$

$$g' - a' - a = 1 + \text{exp-}a' = 1 + 1/2(d-1) \tag{24}$$

$$\begin{aligned} s'' &= \ln 4 - 1 - \text{exp-}a' \\ &= 0.386 - \text{exp-}a' \\ &= 0.386 - 1/2(d-1) \end{aligned} \tag{25}$$

In the limit of "very wide" strips,

$$\max s'' = \ln 4 - 1 = 0.386;$$

$$\frac{\max s''}{\max s'} = \frac{0.386}{0.903} = 0.427 \tag{26}$$

In this extreme, the parallel component is less than one-half the total shaded area in Figs. 3 and 5.

Formula (25) may be used to compute the parallel component s''/a' of the area fraction in the "wide" range. It is found to have a maximum value near the "square" shape, as shown in Fig. 7.

$$\max s''/a' = 0.125;$$

$$\frac{\max s''/a'}{\max s'/a'} = \frac{0.125}{0.201} = 0.622 \tag{27}$$

In this range, the parallel component is greater than one-half the total.

We now have formulas for the area fraction s'/a', resolved into its parallel and series components s''/a' and $(s'-s'')/a'$, for the range of "wide" strips. This includes the marginal range near the "square" shape. These formulas are graphed in Fig. 7, which will be explained further after the derivation for narrow strips. This concludes the conformal-mapping operations for wide strips, so we are ready for the application to mixed dielectric. The objective is the evaluation of the effective dielectric constant k' which is a mean between free-space (1) and the dielectric material k.

With reference to Figs. 3 and 5, the previous discussion has led to formulas for k' as weighted average between k and 1, in terms of the filling fraction. For wide strips, the derived values can be inserted in the general formulas such as (4) and (6) for computing k'; all the quantities needed in these formulas are evaluated in [1] or herein.

An example is given to indicate a procedure for computation. It is based on the "square" shape and a certain dielectric constant of the material. (See [1], Table I.)

	$a/b = 1; \quad a = \pi$
	$g' = d = 6.65$
	$R_1 = 178$ ohms
	$a' = 2.416$
	$a'/g' = 0.363$
(16)	$s' = 0.482$
(25)	$s'' = 0.297$
	. . . .
	$k = 2$
(1)	$s = 0.390 = \frac{1}{2}s' + \frac{1}{2}s''$
	$a' - s = 2.026 = 0.305\, g'$
(4)	$q = 0.695$
(6)	$k' = 1.695$ (between 1.5 and 2)
(9)	$R = 137$ ohms

This case is the center point on the square chart in Fig. 4.

Here we may compare the simple computation based on the lower bound of the filling fraction.

$$\tfrac{1}{2}g'' = \pi + \ln 4 = 4.528 \tag{21}$$

$$q = q'' = 0.681 \tag{7}$$

$$k' = 1.681 \tag{6}$$

This differs from the complete computation by only 0.01 of k' or 0.005 of R.

The present derivation enables a close approximation of the effective dielectric constant, over the range of "wide" strips, including the marginal range near the "square" shape. It is based on the conformal-mapping approximation of the preceding paper [1], from which the effective dielectric constant is here evaluated for the same shape. There remain to be given some explicit formulas for synthesis.

A formula for synthesis must be directed toward one of the properties which involve the dielectric, such as capacitance ($C \propto k'$), wave resistance ($R \propto 1/\sqrt{k'}$) or wave velocity ($\propto 1\sqrt{k'}$). The inductance L is independent of the dielectric. The wave resistance is selected as the most useful objective for synthesis; the shape ratio is to be computed after specifying the dielectric material k.

Therefore an approximate formula has been derived, which gives the shape ratio in terms of the wave resistance R and the dielectric constant k of the sheet material. From (8) and (9), the wave resistance is

$$R = \frac{R_c}{\sqrt{k'}}\frac{\pi}{g'} = \frac{R_c}{\sqrt{k'}}\frac{\pi}{d}; \qquad d = \frac{\pi}{\sqrt{k'}}\frac{R_c}{R} \tag{28}$$

We continue to ignore the small difference between g' and d. Here we define another parameter d_k which is the lesser value of effective half-width that would be required if the space were entirely filled with the dielectric material k.

$$R = \frac{R_c}{\sqrt{k}}\frac{\pi}{d_k}; \qquad d_k = \frac{\pi}{\sqrt{k}}\frac{R_c}{R} \tag{29}$$

The formula is to be a close approximation for "wide" strips (in this case $d_k > 2\pi$).

The required shape is between the limits that would be required for free space and for complete dielectric filling. For "wide" strips, it is much closer to the latter, which therefore is taken as a starting point. The derivation has been based on "very wide" strips, giving the exact terms for second approximation for both extremes of dielectric constant. The resulting approximate formula for synthesis is given in a form convenient for computation.

$$\frac{a}{b} = \frac{1}{\pi}(d_k - 1) - \frac{1}{\pi}\ln(2d_k - 1) + \frac{k-1}{2\pi k}\left[\ln(d_k - 1) + 0.293 - 0.517/k\right] \tag{30}$$

If $d_k > 2\pi$, estimated relative error < 0.01 of R.

In the limit of "very wide" strips, this formula is exact in the first two terms for the lo-k and hi-k extremes. It is an interpolation formula for intermediate conditions.

To obtain an explicit formula for analysis, (A-71) may be modified for mixed dielectric. The result is a second approximation for wide strips.

78

$$\frac{R}{R_c} = \frac{\sqrt{1/k}}{\dfrac{a}{b} + \dfrac{1}{\pi}\ln 4 + \dfrac{k+1}{2\pi k}\ln\dfrac{\pi\epsilon}{2}\left(\dfrac{a}{b} + 0.94\right) + \dfrac{k-1}{2\pi k^2}\ln\dfrac{\epsilon\pi^2}{16}} = \frac{\sqrt{1/k}}{\dfrac{a}{b} + 0.441 + \dfrac{k+1}{2\pi k}\left[\ln\left(\dfrac{a}{b} + 0.94\right) + 1.451\right] + \dfrac{k-1}{k^2}(0.082)} \tag{31}$$

In the denominator, the last term introduces the interpolation between the lower and upper bounds of the effective filling fraction; it has most effect for intermediate values of dielectric constant. It is estimated that the relative error is less than 0.01 for all conditions of "wide" strips.

V. Derivation for Narrow Strips

This derivation for narrow strips is based on free-space formulas and the principles of the foregoing derivation for wide strips. The free-space formulas for narrow strips are found in the literature, to the first approximation. The writer's unpublished derivation to the second approximation, including some properties of the free-space field, will be stated as required, without proof.

As an extension of the parameters already defined, some other are introduced which are better suited for narrow strips. These may be regarded as "separation" parameters h rather than "width" parameters g.

$$R_1 = R_c h'/\pi = 120\,h';$$
$$h' = \pi R_1/R_c = R_1/120 = \pi^2/g' \tag{32}$$

There are some related parameters that are defined here for reference. The average value between unity and the dielectric constant k of the sheet material is

$$k_a = \frac{k+1}{2} \tag{33}$$

If the space were filled with k_a, the wave resistance R would be given by the separation parameter

$$h_a = \pi\sqrt{k_a}\, R/R_c = \sqrt{k_a}\, R/120 \tag{34}$$

This will be used as a reference in some formulas.

Referring back to Fig. 1, the free-space wave resistance R_1 is given by the following formula to a second approximation for narrow strips:

$$\begin{aligned} h' &= \pi R_1/R_c \\ &= \ln\frac{4b}{a} + \frac{1}{8}\left(\frac{a}{b}\right)^2 \pm (\cdots)\left(\frac{a}{b}\right)^4 \pm \cdots \end{aligned} \tag{35}$$

The first approximation, for "very narrow" strips, may be derived by substituting for each strip a round wire which is equivalent in the first-order far field. The second approximation, for "narrow" strips, may be derived by substituting for each strip a spaced pair of round wires which is equivalent in the second-order far field. The later derivation, which is found to be rather simple, has been developed by the writer and verified by several alternative approaches.

On the same basis, the free-space formula for synthesis, to the second approximation, is also simple in explicit form.

$$\begin{aligned} b/a &= \tfrac{1}{4}\exp h' - \tfrac{1}{2}\exp\text{-}h' \\ &\quad \pm (\cdots)\exp\text{-}3h' \pm \cdots \end{aligned} \tag{36}$$

If $b/a > 1$, $h' > 1.485$ and relative error < 0.005 of R.

The marginal condition for narrow strips, as for wide strips, corresponds roughly to $R = \frac{1}{2}R_c/\sqrt{k'}$, for which $h' = \pi/2$.

Referring to Fig. 2, the fields around "very narrow" strips can be computed to a first approximation by replacing each strip by an equivalent small round wire. Several simple relations become apparent, as will be stated.

In Fig. 2, there is a dashed curve between points ③ and ⑦ which represents the free-space flux line terminating at the edge of the strip. This flux line approaches the shape of a circular arc (quadrant). As for the effect of the dielectric, we note that most of the electric energy is close to the strip, and is divided nearly equally between both faces of the strip. Therefore the effective dielectric constant, for "very narrow" strips, approaches the lower limit,

$$k' = k_a = \frac{k+1}{2} \tag{37}$$

The excess over this value is caused by the shaded area, whose effect increases with strip width.

Referring to Fig. 3(a), the flux fraction on the outer faces may be based on the free-space field. It is found to have the following value, a second approximation for "narrow" strips.

$$\begin{aligned} \frac{a'}{g'} &= \frac{1}{2} - \frac{1}{2\pi}\operatorname{antitan}\frac{a}{b} \\ &= \frac{1}{2} - \frac{2}{\pi}\exp\text{-}h' + \frac{20}{3\pi}\exp\text{-}3h' \pm \cdots \end{aligned} \tag{38}$$

In this series, the first two terms are exact. The deficiency from $\frac{1}{2}$ decreases rapidly with decreasing width or with increasing separation and wave resistance. While this flux fraction is of interest for narrow strips, it will not be needed for the first approximation to be developed here.

In Fig. 3(a), the dielectric boundary is represented by the curve between points ③ and ⑧ .From the small-strip or small-wire field in free space, this curve is found to be

$$y' = \pi - \frac{g'}{\pi}\ln\frac{1}{\cos\dfrac{\pi x'}{g'}} \tag{39}$$

79

In the entire rectangle $(0 < x' < g';\ 0 < y' < \pi)$, the area fraction above the curve is the filling fraction, and may be formulated as follows:

$$q' = \frac{1}{2} + \frac{\ln 2}{2\pi^2}g' = \frac{1}{2} + \frac{\ln 2}{2h'} \tag{40}$$

The "parallel" component of this filling fraction is to be evaluated with reference to Figs. 3, 5, and 6, as was done for wide strips. The properties of a narrow strip between two planes are given in the Appendix. The width parameter may be formulated from (56) and (57), as follows, to a first approximation for very narrow strips.

$$\begin{aligned} g'' &= \frac{\pi^2}{\ln\dfrac{4b}{a} - \ln\dfrac{\pi}{2}} = \frac{g'}{1 - \dfrac{\ln \pi/2}{\pi^2}g'} \\ &= g' + \frac{\ln\pi/2}{\pi^2}g'^2 \end{aligned} \tag{41}$$

From this, the parallel part of the filling fraction may be evaluated.

$$q'' = \frac{1}{2}g''/g' = \frac{1}{2} + \frac{\ln \pi/2}{2h'} \tag{42}$$

As before, the excess of the entire filling fraction over the parallel component is taken to be the series component, and may be evaluated as follows:

$$q' - q'' = \frac{\ln 4/\pi}{2h'} \tag{43}$$

From these formulas, by interpolation as in (4), the effective filling fraction is

$$q = q'' + \frac{1}{k}(q' - q'')$$
$$= \frac{1}{2} + \frac{1}{2h'}\left(\ln \pi/2 + \frac{1}{k}\ln 4/\pi\right) \tag{44}$$

Then the effective dielectric constant is

$$k' = \frac{k+1}{2} + \frac{k-1}{2h'}\left(\ln \pi/2 + \frac{1}{k}\ln 4/\pi\right)$$
$$= k_a\left\{1 + \frac{k-1}{k+1}\,\frac{\ln \pi/2 + \dfrac{1}{k}\ln 4/\pi}{\ln 4b/a + \cdots}\right\} \tag{45}$$

Over the range of narrow strips ($b/a>1$) the second term is greatest for hi-k, but is still less than $\frac{1}{4}$ of the total.

80 In the limit of very narrow strips (so $a' = \frac{1}{2}g'$) these relations may be expressed in terms of effective width as was done for wide strips.

$$s' = \frac{2\ln 2}{\pi^2}a'^2; \qquad s'' = \frac{2\ln \pi/2}{\pi^2}a'^2 \tag{46}$$

This form will be used for computing the slope at the "narrow" end of the curves in Fig. 7.

Here again, an approximate formula for synthesis has been derived to give the shape ratio for specified wave resistance R and dielectric constant k. The required separation parameter h' is formulated in terms of the hypothetical parameter h_a for k_a.

$$h' = h_a + \frac{1}{2}\frac{k-1}{k+1}\left(\ln\frac{\pi}{2} + \frac{1}{k}\ln\frac{4}{\pi}\right)$$
$$= h_a + \frac{\ln 2 + (k-1)^2 \ln \pi/2}{2k(k+1)}$$
$$= h_a + \frac{k-1}{k+1}(0.226 + 0.120/k) \tag{47}$$

Then the shape ratio can be computed by (36). If $b/a>1$, the second term is less than $\frac{1}{10}$ as great as the first, and the estimated relative error is within 0.005 of h' or R.

As in the case of wide strips, we can obtain an implicit formula for analysis, which is a second approximation for narrow strips.

$$\frac{R}{R_c} = \frac{1}{\sqrt{k_a}}\,\frac{1}{\pi}\left[\ln\frac{4b}{a} + \frac{1}{8}\left(\frac{a}{b}\right)^2 - \frac{1}{2}\frac{k-1}{k+1}\left(\ln\frac{\pi}{2} + \frac{1}{k}\ln\frac{4}{\pi}\right)\right] \tag{48}$$

It is estimated that the relative error is less than 0.01 for all conditions of narrow strips.

VI. Relations for Wide and Narrow Strips

Having derived various formulas for wide strips or narrow strips, it is interesting and instructive to note the behavior over the entire range of the shape ratio a/b or b/a. The end objective is to present useful design information in convenient form. Figure 9 is intended to serve this purpose; it is introduced here for reference in the discussion within this section. Figure 9(a), (b), and (c) give in three ranges a graphical presentation of the relation between wave resistance and shape ratio, with dielectric constant as a parameter. Figure 9(d) gives the relation between the shape ratio and the effective filling fraction, which will be discussed further on in this section. [The four parts of this figure are arranged in clockwise order for showing (a), (b), and (c) in logical sequence, preserving the proximity of adjoining regions.]

Figure 4 has been introduced as a map of the entire range of parameters. Each scale is chosen to locate the transition condition in the vicinity of the center. The scale of ordinates is

$$\frac{k-1}{k} = 1 - 1/k \tag{49}$$

A scale of abscissas is chosen to locate the "square" shape on the centerline.

The effective filling fraction q is a significant parameter for distinguishing the various limits or bounds over the map in Fig. 4. It has a particular designation and definition on each of the four sides of the square, and it determines the effective dielectric constant k'.

On the lower border, the effective filling fraction is equal to the actual filling fraction q' based on Figs. 3(a) and 5. On the upper border, the effective filling fraction is equal to the parallel part q'' based on Figs. 3 and 5. These provide upper and lower bounds which are rather close together. Each of these is susceptible of exact description and is susceptible of approximate or exact formulation. These are related to the bounds of area fraction graphed in Fig. 7.

Contours of effective dielectric constant k' may be mapped over the region shown in Fig. 4. One contour is shown, the one through the center.

The center point is particularly interesting, as a midway interpolation between the extremes. At the center, where $k=2$, it is found that $k'=1.70$. The corresponding values of k at the ends of this contour (2.4 and 1.7) have substantially equal displacement above and below the horizontal centerline. Therefore, this contour can be drawn as a straight line, which then determines a particular scale of abscissas.

In this family of contours k', the bottom and top contours are horizontal straight lines. Therefore, it is likely that all the contours would be nearly straight lines having slopes between zero and that of the contour through the center.

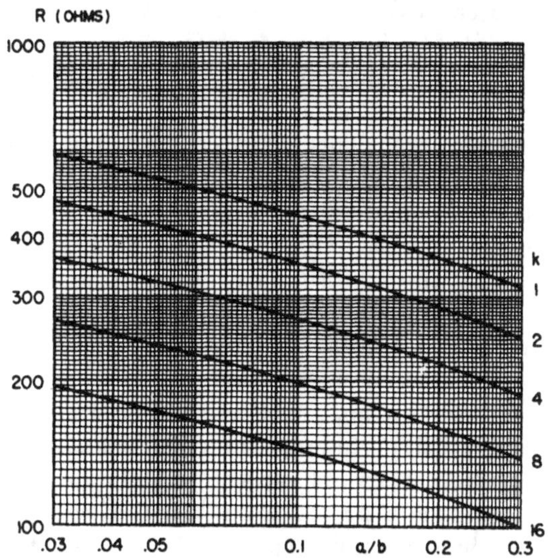

(a) Narrow strips.

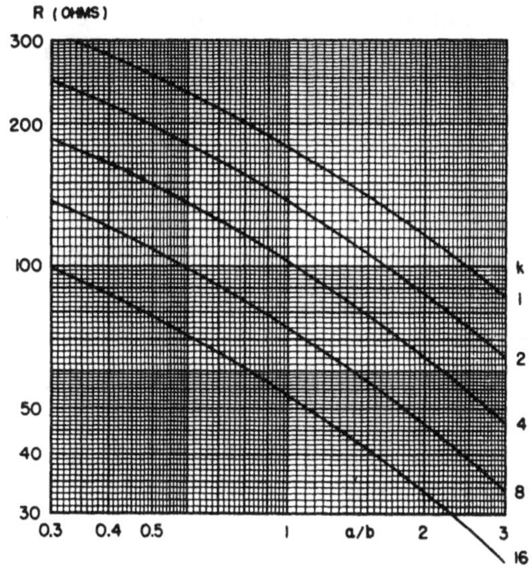

(b) Intermediate strips.

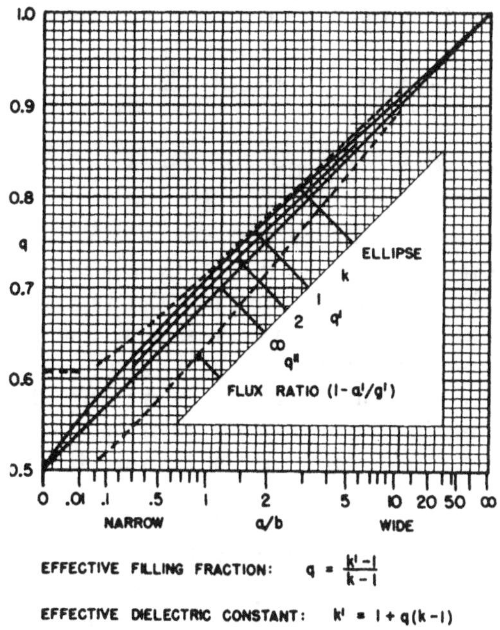

(d) Filling fraction.

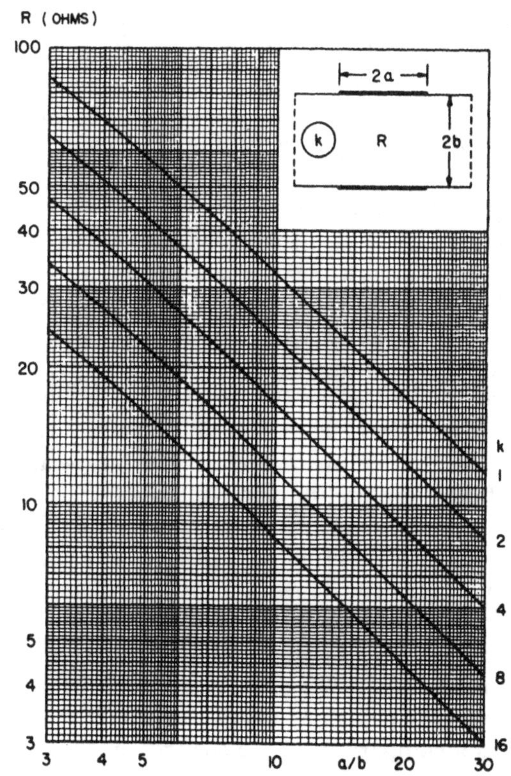

(c) Wide strips.

Fig. 9. Graphical charts for design.

The limits of the effective filling fraction are here determined by evaluation and interpretation of the shaded area in Figs. 3 and 5. This has been formulated in terms of the area fraction graphed in Fig. 7. Referring to the dielectric-boundary curve and to the rectangle in which it is inscribed, the area fraction s'/a' is the ratio of the shaded area (outside of the curve) over the area of the rectangle, as previously defined. This fraction has been divided into two components, the parallel component s''/a' and the series component $s'/a-s''/a'$. Furthermore, the interpolation formula (1) gives the effective area fraction s/a' which depends on the dielectric constant k. These are graphed on a scale of abscissas t which has been chosen for reasons to be developed here.

In general terms, it has been found that the area fraction s'/a' approaches zero at both extremes of shape. Furthermore, it is found that the area fraction is simply related to the parameter a', by direct and inverse proportionalities for the narrow and wide extremes.

Referring to Fig. 7, the following table identifies the equations for evaluating the end slopes and the maximum value of each of the extreme curves.

Shape	Graph	Area Fraction	
		s'/a'	s''/a'
Very narrow strips	Slope	(46)	(46)
Intermediate strips	Maximum	(20)	(27)
Very wide strips	Slope	(15)	(26)

It is desired to graph these area fractions on a scale that will give the curves a simple shape. The following variable for the scale of abscissas gives each curve a shape similar to a sine wave:

$$t = \frac{2}{\pi} \text{ antitan } a'/a^* \qquad (50)$$

82 The center point is located at the "square" shape by choosing $a^* = 2.416$ [1], (Table I). The peak of each curve is then near this centerline.

The effective area fraction s/a' is obtained by interpolation in accordance with (1), depending on k. It is graphed for several values of k giving equal steps of $1/k$.

In Fig. 7, it is seen that the area fraction has a maximum value slightly less than that corresponding to an elliptic quadrant inscribed in the same rectangle. Toward the extremes, the boundary curve approaches the rectangle so the area fraction approaches zero. The scale of the shape ratio a/b is seen to be crowded towards the center so that, over the usual range of shape, the area fraction is near the maximum. Therefore the maximum values, which are noted on the right-hand side, may be taken as a useful approximation.

The effective filling fraction q is particularly significant in that it is determined mainly by the shape ratio and is only slightly dependent on the dielectric constant. From its value, it is easy to compute the effective dielectric constant. For reference, the effective filling fraction is graphed in Fig. 9(d). The scale of shape ratio is determined to give one limit q'' as a straight line. This facilitates a comparison of the slightly different curves.

In Fig. 9(d), the principal (solid-line) graphs are those for the extremes of k and one intermediate value. The lowest (dashed-line) graph is a lower bound set by the free-space flux ratio. The uppermost (dashed-line) graph is an upper bound set by the ellipse, as seen in Fig. 7. The problems and solutions presented herein are best appreciated and understood by reference to this set of graphs.

It would be convenient if the effective filling fraction were dependent entirely on the shape ratio and independent of the dielectric constant, so a single relationship would suffice. The lowest graph of q is q'', which is effective in the limit of hi-k. This comes near to the ideal of a single graph independent of k. If q'' is taken for q, this causes a relative error less than 0.01 of k' or 0.005 of R. This is close enough for practical purposes. Also it happens that this case is based on the strip between two planes, which is the easiest to formulate. As previously mentioned, this concept has been presented by Dukes [3], [4] but it seems that he did not correctly apply it to the problem at hand.

The dashed curves in Fig. 9(d) show the brackets that could be most simply formulated by conformal mapping of the dielectric boundary. The interpolation between these brackets is the subject of Fig. 7.

While the present derivations are limited to thin strips, a small thickness can be compensated by a reduction of width. While retaining the same separation $2b$, each strip is taken to have a rectangular cross section of small thickness Δb less than the half-width a and much less than the half-separation b. Each edge recedes by a small amount Δa, which is the edge correction to retain the same properties.

This edge correction is simply formulated for free-space ($k=1$) as follows, for wide and narrow strips:

$$\text{If } a > b/4\pi > \Delta b: \quad \Delta a = \frac{\Delta b}{2\pi}\left(\ln \frac{2b}{\Delta b} + 1\right) \qquad (51)$$

$$\text{If } b/4\pi > a > \Delta b: \quad \Delta a = \frac{\Delta b}{2\pi}\left(\ln \frac{8\pi a}{\Delta b} + 1\right) \qquad (52)$$

In the range of validity, $\Delta a > \frac{2}{3}\Delta b$.

With the dielectric sheet $k>1$, the edge recession has more effect because it decreases the area in contact with dielectric while the thickness merely increases the area exposed to empty space. In the extreme of hi-k, the edge correction approaches zero. It is proposed that the edge correction Δa in these formulas be multiplied by $1/k$, as a sensible interpolation between the extremes.

VII. Procedures for Computation

The shape ratio determines the free-space wave resistance R_1 and is the principal factor in determining the effective filling fraction q. The latter and the dielectric constant k determine the effective dielectric constant k'. Then R_1 and k' determine the actual wave resistance R. These principles underlie any computation in the present configuration including mixed dielectric. For each case, a procedure is sought which will include a sequence of explicit formulas.

The effective filling fraction q'' for hi-k is particularly significant and is simply defined by (7). For practical purposes, this parameter determines the effect of the dielectric sheet. It may be included in various sequences of computation.

Closer computation for theoretical purposes requires also the actual filling fraction q' which is effective for lo-k. This parameter may be computed from various approximations given for wide or narrow strips. Then (4) gives the effective filling fraction q for any k.

From the derivation for either wide or narrow strips, a complete computation procedure can be gleaned for the closest approximation. There may not be a sequence of explicit formulas, starting with any two of the three essential parameters R, a/b, k, in which case there is the alternative of starting with an intermediate parameter such as R_1. The latter may be required for wide strips, as in [1].

The transition between the conditions of wide and narrow strips, for the purpose of selecting the closer formulas, is usually on the narrow side of the square condition. As a rule, this transition may be associated with the shape, $a/b=\frac{1}{2}$. In particular, this is near the cross-over of the approximate formulas for a strip between two planes.

There have been presented some explicit formulas for R of a/b, or the reverse, including k as a parameter. These are adequate for practical purposes, so they are listed here.

Range of Shape	Synthesis	Analysis
Wide strips	(29)(30)	(31)
Narrow strips	(34)(47)(36)	(48)

VIII. Design Charts

Figure 9 is a graphical presentation of the results of this study, in a form adequate for practical purposes.

Figure 9(a), (b), and (c) gives, in three sections, the relation between R and a/b for k in binary steps from 1 to 16. Figure 9(a) includes the range covered by the complete formulas for narrow strips; Fig. 9(b) and (c), the range covered by the complete formulas for wide strips.

Figure 9(d) gives the relation between the effective filling fraction q and the shape ratio a/b, including the small dependence on k. From this ratio, (6) enables simple computation of k'.

The velocity ratio for any shape is equal to $1/\sqrt{k'}$. It may be taken off the R curves, since it is equal to the resistance ratio R/R_1 for any shape.

As previously mentioned, the form of the R curves is patterned after Dukes [4], but his curves have discrepancies up to 0.15 of R. The present curves cover a wider range of shape, on a more expanded scale.

IX. Conclusion

In this problem of mixed dielectric, it has been found possible to compute the effective dielectric constant to a remarkably close approximation. It is placed between rather close limits that are exactly defined and approximately formulated. Then it is evaluated by interpolation between these limits, using a rule that has a logical basis in principle and seems sensible for this application.

This approach has been applied to the case of parallel-strip conductors on the faces of a dielectric sheet. The result is a formulation of this case over the entire range of the shape ratio and the dielectric constant of the sheet material. For synthesis or analysis, there are given some simple explicit formulas capable of approximation within about one per cent of wave resistance.

X. Symbols

k = dielectric constant of sheet of material separating the pair of strips.

k' = effective dielectric constant of all space around the pair of strips.

$k_a=(k+1)/2$ = average of dielectric constants of sheet and free space.

q' = entire filling fraction of dielectric material.

q'' = parallel part (nearly all) of filling fraction.

$q=(k'-1)/(k-1)$ = effective filling fraction.

R_c = 377 ohms = wave resistance of square area of free space.

R = wave resistance of symmetrical pair of strips on dielectric sheet k; or of one quadrant of its cross section.

R_1 = wave resistance of pair of strips in free space ($k=1$).

C = capacitance.

L = inductance.

ϵ_0 = electrivity (electric permittivity) of free space. 83

μ_0 = magnetivity (magnetic permeability) of free space.

l = length of pair of strips (for C or L).

$x+jy=z$ = complex plane of space coordinates, the cross section of the pair of strips.

$x'+jy'=z'$ = complex plane of flux-potential coordinates.

a = half width of strip conductor.

b = half separation of parallel strips.

a/b = shape ratio.

a' = effective half width of outer face of strip.

g' = effective half width of strip, including flux on both outer and inner faces.

g'/π = ratio of effective width over separation.

a'/g' = fraction of total flux that terminates on outer face of strip.

g'' = effective half width of strip between parallel planes.

$\Delta a=g''-a$ = effective increment of half width of "wide" strip between parallel planes.

$a^*=2.416=a'$ for "square" shape ($a/b=1$).

d = parameter defined in [1], differing slightly from g' for "wide" strips.

d_k = parameter d for same R but all space filled with dielectric k.

$h'=\pi^2/g'$ = separation parameter.

$h'/\pi=\pi/g'$ = ratio of half separation over effective half width.

h_a = parameter h' for same R but all space filled with average dielectric k_a.

s' = increment of effective half width, including parallel and series parts.

s'' = parallel part of s'.

s = interpolated increment of effective half width for any dielectric constant k.

t = special scale of abscissas, related to shape ratio.

Appendix

A Strip Between Parallel Planes

An essential part of this presentation is the relation between the subject shape and a slightly different shape, compared with reference to (a) and (b) in Figs. 3 and 6. The different shape is a thin strip midway between two parallel planes, for comparison with a strip spaced from one plane. The properties of this different shape are summarized here for reference where needed in the text.

Figure 6(a) shows the upper two quadrants of the subject shape, including one of the two strips and the neutral plane below. Figure 6(b) shows the same region around one strip, but with the addition of a parallel second plane equally spaced above the strip. The free-space properties of the latter are to be formulated here.

The derivation for this shape is well known, in terms of complete elliptic integrals. These can be closely approximated for the cases of wide or narrow strips by simple formulas in terms of slide-rule functions.

84

Referring to Fig. 6(b), the effective width of the strip is greater than twice the actual width because both sides are exposed to nearby planes. In one quadrant, the effective width g'' is greater than twice the half-width a by twice the increment of width Δa. The well-known derivation, applied to the one quadrant, gives

$$g''/b = g''/\pi = \pi/h'' = R_c/R'' = 2K/K' \tag{53}$$

in which

$$K = K(\sin\alpha); \qquad K' = K(\cos\alpha) \tag{54}$$

$$\sin\alpha = \tanh a/2; \quad \cos\alpha = \operatorname{sech} a/2 \tag{55}$$

$$\tan\tfrac{1}{2}\alpha = \tanh\tfrac{1}{4}a = \tanh\frac{\pi a}{4b} \tag{56}$$

Here, the free-space wave resistance in one quadrant is R''. The complete elliptic integral K is defined and tabulated in Dwight.

The ratio K/K' can be closely approximated over each of two ranges, corresponding to narrow and wide strips. Applying these approximations, we obtain the following simple formulas.

Narrow strips; $\alpha < \tfrac{1}{4}\pi$; $a/b < 0.56$;

$K/K' < 1$; $R'' > \tfrac{1}{2}R_c$:

$$g'' = \frac{\pi^2}{\ln\dfrac{2}{\tan\frac{1}{2}\alpha}} = \frac{\pi^2}{\ln\dfrac{2}{\tanh\frac{1}{4}a}} < 2\pi \tag{57}$$

Wide strips; $\alpha > \tfrac{1}{4}\pi$; $a/b > 0.56$;

$K/K' > 1$; $R'' < \tfrac{1}{2}R_c$:

$$g'' = 4\ln\frac{2}{\tan\left(\frac{1}{4}\pi - \frac{1}{2}\alpha\right)} = 2(a + \ln 4) > 2\pi \tag{58}$$

Relative error <0.003 of g'' (or 0.005 of a).

The approximation for wide strips is in a well-known form, including $\Delta a = \ln 4$, as given by Maxwell [2]. The corresponding approximation for narrow strips has the same mathematical basis, but is not so well known.

The transition between the two cases is not critical; it may be made at $a/b = 0.5$, which is near $g'' = 2\pi$. In terms of a strip and one plane, this is near $g' = \tfrac{3}{2}\pi$.

While not essential to this presentation, there are given here for reference the relations between wave resistance and shape between two planes.

Narrow strips (as before):

$$R_1 = R_c\frac{1}{2\pi}\left[\ln\frac{8b}{\pi a} + \frac{\pi^2}{48}\left(\frac{a}{b}\right)^2 - \cdots\right] \tag{59}$$

$$\frac{b}{a} = \frac{\pi}{8}\exp 2\pi R_1/R_c - \frac{\pi}{6}\exp - 2\pi R_1/R_c + \cdots \tag{60}$$

Wide strips (as before):

$$R_1 = R_c\frac{\frac{1}{2}}{\dfrac{a}{b} + \dfrac{1}{\pi}\ln 4} + \cdots \tag{61}$$

$$\frac{a}{b} = \frac{1}{2}R_c/R_1 - \frac{1}{\pi}\ln 4 + \cdots \tag{62}$$

Each of these formulas includes the first and second terms of a rapidly convergent series. They are given in explicit form for analysis and for synthesis.

Acknowledgment

This study and presentation have been stimulated by the needs of a printed-circuit array antenna being designed for Hazeltine Corporation. The writer is grateful to Mrs. Nancy M. Staron for the careful typing of the extensive mathematical content of this manuscript, and to Henry J. Newton for the skillful preparation of the graphs.

References

[1] Wheeler, H. A., Transmission-line properties of parallel wide strips by a conformal-mapping approximation, *IEEE Trans on Microwave Theory and Techniques*, vol MTT-12, May 1964, pp 280–289.

[2] Maxwell, J. C., *Electricity and Magnetism*, 3 ed. London: Oxford University Press, 1892. (Edge correction for a wide strip between parallel planes, p 303.)

[3] Dukes, J. M. C., Properties of strip transmission lines with the aid of an electrolytic tank, *Proc. IEE*, vol 103, pt. B, May 1956, pp 319–333. (Rough computation for mixed dielectric in case of strip and plane separated by sheet.)

[4] ——, The application of printed-circuit techniques to the design of microwave components, *Proc. IEE*, vol 105, pt. B, Aug 1958, pp 155–172. (Graphs of wave resistance and velocity ratio for parallel strips separated by dielectric sheet, subject to errors.)

[5] Frazita, R. F., Transmission line properties of coplanar parallel strips on a dielectric sheet, M.S. thesis, Polytech. Inst. of Brooklyn, N. Y., 1965.

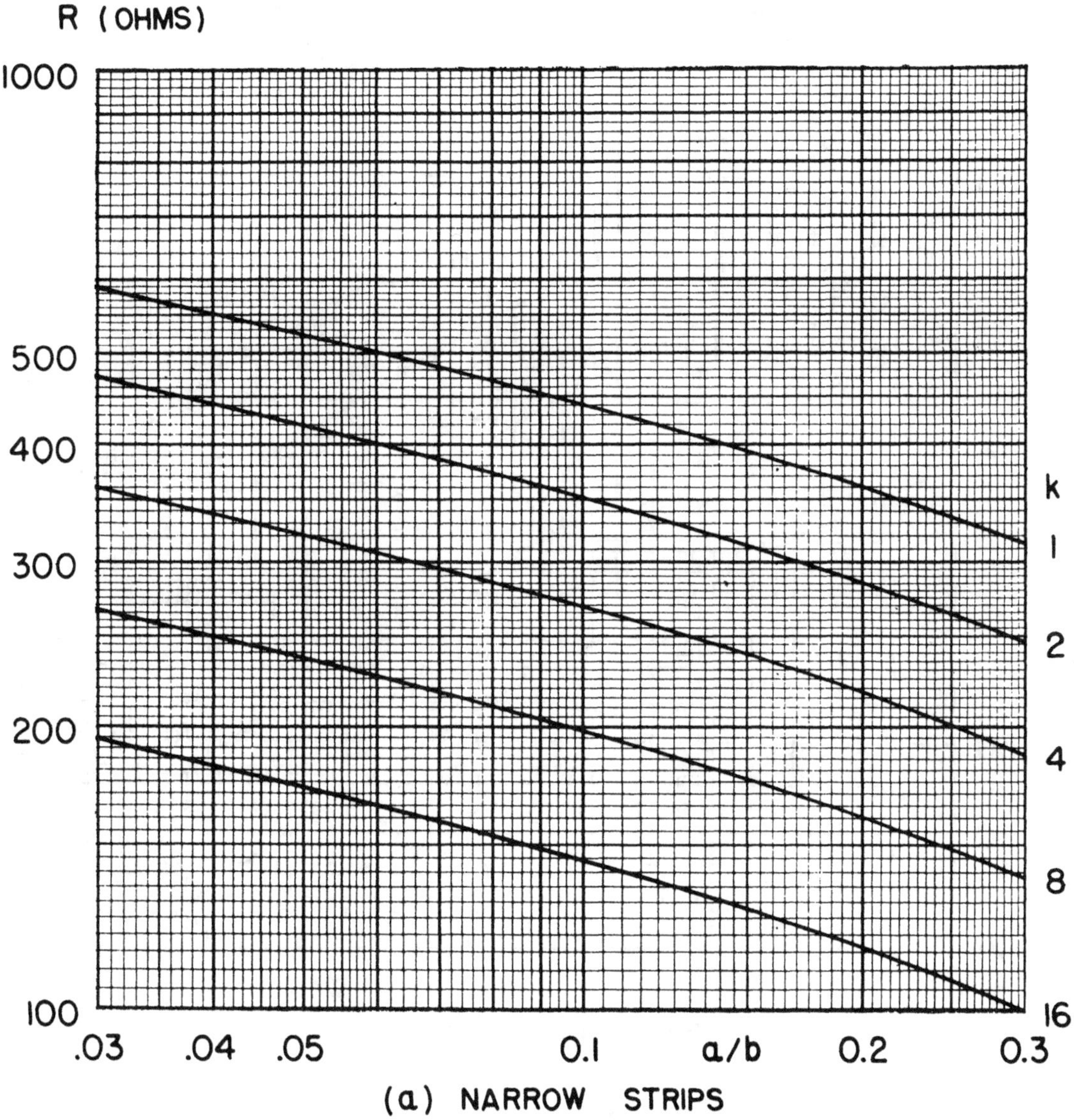

FIG. 9 - GRAPHICAL CHARTS FOR DESIGN.

(Courtesy of H.A. Wheeler-- expanded version of chart on p. 81 fig. 9a)

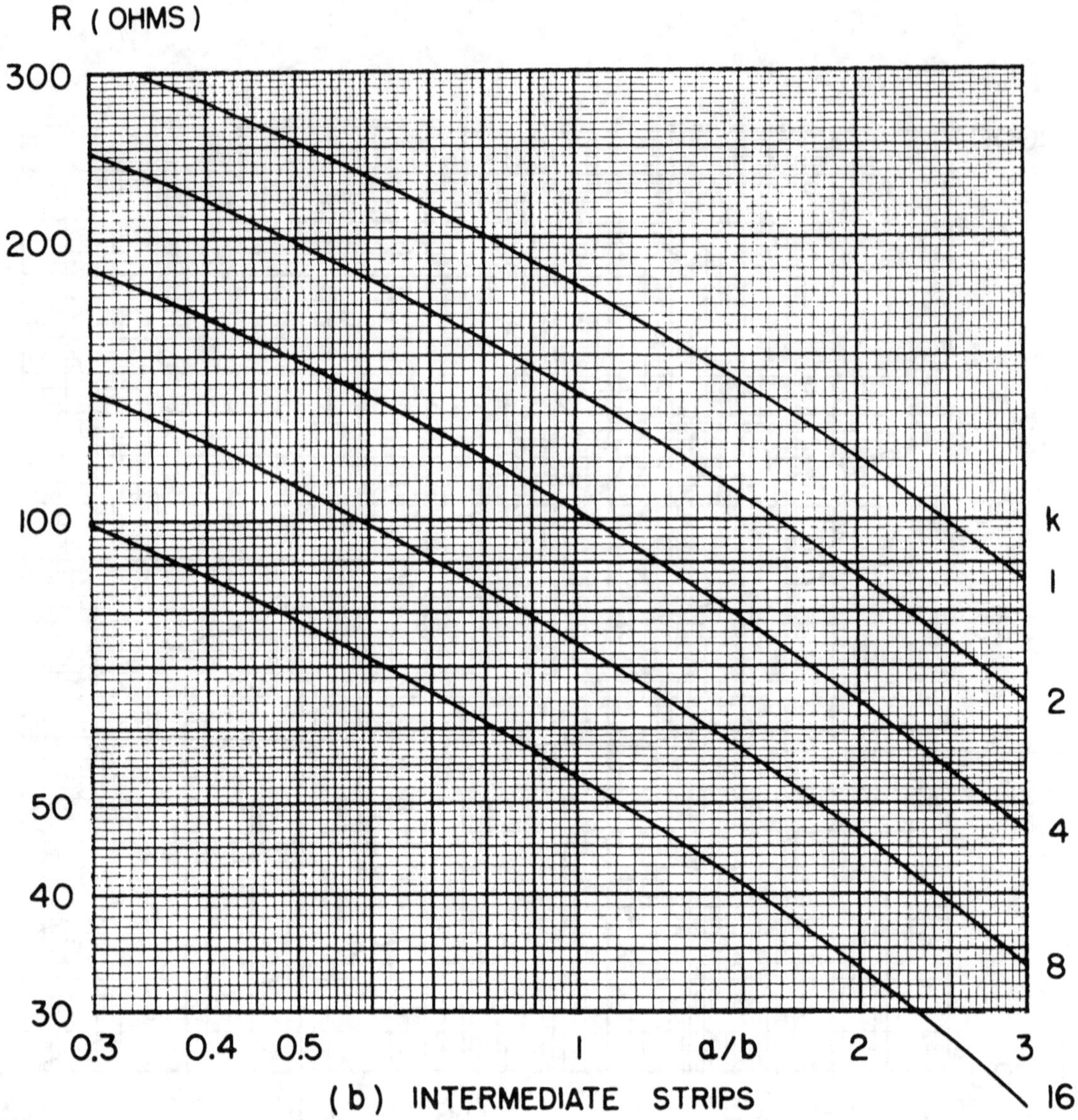

FIG. 9- GRAPHICAL CHARTS FOR DESIGN.

(Courtesy of H.A. Wheeler--
expanded version of chart on
p. 81 fig. 9b)

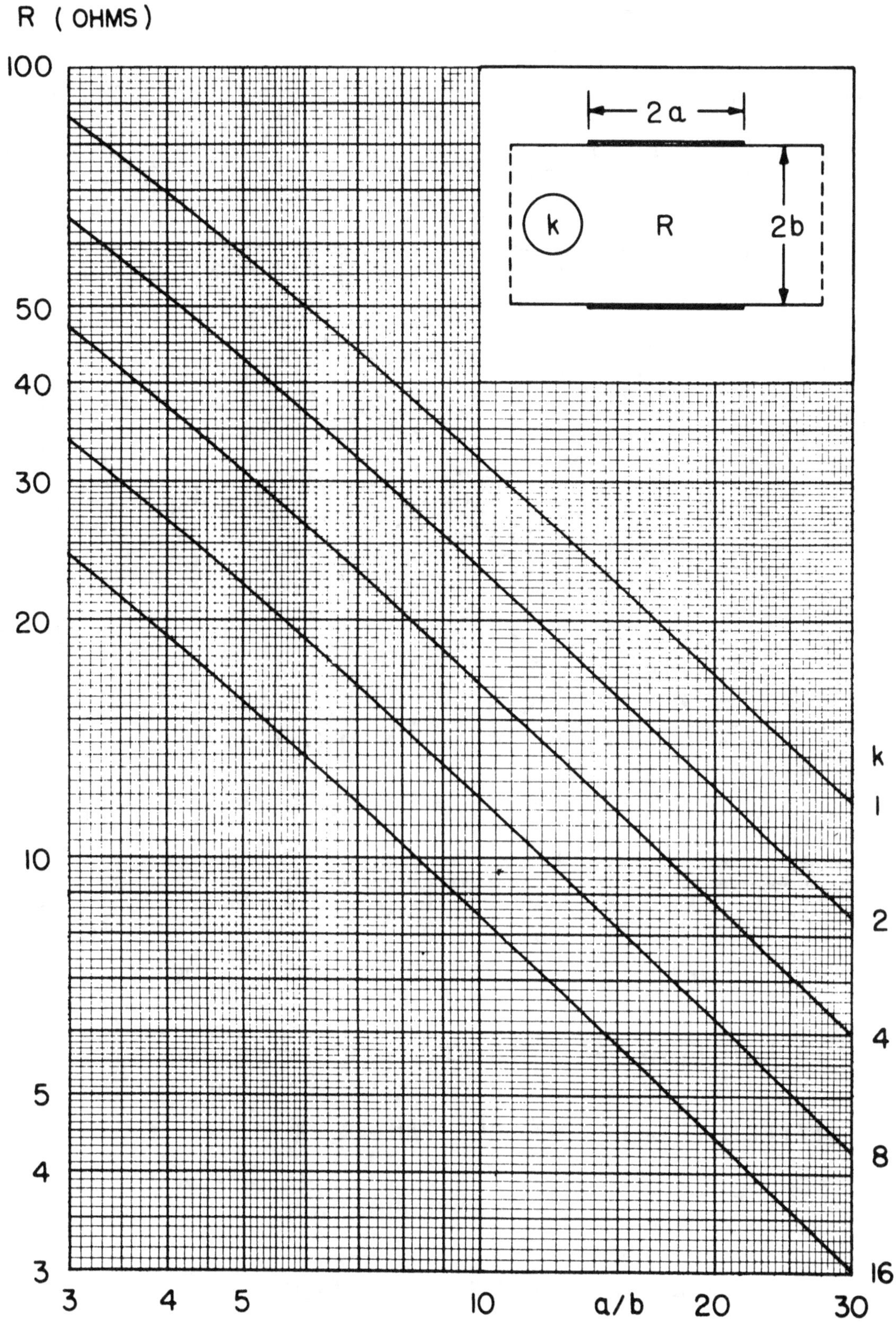

(c) WIDE STRIPS

FIG. 9- GRAPHICAL CHARTS FOR DESIGN.
(Courtesy of H.A. Wheeler--
expanded version of chart on
p. 81 fig. 9c)

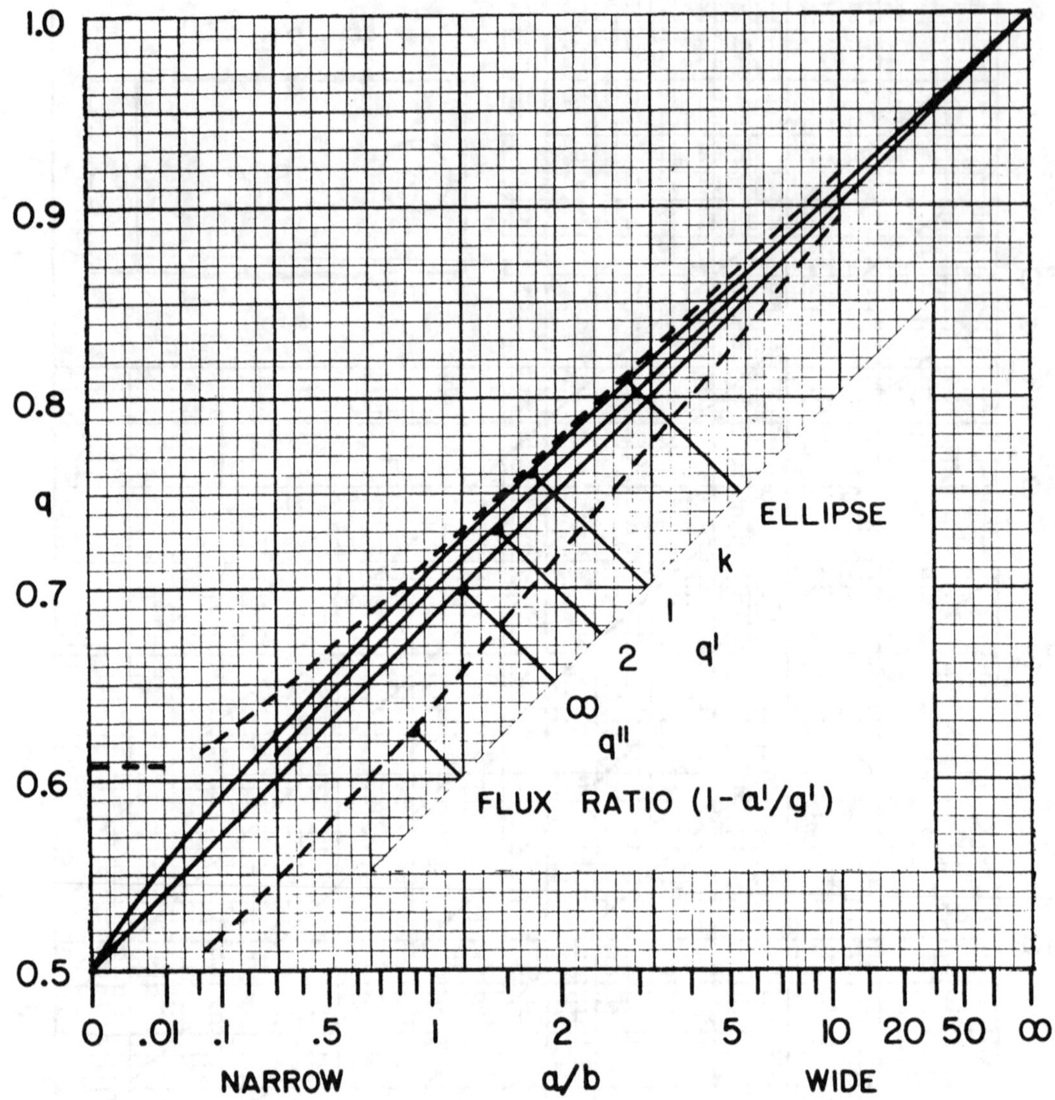

EFFECTIVE FILLING FRACTION: $q = \frac{k' - 1}{k - 1}$

EFFECTIVE DIELECTRIC CONSTANT: $k' = 1 + q(k - 1)$

(d) FILLING FRACTION

FIG. 9 - GRAPHICAL CHARTS FOR DESIGN.

(Courtesy of H.A. Wheeler- expanded version of chart on p. 81 fig. 9d)

Reprinted from IEEE TRANSACTIONS
ON *MICROWAVE THEORY AND TECHNIQUES*
Volume MTT-16, Number 12, December, 1968
pp. 1021-1027

Parameters of Microstrip Transmission Lines and of Coupled Pairs of Microstrip Lines

THOMAS G. BRYANT, STUDENT MEMBER, IEEE, AND JERALD A. WEISS, SENIOR MEMBER, IEEE

Abstract—**A theoretical analysis is presented of microwave propagation on microstrip, with particular reference to the case of coupled pairs of microstrip lines. Data on this type of transmission line are needed for the design of directional couplers, filters, and other components in microwave integrated circuits. The inhomogeneous medium, consisting of the dielectric substrate and the vacuum above it, is treated in a rigorous manner through the use of a "dielectric Green's function" which expresses the discontinuity of the fields at the dielectric-vacuum interface. Results are presented in graphical form for substrate dielectric constants of 1, 9, and 16, and a range of values of width and spacing of the strips. Numerical tables for these and other cases are also available. The tables present capacitance, characteristic impedance, and velocity of propagation of the even and odd normal modes. The method lends itself to the treatment of other geometries which are of practical interest, such as "thick" strips, presence of an unsymmetrically located upper ground plane, etc.**

Manuscript received May 28, 1968; revised August 21, 1968. This work was performed under the sponsorship of the Array Radars Group at M.I.T. Lincoln Laboratory, in part through contracts with the University of Maine and Worcester Polytechnic Institute.

T. G. Bryant was with the University of Maine, Orono, Me. He is now with the Array Radars Group, M.I.T. Lincoln Laboratory, Lexington, Mass. (Operated with support from the U. S. Advanced Research Projects Agency.)

J. A. Weiss is with the Department of Physics, Worcester Polytechnic Institute, Worcester, Mass., and the Array Radars Group, M.I.T. Lincoln Laboratory, Lexington, Mass.

I. INTRODUCTION

FOR GUIDANCE in the design of integrated microwave circuit components, data are required on the parameters of symmetrical coupled pairs of microstrip transmission lines. The parameters needed to characterize this structure are the characteristic impedances and velocities of propagation of the two normal modes. In addition, for certain purposes such as investigation of spurious coupling, peak power capability, and gyromagnetic interaction (in the case of nonreciprocal substrate materials), information is also required on the RF field configuration.

The term "microstrip" is a nickname for a microwave circuit configuration which is constructed by printed-circuit techniques—modified where necessary to reduce loss, reflections, and spurious coupling, but retaining advantages in size, simplicity, reliability, and cost which such production techniques afford. Microstrip shares some of the troublesome properties of dispersive waveguide, in that conductor dimensions influence not only characteristic impedance, but also the velocity of propagation; also in that, on structures which support propagation in more than one mode, the velocities of the modes are in general unequal.

We present a solution of the microstrip problem in the "quasi-static" limit; i.e., for the frequency range in which propagation may be regarded as approximately TEM. Such a solution is valid in the range extending into the low gigahertz region, in the case of microstrip dimensions and substrate materials frequently used in integrated microwave circuit technology. For higher frequencies, the solution provides design guidance and can serve also as a basis for solution of the full propagation problem. The method developed for this work involves the determination of a "dielectric Green's function" which characterizes the effect on the field configuration due to the presence of the dielectric-vacuum interface. Such a function has potentially wide applicability to a variety of problems involving inhomogeneous dielectric media.

90 The physical construction of a coupled pair of microstrip lines is shown in Fig. 1. The configuration is conventionally specified by the parameters W/H and S/H, together with K, where W is the strip width, H is the substrate height, and S is the spacing between adjacent edges of the strips; K is the relative dielectric permittivity of the substrate material. The problem of a single microstrip may be identified with that of coupled strips in the limit $S/H \rightarrow \infty$. In the present report the strip thickness T is assumed to be negligibly small. This assumption is not intrinsic to the method, however, and application to the case of nonzero T/H will be presented in a future publication.

From the symmetry of the structure, the normal modes of propagation on a pair of parallel strips of equal dimensions have even and odd symmetry, respectively, with respect to reflection in a central bisecting plane. In the quasi-static limit, where propagation is approximately TEM, the characteristic impedances of the two modes can be determined from their respective dc capacitances and low-frequency velocities. The difference in impedances becomes large as the coupling between the strips is increased by reducing the spacing between them. The analogous problem of coupled strips in the case of balanced stripline has been treated by Cohn [1]. The microstrip problem presents additional difficulties due to the lower symmetry and the presence of the dielectric–vacuum boundary. Following the first thorough treatment by Wheeler [2], there have been calculations of capacitance, velocity, and impedance of single strips, and also some work on coupled pairs, using various approximate methods, by Cristal [3], Wolters and Clar [4], Policky and Stover [5], and no doubt by others. During the course of this study, we were apprised by a paper by Silvester [6] which treats the dielectric–vacuum boundary by means of a Green's function, similar in some respects to that employed in the present paper. A recent paper by Yamashita and Mittra [7] presents an analysis based on the variational principle.

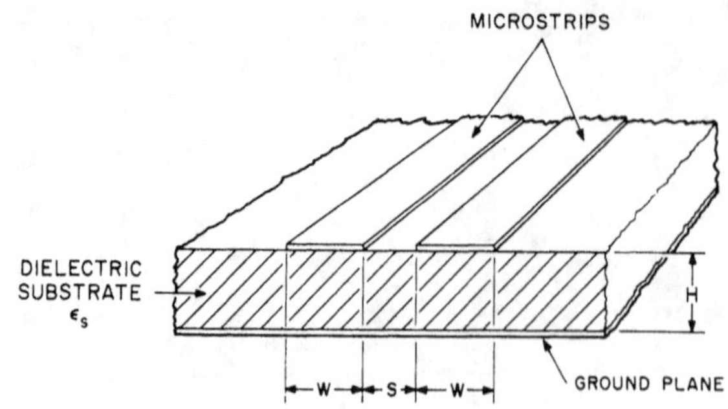

Fig. 1. A coupled pair of microstrip transmission lines.

II. The Dielectric Green's Function

We perform a rigorous solution of the electrostatic problem; namely, the determination of capacitance for a single strip, and for coupled strips, on a dielectric substrate. We then apply this solution to the propagation problem, incorporating the quasi-static approximation in which the longitudinal components of the fields are neglected.

Consider an arrangement of conducting boundaries, denoted collectively by S, and a harmonic function (potential) $\Phi_0(r)$ which satisfies given Dirichlet boundary conditions on S, and thus solves the electrostatic problem for the case of a vacuum or homogeneous dielectric medium. For the microstrip problem, S is simply the conducting ground plane, assumed to be infinite in extent. The potential is created by an elementary source, which we might take to be a uniformly charged line lying parallel to the plane at the position r_0. A more suitable source is a narrow elementary strip, uniformly charged, since its potential is finite everywhere; the element actually used in this work is of that form.

Dielectric material, of relative permittivity K, is now introduced into this field, so as to occupy a region bounded by the surface T. In the following, we shall term "interface" that part of T which does not coincide with S. We seek the solution $\Phi(r)$ of the new electrostatic problem in the presence of the inhomogeneous dielectric medium.

Suppose we were to assume, as a first approximation, that the solution $\Phi_0(r)$ of the vacuum problem solves the dielectric problem also. This function is, indeed, the unique solution—not of the desired problem, but of a related one: namely, that in which, in addition to the charge distributions required to establish the specified values of potential on S, there is also a certain distribution of "free" charge $\sigma_0(r_T)$ on the dielectric–vacuum interface. For, since Φ_0 and grad Φ_0 are continuous at T, the surface divergence of dielectric displacement $\boldsymbol{D}$ is

$$\epsilon_0(K - 1)\boldsymbol{n}_T \cdot \text{grad } \Phi_0 = \sigma_0(\boldsymbol{r}_T) \tag{1}$$

where $\boldsymbol{n}_T$ is the "outward" unit normal vector (directed from T into the vacuum).

Intuitively, we may imagine placing a compensating distribution of "bound" charge $\sigma(r_T)$ on the interface so as to cancel $\sigma_0(r_T)$. Let $\Phi(r)$ denote the new potential, including the effect of this additional charge. Such bound charge, corresponding to a surface divergence of electric field $\boldsymbol{E}$, creates a discontinuity in E such that

$$\epsilon_0 \boldsymbol{n}_T \cdot [\text{grad } \Phi(\boldsymbol{r}_{T-}) - \text{grad } \Phi(\boldsymbol{r}_{T+})] = \sigma(\boldsymbol{r}_T) \tag{2}$$

where r_{T+} and r_{T-} refer to the + (vacuum) and − (dielectric) sides of T, respectively.

Assuming that the correct bound charge distribution $\sigma(r_T)$ has been found, its effect is to annihilate the free charge distribution $\sigma_0(r_T)$:

$$\epsilon_0 \boldsymbol{n}_T \cdot [K \text{ grad } \Phi(\boldsymbol{r}_{T-}) - \text{grad } \Phi(\boldsymbol{r}_{T+})] = 0. \tag{3}$$

To evaluate the contribution of $\sigma(r_T)$ to the potential, we may employ the (vacuum) Green's function $G(r|r')$ appropriate to the conducting boundary S. Let G satisfy

$$\nabla^2 G(\boldsymbol{r} \mid \boldsymbol{r}') = -4\pi\delta(\boldsymbol{r} - \boldsymbol{r}'), \tag{4}$$

with the boundary condition

$$G(\boldsymbol{r}_S \mid \boldsymbol{r}') = 0. \tag{5}$$

Then the total potential $\Phi(r|r_0)$ due to the source at r_0, including the contribution due to $\sigma(r_T)$, is

$$\Phi(\boldsymbol{r} \mid \boldsymbol{r}_0) = \Phi_0(\boldsymbol{r}) + \frac{1}{4\pi\epsilon_0}\int_T \sigma(\boldsymbol{r}'_T) G(\boldsymbol{r} \mid \boldsymbol{r}'_T) da'. \tag{6}$$

As will be shown, (6) is the dielectric Green's function which plays the role of kernel for the solution of the quasi-static microstrip problem. In terms of this potential, the boundary condition (3) becomes

$$\sigma_0(\boldsymbol{r}_T) + \frac{1}{4\pi}\int_T \sigma(\boldsymbol{r}'_T)\boldsymbol{n}_T \cdot [K \text{ grad } G(\boldsymbol{r}_{T-} \mid \boldsymbol{r}'_T) - \text{grad } G(\boldsymbol{r}_{T+} \mid \boldsymbol{r}'_T)] da' = 0. \tag{7}$$

This constitutes an integral equation for the unknown bound charge distribution $\sigma(r_T)$. By making use of the properties of $G(r|r')$ we can put the relation (7) into a form suitable for computation.

We decompose the factor in brackets in (7) according to

$$\begin{aligned} K \text{ grad } G(\boldsymbol{r}_{T-} \mid \boldsymbol{r}'_T) &- \text{grad } G(\boldsymbol{r}_{T+} \mid \boldsymbol{r}'_T) \\ &= \tfrac{1}{2}(K+1)[\text{grad } G(\boldsymbol{r}_{T-} \mid \boldsymbol{r}'_T) - \text{grad } G(\boldsymbol{r}_{T+} \mid \boldsymbol{r}'_T)] \\ &\quad + \tfrac{1}{2}(K-1)[\text{grad } G(\boldsymbol{r}_{T-} \mid \boldsymbol{r}'_T) + \text{grad } G(\boldsymbol{r}_{T+} \mid \boldsymbol{r}'_T)]. \end{aligned} \tag{8}$$

In substituting in (7) according to (8), we note that the two terms on the right side of (8) have the following interpretations. The first term (the one with coefficient $K+1$) yields in effect an integral of the Green's-function field over the surface of a small region enclosing an element of T; with the aid of Gauss's theorem we replace it by a volume integral of $\nabla^2 G$ which, according to (4), becomes $-4\pi\delta(r-r')$. The second term (with coefficient $K-1$) represents an average of the Green's-function fields on the two sides of T. It is therefore continuous at T; in fact, it is the part of the field contributed by the "images" of the elementary source $\delta(r-r')$; or, put another way, it is the field due to the charge distributions induced on S by this source. We denote by $G_I(r|r')$ this "image part" of the Green's function.

With these interpretations, (7) becomes

$$\sigma_0(\boldsymbol{r}_T) + \frac{1}{2}(K+1)\sigma(\boldsymbol{r}_T) + \frac{1}{4\pi}(K-1)\int_T \sigma(\boldsymbol{r}'_T)\boldsymbol{n}_T \cdot \text{grad } G_I(\boldsymbol{r}_T \mid \boldsymbol{r}'_T) da' = 0. \tag{9}$$

Equation (9) is an inhomogeneous Fredholm's integral equation of the second kind [8] for the unknown bound charge distribution $\sigma(r_T)$, in terms of the known initial free charge distribution $\sigma_0(r_T)$, given by (1), and the image part of the Green's function for S, which is known, or in any case is derivable with more or less difficulty depending on the shape of the conducting boundary S. For the case of the flat microstrip ground plane, G_I takes a simple form. Of the various methods available for solution of such equations, one which we have used successfully in the present work is an iterative method based on a slight rearrangement of (9):

$$\sigma(\boldsymbol{r}_T) = -\frac{2}{K+1}\sigma_0(\boldsymbol{r}_T) - \frac{1}{2\pi}\frac{K-1}{K+1}\int_T \sigma(\boldsymbol{r}'_T)\boldsymbol{n}_T \cdot \text{grad } G_I(\boldsymbol{r}_T \mid \boldsymbol{r}'_T) da'. \tag{10}$$

As a first approximation, set the function $\sigma(r'_T)$ appearing under the integral sign equal to zero; (10) gives the second approximation as simply $2/(K+1)$ times the negative of the initial free charge σ_0. Using this in the integral, we obtain the third approximation of σ, and so on. The method is not guaranteed to converge under all conditions, but in any case we can directly determine the success of the process by evaluating the left side of (9) after each stage of iteration. For the microstrip problem contemplated here, convergence has been shown to take place without incident. (In experiments with other surfaces S, we have acquired some experience which enables us often to anticipate convergence difficulties.)

With σ known, the desired dielectric Green's function $\Phi(r|r_0)$ may now be generated according to (6). An illustration of the result is shown in Fig. 2. The graph refers to substrate $K=16$; it shows the potential due to a strip element of $W/H=0.025$ carrying unit charge. Also shown in the figure are the distributions of bound charge σ and of surface charge on the ground plane. The vacuum image Green's function G_I used for this calculation is obtained by elementary methods. The potential due to an element of width 2ξ located at height H is

$$\begin{aligned} G(x, y \mid 0, H) &= \frac{1}{2}\left(\frac{x}{\xi} - 1\right)\ln\frac{(x-\xi)^2 + (y-H)^2}{(x-\xi)^2 + (y+H)^2} \\ &\quad - \frac{1}{2}\left(\frac{x}{\xi} + 1\right)\ln\frac{(x+\xi)^2 + (y-H)^2}{(x+\xi)^2 + (y+H)^2} \\ &\quad + \frac{y+H}{\xi}\left(\arctan\frac{x+\xi}{y+H} - \arctan\frac{x-\xi}{y+H}\right) \\ &\quad - \frac{y-H}{\xi}\left(\arctan\frac{x+\xi}{y-H} - \arctan\frac{x-\xi}{y-H}\right). \end{aligned} \tag{11}$$

For use in (10), we require the component of field normal to T, due to the image part of (11):

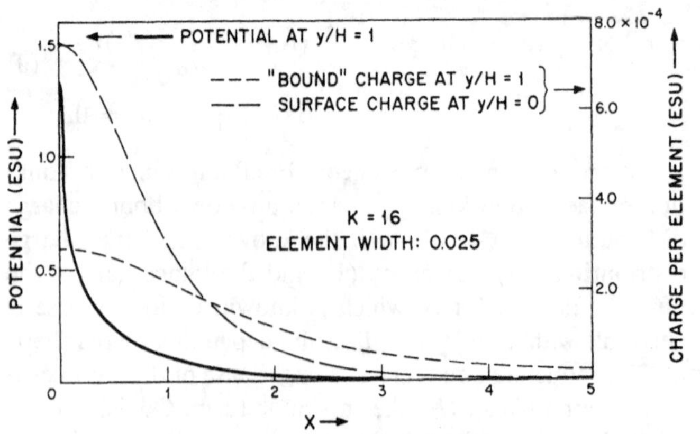

Fig. 2. The potential and charge distributions of the dielectric Green's function.

$$E_y(x, H) = \frac{1}{\xi}\left(\arctan\frac{x-\xi}{2H} - \arctan\frac{x+\xi}{2H}\right). \quad (12)$$

III. Calculation of Normal Mode Parameters

Consider a coupled-strip configuration composed of two strips, spacing S/H and equal widths W/H. One of these strips is at potential V; the other is at potential $\pm V$ for the even or odd mode, respectively. We divide the strips and the space between them into elements of width 2ξ (12); let $W/H=2\xi M$, $S/H=2\xi N$, and let $x_{j-i}=2\xi(j-i)$. The potentials on the strips are to be produced by the placement of charge λ_i on the ith element, $i=1, \cdots, M$, and $\pm\lambda_i$ on the symmetrically located element in the case of the even or odd mode, respectively. We determine the charges λ_i. Referring to the dielectric Green's function (6), let

$$\Phi_{ij} = \Phi(x_i, H \mid x_j, H) = \Phi(x_{j-i}, H \mid 0, H) \quad (13)$$

denote the potential at x_i due to the element at x_j carrying unit charge. Then the resultant potential at the center of the ith element is

$$\sum_{j=1}^{M} \lambda_j(\Phi_{ij} \pm \Phi_{i,2M+N+1-j}) = V. \quad (14)$$

This system is to be solved for the M values of charge λ_i. The solution yields the charge distribution (and therefore also the current distribution in the propagation problem) and the total charge

$$Q = \sum_{i=1}^{M} \lambda_i \quad (15)$$

from which the capacitance C_K per strip follows as $C_K = Q/V$. The method of calculation outlined above is obviously not limited to symmetrical pairs of strips, but may be applied to any strip configuration.

To determine the effective relative permittivity K_{eff}, the calculation must be performed for the vacuum $K=1$ as well as for the value of substrate K contemplated. The vacuum problem can, of course, be solved without iteration, since there is no bound charge in that case. Denoting the resulting capacitance per strip by C_1, we have

$$K_{eff} = C_K/C_1. \quad (16)$$

The velocity of propagation v is

$$v = \frac{c}{\sqrt{K_{eff}}} \quad (17)$$

where $c=2.997925\times10^8$ m·s^{-1} is the characteristic velocity of the vacuum. Within the quasi-static approximation, we may calculate the characteristic impedance per strip from

$$Z_0 = \frac{1}{vC_K} = \frac{1}{cC_1\sqrt{K_{eff}}}. \quad (18)$$

The wavelength ratio, which is occasionally convenient in design work, is

$$\frac{\lambda(K_{eff})}{\lambda(K)} = \sqrt{\frac{K}{K_{eff}}}. \quad (19)$$

IV. Data for Single and Coupled Strips

The method outlined above has been carried out for various values of K; in particular, for $K=9$, appropriate for an alumina substrate, and $K=16$, for the magnetic substrate material yttrium iron garnet. Fig. 3 shows the characteristic impedance of a single strip as a function of W/H for values of K from unity to 16; Fig. 4 shows the corresponding velocities.

The computational procedure contains its own means for determining accuracy, through the assessment of errors in the fulfillment of the boundary conditions, as will be discussed in Section VI. In addition, it is illuminating to compare these results with those which Wheeler [2] obtained analytically by means of a conformal transformation. The comparison is shown in Fig. 5, where the differences between our values of Z_0 and those of Wheeler are shown on a greatly expanded scale. Wheeler offers two approximations, appropriate for wide (Wheelw) and narrow (Wheeln) strips, respectively. Agreement between those two is best at about $W/H=1$. The figure shows that our results are intermediate between the two, the disagreement being well within one percent everywhere except for $W/H<0.2$. In that range, our values are higher. This is an error in our results, due to the fact that we have employed an elementary strip width of $W/H=0.025$, which means that for $W/H=0.2$ our representation of the strip is composed of only eight elements. With such coarse division of the strip, the steps in the charge distribution become a source of appreciable error, particularly at the edges where our discrete representation misses the sharp rise in concentration of charge. As the need arises for data on very narrow strips, this error can be made as small as desired by a straightforward modification of the program.

Results for coupled strips are shown graphically in Figs. 6–8, covering the ranges $W/H=0.2, \cdots, 2.0$, and $S/H=0.3$, 0.6, and 1.0 for values of substrate dielectric constant $K=1$, 9, and 16. Table I shows a sample page of tabular data. For given K and spacing parameter S/H, the table lists capacitance C_K, effective dielectric constant K_{eff} (16), char-

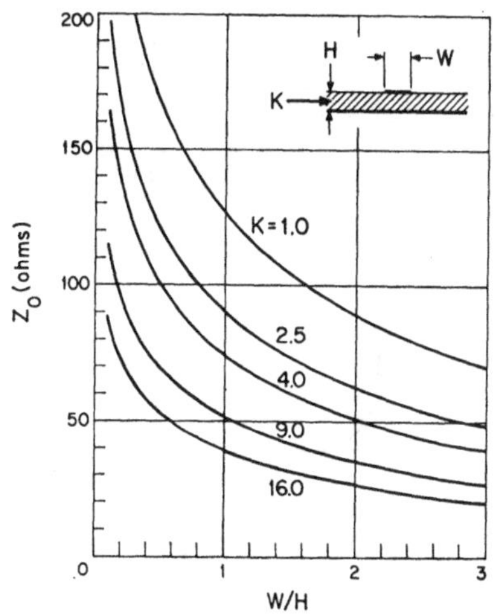

Fig. 3. Characteristic impedance of microstrip—single strip.

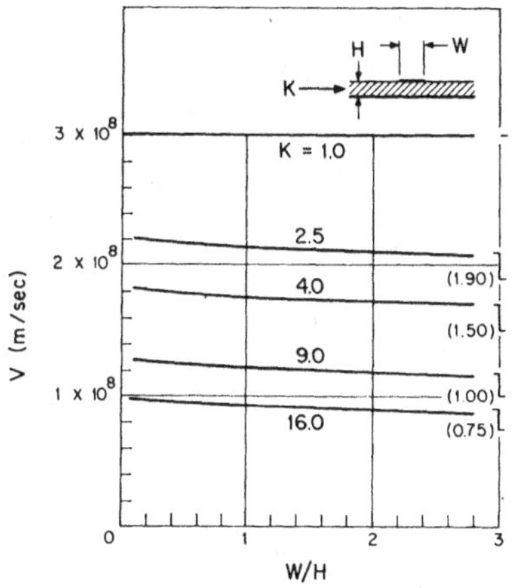

Fig. 4. Velocity of propagation on microstrip—single strip.

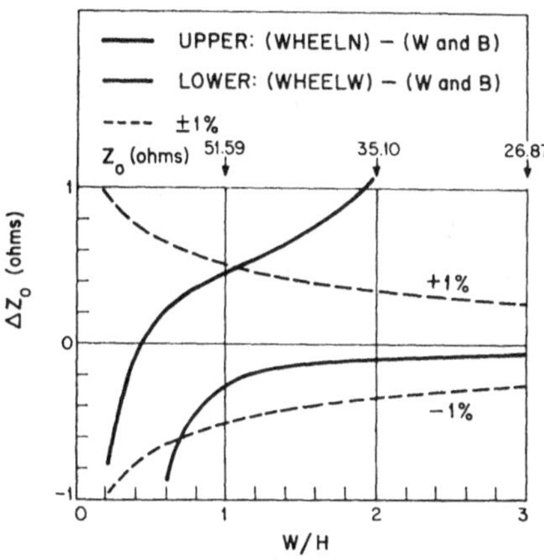

Fig. 5. Comparison of three values of characteristic impedance of microstrip.

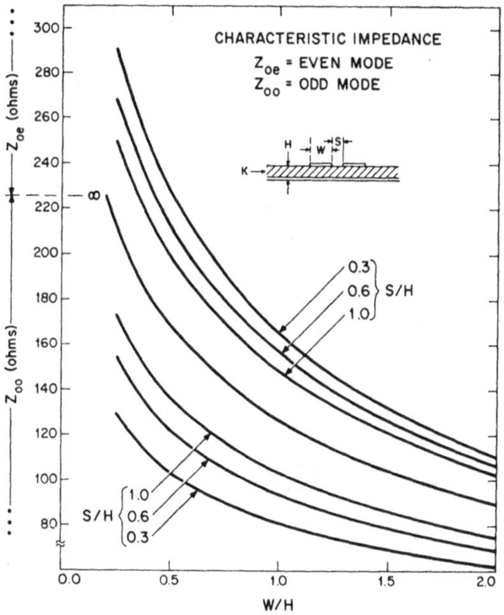

Fig. 6. Characteristic impedance of coupled pairs of microstrip transmission lines—$K=1$.

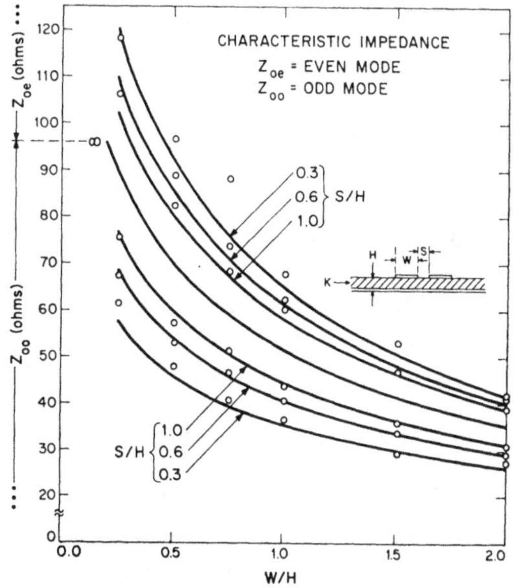

Fig. 7. Characteristic impedance of coupled pairs of microstrip transmission lines—$K=9.0$.

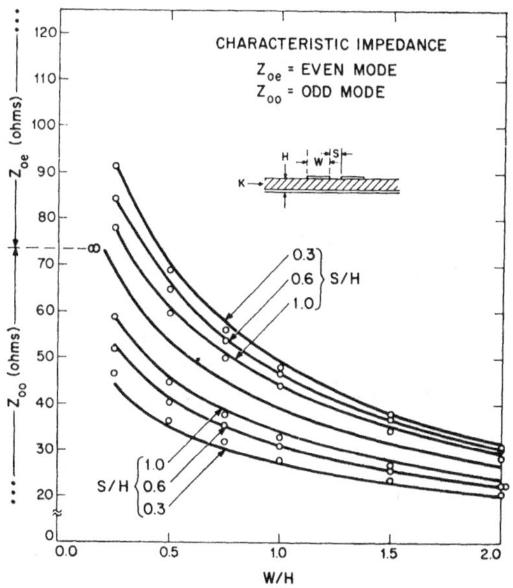

Fig. 8. Characteristic impedance of coupled pairs of microstrip transmission lines—$K=16.0$.

93

TABLE I

PARAMETERS OF MICROSTRIP

$S/H=0.2$, $K=10.0$

W/H	C_K (pF/m)	K_{eff}	Z_0 (ohms)	V (10^8 m/s)	$L(K_{eff})/L(K)$
		Coupled Strips—Even Mode			
0.10	54.512	6.2506	152.984	1.1991	1.2648
0.20	66.463	6.4106	127.072	1.1841	1.2490
0.30	76.813	6.5429	111.079	1.1720	1.2363
0.40	86.558	6.6611	99.459	1.1616	1.2253
0.50	96.009	6.7695	90.395	1.1522	1.2154
0.60	105.298	6.8700	83.030	1.1438	1.2065
0.70	114.490	6.9635	76.882	1.1361	1.1984
0.80	123.621	7.0508	71.649	1.1290	1.1909
0.90	132.709	7.1326	67.128	1.1225	1.1841
1.00	141.767	7.2094	63.176	1.1165	1.1777
1.10	150.803	7.2815	59.687	1.1110	1.1719
1.20	159.822	7.3495	56.581	1.1058	1.1665
1.30	168.826	7.4135	53.796	1.1011	1.1614
1.40	177.819	7.4741	51.284	1.0966	1.1567
1.50	186.801	7.5313	49.004	1.0924	1.1523
1.60	195.775	7.5856	46.926	1.0885	1.1482
1.70	204.742	7.6372	45.023	1.0848	1.1443
1.80	213.701	7.6861	43.274	1.0814	1.1406
1.90	222.656	7.7328	41.659	1.0781	1.1372
2.00	231.604	7.7772	40.165	1.0750	1.1339
		Coupled Strips—Odd Mode			
0.10	120.499	5.4993	64.916	1.2784	1.3485
0.20	150.072	5.5119	52.183	1.2769	1.3469
0.30	170.603	5.5275	45.968	1.2751	1.3450
0.40	186.988	5.5458	42.010	1.2730	1.3428
0.50	201.035	5.5666	39.148	1.2706	1.3403
0.60	213.624	5.5896	36.917	1.2680	1.3375
0.70	225.252	5.6146	35.089	1.2652	1.3346
0.80	236.220	5.6412	33.539	1.2622	1.3314
0.90	246.724	5.6693	32.191	1.2591	1.3281
1.00	256.892	5.6986	30.997	1.2558	1.3247
1.10	266.817	5.7289	29.923	1.2525	1.3212
1.20	276.560	5.7600	28.947	1.2491	1.3176
1.30	286.167	5.7917	28.052	1.2457	1.3140
1.40	295.671	5.8238	27.225	1.2423	1.3104
1.50	305.095	5.8563	26.458	1.2388	1.3067
1.60	314.457	5.8889	25.742	1.2354	1.3031
1.70	323.769	5.9217	25.071	1.2320	1.2995
1.80	333.042	5.9544	24.440	1.2286	1.2959
1.90	342.281	5.9870	23.845	1.2252	1.2924
2.00	351.494	6.0195	23.283	1.2219	1.2889

acteristic impedance Z_0, velocity of propagation v and the wavelength ratio (19). The table refers to the even and odd modes and covers a range of the width parameter W/H. Complete tables are available[1] covering the following ranges: $W/H=0.1(0.1)2.0$, single strip and even and odd modes for $S/H=0.2(0.2)1.0$, for substrate dielectric constant $K=9$, 10, 15, 16; also, a table for wide spacing: $S/H=1.0(0.5)4.0$, for $K=9.6$.

[1] Tables have been deposited as Document No. NAPS-00087 with ADI Auxiliary Publications Service, American Society for Information Sciences, c/o CCM Information Sciences, Inc., 22 West 34 Street, New York, N. Y. 10001. Microfiche copies are available at $1.00 or 8½ by 11-inch photocopies at $3.00. To order, include the document number and the advance payment.

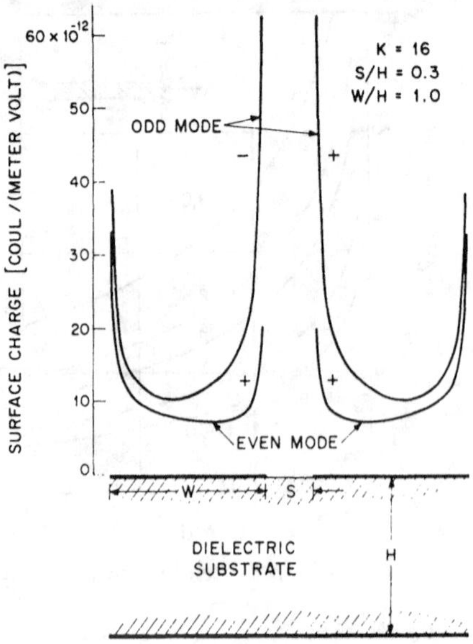

Fig. 9. Charge distributions for the even and odd modes on coupled strips.

The programs also furnish other useful information, including the determination of errors and the values of charge, potential, fields over the complete cross-section of the transmission line. An example of the charge distributions on the strips is shown in Fig. 9.

The data points in Figs. 7 and 8 are values measured by MacFarland [9].

V. PROGRAMS

The computations were performed at the computer facility at the University of Maine and on the CP/CMS system at M.I.T. Lincoln Laboratory. In the present form, the program is composed of two parts: the first computes the dielectric Green's function according to (10) for any specified K and stores these data. This computation has been used to generate data for a sequence of seven values of K in the range from 6 to 18. The dielectric Green's function for intermediate values of K is determined by interpolation. The second program uses these data to compute the single- and coupled-strip parameters as illustrated in Table I, according to (14).

VI. DETERMINATION OF ACCURACY

The calculation of the dielectric Green's function includes means for verification of its accuracy, through determination of errors in the fulfillment of the required boundary conditions. Using (9), we evaluate the surface divergence of electric displacement on the dielectric–vacuum interface. Typically, the residual free charge is less than one part in 10^4 or 10^5 relative to the bound charge at all points near the source, remaining roughly constant in magnitude—but therefore rising, relatively, to a few parts in 10^2 at distances about 10H from the source. We can assess the effect of this error by determining the total amount of "missing" bound charge. This estimate includes also the error resulting from the fact that the calculation was not normally carried to points beyond ±10H. As discussed below, this missing charge amounted to less than five percent of the required bound charge. If all the missing charge were lumped at the closest possible position, namely at ±10H, its contribution to the potential at points near the source would be less than one part in 10^4. This value is, of course, a great overestimate of the error, since in fact the missing charge would have been distributed over the entire infinite breadth of the interface.

All of the electric field originating on the source must, of course, terminate somewhere. The three locations of the terminating charge are 1) on the dielectric surface directly in contact with the source strip, 2) on the exposed area of the interface, and 3) on the ground plane. The accounting of these charges is as follows, for unit charge on the source: under the source, $(1-K)/(1+K)$—leaving net charge $2/(1+K)$ at the position of the strip; on the exposed interface, $(1-K)/K(1+K)$; on the ground plane, $-1/K$. We find that the calculated total of each of these agrees with requirements to within 1.6 percent at worst, and generally to within 0.7 percent or better, depending on the value of K. The greatest error is in the "missing" bound charge at the interface, as discussed above. It is noteworthy that the value of K for which the greatest deficiency is found to occur is in the vicinity of $K=2.5$, consistent with the fact that the maximum value of $(K-1)/K(K+1)$ occurs at $K=1+\sqrt{2}\cong 2.4$.

The conclusion from our various tests is that our calculation of the dielectric Green's function leads to errors in the microstrip parameters which are of the order of one part in 10^4 or less; completely adequate for all practical purposes.

As pointed out in Section IV, there is another type of error which enters in calculations involving values of W/H or S/H in the range less than 0.2. This is the consequence of the finite width of the discrete elementary strip, namely $W/H=0.025$ in the data reported here. As Fig. 5 shows, the effect is that the characteristic impedance is falsely high in that range: about one percent high at $W/H=0.2$ and becoming worse fast. As previously mentioned, this error can be easily corrected when the need arises.

Acknowledgment

We wish to acknowledge the contributions of Prof. F. Irons of the University of Maine and N. W. Cook of WPI. We are grateful to C. Blake and D. H. Temme of Lincoln Laboratory for their support and encouragement, and to H. T. MacFarland and J. D. Welch for the benefit of many discussions and other assistance. We also thank MacFarland for permission to quote his experimental results in advance of publication. Finally, we thank Mrs. J. Reid for her valuable instruction and assistance in the use of the CP/CMS time-sharing computer system.

References

[1] S. B. Cohn, "Shielded coupled-strip transmission line," *IRE Trans. Microwave Theory and Techniques*, vol. MTT-3, pp. 29–38, October 1955.

[2] H. A. Wheeler, "Transmission-line properties of parallel strips separated by dielectric sheet," *IEEE Trans. Microwave Theory and Techniques*, vol. MTT-13, pp. 172–185, March 1965.

[3] E. G. Cristal, Tech. Rept. USAEL Contract DA-28-043 AMC-02266 (E).

[4] K. C. Wolters and P. L. Clar, *Proc. 1967 G-MTT Internat'l Microwave Symp.*, paper V-2.

[5] G. Policky and H. L. Stover, "Parallel-coupled lines on microstrip," Texas Instruments, Inc., Rept. 03-67-61.

[6] P. Silvester, "TEM wave properties of microstrip transmission lines," *Proc. IEE* (London), vol. 115, pp. 43–48, January 1968.

[7] E. Yamashita and R. Mittra, "Variational method for the analysis of microstrip lines," *IEEE Trans. Microwave Theory and Techniques*, vol. MTT-16, pp. 251–256, April 1968.

[8] See, for example, L. P. Smith, *Mathematical Methods for Scientists and Engineers.* Englewood Cliffs, N. J.: Prentice-Hall, 1953.

[9] H. T. MacFarland, private communication.

A HIGH DIRECTIVITY MICROSTRIP COUPLER TECHNIQUE

Allen Podell
Anzac Electronics

Typically loosely coupled backward wave microstrip couplers have directivities which decrease with increasing frequency. High directivity becomes more difficult to obtain as the coupling is loosened. The problem appears to be that the propagating velocities of the odd and even modes are not equal. This paper deals with a technique which equalizes these velocities up to an arbitrarily high frequency.

Techniques such as capacitive tabs and lumps at the ends of the coupler may be used to increase its directivity.[1] Generally the techniques which increase the directivity seem to decrease the overall bandwidth of the coupler. The overlay coupler has equal even and odd mode velocities, and may be used in broadband couplers. The technique described herein is also a fundamental velocity equalizer, but without any additional processing beyond the normal printing and etching process of a microstrip coupler.

Figure 1 shows a typical wiggly line coupler. Wiggling the adjacent edges of the lines in the manner pictured slows the odd mode wave without much affecting the even mode. If the outsides of the lines were wiggled in unison with the insides then the even mode would be slowed down, and the wiggling would have to be more severe before velocity equalization would occur.

Figure 2 shows a typical coupling and directivity versus frequency plot for a wiggly coupler. Note that the directivity does not slope severely upward with frequency, demonstrating the fundamental velocity equalization.

Wiggly couplers have been built with couplings ranging from 7.8 to 27 dB on glass teflon, glass epoxy, and alumina. Achieving high directivity seems mainly a result of patience in determining the necessary wiggle depth, and care in fabricating the photographic mask.

Five tandem pair 3 dB wiggly couplers were built as shown in Figure 3 to check their reproducibility. Their performance is shown in Table I. A welded strip, a bond wire or a beam lead may be used to connect the lines at the crossover. In our case, the jumper was made with a pair of 1 mil bond wires. For just 3 dB, one would guess that the single section interdigitated coupler would be as easy to construct and be less lossy.[2] Above 7 dB coupling, however, there is no need for the crossover; a singly coupler can be etched with the necessary tolerances for reproducibility.

In order to exploit the broadband possibilities of a wiggly coupler, a 54 to 290 MHz 3 section 18 dB coupler was built. The 3 sections were individually built and tested before being patched together. Figure 4 shows the coupling and directivity performance of this coupler, built on glass epoxy circuit board.

The side effects of wiggling seem to be to increase the coupling for a given line spacing, and to slightly increase the loss. When the wiggle spacing becomes an appreciably fraction of a wavelength, the coupler will no longer appear continuous, but this does not appear to be a serious limitation until well above X band.

What has been described here is a technique for the broadband equalization of the even and odd mode velocities of a microstrip coupler. This technique makes possible simple, multi-octave microstrip couplers of more than 20 dB directivity.

REPRODUCED FROM
G-MTT 1970 INTERNATIONAL MICROWAVE SYMPOSIUM,
DIGEST, pp. 33-36
IEEE Cat. No. 70 C 10-MTT

1. H.E. Brenner, "Perturbations of the critical parameters of quarter-wave directional couplers," IEEE MTT and J.E. Morris, IEEE TMTT, July 1968. Overlay Coupler

2. Julius Lange, IEEE TMTT, December 1969

TABLE I

Frequency	1-2 GHz
Coupling	3 dB Nominal
Insertion Loss	.35 dB Max
Unbalance	1.0 dB Max
Isolation	27 dB Min

Summary, 5 units worst case

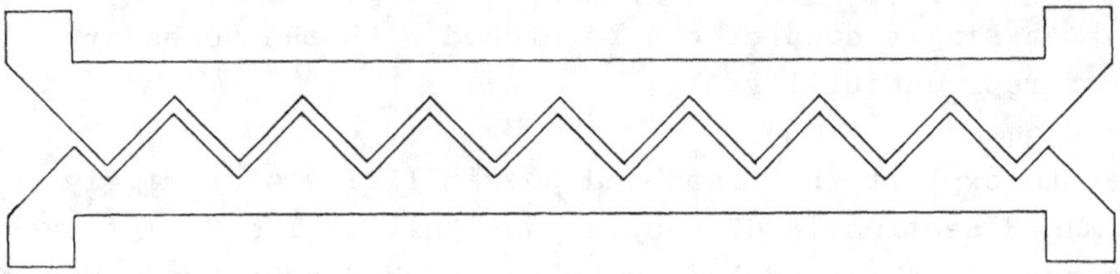

FIGURE 1 - TYPICAL WIGGLY LINE COUPLER

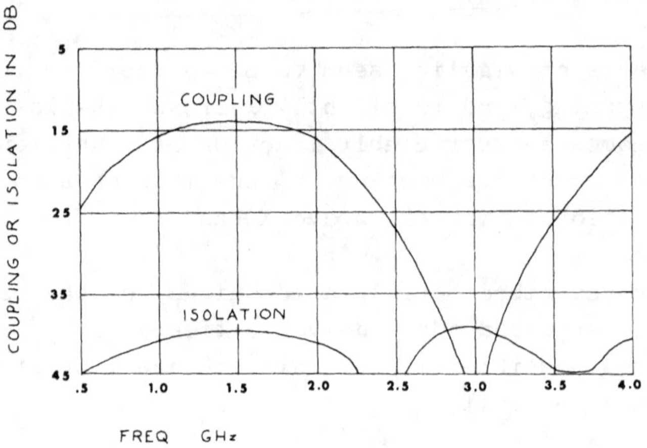

FIGURE 2 - COUPLING AND DIRECTIVITY

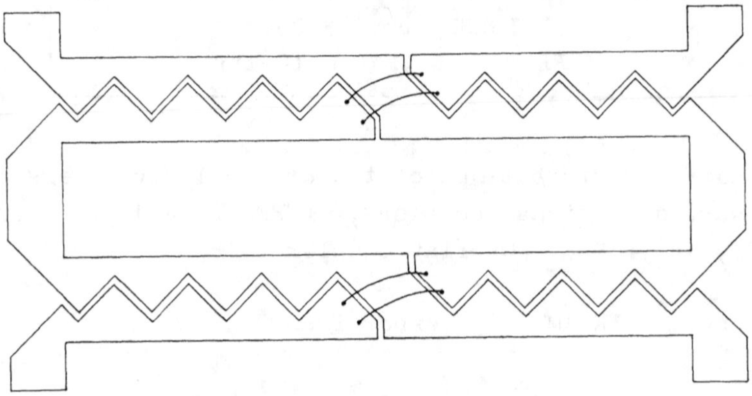

FIGURE 3 - 3 DB COUPLER

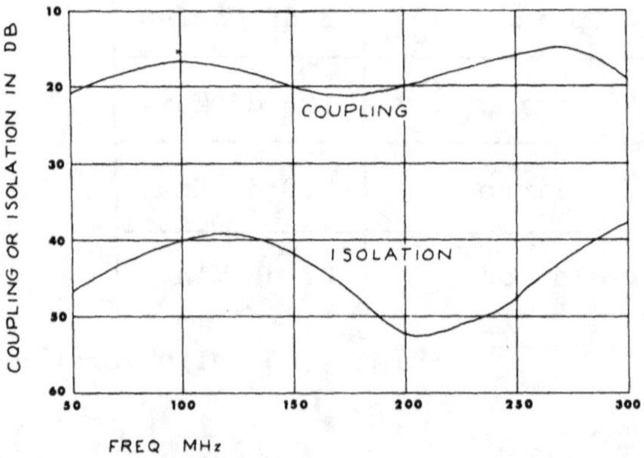

FIGURE 4 - COUPLING AND DIRECTIVITY OF 54 TO 290 MHz COUPLER

BIBLIOGRAPHY ON DIRECTIONAL COUPLERS IV

101

The Directional Coupler—1966

I Introduction

The directional coupler is a well-known device used as an attenuator, power splitter, hybrid junction, local oscillator injection device, and, most commonly, a sampling device for measuring separately the forward and backward waves on a transmission line. Over a period of twenty-five years many papers have appeared dealing with analysis, design, and application of the device.

In 1954, R. F. Schwartz prepared a bibliography of 91 papers which had appeared up to that time relating to the directional coupler field and, in 1955, Schwartz and Medhurst prepared a supplement of 41 additional papers. The present correspondence focuses attention on those problems which were treated up to 1954–55, highlights the significant advances to date, and points out some of the fruitful areas for further work.

While a considerable number of bibliographical entries are presented here, it is almost impossible to be exhaustive. Some restricted circulation reports and theses have been noted, for they help to indicate certain areas of interest as well as organizations and people involved. It is recognized, however, that there are probably many more entries that could be added.

II. Pre-1955 Directional Couplers

The directional coupler became a common transmission line and microwave device during World War II. It was early appreciated that two distinct mechanisms could be exploited: 1) the constructive and destructive interference of waves coupled by two superimposed, but different type, couplings as in the Bethe-hole and loop-type directional couplers, and 2) the constructive and destructive interference produced by waves arriving

Manuscript received May 23, 1966; revised September 12, 1966.

at a given point by different paths as in the multihole directional coupler. During the 1940's many different configurations of couplers utilizing these principles were developed and couplings in the range of 15 to 60 dB with directivities of 20 dB or greater were common. Bandwidths in most types of couplers were rather narrow, even though the principles which could lead to broader band performance were known. Many of the early couplers were severely limited by other defects in the transmission systems, e.g., connectors.

In the early 1950's refinements were beginning to be made. Lower dB values of coupling were sought with greater directivity over wide bands, 40 dB in waveguide and 30 dB in coaxial line being typical. The trend in radars was toward higher and higher powers and therefore power handling capability became an area of concern. During this period the awareness of the coupled-mode viewpoint and the concept of odd and even mode distributions grew and with it developed techniques for design. The application of Fourier transforms in the approximate analysis and the carry over of information from modern circuit theory was also accomplished. The loop-type and coupled transmission line coupler were reinvestigated to develop quantitative criteria for design. These techniques and principles were subsequently applied to coaxial, open wire, and strip transmission lines, as well as to waveguide. New applications were
102 also constantly being developed in operational and measurement systems.

III. 1955–1965 Directional Couplers

In the most recent decade the problem emphasis has broadened. Further improvement of conventional directional couplers has continued. The 3 dB broadband coupler has become a reality, and octave or more bandwidths in some types of directional couplers have been attained.

Directional couplers for other transmission systems such as, for example, ridged waveguide and strip line, have been developed. The latter in particular has led to new techniques of design which were impossible to utilize with other TEM transmission lines. The improvement of other components such as loads and connectors, and the design of broad-banded transitions between systems have made it possible in many cases to realize the full benefit of coupler design improvements. The availability of good miniature connectors has allowed some types of directional couplers to be shrunk in size.

Modern network theory has continued to influence directional coupler design and a general body of knowledge on coupling structures has evolved. At the same time the digital computer has been put to work calculating parameter tables so that for certain types of directional couplers the design task has become almost routine.

The marriage of directional couplers with filters has been consummated both in design and application. Directional coupling in multimode systems for the purpose of measurement, filtering, and multiplexing has been accomplished.

Directional coupling structures utilizing ferrite and YIG materials have also received some attention. Quasi-optical couplers in the mm range and also at optical wavelengths have been realized.

IV. The Future

Undoubtedly, the influence of modern network theory and the trend toward computer-generated design tables will continue in the directional coupler field. Probably formalized synthesis procedures will also be developed. Continued gains from these techniques as well as from improved connectors will make possible tight couplings over extremely wide bands. Perhaps the log periodic principle will be applied in some of these designs. While conventional waveguide will continue to be limited by dominant-mode bandwidth, there will be applications where ridged waveguide will be further exploited for multiband operation. One could also expect continued efforts in over moded systems, and as mm and sub-mm waves find more applications, quasi-optical directional coupler design will receive more attention. Constructions using prisms will probably become quite familiar.

Finally, it would be expected that a great deal more will be done in applying anisotropic and gyrotropic materials to directional couplers. What will evolve from this is a matter for speculation.

W. E. Caswell[1]
Electronic Materiels of Canada, Ltd.
Ottawa, Canada
R. F. Schwartz
The Moore School of Elec. Engrg.
University of Pennsylvania
Philadelphia, Pa.

Bibliography

Reference Articles

[1] R. F. Schwartz, "Bibliography on directional couplers," *IRE Trans. on Microwave Theory and Techniques*, vol. MTT-2, pp. 58–63, July 1954.
[2] R. F. Schwartz and R. G. Medhurst, "Addenda to bibliography on directional couplers," *IRE Trans. on Microwave Theory and Techniques*, vol. MTT-3, pp. 42–43, April 1955.

Additional Pre-1955 Publications

1951

[3] F. Bolinder, "Approximate theory of the directional coupler," *Proc. IRE*, vol. 39, p. 291, March 1951.
[4] Tsuchey and Takahashi, "CM type directional coupler," *The 1951 Three Electricity Societies (Japan)*, pp. 10–12, 1951.

1952

[5] F. Assadourian and E. Rimai, "Simplified theory of microstrip transmission systems," *Proc. IRE*, vol. 40, pp. 1651–1657, December 1952.
[6] D. Grieg and H. Englemann, "Microstrip—a new transmission technique for the kilomegacycle range," *Proc, IRE*, vol. 40, pp. 1644–1650, December 1952.
[7] J. Kostriza, "Microstrip components," *Proc. IRE*, vol. 40, pp. 1658–1663, December 1952.
[8] H. Riblet, "Super directivity with directional coupler arrays," *Proc. IRE*, vol. 40, pp. 994–995, August 1952.

1953

[9] "Unidirectional coupler research and development program," Sperry Gyroscope Co., Great Neck, N. Y., Final Rept. 5224-1240-8, 1953.
[10] S. Kobayashi, "Directional couplers of the bridged-T type," *J. Inst. Elect. Commun. Engrs. (Japan)*, vol. 36, pp. 475–480, September 1953.
[11] S. Kumagai and K. Itakura, "Theoretical and experimental researches on new type directional couplers," Osaka Univ., Japan, Tech. Rept., p. 257, 1953.
[12] S. Tanaka and T. Okakura, "Characteristics and improvement of Bethe-hole directional couplers for waveguides," *J. Inst. Elect. Commun. Engrs. (Japan)*, vol. 36, pp. 489–493, September 1953.

1954

[13] F. Barnett, P. Lacy, and B. Oliver, "Principles of directional coupling in reciprocal systems," *Proc. on Mod. Advances in Microwave Techniques*, pp. 251–269, November 1954; also, New York: Interscience, 1955.
[14] L. I. Kent, "The optimum design of multi-element directional couplers," M.S. thesis, Polytechnic Inst. of Brooklyn, N. Y., 1954.
[15] B. Oliver, "Directional electromagnetic couplers," *Proc. IRE*, vol. 42, pp. 1686–1692, November 1954.
[16] Weissfloch, "Use of short circuiting pistons in the study of junctions and directional couplers," *Ann. Telecomm. (France)*, vol. 9, pp. 81–92, March 1954.
[17] E. M. Wells, "A short-slot hybrid for 9 mm," *Marconi Rev.*, vol. 17, pp. 86–87, 3rd Quarter 1954.

Post-1954 Publications

1955

[18] P. Andrew, "A simple waveguide directional coupler," *J. Instn. Radio Engrs. (Britain)*, vol. 15, pp. 112–116, February 1955.
[19] E. Belohonbek, "Investigations on a broad-band coupling between a coaxial line and a waveguide; I. theoretical part," *Arch. elekt. Ubertragung (Germany)*, vol. 9, pp. 432–440, September 1955 (in German).
[20] ——, "Investigations on a broad-band coupling between a coaxial line and a waveguide; II, mechanical part," *Arch. elekt. Übertragung (Germany)*, vol. 9., pp. 469–474, October 1955 (in German).
[21] S. Cohn, "Shielded coupled-strip transmission line," *IRE Trans. on Microwave Theory and Techniques*, vol. MTT-3, pp. 29–38, October 1955.
[22] J. Cook, "Tapered velocity couplers," *B.S.T.J.*, vol. 34, pp. 807–822, July 1955.
[23] R. Damon, "Magnetically controlled microwave directional coupler," *J. Appl. Phys.*, vol. 26, pp. 1281–1282, October 1955.
[24] A. Fox, "Wave coupling by warped normal modes," *Bell Sys. Tech. J.*, vol 34, pp. 823–852, July 1955.
[25] J. K. Hunton, H. C. Poulter, and C. S. Reis, "High directivity coaxial directional couplers and reflectometers," *Hewlett-Packard J.*, vol. 7, October 1955.
[26] R. Knechtli, "Further analysis of transmission-line directional couplers," *Proc. IRE*, vol. 43, pp. 867–869, July 1955.
[27] R. L. Kyle, "A new broad band coax directional coupler," *Proc. NEC*, pp. 402–414, October 1955.
[28] W. Louisell, "Analysis of the single tapered mode coupler," *Bell Sys. Tech. J.*, vol. 34, pp. 853–870, July 1955.
[29] G. Monteath, "Coupled transmission lines as symmetrical directional couplers," *Proc. IEE (London)*, vol. 102B, pp. 383–392, May 1955.
[30] H. Wolf, "Anwendung theorie des reflectometers," *Arch. elekt. Übertragung*, vol. 9, pp. 221–227, May 1955.

1956

[31] A. Berk and E. Strumwasser, "Ferrite directional couplers," *Proc. IRE*, vol. 44, pp. 1439–1445, October 1956.
[32] F. S. Coale, "A traveling wave directional filter," *IRE Trans. on Microwave Theory and Techniques*, vol. MTT-4, pp. 256–260, October 1956.
[33] E. Jones and J. Bolljahn, "Coupled-strip-transmission-line filters and directional couplers," *IRE Trans. on Microwave Theory and Techniques*, vol. MTT-4, pp. 75–81, April 1956.
[34] P. P. Lombardini, R. F. Schwartz, and P. J. Kelly, "Criteria for the design of loop-type directional couplers for the L band," *IRE Trans. on Microwave Theory and Techniques*, vol. MTT-4, pp. 234–239, October 1956.
[35] ——, "Multi-output directional couplers," Moore School of Elec. Engrg., University of Pennsylvania, Philadelphia, Final Rept., Contract AF 30(635)-2801, January 31, 1956.
[36] J. Reed and G. Wheeler, "A method of analysis of symmetrical four-port networks," *IRE Trans. on Microwave Theory and Techniques*, vol. MTT-4, pp. 246–252, October 1956.
[37] H. Wolf, "Coupled high frequency transmission lines for use as directional couplers," *Nachrichtentech. Z. (Germany)*, vol. 9, pp. 375–382, August 1956 (in German).
[38] L. Young, "Branch guide couplers," *Proc. NEC*, vol. 12, pp. 723–732, 1956.

[1] Presently with Pan American World Airways, Guided Missiles Range Division, Fort Churchill, Manitoba, Canada.

1957

[39] S. B. Cohn, P. M. Sherk, J. K. Shimizu, and E. M. T. Jones, "Strip transmission line and components," Stanford Research Inst., Menlo Park, Calif., Final Rept., Contract DA 36-039-SC-63232, March 1957.

[40] J. Crompton, "A contribution to the design of multi-element directional couplers," *Proc. IEE (London)*, vol. 104C, pp. 398–402, September 1957.

[41] R. Koike, T. Asano, and A. Morishima, "Coaxial directional couplers," *J. Inst. Elect. Commun. Engrs. (Japan)*, vol. 41, pp. 626–631, June 1958 (in Japanese).

[42] P. P. Lombardini, and R. F. Schwartz, "A new type of directional coupler for coupling coaxial line to TE_{10} waveguide," *IRE WESCON Conv. Rec.*, vol. 1, pt. 1, pp. 22–29, 1957.

[43] P. D. Lomer and J. W. Crompton, "A new form of hybrid junction for microwave frequencies," *Proc. IEE (London)*, vol. 104B, pp. 261–263, May 1957.

[44] H. Perini and P. Sferrazza, "Rectangular waveguide to strip-transmission-line directional couplers," *IRE WESCON Conv. Rec.*, vol. 1, pt. 1, pp. 16–21, 1957.

[45] J. K. Shimizu, "A strip line 3 db directional coupler," Stanford Research Inst., Menlo Park, Calif., Sci. Rept. 1, Contract AF 19(604)-1571; also, *IRE WESCON Conv. Rec.*, vol. 1, pt. 1, pp. 4–15, 1957.

1958

[46] G. Broussaud, "Waveguide coupling—theory of thin slot directional couplers," *Ann. Radioelect. (France)*, vol. 13, pp. 187–199, July 1958 (in French).

[47] S. B. Cohn et al., "Design criteria for microwave filters and coupling structures," Stanford Research Inst., Menlo Park, Calif., Tech. Rept. 3, Contract DA36-039-SC-74862, August 1958.

[48] J. M. C. Dukes, "Broad-band slot coupled microstrip directional couplers," *Proc. IEE (London)*, vol. 105B, pp. 147–154, March 1958.

[49] J. Figanier and E. A. Ash, "Intrinsic directional coupler using elliptical coupling apertures," *Proc. IEE (London)*, vol. 105C, pp. 432–437, March 1958.

[50] A. Jaumann, "Directional couplers for the production of the H_{01} wave in circular waveguide," *Arch. elekt. Übertragung (Germany)*, vol. 12, pp. 440–446, October 1958 (in German).

[51] S. Kurokawa, T. Takahashi, and M. Arai, "Mutually coupled CR type directional coupler," *J. Rad. Res. Lab. (Japan)*, vol. 5, pp. 127–133, April 1958.

[52] I. Lucas, "Matching elements for the coupling of periodic structure waveguide directional couplers," *Arch. elekt. Übertragung (Germany)*, vol. 12, pp. 91–96, February 1958 (in German).

[53] L. Milosevic, "Frequency selective directional couplers for hyperfrequencies," *Revue-Tech. de la Compagnie Francaise Thomson-Houston*, no. 29, pp. 49–54, September 1958.

[54] J. Reed, "The multiple branch waveguide coupler," *IRE Trans. on Microwave Theory and Techniques*, vol. MTT-6, pp. 398–403, October 1958.

[55] E. Schuon, "Properties and design of long slot directional couplers," *Arch. elekt. Übertragung (Germany)*, vol. 12, p. 237, 1958.

[56] J. Shimizu and E. Jones, "Coupled-transmission-line directional couplers," *IRE Trans. on Microwave Theory and Techniques*, vol. MTT-6, pp. 403–410, October 1958.

[57] D. C. Stinson, "Ferrite directional couplers with off-center apertures," *IRE Trans. on Microwave Theory and Techniques*, vol. MTT-6, pp. 332–333, July 1958.

1959

[58] G. L. Allerton, "Microwave directional couplers," *Electronics*, vol. 32, p. 40, September 18, 1959.

[59] T. Anderson, "Directional coupler design nomograms," *Microwave J.*, vol. 2, pp. 34–38, May 1959.

[60] D. J. Angelakos, "An image line coupler," *IRE Trans. on Microwave Theory and Techniques*, vol. MTT-7, pp. 391–392, July 1959.

[61] E. S. Hensperger, "The design of multi-hole coupling arrays," *Microwave J.*, vol. 2, pp. 38–42, August 1959.

[62] D. J. Lewis, "Mode couplers and multimode measurement techniques," *IRE Trans. on Microwave Theory and Techniques*, vol. MTT-7, pp. 110–116, January 1959.

[63] V. V. Meriakri, "Closely coupled directional couplers in multimode circular waveguides," *Radiotekhnika i Elektronika (U.S.S.R.)*, vol. 4, pp. 1929–1931, November 1959 (in Russian).

[64] D. J. Nigg, "Shrinking the directional couplers," *Electronic Industries*, vol. 18, pp. 92–96, September 1959.

[65] H. Pascher, "Long-slot directional coupler for H-mode waves," *Arch. elekt. Übertragung (Germany)*, vol. 13, pp. 76–82, February 1959 (in German).

[66] K. G. Patterson, "A method for accurate design of a broad-band multibranch waveguide coupler," *IRE Trans. on Microwave Theory and Techniques*, vol. MTT-7, pp. 466–473, October 1959.

[67] V. V. Tyazhelov, "An experimental investigation of the interaction of single wire transmission lines," *Radiotekhnika i Elektronika (U.S.S.R.)*, vol. 4, pp. 592–598, 1959 (in Russian).

[68] T. A. Vereschnakova and V. V. Tyazhelov, "Experimental investigation of space-beats in a two-wire decimetric waves," *Bulletin of Higher Education Institutions, Radio Engineering (U.S.S.R.)*, vol. 2, p. 217, 1959 (in Russian).

[69] W. G. Voss, "Modified two-hole directional coupler," *Electronic Rad. Eng.*, vol. 36, p. 28, June 1959.

1960

[70] F. Andeceza, "Coupling of coaxial transmission lines and waveguide," *Rev. Telecom. (Madrid)*, vol. 15, pp. 20–25, March 1960 (in Spanish).

[71] A. N. Akhiezer, "Coupling of rectangular waveguides by means of a hole in the wide wall," *Zh. tekh. phys. (U.S.S.R.)*, vol. 30, pp. 851–854, July 1960 (in Russian).

[72] R. Mashkovtsev, L. Bensman, and A. A. Khokhrev, "A wide band waveguide directional coupler," *Radiotekhnika (U.S.S.R.)*, vol. 15, pp. 8–17, April 1960 (in Russian).

[73] D. I. Mirovitsky and G. G. Valieev, "Surface wave directional couplers," *Radiotekhnika i Electronika (U.S.S.R.)*, vol. 5, pp. 1078–1084, 1960 (in Russian).

[74] W. Nonak, "Application of unsymmetrical coupling for waveguide directional couplers with frequency independent coupling factor," *Hochfrequenztech. u. Elektakust. (Germany)*, vol. 69, pp. 179–188, October 1960 (in German).

[75] B. Oguchi, "Circular electric mode coupler," *IRE Trans. on Microwave Theory and Techniques*, vol. MTT-8, pp. 660–666, November 1960.

[76] L. Solymar, "Some properties of three coupled waves," *IRE Trans. on Microwave Theory and Techniques*, vol. MTT-8, pp. 284–291, May 1960.

[77] W. Stosser, "The 3-db coupler," *Frequenz (Germany)*, vol. 14, pp. 117–121, April 1960 (in German).

1961

[78] V. J. Albanese and H. Kagan, "The cross-over directional coupler," *Microwave J.*, vol. 4, pp. 88–91, September 1961.

[79] K. Araki, "Study of the design for directional couplers," Rept. of the Elec. Commun. Lab. (Japan), no. 1311, 1961.

[80] S. B. Cohn and S. L. Wehn, "Microwave hybrid coupler study program," Rantee Corp., Calabasas, Calif., 2nd Quart. Rept., Contract DA 36-2395c-87435, November 1961.

[81] A. L. Feldshtein, "Synthesis of stepped directional couplers," *Radiotekhnika i Electronika (U.S.S.R.)*, vol. 6, pp. 234–240, February 1961 (in Russian); also, New York: Pergamon Press, pp. 74–85 (in English).

[82] R. Z. Gerlack, "10-dB X_L cross-guide coupler," *IRE Trans. on Microwave Theory and Techniques*, vol. MTT-9, p. 571, November 1961.

[83] W. J. Getsinger, "Ridge waveguide directional couplers," Stanford Research Inst., Menlo Park, Calif., Contract AF 19(604)3502 AD 260, 246, May 1961.

[84] W. J. Getsinger, "A coupled strip line configuration using printed circuit construction that allows very close coupling," *IRE Trans. on Microwave Theory and Techniques*, vol. MTT-9, pp. 535–544, November 1961.

[85] R. Koike, "Recent advances in the use of coupled transmission lines as directional couplers," *Proc. IEE (London)*, vol. 108B, pp. 120–124, January 1961.

[86] Z. Krzycki, "Cross-guide coupler with a ferrite filled coupling hole," *Rozprawy elektrotech. (Poland)*, vol. 7, pp. 355–364, 1961 (in Polish).

[87] B. Maher, "An *L*-band loop type coupler," *IRE Trans. on Microwave Theory and Techniques*, vol. MTT-9, pp. 362–363, July 1961.

[88] Z. Novotny, "Directional coupler of optimum directivity," *Slaboproudy Obzor (Czechoslavakia)*, vol. 22, pp. 653–659, 1961 (in Czechoslovakian).

[89] W. Nowak, "Two new waveguide directional couplers with particularly frequency independent coupling," *Wiss. Z. Tech. Hochsch. Dresden (Germany)*, vol. 10, pp. 743 and 747, 1961 (in German).

[90] L. L. Oh and C. D. Lunden, "Zig zag-line couplers transfer microwave power," *Electronics*, vol. 34, pp. 58–59, July 14, 1961.

[91] C. Y. Pon, "Hybrid-ring directional coupler for arbitrary power division," *IRE Trans. on Microwave Theory and Techniques*, vol. MTT-9, pp. 529–535, November 1961.

[92] B. Richard, "High performance directional couplers," *Ann. Radioelect. (France)*, vol. 16, pp. 293–301, October 1961 (in French).

[93] W. Shelton, "Compact multihole directional couplers," *Microwave J.*, vol. 4, pp. 89–92, July 1961.

[94] S. Tschiya, "On the action of dipoles in waveguide circuits," *J. Inst. Elect. Commun. Engrg. (Japan)*, vol. 44, pp. 963–973, June 1961 (in Japanese).

[95] C. W. van Es, M. Gevers, and F. C. deRonde, "Waveguide equipment for 2mm microwaves," *Philips Tech. Rev.*, vol. 22, p. 122, January 16, 1961.

[96] A. I. Zykov and I. N. Dudkina, "Coupling of matching elements in diaphragm-type waveguides," *Pribory i Tekh. Eksper. (U.S.S.R.)*, no. 2, p. 191, March–April 1961 (in Russian); also, *Instrum. Exper. Tech. (U.S.A.)*, no. 2, p. 390, December 1961 (in English).

1962

[97] K. Araki, "The characteristic impedance of the pair cable type directional coupler," *Rev. of the Elect. Commun. Lab. (Japan)*, vol. 10, pp. 607–625, November–December 1962.

[98] C. R. Boyd, "On a class of multiple line directional couplers," *IRE Trans. on Microwave Theory and Techniques*, vol. MTT-10, pp. 287–294, July 1962.

[99] S. B. Cohn and R. H. Koontz, "Microwave hybrid coupler study program," Rantee Corp., Calabasas, Calif., 3rd Quart. Rept., Contract DA36-239-SC-87435, February 1962.

[100] A. L. Feldshtein and E. S. Zhavoronka, "The design of loosely coupled Chebyshev directional couplers," *Radiotekhnika (U.S.S.R.)*, vol. 17, pp. 40–50, January 1962 (in Russian); also, New York: Pergamon Press, pp. 39–50 (in English).

[101] W. J. Getsinger, "Coupled rectangular bars between parallel plates," *IRE Trans. on Microwave Theory and Techniques*, vol. MTT-10, pp. 479–491, November 1962.

[102] H. Huang, "Generalized theory of coupled local normal modes in multiwave guides," *Acta Phys. Sinica (China)*, vol. 18, pp. 325–333, July 1962 (in Chinese).

[103] H. J. Hinden and J. J. Taub, "Oversize waveguide directional coupler," *IRE Trans. on Microwave Theory and Techniques*, vol. MTT-10, pp. 394–395, September 1962.

[104] E. A. Marcatili and D. H. Ring, "Broad-band directional couplers," *IRE Trans. on Microwave Theory and Techniques*, vol. MTT-10, pp. 251–257, July 1962.

[105] B. M. Mashkovstev, "A method of analyzing multicircuit directional couplers with rotating field polarization," *Radiotekhnika (U.S.S.R.)*, vol. 17, pp. 3–10, 1962 (in Russian).

[106] L. O. Sweet, "Analysis of a two-section coupler," *IRE Trans. on Microwave Theory and Techniques*, vol. MTT-10, p. 295, July 1962.

[107] J. Willis, "A plasma controlled directional coupler," *IRE Trans. on Microwave Theory and Techniques*, vol. MTT-10, pp. 383–389, September 1962.

[108] L. Young, "Synchronous branch guide directional couplers for low and high power applications," *IRE Trans. on Microwave Theory and Techniques*, vol. MTT-10, pp. 459–475, November 1962.

1963

[109] M. E. Brodwin and V. Ramaswamy, "Continuously variable directional couplers in rectangular waveguide," *IEEE Trans. on Microwave Theory and Techniques*, vol. MTT-11, pp. 137–142, March 1963.

[110] S. B. Cohn, "The re-entrant cross section and wide-band 3 dB hybrid coupler," *IEEE Trans. on Microwave Theory and Techniques*, vol. MTT-11, pp. 254–258, July 1963.

[111] A. F. Harvey, *Microwave Engineering*. New York: Academic Press, 1963, ch. 3., sec. 3.3.

[112] H. Huang, "Theory of coupled waveguides," *Acta Phys. Sinica* (*China*), vol. 18, pp. 27–55, January 1963 (in Chinese).
[113] H. Huang, "Notes on discontinuity problems in coupled wave theory," *Acta Phys. Sinica* (*China*), vol. 18, pp. 56–62, January 1963 (in Chinese).
[114] J. G. Humphreys, "Microwave coaxial line components," *Brit. Commun. and Electron.*, vol. 10, pp. 362–367, May 1963.
[115] R. Levy, "General synthesis of asymmetric multi-element coupled-transmission-line directional couplers," *IEEE Trans. on Microwave Theory and Techniques*, vol. MTT-11, pp. 226–237, July 1963.
[116] R. J. Mohr and J. E. McFarland, "Exact analysis of asymmetric couplers," *Microwaves*, vol. 2, pp. 90–93, March 1963.
[117] L. L. Oh, "Serpentine line directional couplers," *Microwaves*, vol. 2, p. 32, December 1963.
[118] J. Reed, "Branch waveguide coupler design charts," *Microwave J.*, vol. 6, p. 103, January 1963.
[119] H. Smith, "Tables for the design of aperture type waveguide couplers," *Microwave J.*, vol. 6, pp. 91–94, June 1963.
[120] L. Sweet, "A method of improving the response of waveguide directional couplers," *IEEE Trans. on Microwave Theory and Techniques*, vol. MTT-11, p. 554, November 1963.
[121] W. A. G. Voss, "Optimized cross slot directional coupler," *Microwave J.*, vol. 6, pp. 83–87, May 1963.
[122] L. Young, "The analytical equivalence of TEM-mode directional couplers and transmission-line stepped impedance filters," *Proc. IEE* (*London*), vol. 110, pp. 275–281, February 1963.

1964

[123] E. G. Cristal, "Coupled circular cylindrical rods between parallel ground planes," *IEEE Trans. on Microwave Theory and Techniques*, vol. MTT-12, pp. 428–439, July 1964.
[124] R. G. Fellers and J. Taylor, "Internal reflections in dielectric prisms," *IEEE Trans. on Microwave Theory and Techniques*, vol. MTT-12, pp. 584–587, November 1964.
[125] V. D. Kuznetsov and V. K. Paramonov, "Stepped directional couplers," *Telecom. and Radio Engrg.* (*U.S.S.R.*), vol. 19, January 1964 (in Russian); also, New York: Pergamon Press, pp. 100–113 (in English).
[126] R. Levy, "Tables for asymmetric multi-element coupled tranmission-line directional couplers," *IEEE Trans. on Microwave Theory and Techniques*, vol. MTT-12, pp. 275–279, May 1964.
[127] W. R. Lind, "A TE mode selective coaxial directional coupler," M.S. thesis, Moore School of Elect. Engrg., University of Pennsylvania, 1964.
[128] G. L. Matthaei, L. Young, and E. M. T. Jones, *Microwave Filters, Impedance Matching Networks and Coupling Structures.* New York: McGraw-Hill, 1964.
[129] L. Young, "Waveguide 0-db and 3-db directional couplers as harmonic pads," *Microwave J.*, vol. 7, p. 79, March 1964.

1965

[130] E. G. Cristal and L. Young, "Theory and tables of optimum symmetrical TEM-mode coupled-transmission-line directional couplers," *IEEE Trans. on Microwave Theory and Techniques*, vol. MTT-13, pp. 544–558, September 1965.
[131] H. Berger, "Nonreciprocal directional couplers," *IEEE Trans. on Microwave Theory and Techniques*, vol. MTT-13, p. 474, July 1965.
[132] J. Daglian, "A contradirectional waveguide coupler with high directivity and tight coupling," M.S. thesis, Moore School of Elect. Engrg., University of Pennsylvania, Philadelphia, 1965.
[133] L. C. Gunderson and A. Guida, "Stripline coupler design," *Microwave J.*, vol. 8, pp. 97–101, June 1965.
[134] H. Jones and R. Norris, "Plated dielectric waveguide components," *Microwaves*, vol. 4, pp. 14–18, July 1965.
[135] R. J. Kalagher, "A TEM mode-selective coaxial directional coupler," M.S. thesis, Moore School of Elect. Engrg., University of Pennsylvania, Philadelphia, 1965.
[136] L. Lavendol and J. J. Taub, "Re-entrant directional coupler using strip transmission line," *IEEE Trans. on Microwave Theory and Techniques*, vol. MTT-13, pp. 700–701, September 1965.
[137] R. Levy, "Transmission line directional coupler for very broadband operation," *Proc. IEE* (*London*), vol. 112, pp. 469–476, April 1965.
[138] E. Martin, "Calibrating directional couplers en masse," *Microwaves*, vol. 4, pp. 40–43, June 1965.
[139] J Shelton, J. Wolfe, and R. Van Wagoner, "Tandem couplers and phase shifter for multi-octave bandwidth," *Microwaves*, vol. 4, pp. 14–19, April 1965.
[140] P. P. Toulios and A. C. Todd, "Synthesis of symmetrical TEM-mode directional couplers," *IEEE Trans. on Microwave Theory and Techniques*, vol. MTT-13, pp. 536–544, September 1965.

104

As can be seen, the directional property of parallel coupled lines has been recognized in print for many years. See the references in papers by Olivers (p. 107) and Monteath (p. 114), and also a contemporary paper by Firestone.[18]

The paper by Jones and Bolljahn (p. 124) detailed the theory of even and odd modes and gave ten canonical circuit forms for parallel coupled lines, subsequently extended in Japan: see Uchida[15] and Young.[12]

GENERAL THEORY (AND SOME HISTORY) V

The paper by Young (p. 131) was the first publication in English showing that multi-section couplers can be designed from stepped-impedance filter prototypes. Perhaps a little bit of history might be enlightening. I discovered this concept independently in early 1962, but subsequently saw a Russian paper by Feldshtein (Ref. 9 in my paper) with a similar approach. Henry Riblet visited me about mid-1962, after he had also come to the same conclusion. In September 1962 I met Levy in London who was also working independently on the same problem, which resulted in two papers (p. 182), (p. 194), reprinted in Section VI. It was not till this book of reprints was assembled that we discovered that Kuroda (p. 181), whose short paper is also reproduced in Section VI, had designed a two-section asymmetrical coupler by 1960. Such a coupler (of two sections) has also been built by Stuart D. Casper and James E. McFarland[19] of the Narda Microwave Corporation by 1960. (Their patent disclosure also mentioned asymmetrical couplers of more than two sections.) Commercial products have often advanced the state of the art, and much ingenuity has been shown by manufacturers (like Sage Laboratories under T.S. Saad, who made a forward coupler out of a backward coupler by a clever twist[20]). There are probably numerous other examples not known to us.

The paper by Sharpe (p. 148) is of general interest and gives an example on a tapered-line magic T, which will be the subject of Section VIII.

Directional Electromagnetic Couplers*

BERNARD M. OLIVER†, FELLOW, IRE

Summary—**The natural coupling between parallel transmission lines is inherently directional. Very simple and cheap directional couplers can be made which utilize this effect. By introducing appropriate variation of coupling with distance a wide variety of transmission characteristics may be realized, including high-pass (ideally, infinite bandwidth) characteristics. The coupled lines may be of equal or different impedance. Thus, a simple configuration of four wires (or two wires and ground) can serve simultaneously as a directional coupler, filter, and transformer.**

Introduction

The propagation of waves on systems of parallel conductors was investigated many years ago[1] in connection with the problem of crosstalk between open wire lines and between cable pairs. These early investigations were directed toward eliminating the natural coupling rather than using it. Later investigations[2–4] using matrix algebra provided elegant mathematical treatments of the general multiconductor transmission problem. The generality of these treatments obscures the simplicity of the special cases we are interested in here. When directional couplers became popular a host of papers[5–7] appeared on the subject. With few exceptions[5] the emphasis in these papers was on waveguide devices. The purpose of this paper is to provide a reasonably simple and not unnecessarily general treatment of the directional coupling action of ordinary transmission lines.

The following assumptions will be made:

1. The loss is negligible; i.e. $\alpha l \ll 1$ for any l considered.
2. The conductor sizes and spacings are infinitesimal compared with the wavelength.
3. All currents are confined to the conductor surfaces.
4. The medium is homogeneous and isotropic (except where controlled anisotropy is introduced to give co-directional coupling).
5. The lines are balanced and the coupling is symmetrical.

Assumptions (1) to (4) are readily met in the 1 to 1,000 mc range; and assure that only TEM waves need be considered. All such waves propagate at the velocity of light in the medium.

Assumption (5) means that only balanced modes will exist; and limits the configurations to those shown in Fig. 1 (opposite page). The coupling between two unbalanced (single-wire-to-ground) circuits is a degenerate case of Fig. 1(a) so all results save those involving cross-coupling apply to this case as well.

Normal Modes and Line Constants

The propagation of waves on two parallel lines such as shown in Fig. 1 is described by the set of differential equations:

$$\left.\begin{aligned}
\frac{\partial e_1}{\partial z} + L_{11}\frac{\partial i_1}{\partial t} + L_m\frac{\partial i_2}{\partial t} &= 0\\
\frac{\partial i_1}{\partial z} + C_{11}\frac{\partial e_1}{\partial t} - C_m\frac{\partial e_2}{\partial t} &= 0\\
\frac{\partial e_2}{\partial z} + L_{22}\frac{\partial i_2}{\partial t} + L_m\frac{\partial i_1}{\partial t} &= 0\\
\frac{\partial i_2}{\partial z} + C_{22}\frac{\partial e_2}{\partial t} - C_m\frac{\partial e_1}{\partial t} &= 0
\end{aligned}\right\} \tag{1}$$

where:

L_{jj}, C_{jj} $(j=1, 2)$ are the self inductance and capacitance per unit length of line j, in the presence of line k $(k=2, 1)$ taking e_k and i_k to be zero,
L_m is the mutual inductance per unit length,
C_m is the mutual capacity per unit length.

As in any system involving coupled linear circuits, we have a set of simultaneous linear differential equations which admits a particularly simple set of solutions: the *normal modes* of the system. In this case we have four normal modes (two in each direction) characterized by fixed ratios of e_2/e_1 and i_2/i_1 which do not change as the wave propagates.[8] The modes are characterized by:

$$\frac{e_2}{e_1} = \pm\sqrt{\frac{L_{22}}{L_{11}}} = \pm\sqrt{\frac{C_{11}}{C_{22}}} \tag{2}$$

$$e_1 i_1 = e_2 i_2. \tag{3}$$

The choice of the plus sign corresponds to "lateral" excitation in Fig. 1(a), the minus sign to "diagonal" excitation. In either normal mode equal power is associated with each line.

* Decimal classification: R117.1. Original manuscript received by the IRE, May 7, 1952; revised manuscript received, August 12, 1954.
† Formerly Bell Telephone Labs., Inc., Murray Hill, N. J.; now Hewlett-Packard Company, Palo Alto, Calif.

[1] J. R. Carson and R. S. Hoyt, "Propagation of periodic waves over a system of parallel wires," *Bell Sys. Tech. Jour.*, vol. IV, no. 3, pp. 495–545; July, 1927.

[2] L. A. Pipes, "Matrix theory of multiconductor transmission lines," *Phil. Mag.*, vol. 24, pp. 97–100; 1937.

[3] S. Koizumi, "Mehrpolleitungstheorie," *Arch. für Electrotech.*, vol. 33, pp. 171–188, 609–622; 1939.

[4] S. O. Rice, "Steady state solution of transmission line equations," *Bell Sys. Tech. Jour.*, vol. XX, no. 2, pp. 131–178; April, 1941.

[5] A. A. Pistolkors and M. S. Neumann, "An instrument for direct measurement of the travelling wave coefficient in feeders," *Electrosvyaz*, vol. IX, no. 4; April, 1941. (RTP trans. no. 1525.)

[6] F. Bolinder, "Approximate theory of the directional coupler," Proc. I.R.E., vol. 39, p. 291; March, 1951.

[7] For a listing of about ninety directional coupler references see R. F. Schwartz, "Bibliography on directional couplers," *Trans. I R.E.*, vol. MTT-2, no. 2, pp. 58-63; July, 1954.

[8] That is to say there is no "crosstalk" between normal modes. As Carson and Hoyt, *loc. cit.*, point out, if each telephone channel were assigned a normal mode, transposition of telephone wires would be unnecessary. Transposition merely confines each normal mode to a particular pair.

Upon making these substitutions and letting

$$k_C = \frac{C_M}{\sqrt{C_{11}C_{22}}}, \qquad k_L = \frac{L_M}{\sqrt{L_{11}L_{22}}} \tag{4}$$

(1) reduces to the familiar transmission line equations:

$$\begin{aligned} \frac{\partial e_j}{\partial z} + L_{jj}(1 \pm k_L)\frac{\partial i_j}{\partial t} &= 0 \\ \frac{\partial i_j}{\partial z} + C_{jj}(1 \mp k_C)\frac{\partial e_j}{\partial t} &= 0 \end{aligned} \qquad j = 1, 2.$$

The velocity of propagation

$$v = \frac{1}{\sqrt{L_{jj}C_{jj}(1 \pm k_L)(1 \mp k_C)}}$$

must be the same for all modes, i.e. whether the upper signs or the lower signs are chosen. This can only be true if $k_L = k_C = k$. Hence for natural isotropic coupling

$$\frac{L_M}{C_M} = \sqrt{\frac{L_{11}L_{22}}{C_{11}C_{22}}} \tag{5}$$

and

$$v = \frac{1}{\sqrt{L_{jj}C_{jj}(1 - k^2)}}.$$

Further, since v must be independent of k, $L_{jj}C_{jj}$ must vary as $(1-k^2)^{-1}$. Both L and C will change with k, but the essential variation will be due to C alone, the remaining variations being mutually reciprocal and negligible except when the conductors are in close proximity

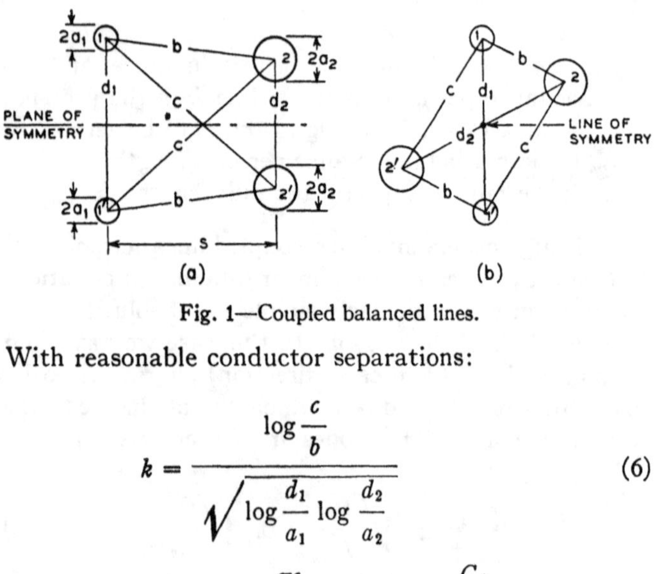

Fig. 1—Coupled balanced lines.

With reasonable conductor separations:

$$k = \frac{\log \dfrac{c}{b}}{\sqrt{\log \dfrac{d_1}{a_1} \log \dfrac{d_2}{a_2}}} \tag{6}$$

$$C_{jj} = \frac{\pi\epsilon}{(1 - k^2)\log \dfrac{d_j}{a_j}} = \frac{C_j}{(1 - k^2)} \tag{7}$$

$$L_{jj} = \frac{\mu}{\pi}\log\frac{d_j}{a_j} = L_j. \tag{8}$$

C_j and L_j $(j=1, 2)$ are the self-capacity and inductance of either line in the absence of the other.

When the lines are identical (i.e. if $a_1=a_2$, $d_1=d_2$) (6) may be written:

$$k = \frac{\log \sqrt{1 + \left(\dfrac{d}{s}\right)^2}}{\log \dfrac{d}{a}}. \tag{6a}$$

For small coupling $[(d/s) \ll 1]$ this simplifies to

$$k \cong \frac{d^2}{2s^2} \frac{1}{\log \dfrac{d}{a}} \tag{6b}$$

which, for lines in air becomes

$$k \cong \frac{60}{Z_0}\left(\frac{d}{s}\right)^2.$$

Natural Directional Coupling

Assume a wave is traveling from left to right in line 1 and consider the induced voltages and currents produced in an elemental length dz of line 2 terminated in impedances Z_a and Z_b as shown in Fig. 2. The rising voltage

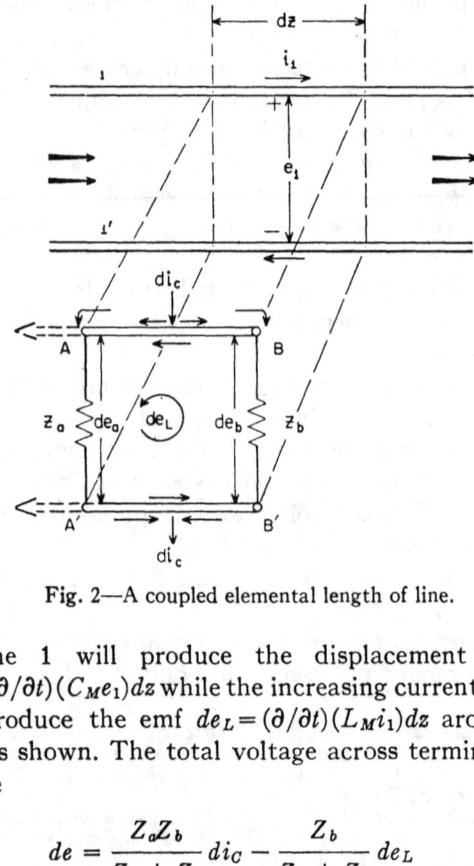

Fig. 2—A coupled elemental length of line.

on line 1 will produce the displacement current $di_C = (\partial/\partial t)(C_M e_1)dz$ while the increasing current in line 1 will produce the emf $de_L = (\partial/\partial t)(L_M i_1)dz$ around the loop as shown. The total voltage across terminals BB' will be

$$\begin{aligned} de &= \frac{Z_a Z_b}{Z_a + Z_b} di_C - \frac{Z_b}{Z_a + Z_b} de_L \\ &= \frac{Z_b}{Z_a + Z_b}\frac{\partial}{\partial t}(Z_a C_M e_1 - L_M i_1)dz. \end{aligned}$$

This voltage will be zero if

$$Z_a = \frac{L_M}{C_M}\frac{i_1}{e_1}.$$

108

But

$$\frac{i_1}{e_1} = \sqrt{\frac{C_{11}}{L_{11}}},$$

while from (5)

$$\frac{L_M}{C_M} = \sqrt{\frac{L_{11}L_{22}}{C_{11}C_{22}}},$$

so that we require

$$Z_a = \sqrt{\frac{L_{22}}{C_{22}}}.$$

Since this is the characteristic impedance of line 1 in the presence of line 2, the voltage and current at BB' will still be zero if the conductors AB and $A'B'$ are extended an arbitrary distance to the left and then terminated in Z_a. This follows since the induction in each added elemental section produces no voltages or currents at its right-hand terminals. The entire induced voltage and current in each elemental section appear at the left and are of the proper relative polarity to launch a wave traveling to the left. Any voltage or current appearing at the extreme right is a result of reflection. We may conclude, therefore, that the natural coupling between parallel lines is inherently directional. In particular it is *contra-directional:* the induced wave travels in the opposite direction to the inducing wave.

Obviously if either the electric or magnetic coupling were eliminated the coupling would be nondirectional, while if either were reversed the coupling would be *co-directional*. This latter case will be considered later.

If $Z_a = Z_{22}$, we find, since all the voltage and current appear at $A-A'$:

$$\left.\begin{aligned} de_A &= de_L = \frac{\partial}{\partial t}(L_M i_1)dz = ik\sqrt{LC}\,\frac{\partial e_1}{\partial t}dz \\ di_A &= di_C = \frac{\partial}{\partial t}(C_M e_1)dz = ik\sqrt{LC}\,\frac{\partial i_1}{\partial t}dz \end{aligned}\right\}. \quad (9)$$

Contra-Directional Coupler with Constant Coupling

From here on we will assume that the two coupled lines are of equal impedance. Little will be lost by this since all results may be converted to the unequal case merely by multiplying by the appropriate impedance ratio. Let:

L_0, C_0, Z_0 = Inductance, capacitance, impedance of either line in the absence of the other ($k=0$).

L, C, Z = Corresponding quantities in the presence of other line.

L_m, C_m = Mutual inductance and capacitance per unit length.

Z^+ = Ratio of voltage to current in one line for "plus" or "lateral" mode.

Z^- = Ratio of voltage to current in one line for "minus" or "diagonal" mode.

k = Coupling coefficient.

We then have the following relations

$$\left.\begin{aligned} &L \cong L_0, \quad C \cong \frac{C_0}{1-k^2}, \quad Z = Z_0\sqrt{1-k^2} \\ &Z^+ = Z\sqrt{\frac{1+k}{1-k}} = Z_0(1+k) \\ &Z^- = Z\sqrt{\frac{1-k}{1+k}} = Z_0(1-k) \end{aligned}\right\}. \quad (10)$$

Now consider two lines with constant coupling over a length l, as shown in Fig. 3. If the connecting lines are of impedance $Z=\sqrt{L/C}$ (*not* $Z_0=\sqrt{L_0/C_0}$) and are properly terminated, then from the previous results, we should expect a wave incident on terminals $a-a'$ to produce an output at $b-b'$ and at $c-c'$, but none at $d-d'$. In analyzing the behavior of this coupler it is convenient to represent the input E_0 at $a-a'$ as the sum of an input $E_0/2$ applied to $a-a'$ and $c-c'$ in the $+$mode, and an input $E_0/2$ applied to these same terminals in the $-$ mode. The two normal mode inputs cancel at $c-c'$ but add at $a-a'$ to give the actual input.

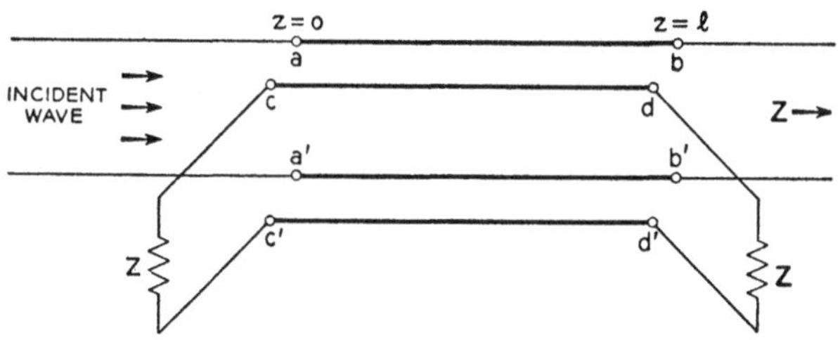

Fig. 3—A directional coupler.

109

If both normal mode inputs are supplied over uncoupled lines of impedance Z then at the input terminals a reflection will be produced. If

$$\zeta = \sqrt{\frac{1+k}{1-k}}$$

then for the $+$ mode the reflection coefficient will be

$$\rho^+ = \frac{\zeta - 1}{\zeta + 1} = \rho$$

while for the $-$ mode

$$\rho^- = \frac{\dfrac{1}{\zeta} - 1}{\dfrac{1}{\zeta} + 1} = \frac{1-\zeta}{1+\zeta} = -\rho.$$

The two reflections cancel at the terminals $a-a'$, add to produce an output at $c-c'$. A complete picture of the

110

TABLE I

Time and Place	Wave	$\overset{+}{\text{Mode}}$	$\overset{-}{\text{Mode}}$	Total: Line 1	Total: Line 2
$t=0$, $z=0$	Inc	$\frac{1}{2}$	$\frac{1}{2}$	1	0
	Refl	$\frac{1}{2}\rho$	$-\frac{1}{2}\rho$	0	**ρ**
	Trans	$\frac{\zeta}{\zeta+1}$	$\frac{1}{\zeta+1}$	1	ρ
$t=\tau$, $z=l$	Inc	$\frac{\zeta}{\zeta+1}$	$\frac{1}{\zeta+1}$	1	ρ
	Refl	$-\rho\frac{\zeta}{\zeta+1}$	$\rho\frac{1}{\zeta+1}$	$-\rho^2$	$-\rho$
	Trans	$\frac{2\zeta}{(\zeta+1)^2}$	$\frac{2\zeta}{(\zeta+1)^2}$	**$\frac{4\zeta}{(\zeta+1)^2}$**	0
$t=2\tau$, $z=0$	Inc	$-\rho\frac{\zeta}{\zeta+1}$	$\rho\frac{1}{\zeta+1}$	$-\rho^2$	$-\rho$
	Refl	$\rho^2\frac{\zeta}{\zeta+1}$	$\rho^2\frac{1}{\zeta+1}$	ρ^2	ρ^3
	Trans	$-\rho\frac{2\zeta}{(\zeta+1)^2}$	$\rho\frac{2\zeta}{(\zeta+1)^2}$	0	**$-\rho\frac{4\zeta}{(\zeta+1)^2}$**
$t=3\tau$, $z=l$	Inc	ρ^2 times row 4. Hence geometric series of outputs produced from here on.			

Fig. 4—Sequence of reflections and transmissions with unit impulse input. Quantities in heavy boxes are first direct output and first two coupled outputs.

Note: $\zeta=\sqrt{\frac{1+k}{1-k}}$, $\rho=\frac{\zeta-1}{\zeta+1}$.

action may be had by taking the component normal mode inputs to be impulses of area $\frac{1}{2}$. An impulse thus appears at $c-c'$ the moment the actual input enters the coupled region. The transmitted impulses propagate down the coupler to produce a combined output at $b-b'$ but none (total) at $d-d'$. The reflections from this end propagate back to $z=0$ to produce an output at $c-c'$ of reversed polarity, together with further reflections to the right. At this point the conditions duplicate those just after the impulse first entered the coupler, except that all amplitudes are reduced by the factor

$$\left(\frac{\zeta-1}{\zeta+1}\right)^2.$$

As a result there now follows a geometric series of impulse outputs at $b-b'$ and $c-c'$, each impulse smaller than the last by the factor

$$\left(\frac{\zeta-1}{\zeta+1}\right)^2$$

and delayed an additional time $2\tau=2l/v$.

The initial series of reflected and transmitted impulse amplitudes for both normal modes and for their sums are given in Table I (Fig. 4) left, while the actual outputs are shown in Fig. 5 for various values of k.

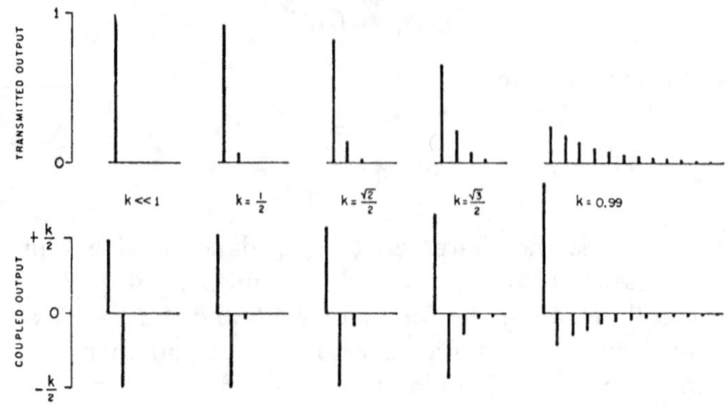

Fig. 5—Impulse response of contra-directional coupler with constant coupling.

Since the spectrum of a unit impulse occurring at a time τ is $e^{-i\omega t}$, the spectrum of the output for a unit impulse input (*i.e. the frequency response*) may readily be found from the preceding. Thus, for the direct, or transmitted, output (at $b-b'$) we have

$$G=\frac{4\zeta}{(\zeta+1)^2}e^{-i\omega\tau}\left[1+\rho^2e^{-i2\omega\tau}+\rho^4e^{-i4\omega\tau}+\cdots\right]$$
$$=\frac{4\zeta}{(\zeta+1)^2}\frac{e^{-i\omega\tau}}{1-\rho^2e^{-i2\omega\tau}}$$
$$=\frac{\sqrt{1-k^2}}{\sqrt{1-k^2}\cos\beta l+i\sin\beta l}. \tag{11}$$

Likewise the series of impulses at $c-c'$ gives for the coupled output

$$F=\rho-\rho\frac{4\zeta}{(\zeta+1)^2}e^{-i2\omega\tau}\left[1+\rho^2e^{-i2\omega\tau}+\rho^4e^{-24\omega\tau}+\cdots\right]$$
$$=ik\frac{\sin\beta l}{\sqrt{1-k^2}\cos\beta l+i\sin\beta l}. \tag{12}$$

From (11) and (12) we note that

$$\frac{F}{G}=i\frac{k}{\sqrt{1-k^2}}\sin\beta l,$$

so that for any k the two outputs are in quadrature and their ratio varies sinusoidally with ω or l. This quadrature property is characteristic of all directional couplers. We further note that $|G|^2+|F|^2=1$ as must be true since the coupler is lossless.

Fig. 6 (page 1690) shows the magnitude and phase of F as a function of βl for various values of k. It will be seen that for small k the transmission varies sinusoidally with frequency (or length) and the phase linearly. With

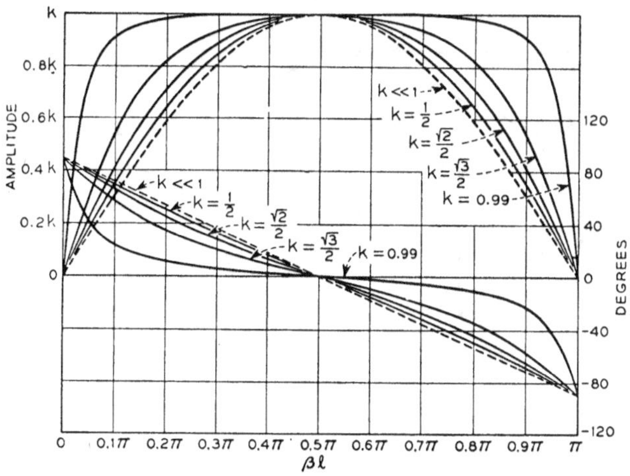

Fig. 6—Contra-directional coupler with constant coupling, k. With tighter coupling the transmission is flatter and the midband delay is less.

$k \ll 1$ the coupler covers a 3 to 1 band between 3 db points. With larger values of k the transmission is even broader; for example $k=\sqrt{2}/2$ (which produces an equal power split at $\beta l=\pi/2$) gives a characteristic covering a 4 to 1 band. *The maximum transmission* to the coupled line always occurs when $l=(2n-1)(\lambda/4)$, and *has the magnitude* k. With a harmonic input of amplitude E_0 applied to terminals $a-a'$, the voltages as a function of distance along the coupler are

$$e_1 = E_0 \frac{\sqrt{1-k^2}\cos\beta(l-z) + i\sin\beta(l-z)}{\sqrt{1-k^2}\cos\beta l + i\sin\beta l}$$

for the voltage on line 1, and

$$e_2 = E_0 ik \frac{\sin\beta(l-z)}{\sqrt{1-k^2}\cos\beta l + i\sin\beta l}$$

for the voltage in line 2 (the branch line). Fig. 7 shows the magnitudes of e_1 and e_2 for $k=1/\sqrt{10}$.

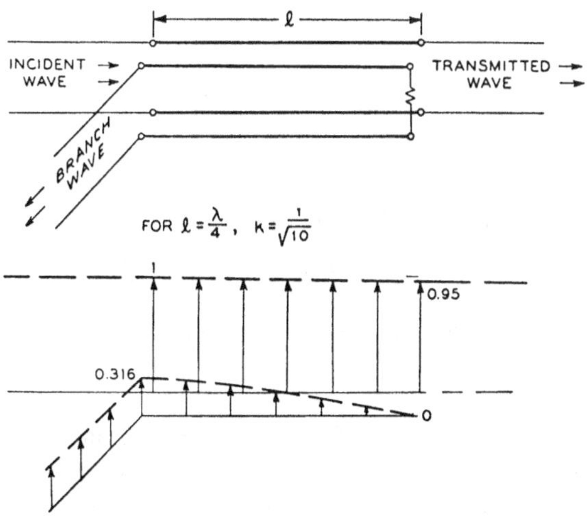

Fig. 7—Illustrating the voltage magnitudes in a contra-directional coupler.

Variable Coupling—Filter Properties

If the coupling coefficient, k, is varied with distance along the coupler, a wide variety of transmission characteristics may be realized. In order to avoid reflections which would destroy the directivity, the impedance, Z, of the lines must be kept constant as $k(z)$ changes. This requires that Z_0 vary as $[1-k^2(z)]^{-1/2}$. However, if the maximum coupling is such that $k^2(z) \ll 1$ and if extremely high directivities are not required, this complication is unnecessary.

If $|F(\omega)|^2 \ll 1$ for all frequencies, the reaction of the induced waves on the inducing waves may be neglected and the transmission of the coupler is readily calculated. By integrating (9) after including the delay from the input terminals to z and back to the output terminals, we find:

$$F = i\beta \int k(z) e^{-i2\beta z} dz.$$

Upon substituting vt/z for z and ω/v for β, this becomes

$$F(\omega) = i\omega \int \frac{1}{2} k\left(\frac{vt}{2}\right) e^{-i\omega t} dt. \tag{13}$$

If both sides of (13) are divided by $i\omega$, the left-hand side will be recognized as the spectrum of the response to a step function, and the right-hand side as the Fourier transform (spectrum) of $\frac{1}{2}k(vt/2)$. Thus, *the step function response is one half the coupling function $k(z)$ with $vt/2$ substituted for z*, i.e.

$$s(t) = \frac{1}{2} k\left(\frac{vt}{2}\right). \tag{14}$$

Further, since the impulse response, $f(t)$, is the time derivative of the step response we have

$$f(t) = \frac{v}{4} k'\left(\frac{vt}{2}\right). \tag{15}$$

Suppose now we wish to realize a coupler having a specified transmission $F(\omega)$ between input and coupled output. The corresponding impulse and step responses are

$$f(t) = \frac{1}{2\pi} \int F(\omega) e^{i\omega t} d\omega.$$

$$s(t) = \frac{1}{2\pi} \int \frac{F(\omega)}{i\omega} e^{i\omega t} d\omega.$$

From (14) and (15) we then find

$$k'(z) = \frac{4}{v} f\left(\frac{2z}{v}\right). \tag{16a}$$

$$k(z) = 2s\left(\frac{2z}{v}\right). \tag{16b}$$

If $F(0) \neq 0$ then $s(\infty) \neq 0$ and from (16b) a coupler of infinite length is required. Aside from this, any $F(\omega)$ which can be realized as an ordinary filter characteristic can be realized in a directional coupler. If $s(t)$ reverses sign the same must be true of $k(z)$. This requires the conductors of either line to be transposed.

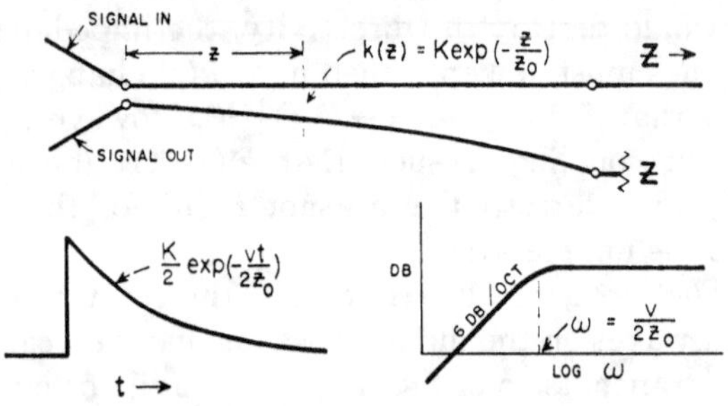

Fig. 8—Exponential variation of coupling with distance. Reversing coupler end-for-end reverses transient response and phase characteristic; leaves gain unchanged.

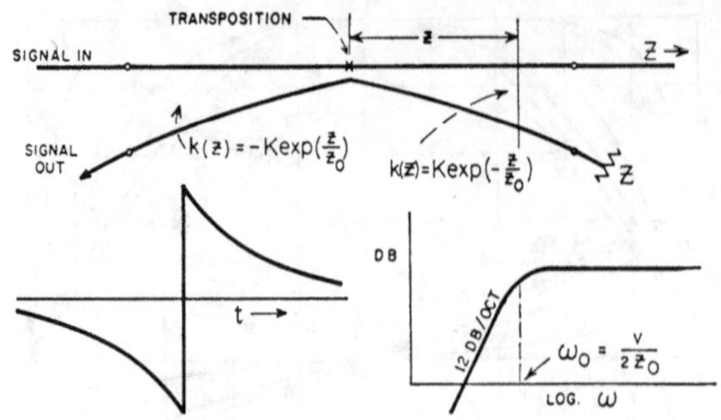

Fig. 9—Symmetrical high-pass coupler.

Thus filter characteristics whose step-function responses reverse sign can be realized only in balanced-line structures. Sharp cut-off filters cannot be realized in two dimensional coaxial structures where such transposition is impossible. Figs. 8 and 9 show two high-pass couplers together with their corresponding step-function responses and steady-state transmissions. Actually, high-frequency cutoffs will be present because the coupling cannot in practice be initiated or reversed infinitely sharply. However, with small lines closely spaced, 100 to 1 bandwidths should be feasible.

Besides changes in Z (which can be avoided) as k changes there is another source of backward transmission (i.e. output at terminals $d-d'$) in a contra-directional coupler. As the conductor separation is changing so that the lines are at an angle θ with respect to each other, the electric coupling in the length dz is the same as for $\theta=0$, but the magnetic coupling is decreased by the factor $\cos\theta$. The two no longer cancel in the backward direction, and a small backward transmission proportional to $\frac{1}{2}(1-\cos\theta)\approx\theta^2/4$ will result. This is most apt to happen at the ends, where the connecting lines diverge; thus, whenever possible, the coupling should gradually be reduced to low values at the ends.

Co-Directional Couplers

By adding excess capacitance either as capacitors along the diagonals c in Fig. 1(a), or as dielectric slabs along the sides c in Fig. 1(b) the capacitive coupling may be reversed, to give $k_C=-k_L$. The required capacitance per unit length, ΔC, for each diagonal (or side) is

$$\Delta C = \frac{4k}{1-k}C = \frac{4k}{1-k}\,\frac{C_0}{1-k^2}\,.$$

The resulting effective capacitance per unit length for each line is

$$C' = C + \frac{\Delta C}{2}$$

$$C' = C\frac{1+k}{1-k} = \frac{C_0}{(1-k)^2}\,.$$

With this loading the impedance for both normal modes is the same

$$Z = Z^+ = Z^- = \sqrt{\frac{L}{C'}} = (1-k)\sqrt{\frac{L_0}{C_0}},$$

but the velocities of propagation are different. The phase constants are

$$\beta^+ = \beta(1+k),$$
$$\beta^- = \beta(1-k),$$

where

$$\beta = \omega\sqrt{LC'} = \frac{\omega\sqrt{L_0C_0}}{1-k}\,.$$

Thus the $-$ mode (which is unloaded) propagates at the velocity of light while the $+$ mode is slowed by the factor $(1-k)/(1+k)$. An input applied to $a-a'$ can be decomposed as before into two equal normal-mode components which add at $a-a'$, cancel at $c-c'$. As these propagate down the lines the modes will slip in relative phase till the cancellation takes place in line 1 and the addition in line 2. The power simply transfers back and forth from line to line. With an input $E_0e^{i\omega t}$ to line 1 we find

$$\frac{e_1}{E_0} = e^{-i\beta z}\cos k\beta z$$

$$\frac{e_2}{E_0} = -\,ie^{-i\beta z}\sin k\beta l.$$

With a coupler of length l, and constant coupling, the direct transmission (terminals $b-b'$) is therefore

$$G = e^{-i\beta l}\cos k\beta l \tag{17}$$

while the coupled transmission (terminals $d-d'$) is

$$F = -\,ie^{-i\beta l}\sin k\beta l. \tag{18}$$

In co-directional couplers, variation of k merely alters the rate at which energy transfer takes place with distance. The output still varies sinusoidally with frequency, k in (17) and (18) being replaced by the average coupling

$$\bar{k} = \frac{1}{l}\int_0^l k(z)dz.$$

No interesting filter properties can be produced, for the

increments from each elemental length merely add without relative delay, and the Fourier relations do not apply. The chief point of interest about co-directional couplers is that *complete* power transfer takes place when $k\beta l = (2n-1)\pi/2$, no matter how small k itself may be.

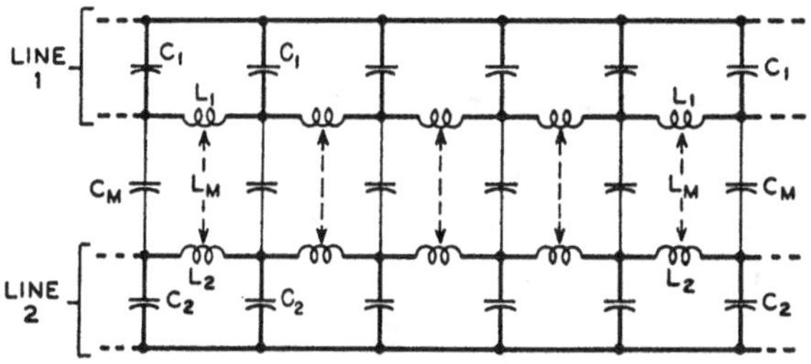

Fig. 10—Coupled lumped lines. With such lines directional couplers can be made to operate at video frequencies.

Coupled Lumped Lines and Helixes

By coupling the coils of two lumped constant delay lines, and by adding capacitance between the nodes as shown in Fig. 10, directional coupling may be achieved at video frequencies. All that is required is that

$$L_M = \pm k\sqrt{L_1 L_2}$$

where

$$k = \frac{C_M}{\sqrt{(C_1 + C_M)(C_2 + C_M)}}\,.$$

In lumped lines the magnetic coupling can be reversed by reversing the winding sense of the coils on either line, so that here either co-directional or contra-directional coupling may be realized. The same situation exists in the case of coupled helixes. By winding one helix over the other in the opposite sense, co-directional coupling with complete power transfer may be achieved. The velocities of propagation differ because the magnetic fields add for the $-$ mode to give greater inductance than for the $+$ mode for which the fields cancel. (Or one can say that while the natural coupling is contra-directional in the fine, i.e. wire-to-wire, the opposite winding sense makes both induced and inducing waves travel in the same axial direction.)

An Experimental VHF Coupler

Fig. 11 shows a double contra-directional coupler designed for reflectometer measurements in the vhf band. The conductors are No. 12 wire, spaced $\frac{1}{2}$ inch on centers, and the two lines are separated $\frac{1}{2}$ inch in the center sections, 1 inch at the ends. Each coupler is 2 feet long. The computed forward loss at four points is given below, in Table II. These figures agree fairly closely with the measured results shown in Fig. 12. The directivity in

TABLE II

Freq.	TV Channel	Loss
54 mc	2	23 db
88 mc	5	20 db
177 mc	7	17.5 db
216 mc	13	19 db

Fig. 11—Experimental vhf directional couplers.

this case was limited by the terminations which were ordinary $\frac{1}{2}$ w 300Ω resistors.

If high directivities are not imperative, and if forward losses on the order of 20 db are permissible, merely running two lengths of twin lead side-by-side for a quarter-wavelength will provide a satisfactory coupler. The directivity is impaired by the dielectric which, by increasing C but not C_m, makes $k_C < k_L$. 113

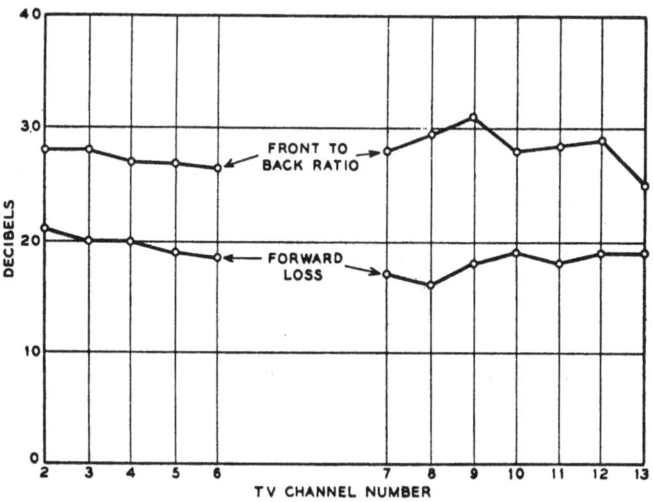

Fig. 12—The performance of a simple balanced 300Ω directional coupler in the vhf TV bands.

Co-directional couplers can also be made with twin lead using the configuration of Fig. 1(b), with the dielectric along sides c. When adjusted to make $k_C = -k_L$, sides b will be slightly shorter than sides c; i.e. the two twin leads will be separated slightly less than their width. The input and output connections at both ends are made across the two diagonals.

Reprinted from Proc. IEE, Vol. 102, Part B, No. 3, May 1955

621.372.512

Paper No. 1833 R
May 1955

COUPLED TRANSMISSION LINES AS SYMMETRICAL DIRECTIONAL COUPLERS

By G. D. MONTEATH, B.Sc., Associate Member.

(*The paper was first received 7th October,* 1954, *and in revised form* 18*th January,* 1955.)

SUMMARY

If two similar unbalanced transmission lines are coupled together by sharing a common outer conductor for part of their length, so as to form a 3-conductor line, they constitute a symmetrical directional coupler. The output ratio of this varies sinusoidally with frequency, being greatest for lengths equal to an odd number of quarter-wavelengths; the characteristic impedance is independent of frequency, being determined by the cross-section of the 3-conductor line.

Two forms of the directional coupler are described. One of these, intended for laboratory power measurement at frequencies up to 1 000 Mc/s, consists of two parallel strips enclosed in a square-section outer conductor. The other, a very simple device for monitoring transmitter power, consists of two lead-covered coaxial cables grafted together.

Among possible uses suggested are all-pass filters and applications to aerial systems consisting of two parts fed in phase quadrature.

LIST OF PRINCIPAL SYMBOLS

A_1, A_2, B_1, B_2 = Four pairs of input terminals of a symmetrical directional coupler.
C = Capacitance in Maxwell bridge, farads.
c = Velocity of light, m/sec.
K = Output ratio of directional coupler.
114 $k = \sqrt{(R_u/R_b)}$.
L = Inductance in Maxwell bridge, henrys.
R_0 = Characteristic impedance of directional coupler, ohms.
R_0' = Characteristic impedance of cable from which the simple directional coupler has been made.
R_b, R_u = Characteristic impedance of the 3-conductor line excited respectively as a balanced and an unbalanced two-conductor line (Fig. 3), ohms.
βl = Angular length of 3-conductor line, radians.
θ = Angle subtended by the slot at the axis of either inner conductor of the simple directional coupler, radians.
ω = Angular frequency, rad/sec.

(1) INTRODUCTION

A directional coupler is a device for sampling waves travelling in one direction along a transmission line or waveguide. (Waveguides will not be considered in the paper.) In its more general form it has two pairs of terminals for connection to the transmission line and two auxiliary outputs. When the auxiliary outputs are correctly terminated in resistive loads, a fraction of the energy travelling in one direction along the transmission line is diverted into one load, and the same fraction of the energy travelling in the opposite direction is diverted into the other load. It follows that if the loads are replaced by detectors of the appropriate impedance, a simultaneous indication is obtained of the powers flowing in both directions. The net power transfer may then be obtained as the difference of the powers; their ratio is equal to the square of the reflection coefficient at the load in which the transmission line is terminated.

For many applications it is essential that a directional coupler should not cause reflection in the transmission line into which it is inserted (this will be termed the main transmission line), and that the two auxiliary outputs should be well matched to transmission lines of the same characteristic impedance. The simplest method of satisfying these requirements simultaneously, particularly when close coupling to the main transmission line is required, is to use a symmetrical directional coupler in which there is no distinction between the terminals intended for connection to the main transmission line and those of the auxiliary outputs.

It is convenient to represent a symmetrical directional coupler by the symbol shown in Fig. 1(*a*), in which A_1, A_2, B_1, B_2 each

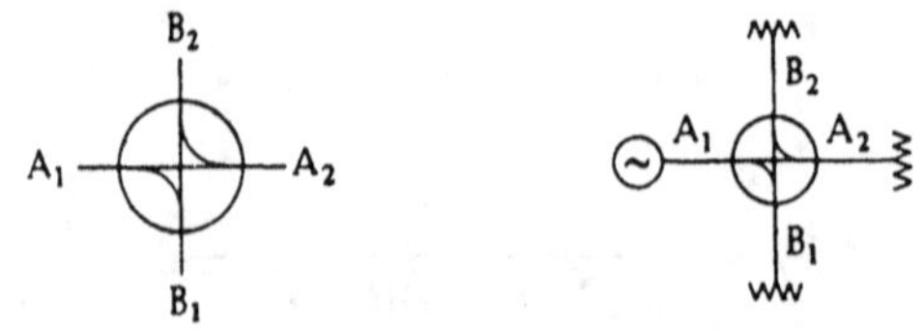

(*a*) (*b*)

Fig. 1.—(*a*) Symbol representing a symmetrical directional coupler.
(*b*) Action of a directional coupler.

represent a pair of terminals suitable for connection to a transmission line. As suggested by the symbol there are two planes of symmetry, so that each pair of terminals is indistinguishable from any other. In Fig. 1(*b*) three of the pairs of terminals are terminated in pure resistances equal to the characteristic impedance of the transmission line for which the directional coupler has been designed (this will be termed the characteristic impedance of the directional coupler), and the fourth, A_1, is connected to a generator. Part (usually the greater part) of the r.f. power passes along the straight line A_1A_2 into the load at A_2, but a proportion of it passes along the curved line A_1B_1 into the auxiliary load at B_1. If the ratio of the power supplied to the load at B_1 to that supplied to the load at A_2 is denoted by K^2, K may be termed the "output ratio" of the directional coupler. If the directional coupler is perfect and if the loads at A_2 and B_1 are perfectly matched to its characteristic impedance, no power is supplied to the load at B_2. Imperfection of the directional coupler in this respect may be expressed in terms of the directivity (the ratio of the output at B_1 to that at B_2) which is usually expressed in decibels.

A symmetrical directional coupler having perfect directivity has the following properties:[1]

(*a*) Referring to Fig. 1(*b*), the input impedance at A_1 is perfectly matched, provided that the loads at A_2 and B_1 are matched. By symmetry it follows that if the source impedance of the generator at A_1 is matched, the source impedances of the outputs at A_2 and B_1 are matched.

(*b*) The outputs at A_2 and B_1 are in phase quadrature (in view of the symmetry it is not necessary to specify the reference planes of the terminals A_2 and B_1 in making this statement).

The principal object of the work described in the paper was to extend the range of power-measuring equipment for frequencies between 100 and 1 000 Mc/s. For this purpose a range of

Written contributions on papers published without being read at meetings are invited for consideration with a view to publication.
Mr. Monteath is with the British Broadcasting Corporation.

directional couplers was required having various output ratios from about $\frac{1}{10}$ (-10dB) downwards. A minimum directivity of about 30dB was required, and like other components of the equipment, the directional couplers were to be matched to a standing-wave ratio of not less than 0·9. It was clearly desirable to adopt a symmetrical design if possible, since the matching would then be a necessary consequence of the directivity, instead of being a separate requirement. The directional coupler produced is described in Section 4. A simpler form, which has been incorporated in transmitters for monitoring power, is described in Section 5.

(2) TYPES OF DIRECTIONAL COUPLER

(2.1) The Maxwell Bridge

Most transmission-line directional couplers behave at sufficiently low frequencies like the Maxwell bridge illustrated in Fig. 2(*a*), which has been rearranged in Fig. 2(*b*) to correspond

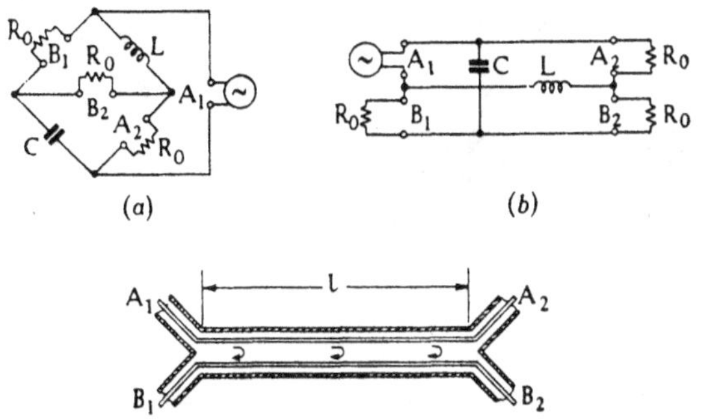

Fig. 2.—(*a*) Maxwell bridge.
(*b*) Maxwell bridge circuit rearranged as a directional coupler.
(*c*) Coupled transmission lines (arrows show direction of coupling).

to the physical configuration of practical directional couplers. The two circuits A_1A_2 and B_1B_2 are coupled by the inductance L and the capacitance C.

Provided that $L/C = R_0^2$ the bridge is balanced at all frequencies, and it therefore behaves as a symmetrical directional coupler of characteristic impedance R_0. The output ratio is given by

$$K = \omega L/R_0 = \omega C R_0 \quad . \quad . \quad . \quad . \quad . \quad (1)$$

The fact that the output ratio is proportional to the frequency may be inconvenient for power measurement, since it accentuates harmonics, whereas usually the fundamental power is required.

The simple circuit of Figs. 2(*a*) and 2(*b*) becomes modified by stray inductances and capacitances at high frequencies, particularly if strong coupling (output ratios greater than about -30dB) is required, since the dimensions become comparable with the wavelength. Inductance in series with C may be compensated for by capacitance in parallel with L, but other stray reactances add to the difficulty of achieving high directivity over a wide band of frequencies.

(2.2) The Reflectometer

One of the best-known types of directional coupler is the reflectometer,[2, 3] which consists of a small loop coupled to a coaxial feeder so that capacitance and mutual inductance between the loop and the inner conductor perform the functions of C and L in Fig. 2(*b*). One of the load resistors [at B_1 or B_2 in Fig. 2(*b*)] is usually built in, so that only one auxiliary output is available. The reflectometer cannot readily be made to give an output ratio greater than 0·05 (-26dB), but if a lower output ratio can be accepted, directivities up to at least 60dB are possible. It is not symmetrical.

(2.3) The Bethe-Hole Directional Coupler

This is a symmetrical directional coupler consisting of two coaxial lines coupled through a common hole in their outer conductors. The diversion of the current path round the hole provides the common inductance L [Fig. 2(*b*)], while the capacitance between the inner conductors through the hole corresponds to C. Ginzton and Goodwin[5] have obtained output ratios up to 0·1, but the diameter of each coaxial line was then about one-third of the wavelength; the dimensions would be inconveniently large except at microwave frequencies.

(2.4) Coupled Transmission Lines

The directional coupler forming the subject of the paper consists of two unbalanced transmission lines, which are coupled together by sharing a common outer conductor [Fig. 2(*c*)] so as to form, in effect, a 3-conductor transmission line. It is convenient to ensure symmetry by making the two inner conductors identical. This arrangement may be regarded as a development of the reflectometer; one of the two inner conductors corresponds to the inner conductor of the main transmission line, and the other to the coupling loop. Alternatively it may be considered as a Bethe-hole directional coupler in which the hole has been extended longitudinally. It will be shown that the 3-conductor line acts as a directional coupler even when its length is comparable with the wavelength, and that it has other advantages.

Firestone[6] has investigated directional couplers in which balanced transmission lines consisting of thin wires are coupled together, and has pointed out their relationship to the 3-conductor line, which is the unbalanced equivalent. A more general treatment of balanced coupled transmission lines has recently been given by Oliver.[7]

(3) THEORETICAL CONSIDERATIONS

Figs. 3(*a*) and 3(*b*) show a symmetrical 3-conductor transmission line excited as a balanced and an unbalanced 2-conductor line, respectively. Let the characteristic impedance be $2R_b$ for

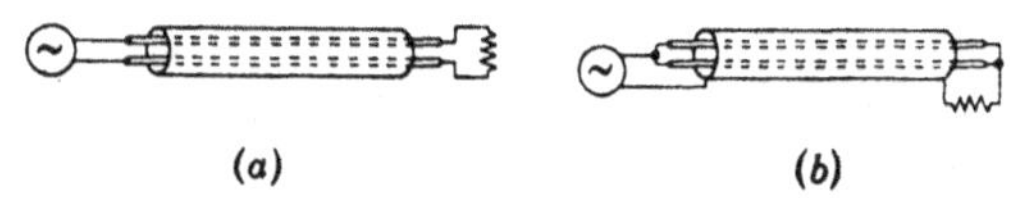

Fig. 3.—Three-conductor line.
(*a*) As a screened balanced transmission line: characteristic impedance, $2R_b$.
(*b*) As an unbalanced transmission line: characteristic impedance, $\frac{1}{2}R_u$.

the balanced line and $\frac{1}{2}R_u$ for the unbalanced line. It will be evident that R_u and R_b differ only by reason of the coupling between the inner conductors. It is shown in Section 11.1 that the 3-conductor line behaves as a symmetrical directional coupler whose characteristic impedance R_0 is given by

$$R_0^2 = R_u R_b \quad . \quad . \quad . \quad . \quad . \quad . \quad (2)$$

provided that the velocity of propagation is the same both for the unbalanced and the balanced modes of excitation shown in Fig. 3. This last condition might not be satisfied if the dielectric were not uniform.

It is convenient, in view of eqn. (2), to express R_u and R_b as

$$R_u = kR_0$$
$$R_b = R_0/k \quad . \quad . \quad . \quad . \quad . \quad . \quad (3)$$

where k is a parameter which must be greater than unity; its departure from unity is a measure of the coupling between the inner conductors. The output ratio K is shown in Section 11.1 to be given by

$$K = \tfrac{1}{2}(k - 1/k) \sin \beta l \quad . \quad . \quad . \quad . \quad . \quad (4)$$

where βl is the angular length. An equivalent analysis of balanced coupled transmission lines has been published independently by Oliver.[7] The parameter ζ in Oliver's paper corresponds to k in this paper.

At low frequencies, such that $\beta l \ll 1$, the mutual inductance and capacitance between the inner conductors may be regarded as lumped rather than distributed, so that the coupled transmission lines behave like the Maxwell bridge shown in Fig. 3. The output ratio may be deduced from eqn. (1), if L and C are taken to be the mutual inductance and capacitance between the inner conductors. The result is identical to that deduced from eqn. (4) if $\sin \beta l$ is replaced by βl. At higher frequencies the output ratio varies sinusoidally with frequency; a qualitative explanation of this effect is outlined below.

Referring to Fig. 2(*c*), the direction of coupling is such that if power flows through the directional coupler from A_1 to A_2, a fraction is diverted to B_1. Each short element of the 3-conductor line may be considered as containing a directional coupler of the Maxwell-bridge type, and the total output at B_1 is the sum of the contributions from different elements. These contributions will take the paths indicated by the arrows in Fig. 2(*c*), and will arrive at B_1 in different phases. If the length is increased from zero the output ratio will increase linearly until the mis-phasing begins to take effect. When the length reaches one half-wavelength, all phases from 0 to 2π will be represented equally, and the output ratio will be zero.

The maximum output ratio, which is obtained for a length equal to one-quarter wavelength, is equal to $\tfrac{1}{2}(k - 1/k)$. It follows that for a high output ratio k must be large; the inner conductors must be close together so as to form a balanced transmission line of low characteristic impedance, while the overall transverse dimensions of the two inner conductors together must be small compared with the outer conductor, so as to form an unbalanced line of high characteristic impedance when connected as in Fig. 3(*b*).

(4) A DIRECTIONAL COUPLER FOR GENERAL LABORATORY USE

This Section is concerned with the prototype of a range of directional couplers for power measurement and general laboratory use at frequencies up to 1 000 Mc/s. The principal application envisaged will be discussed in Section 7.1; for the present it is sufficient to list the following requirements:

(*a*) The directional couplers were to be symmetrical, and to be fitted with four identical r.f. connectors suitable for connection to 71-ohm coaxial cable (Uniradio 21).

(*b*) The directivity was to be at least 30 dB.

(*c*) A range of output ratios up to about −10 dB was required.

In order to simplify the design and construction of a range of directional couplers having different output ratios at the same frequency (or giving the same output ratio at different frequencies), it was decided that the directional couplers were to differ only in the length of the 3-conductor line. The cross-section of this line and the configuration of the junction at each end were to be the same in all cases.

Fig. 4 shows the prototype. The 3-conductor line consists of two parallel strips in a square-section outer shield. At each end the two inner conductors are supported by a polystyrene insulator, and are splayed outwards and tapered down to join the inner conductors of matched r.f. connectors.

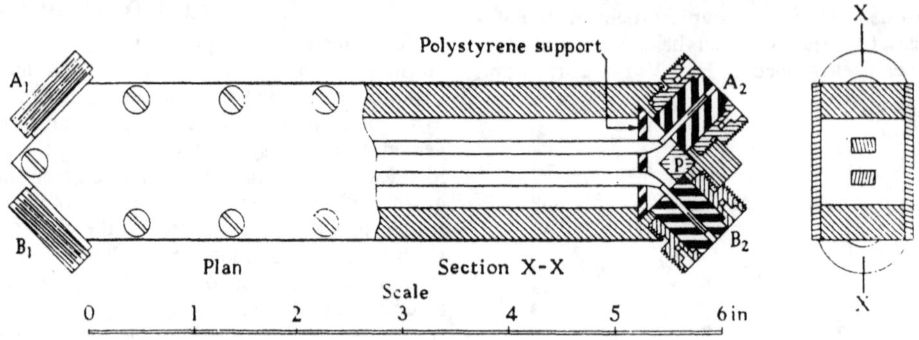

Fig. 4.—Directional coupler for laboratory use.

In the first place it was necessary to choose the cross-section of the 3-conductor line so as to satisfy eqn. (2) and to design the ends so as to ensure that discontinuities and stray coupling would not impair the directivity. The characteristics of the 3-conductor line were determined by means of inductance measurements at 1 Mc/s—a frequency low enough to enable capacitances to be ignored but high enough for skin and proximity effects to be substantially complete. The internal impedance of the conductors, which is negligible at very-high frequencies, could not be neglected completely at 1 Mc/s. A small correction was therefore made for its effect on the measured inductance by observing the measured resistance and using the fact that, when skin effect is substantially complete, the internal resistance and internal reactance are equal. This procedure was applied in the measurement both of self-inductance per unit length and mutual-inductance per unit length.

The 1 Mc/s measurements were made with a medium- and long-wave series impedance bridge* measuring inductance and resistance down to $0{\cdot}001\,\mu$H and 0·01 ohm respectively, and provided with a pair of floating potential terminals. Measurements were made by connecting both inner conductors to the outer at one end, and passing current along one inner conductor from the current terminals of the bridge. Holes drilled through the upper cover-plate enabled a coaxial probe, which was connected by a screened cable to the potential terminals, to be used to sample the potential difference between inner and outer conductors. By measuring this at two points on each inner conductor, the self-inductance and mutual-inductance per unit length were obtained, and R_u and R_b were deduced. Changes were then made in the cross-section until $\sqrt{(R_u R_b)}$ was as near to 71 ohms as could be measured. With the final cross-section the parameter $k = \sqrt{(R_u/R_b)}$ was 1·46. This value indicated a maximum output ratio of 0·39 (−8·2 dB).

After determining the required cross-sectional dimensions of the 3-conductor transmission line in the manner described above, the directional coupler was assembled in the form shown in

* This was developed by C. G. Mayo.

Fig. 4, and measurements of the directivity were made at frequencies in the range 100–900 Mc/s. It was assumed that imperfect directivity could be attributed to discontinuities and stray coupling at the ends, and adjustments were therefore made at the ends to obtain maximum directivity, while preserving symmetry. The directivity was initially determined by observing the apparent reflection coefficient of a 71-ohm resistor fitted into a r.f. connector, but final adjustment and testing could not be carried out in this way because the matching of the load resistors available could not be relied upon to the accuracy required. (Referring to Fig. 1(*b*), if the load at B_2 is replaced by a detector, the observed signal will be the vector sum of components due to reflection at A_2 and B_1 in addition to the component due to imperfect directivity of the directional coupler.) The load resistors were therefore replaced by equal lengths of cable terminated in resistors that were deliberately mismatched, and measurements were carried out over a band of frequencies. Owing to the lengths of cable, the comparatively strong reflections occurring at the resistors varied in phase relative to the apparent reflection introduced by the imperfect directivity. As a result the measured reflection coefficient showed a cyclical variation whose amplitude was a measure of the imperfect directivity.

The end regions in which adjustment is required may be considered as additional short sections of 3-conductor line. In order that these should not impair the directivity they must satisfy the same conditions as the main 3-conductor line. The electrical length must be the same both for balanced and unbalanced modes of propagation, and the characteristic impedances must satisfy eqn. (2). The value of the parameter k applicable to the end sections is unimportant. For example, it could be the same as for the main 3-conductor line, which would in effect be lengthened. Alternatively it would be unity, when the end sections would not contribute to the coupling, but merely act as short lengths of matched coaxial line between the directional coupler and its terminals. Since two independent conditions must be satisfied, two independent adjustments will generally be necessary. For example, these could be equal-value variable capacitors between the inner conductors and the outer, and a variable capacitor between the inners. In fact, the procedure adopted was to design the directional coupler with excessive inductance at each end, and to begin the adjustment by adding the square-section brass blocks marked P in Fig. 4. These were long enough to make contact with both cover-plates, and their cross-section was chosen experimentally for the highest directivity. It happened by chance that the optimum size gave an adequate directivity, and no further change was made.

In the absence of signal generators having accurate attenuators and covering a sufficiently wide band of frequencies, the output ratio was measured with a thermistor milliwattmeter. An oscillator was connected to A_1 through a matched dissipative attenuator, and the powers available at A_2 and B_1 were compared by means of the milliwattmeter, outputs not otherwise connected being terminated in matched load resistors. To reduce the range of powers to be measured, a 10 dB dissipative attenuator, whose attenuation had been measured in a separate experiment, was used to reduce the output from A_2. In Fig. 5 the measured output ratio is plotted against frequency and compared with a sine curve chosen for best fit to the measured points.

Fig. 5 shows that the output ratio satisfies a sinusoidal law within the limits of experimental error. It attains a maximum value at 565 Mc/s, at which frequency one quarter-wavelength is 13·3 cm. This effective length is 0·15 cm longer than the 3-conductor line, measured between the outer surfaces of the polystyrene supports. The maximum output ratio is 0·42 (−7·5 dB), as compared with the value 0·39 deduced from the low-frequency inductance measurements. The discrepancy is attributed to error in measuring the small mutual inductance between the inner conductors.

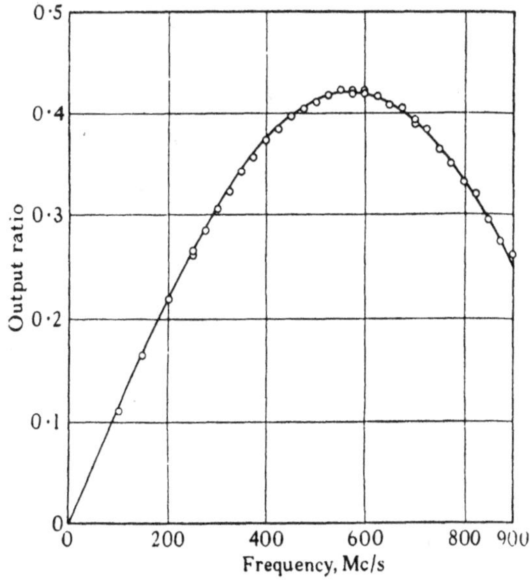

Fig. 5.—Output ratio of directional coupler shown in Fig. 4.

○ Measured values.
—— Sine curve chosen for best fit.

The directivity, measured with the lengths of cable used for the adjustment, exceeded 35 dB at all frequencies in the band 100–900 Mc/s. A lower directivity would probably be observed if different samples of cable were used, owing to the non-uniformity of the characteristic impedance.

(5) A SIMPLE DIRECTIONAL COUPLER FOR MONITORING TRANSMITTER POWER

A different type of directional coupler was required for monitoring transmitter output power in connection with propagation experiments in the v.h.f. and u.h.f. bands. Output ratios of the order of −25 dB were required. A high directivity was not essential, 20 dB being adequate.

Reflectometers could have been designed to give the required performance at the fundamental frequency, but they would have been unduly responsive to harmonics, since the output ratio is proportional to the frequency. This difficulty was overcome by using a 3-conductor transmission line one quarter-wavelength long at the fundamental frequency, which would suppress the even harmonics [$\sin \beta l = 0$ in eqn. (4)], and would couple to the odd harmonics only to the same extent as to the fundamental.

The design of the 3-conductor line was suggested by the fact that if two transmission lines, each having a characteristic impedance R_0' in the absence of the other, are arranged parallel to one another with weak coupling between them, they form a directional coupler whose characteristic impedance is approximately equal to R_0'. Firestone[6] investigated balanced open-wire directional couplers based on this principle, and suggested that their unbalanced equivalents could be devised.

Fig. 6 shows one of the directional couplers. Two lead-covered polythene-filled cables (Uniradio 25) were each milled to form a flat surface, the cut extending into the polythene dielectric, which was exposed in a slot. They were then clamped together with the slots coincidental, and the outer conductors were soldered together. The slots were necessarily tapered at each end where the cables were bent away from the flat surface,

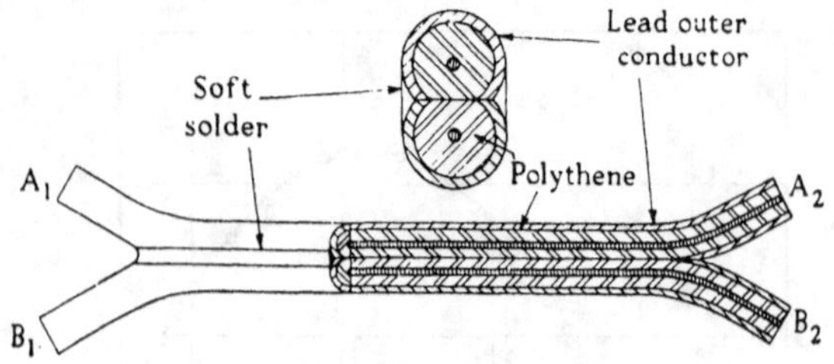

Fig. 6.—Simple directional coupler.

but this was in any case desirable, since the discontinuities caused by rectangular ends would have impaired the directivity. The length of the slot was made slightly greater than one quarter-wavelength, in order to allow for the tapering. One of the cables was arranged to replace an existing cable link in the transmitter; the ends of the other cable were terminated in matched crystal detectors for monitoring the forward and reflected powers.

Section 11.2 contains an approximate theoretical analysis based upon the assumption that the inner conductors are thin compared with their spacing. The result, as applied to 71-ohm polythene-dielectric cables, is illustrated in Fig. 7, which shows the output ratio and the directivity as a function of θ, the angle subtended by the slot at the axis of either inner conductor.

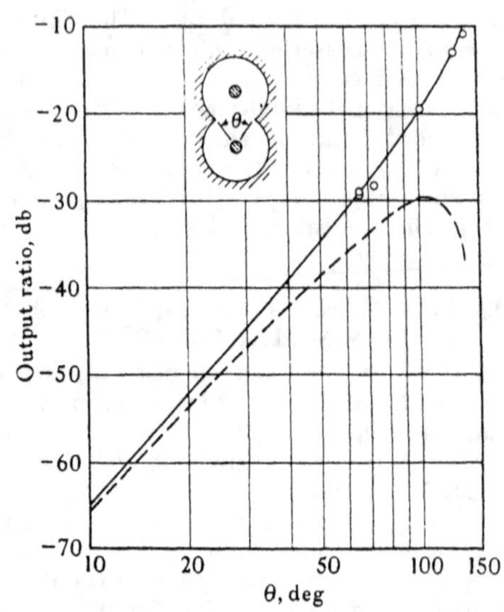

Fig. 7.—Characteristics of directional couplers made from 71-ohm polythene cable.

——— Theoretical maximum output ratio.
○ Measured maximum output ratios.
- - - - Theoretical reciprocal of directivity.

The finite directivity shown in Fig. 7 is a measure of the departure of R_0, which is equal to $\sqrt{(R_u R_b)}$, from R_0', the characteristic impedance of the cable. If θ is increased from zero, R_0 increases from R_0' to a maximum value, the directivity falling from infinity to a minimum. If θ is increased further, R_0 decreases; as it passes through R_0' the directivity again becomes infinite. For values of θ greater than about 100°, at which the minimum directivity occurs, the assumptions upon which the approximate analysis depends are no longer valid, but this qualitative description of the variation of directivity with θ will nevertheless be correct. In any case it would be impracticable to make θ close to 180°, since the inner conductors would be so close together that small mechanical imperfections could have a serious effect.

For small values of θ the output ratio and the reciprocal of the directivity are equal and proportional to θ^2.

A number of directional couplers of the type shown in Fig. 6 have been made for frequencies between 180 and 600 Mc/s. Accurate measurements of the directivity have not been made, but its value exceeds 20 dB in all cases. The maximum output ratios are plotted against θ in Fig. 7; they show good agreement with the theoretical curve.

(6) CASCADE CONNECTION OF SYMMETRICAL DIRECTIONAL COUPLERS

A feature of the directional couplers made from coupled transmission lines is the fact that a high output ratio can be achieved while preserving a high directivity over a wide band of frequencies. Nevertheless practical difficulties may be encountered if output ratios in excess of −10 dB are required, particularly if the frequency and the power to be handled are both high. For example, if the characteristic impedance is 70 ohms and the maximum output ratio is to be unity (0 dB), from eqns. (3) and (4) it is found that R_u and R_b must be 169 and 29 ohms, respectively. These values could be realized by the use of two parallel strips 1 in wide spaced approximately 0·2 in apart in a cylindrical outer conductor approximately 2·5 in in diameter. The narrow gap between the inner conductors would limit the power-handling capacity, but an attempt to overcome this difficulty by increasing the cross-sectional dimensions might impair the directivity at frequencies above about 100 Mc/s owing to end effects. One solution is to obtain a high output ratio by combining together two or more directional couplers of lower output ratio.

Fig. 8(*a*) shows two directional couplers $A_1A_2B_1B_2$ and $A_1'A_2'B_1'B_2'$ of the same characteristic impedance R_0 connected

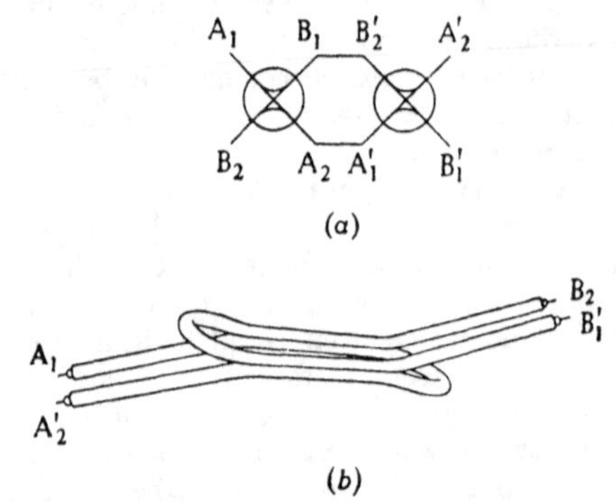

Fig. 8.—Cascade connection of directional couplers.
(*a*) Diagrammatic.
(*b*) A practical arrangement.

together by equal transmission lines A_2A_1' and B_1B_2' to form a single network $A_1A_2'B_1'B_2$. Fig. 8(*b*) shows this method of combination applied to the simple directional couplers described in Section 5. In order to verify that the combination is also a true directional coupler whose characteristic impedance is R_0, suppose that A_2', B_1', and B_2 are each terminated in a resistance R_0 and that a signal is injected at A_1. It is clear that the only possible paths for the signal are $A_1A_2A_1'A_2'$, $A_1A_2A_1'B_1'$, $A_1B_1B_2'B_1'$, and $A_1B_1B_2'A_2'$, and that there can be no output at B_2 and no reflection back towards A_1. Let the input at A_1 be such that the signal reaching A_2' by the path $A_1A_2A_1'A_2'$ is represented in magnitude and phase by unity. Then signals reaching B_1' by the paths $A_1A_2A_1'B_1'$ and $A_1B_1B_2'B_1'$ will respectively be represented by jK' and jK, where K and K' are, respectively, the output ratios of the first and second directional couplers. A signal represented by $-KK'$ will also reach A_2'.

118

It follows that the total outputs at B_1' and A_2' are respectively $j(K - K')$ and $(1 - KK')$. The combination therefore constitutes a directional coupler of output ratio K_0 given by

$$K_0 = (K + K')/(1 - KK')$$

or

$$\text{arc tan } K_0 = \text{arc tan } K + \text{arc tan } K' \quad . \quad . \quad . \quad (5)$$

The lengths of the connecting transmission lines B_1B_2' and A_2A_1' are immaterial, provided that they are equal and matched.

Eqn. (5) may be generalized by induction to give the output ratio of a directional coupler formed by the cascade connection of any number n of directional couplers having output ratios $K_1, K_2 \ldots K_r \ldots K_n$. It is given by

$$\text{arc tan } K_0 = \sum_{r=1}^{n} \text{arc tan } K_r \quad . \quad . \quad . \quad . \quad (6)$$

The band of frequencies in which the output ratio remains within given limits is slightly narrower for a combination of directional couplers of equal length than for a single directional coupler. For example, suppose that the maximum output ratio is unity, and that the bandwidth is determined by the frequencies at which the output ratio falls to 0·9. A single directional coupler would have a bandwidth equal to 57% of the mid-band frequency. Alternatively if three identical directional couplers, each having a maximum output ratio of 0·27 (= tan 15°) were combined, the bandwidth would be reduced to 46% of the mid-band frequency.

A considerable improvement in the constancy of output ratio with frequency may be achieved by combining in cascade directional couplers formed by coupled transmission lines of differing lengths; in this way the bandwidth can be made greater than that of a single directional coupler. The simplest arrangement is to use one or more directional couplers a quarter-wavelength long and one three-quarter-wavelength long at the mid-band frequency; more elaborate combinations are possible.

A method of cascade connection, alternative to that shown in Fig. 8, exists in which B_2 is connected to B_1' and A_2 to A_1'. This method appears to be less useful, since the output ratio depends upon the length of the connecting cables and on the phase delay in the constituent directional couplers. Nevertheless, a pair of coupled transmission lines may be considered as an assembly of elementary Maxwell bridges combined in this way.

(7) APPLICATIONS

(7.1) Power Measurement

Directional couplers are particularly useful for extending the range of r.f. power-measuring devices, such as thermistor milliwattmeters. The most convenient level for power measurement is a few milliwatts, since the thermal time-constant of the thermistor can be made short, and since its physical dimensions can be made small so as to facilitate matching. Lower powers can be measured if steps are taken to compensate for changes in ambient temperature. At frequencies up to 1 000 Mc/s at least, dissipative attenuators can be constructed from rod-and-disc resistors so as to extend the range to powers of the order of 1 watt, but for higher powers the difficulty of matching attenuators increases, since larger resistors are necessary.

A method of using a directional coupler to extend the range of a milliwattmeter to higher powers is indicated in Fig. 9. Although a load capable of absorbing the power to be measured is required, this need not be accurately matched. For many purposes reflection at the load may be ignored, since the error caused by doing so is proportional to the square of the reflection coefficient, but if it must be taken into account, the reflected power may be determined by interchanging the milliwattmeter

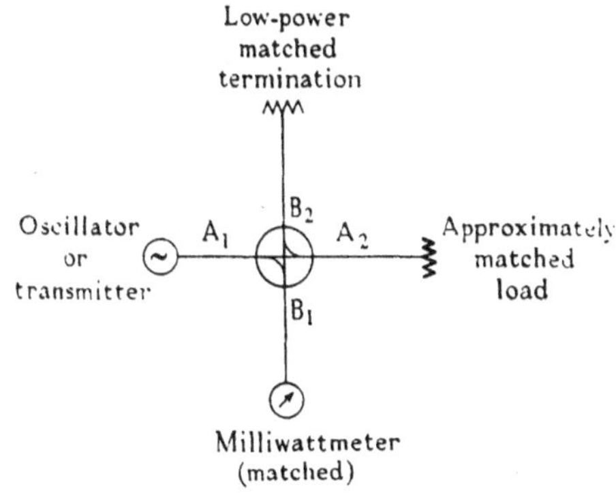

Fig. 9.—Power measurement.

and the low-power matched termination; the net power supplied to the load is then obtained as the difference between the incident and reflected powers. The output ratio of the directional coupler must be determined in a subsidiary experiment, e.g. by using a signal generator having an accurate piston attenuator.

The arrangement of Fig. 9 is particularly convenient when it is required to radiate a known power from an aerial for calibrating field-strength measuring equipment, since the radiated power can be measured at intervals without interrupting the calibration. A similar application to transmitter power monitors has been mentioned in Section 5.

It is sometimes required to generate a known power at too low a level for direct measurement, e.g. to compare with the output of a signal generator. For this purpose the circuit of Fig. 9 may be modified by connecting the milliwattmeter to the directional coupler at A_2 instead of B_1, and taking the low-level output from B_1.

(7.2) All-Pass Networks

An all-pass or constant-resistance network is a lossless four-terminal network whose image impedances are equal resistances independent of the frequency. When this is connected between a source and a load that are matched to the image impedance, the insertion loss is zero at all frequencies, although the phase shift is in general a non-linear function of the frequency. Since all-pass networks are not networks of minimum phase shift, there must always be at least two paths between the input and output.

Various all-pass networks may be made up from directional couplers and lengths of lossless transmission line. Two examples are shown in Figs. 10(*a*) and 10(*b*). The all-pass character of these networks may be verified by tracing out the path of the signal as indicated by the arrows, and noting that if the output is terminated in a matched load no energy is reflected back to the input. The insertion loss is therefore zero if dissipation can be neglected. More complicated networks can be constructed by inserting subsidiary all-pass networks in the closed loop of transmission line.

Fig. 10(*c*) shows the simplest unbalanced all-pass network, which corresponds to the network of Fig. 10(*b*) with the length of the loop of transmission line reduced to zero. The phase shift ϕ may be shown to be given by

$$\tan(\phi/2) = (1/k)\tan\theta \quad . \quad . \quad . \quad . \quad (7)$$

where θ is the angular length of the three-conductor line and k is the parameter defined by eqns. (3). The group delay varies with frequency in a cyclical manner about the mean value; it is a maximum and minimum when θ is an odd multiple and an even multiple of $\pi/2$, respectively. The response to a positive "unit-impulse" signal may be shown to consist of a positive

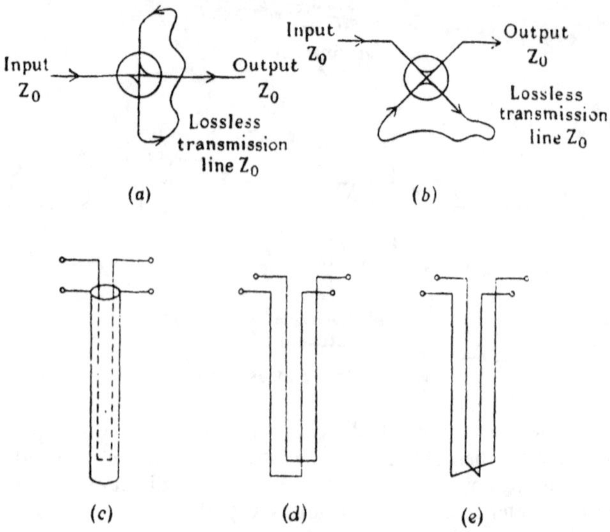

Fig. 10.—All-pass networks.

(a), (b) Networks consisting of a directional coupler and a length of transmission line.
(c) Simplest unbalanced network corresponding to (b).
(d) Balanced equivalent of (c).
(e) Network (d) modified by transposition of cross-connections.

impulse coincident in time with the input but reduced in amplitude, followed by a train of impulses of alternating sign, the first being positive.

The balanced equivalent of the network of Fig. 10(*c*), which is shown in Fig. 10(*d*), requires no comment. It may be modified by transposing the cross-connections, as shown in Fig. 10(*e*), to form a new all-pass network which has no simple unbalanced equivalent. To obtain the phase shift of this network, $1/k$ in eqn. (7) is replaced by k.

The all-pass networks have been discussed mainly as a curiosity, but applications may exist. For example, they could
120 be used for ensuring a quadrature phase relationship between the feeds to two aerials over a wide band of frequencies. Phillips[8] has employed conventional all-pass networks for this purpose at frequencies up to a few megacycles. An alternative method of maintaining a quadrature phase relationship over a wide band of frequencies is discussed in the following Section.

(7.3) Quadrature Hybrid Networks

Referring to eqn. (4) it will be seen that K attains a maximum value at the frequency at which $\theta = \pi/2$, i.e. when the length of the directional coupler is equal to one quarter-wavelength. At neighbouring frequencies K will be approximately constant. By arranging that the maximum value is equal to unity (or slightly greater than unity) K can be made approximately equal to unity over a wide band. (For example if its maximum value is 1·05, K will lie between 0·95 and 1·05 over a band whose width is 56% of the mid-band frequency.) When K is equal to unity the directional coupler belongs to a class of networks which may be termed *quadrature hybrid networks*. These have applications in aerial systems consisting of two parts fed with equal powers in phase quadrature. Other networks of this type are known, but they are more complicated and do not cover such a wide band.

In one application, illustrated in Fig. 11(*a*), the directional coupler, acting as a quadrature hybrid network, could be used to combine the outputs of two transmitters without the use of frequency-selective networks. For example, it could take the place of the diplexer and a quarter-wavelength of transmission line in an aerial system of the type used at the Sutton Coldfield television station, which has been described by Bevan and Page.[9]

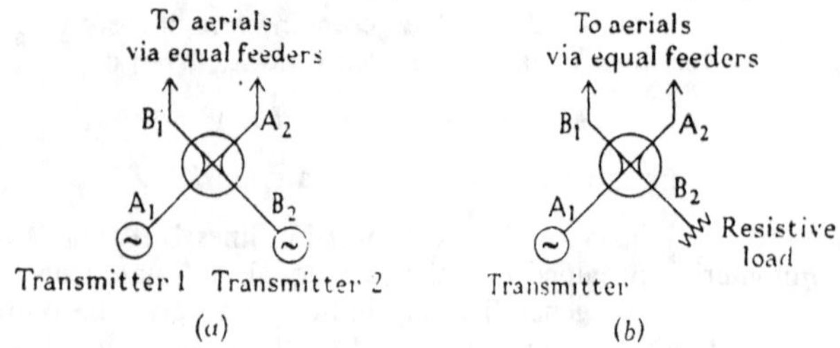

Fig. 11.—Applications to phase-quadrature aerial systems.

(a) Diplexer.
(b) Power-equalizing network.

Fig. 11(*b*) illustrates a similar application, in which the directional coupler, again acting as a quadrature hybrid network, forms, with a resistive load, a *power-equalizing network*. In this arrangement the impedance of the aerials may be allowed to depart more widely from the characteristic impedance of the feeders than would be permissible if the aerials were fed in parallel through feeders differing by one quarter-wavelength. Provided that the two aerials have the same impedance, departure of this impedance from the characteristic impedance of the feeders does not affect the radiation pattern or the input impedance at A_1, since energy reflected owing to mismatch is absorbed in the resistive load at B_2. The use of a power-equalizing network in this way has been described by Masters.[10] The principal advantage of using the directional coupler (with a resistive load) as a power-equalizing network is the wide band of frequencies covered. A possible application would be the radiation of two television programmes from a single aerial of the "superturnstile" type, which has itself a wide bandwidth.

For many transmitting applications, power-handling difficulties would necessitate the use of directional couplers connected in cascade in order to achieve the required output ratio of unity.

(8) CONCLUSIONS

Directional couplers consisting of coupled transmission lines offer advantages over other types for general laboratory use. A directivity exceeding 35 dB has been achieved at frequencies up to 900 Mc/s with a maximum output ratio of −7·5 dB. A simple form of directional coupler made from lead-covered cable has given directivities exceeding 20 dB without any form of experimental adjustment. The sinusoidal variation of output ratio with frequency enables a fairly constant output ratio to be obtained over a band of frequencies by arranging the length to be one quarter-wavelength at the mid-band frequency. This procedure also suppresses even harmonics of the mid-band frequency; the output ratio for odd harmonics is then the same as that for the fundamental.

Up to the present the directional couplers described have been used only for power measurement, but the possibility of other uses exist. These include a range of all-pass networks consisting of directional couplers and lengths of transmission line, and applications to aerial systems consisting of two parts fed in phase quadrature.

(9) ACKNOWLEDGMENTS

The author is indebted to the Chief Engineer of the British Broadcasting Corporation for permission to publish the paper. He also wishes to express his appreciation of the help given by Mr. G. H. Millard who carried out the experimental work described in the paper.

(10) REFERENCES

(1) Montgomery, Dicke, and Purcell: "Principles of Microwave Circuits" (McGraw-Hill, 1948).

(2) Earley, H. C.: "A Wide-Band Directional Coupler for Wave Guide," *Proceedings of the Institute of Radio Engineers*, 1946, **34**, p. 883.

(3) Allan, H. R., and Curling, C. D.: "The Reflectometer," *Proceedings I.E.E.*, Paper No. 767 R, January, 1949 (**96**, Part III, p. 25).

(4) Sosin, B. M.: "Application of Directional Couplers to Amplitude Ratio and Phase Measurement at Very High Frequencies," *Marconi Review*, 1951, **14**, p. 39.

(5) Ginzton, E. L., and Goodwin, P. S.: "A Note on Coaxial Bethe-Hole Directional Couplers," *Proceedings of the Institute of Radio Engineers*, 1950, **38**, p. 305.

(6) Firestone, W. L.: "Directional Coupling with Transmission Lines," *Tele-Tech*, November, 1953, **12**, No. 11, p. 95.

(7) Oliver, B. M.: "Directional Electromagnetic Couplers," *Proceedings of the Institute of Radio Engineers*, 1954, **42**, p. 1686.

(8) Phillips, G. J.: "A Wide-Band Aerial System for Circularly Polarized Waves, Suitable for Ionospheric Research," *Proceedings I.E.E.*, Paper No. 1121 R, May, 1951 (**98**, Part III, p. 237).

(9) Bevan, P. A. T., and Page, H.: "The Sutton Coldfield Television Broadcasting Station," *ibid.*, Paper No. 116 R, April, 1951 (**98**, Part III, p. 416).

(10) Masters, R. W.: "A Power-Equalizing Network for Antennas," *Proceedings of the Institute of Radio Engineers*, 1949, **37**, p. 735.

(11) APPENDICES

(11.1) Derivation of the Results of Section 3

Fig. 12 shows a 3-conductor transmission line characterized by push-pull and push-push characteristic impedance $2R_b$ and

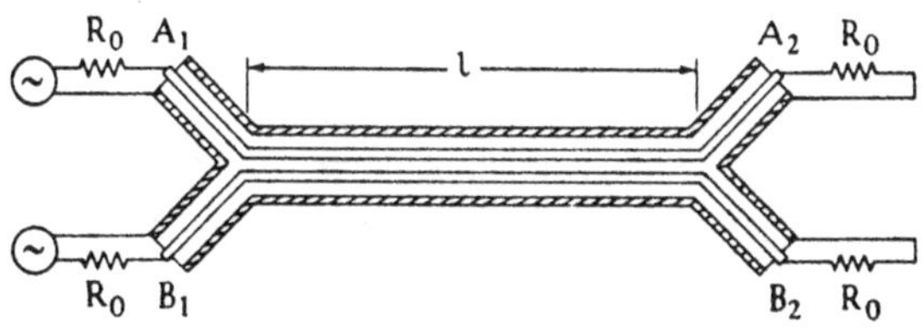

Fig. 12.—Three-conductor line characterized by R_u and R_b (defined in Fig. 3).

R_0 equals the characteristic impedance of the coaxial lines. Source impedance of generators is zero.

$\frac{1}{2}R_u$ as defined in Fig. 3. The inner conductors are connected to coaxial transmission lines terminated in their characteristic impedance R_0, which is assumed to be equal to $(R_bR_u)^{1/2}$. The generators shown connected in series with the loads at A_1 and B_1 are assumed to have zero self-impedance. If the generator at A_1 only is assumed to be operative, this will cause waves to pass along the coaxial line to the 3-conductor line. As a result waves may in general pass from the 3-conductor line towards A_2, B_1 and B_2, and there may also be reflection back to A_1. It is required to find the relative amplitudes and phases of these waves.

The problem is simplified by considering a second generator connected at B_1, as shown in the Figure, and assuming the two generators to operate firstly in phase and secondly in antiphase. By applying the superposition theorem the effect of a single generator at A_1 is obtained. In what follows the amplitudes of waves will be characterized by the voltage, rather than the current.

First it will be assumed that each generator produces a positive impulse, so that a positive unit impulse travels from A_1 and B_1 towards the left-hand junction. Two coaxial lines in parallel have a characteristic impedance $\frac{1}{2}R_0$, and the 3-conductor line behaves like an unbalanced transmission line of characteristic impedance $\frac{1}{2}R_u = \frac{1}{2}kR_0$. The voltage reflection coefficient at the junction is therefore $(k-1)/(k+1)$; positive impulses of this magnitude will be returned to A_1 and B_1. An impulse of magnitude $2k/(k+1)$ will travel from left to right along the 3-conductor line. Since the reflection coefficient at the right-hand junction will be $(1-k)/(k+1)$, an impulse of magnitude $4k/(k+1)^2$ will pass into the loads at A_2 and B_2, and an impulse of magnitude $2k(1-k)/(k+1)^2$ will travel from right to left along the 3-conductor line. The last-mentioned impulse will now be reflected backwards and forwards in the 3-conductor line, and it will be changed in sign at each reflection. A geometrically-decreasing train of positive impulses will arrive at A_2 and B_2, and a train of negative impulses will arrive at A_1 and B_1.

The result may be summarized as follows. If waves characterized by the unit impulse $\delta(t)$ are incident on the left-hand junction from A_1 and B_1, the resulting output at A_2 and B_2 is characterized by the function $H_1(t)$, where

$$H_1(t) = \frac{4k}{(k+1)^2}\sum_0^\infty\left(\frac{k-1}{k+1}\right)^{2n}\delta[t-(2n+1)l/c] \quad . \quad (8)$$

The waves reflected back to A_1 and B_1 are characterized by the function $H_2(t)$, where

$$H_2(t) = \frac{k-1}{k+1}\delta(t) - \frac{4k}{(k+1)^2}\sum_0^\infty\left(\frac{k-1}{k+1}\right)^{2n+1}\delta[t-(2n+2)l/c] \quad . \quad . \quad . \quad . \quad (9)$$

A similar process occurs if the generator at B_1 is reversed, so that the two generators are excited in anti-phase; the reflection coefficient at either junction is found to be of the same magnitude as before but reversed in sign. The outputs at A_1, A_2, B_1, and B_2 are found to be $-H_2(t)$, $H_1(t)$, $H_2(t)$, and $-H_1(t)$, respectively.

If the outputs due to the push-push and push-pull excitations are added and the sum is halved, the output due to a unit impulse incident on the left-hand junction from A_1 is obtained. It is zero at A_1 and B_2, $H_1(t)$ at A_2, and $H_2(t)$ at B_1.

By expressing the outputs at A_2 and B_1 in terms of their frequency spectra, the ratio of the output at B_1 to that at A_2 at an angular frequency ω is found to be

$$\tfrac{1}{2}j(k-1/k)\sin(\omega l/c)$$

where c is the velocity of light.

The fact that the two outputs are in phase quadrature and that there is no output at B_2 establishes that the device is a directional coupler of characteristic impedance R_0. The output ratio as stated in eqn. (4) is confirmed.

(11.2) Analysis of the Directional Coupler described in Section 5

It is required to find the push-push and push-pull characteristic impedances, $\frac{1}{2}R_u$, $2R_b$, of the 3-conductor line. The geometric mean of these quantities is the characteristic impedance of the directional coupler, whose departure from the characteristic impedance of the coaxial cable from which the directional coupler is made determines the finite directivity. The ratio R_u/R_b determines the output ratio.

The method is to change the cross-section of the 3-conductor line into a simpler form by a conformal transformation. Provided that the transformation is 1 : 1 (this excludes transformations which divide one conductor into two or more, and those which superimpose two or more conductors), R_u and R_b will be unchanged by the transformation.

An approximation will be made by assuming that the radius of the inner conductors is small compared with their spacing from each other and from the outer conductor. Expressed in physical terms the approximation is equivalent to neglecting the non-uniformity of the distribution of current round the periphery of each inner conductor. Mathematically, the procedure is as follows. Suppose that in the z-plane the axis of an inner conductor is represented by the complex number z_1, and that its periphery is the locus $z = z_1 + a\varepsilon^{j0}$, a being the radius. If now the transformation $w = f(z)$ is made, the new axis will be at $w = f(z_1)$ and the periphery will be the locus $w = f(z_1 + a\varepsilon^{j0})$. If a is small the latter expression may be expanded by Taylor's theorem and powers of a above the first may be neglected:

$$w \simeq f(z_1) + a\varepsilon^{j0} f'(z_1) \quad . \quad . \quad . \quad . \quad (10)$$

It follows that to a first approximation the periphery in the w-plane remains a circle centred on the point $w = f(z_1)$ corresponding to the centre in the z-plane. The radius is, however, changed from a to $a|f'(z_1)|$. It would not be difficult to obtain a second approximation by taking into account the next term of the Taylor series—the effect is mainly to displace the centre of the periphery slightly—but unwieldy expressions result.

The transformation is illustrated in Fig. 13, in which (*a*) is the cross-section of the 3-conductor line. For convenience the radius of the outer conductor is taken to be unity, and the width of the slot is expressed in terms of θ, the angle subtended at the axis of either inner conductor. The actual cross-section will be regarded as the z-plane.

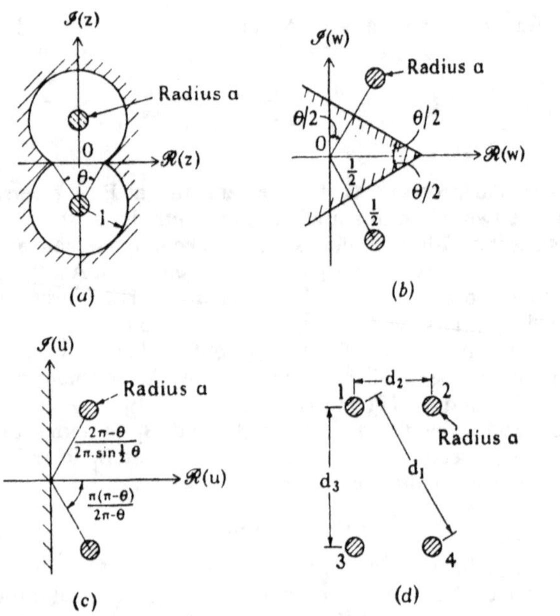

Fig. 13.—Conformal transformation of the cross-section.

(*a*) z-plane (original cross-section).
(*b*) w-plane. $w = (z + \sin\frac{1}{2}\theta)^{-1}$.
(*c*) u-plane. $u = \dfrac{2\pi - \theta}{2\pi\sin\frac{1}{2}\theta}(2w\sin\tfrac{1}{2}\theta - 1)^{\pi/(2\pi-\theta)}$
(*d*) Conducting half-plane replaced by images of conductors in it.
Change of scale.

$$d_1 = \frac{2\pi - \theta}{\pi\sin\frac{1}{2}\theta}$$

$$d_2 = \frac{2\pi - \theta}{\pi\sin\frac{1}{2}\theta}\sin\frac{\pi\theta}{4\pi - 2\theta}$$

$$d_3 = \frac{2\pi - \theta}{\pi\sin\frac{1}{2}\theta}\cos\frac{\pi\theta}{4\pi - 2\theta}$$

The first stage in the transformation is to carry out a geometrical inversion with respect to the left-hand edge of the slot, the radius of inversion being unity. This is done by putting

$$w = (z + \sin\tfrac{1}{2}\theta)^{-1} \quad . \quad . \quad . \quad . \quad . \quad (11)$$

The resulting cross-section in the w-plane is shown in Fig. 13(*b*). It will be seen that the outer conductor now takes the simpler form of a conducting wedge, which partially screens the two inner conductors from one another. If the width of the slot tends to zero in Fig. 13(*a*), the wedge in Fig. 13(*b*) becomes more acute and its apex moves to the right, until eventually the two inner conductors are completely separated by an infinite conducting plane. The radius of each inner conductor is unchanged, since $|dw/dz|$ is unity at the axis.

The cross-section in the w-plane is further simplified by moving the origin to the apex of the wedge and raising w to a fractional power so as to reduce the angle of the sector not occupied by the wedge from 2π–θ to π. At the same time the dimensions are simplified by making a change of scale. This is done by putting

$$u = \frac{2\pi - \theta}{2\pi\sin\frac{1}{2}\theta}(2w\sin\tfrac{1}{2}\theta - 1)^{\pi/(2\pi-\theta)} \quad . \quad . \quad (12)$$

The derivative of u is given by

$$\frac{du}{dw} = (2w\sin\tfrac{1}{2}\theta - 1)^{-(\pi-\theta)/(2\pi-\theta)} \quad . \quad . \quad . \quad (13)$$

At the axis of either inner conductor $|du/dw|$ is equal to unity; the radius of the inner conductors therefore remains unchanged. The resulting cross-section in the u-plane is shown in Fig. 13(*c*).

Instead of considering the cross-section of Fig. 13(*c*) directly, the conducting plane into which the outer conductor has finally been transformed is replaced by the images of the inner conductors in it, as shown in Fig. 13(*d*). We now have to consider a balanced directional coupler formed by two open-wire transmission lines coupled together.

Referring to Fig. 13(*d*), R_u is equal to the characteristic impedance of the balanced transmission line formed by connecting conductors 1 and 3 and conductors 2 and 4. R_b is equal to the characteristic impedance of the balanced transmission line formed by connecting conductors 1 and 4 and conductors 2 and 3. Assuming that the radius of the inner conductors is small compared with the spacing between them the result is found to be

$$R_u = 60\log_\varepsilon\frac{d_1 d_2}{a d_3}$$

$$= 60\log_\varepsilon\left(\frac{2\pi - \theta}{a\pi\sin\frac{1}{2}\theta}\sin\frac{\pi\theta}{4\pi - 2\theta}\right) - 60\log_\varepsilon\cos\frac{\pi\theta}{4\pi - 2\theta} \quad (14)$$

$$R_b = 60\log_\varepsilon\frac{d_2 d_3}{a d_1}$$

$$= 60\log_\varepsilon\left(\frac{2\pi - \theta}{a\pi\sin\frac{1}{2}\theta}\sin\frac{\pi\theta}{4\pi - 2\theta}\right) + 60\log_\varepsilon\cos\frac{\pi\theta}{4\pi - 2\theta} \quad (15)$$

Eqns. (14) and (15) may be interpreted as follows. The first term in the expressions for R_u and R_b is the characteristic impedance of the unbalanced transmission line formed by one inner conductor and the outer conductor, the other inner conductor being removed. If θ tends to zero this term tends to R'_0, the characteristic impedance of the coaxial cable from which the directional coupler has been made. This is given by

$$R'_0 = 60\log_\varepsilon(1/a) \quad . \quad . \quad . \quad . \quad . \quad (16)$$

For finite but small values of θ the presence of the slot increases the characteristic impedance by an amount proportional to θ^2. The second term takes account of the coupling between the inner conductors through the slot; it is positive in the expression for R_u and negative in the expression for R_b, and is proportional to θ^2 for small values of θ.

The output ratio K and the characteristic impedance R_0 of a directional coupler constructed from the 3-conductor line of Fig. 13(*a*) are obtained by substituting for R_u and R_b in eqns. (4) and (2), respectively. In practice it is assumed in using the directional coupler that its characteristic impedance is equal to R_0', which is given by eqn. (16). It can be shown that if the difference between R_0 and R_0' is small, as is the case in practice, the directivity is equal to

$$\frac{R_0'\sqrt{(1 + K^2)}}{|R_0 - R_0'|}$$

In most applications the square root in this expression will be approximately equal to unity.

Coupled-Strip-Transmission-Line Filters and Directional Couplers

E. M. T. JONES† AND J. T. BOLLJAHN†

Summary—**This paper describes the theory of operation of coupled strip line filters and directional couplers, and presents information from which these components may be easily designed. Low-pass, band-pass, all-pass, and all-stop filter characteristics are obtained from these coupled lines either by placing open or short circuits at two of the four available terminal pairs, or by interconnecting two of the terminal pairs. Directional couplers having perfect directivity and constant input impedance at all frequencies, and for all degrees of coupling, are obtained by placing equal resistive loads at each of the four terminal pairs.**

Introduction

THE ELECTROMAGNETIC COUPLING that exists between parallel transmission lines has been used to advantage by a number of investigators to construct filters and directional couplers from coupled open-wire transmission lines, and coupled coaxial lines.[1–5] This natural coupling can be used equally well in the construction of shielded strip transmission line filters and directional couplers. This paper derives formulas from which these coupled-strip-line filters and coupled-strip-line directional couplers may be designed.

Coupled Transmission-Line Impedance and Admittance Matrices

The general physical configuration of coupled strip lines to be considered here is illustrated in Fig. 1. A convenient way to describe the behavior of these coupled strip transmission lines is by means of the impedance matrix which will now be derived. Consider the shielded coupled transmission lines shown in Fig. 1, where the two transmission lines are supported midway between the two ground planes and are driven by a set of five constant-current generators at each end. The four current generators $i_1/2$ energize the even or unbalanced mode on the structure producing voltages on the two conductors of the form

$$v_{a_1}(z) = v_{b_1}(z) = -jZ_{oe}i_1 \frac{\cos k(l-z)}{\sin kl}, \qquad (1)$$

† Stanford Research Institute Menlo Park, Calif.

[1] A. Alford, "Coupled networks in radio-frequency circuits," Proc. IRE, vol. 29, pp. 55–70; February, 1941.

[2] J. J. Karakash and D. E. Mode, "A coupled coaxial transmission-line band-pass filter," Proc. IRE, vol 38, pp. 48–52; January, 1950.

[3] W. L. Firestone, "Analysis of transmission line directional couplers," Proc. IRE, vol. 42, pp. 1529–1538; October, 1954.

[4] B. M. Oliver, "Directional electromagnetic couplers," Proc. IRE, vol. 42, pp. 1686–1692; November, 1954.

[5] R. C. Knechtli, "Further analysis of transmission-line directional couplers," Proc. IRE, vol. 43, pp. 867–869; July, 1955.

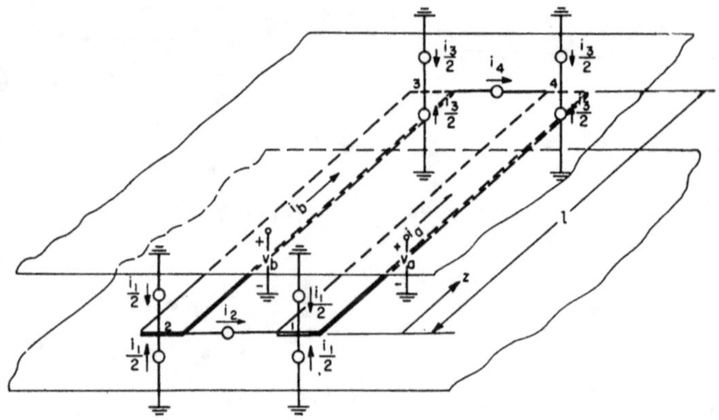

Fig. 1.—Notation used in deriving the impedance matrix of coupled transmission lines.

where

Z_{oe} = characteristic impedance of one wire to ground with equal currents in the same direction.

Likewise, the current generators $i_3/2$ produce voltages on the two conductors

$$v_{a_3}(z) = v_{b_3}(z) = -jZ_{oe}i_3 \frac{\cos kz}{\sin kl}. \qquad (2)$$

The current generators i_2 and i_4 produce odd or balanced voltages on the lines of the form

$$v_{a_2}(z) = -v_{b_2}(z) = -jZ_{oo}i_2 \frac{\cos k(l-z)}{\sin kl}, \qquad (3)$$

and

$$v_{a_4}(z) = -v_{b_4}(z) = -jZ_{oo}i_4 \frac{\cos kz}{\sin kl}, \qquad (4)$$

where

Z_{oo} = characteristic impedance of one wire to ground with equal currents in opposite directions.

The total current and voltage at each of the terminals (1, 2, 3, and 4) are easily related to the currents and voltages of the even- and odd-mode excitation. Assume that the positive current in each case is into a terminal; then the terminal currents are

$$\begin{aligned} I_1 &= i_1 + i_2 \\ I_2 &= i_1 - i_2 \\ I_3 &= i_3 - i_4 \\ I_4 &= i_3 + i_4. \end{aligned} \qquad (5)$$

FILTER	Z_I	α, β and θ_c
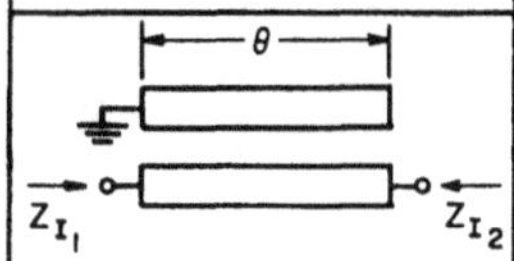 1- LOW PASS	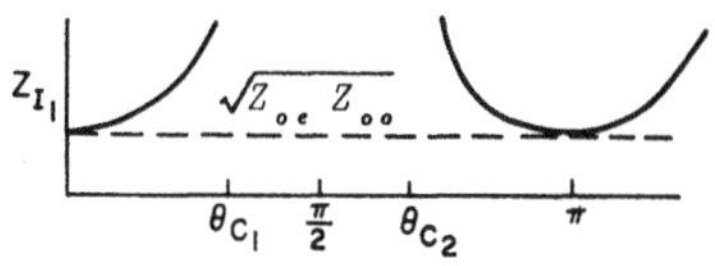 $Z_{I_1} = \dfrac{2 Z_{oe} Z_{oo} \cos\theta}{[-(Z_{oe}-Z_{oo})^2 + (Z_{oe}+Z_{oo})^2 \cos^2\theta]^{1/2}}$ $Z_{I_2} = \dfrac{Z_{oe} Z_{oo}}{Z_{I_1}}$	$\cosh(\alpha + j\beta) = \dfrac{[(Z_{oe}+Z_{oo})^2 \cos^2\theta - (Z_{oe}-Z_{oo})^2]^{1/2}}{2\sqrt{Z_{oe} Z_{oo}}}$ $\cos\theta_{c_1} = -\cos\theta_{c_2} = \left[\dfrac{\frac{Z_{oe}}{Z_{oo}} - 1}{\frac{Z_{oe}}{Z_{oo}} + 1}\right]$
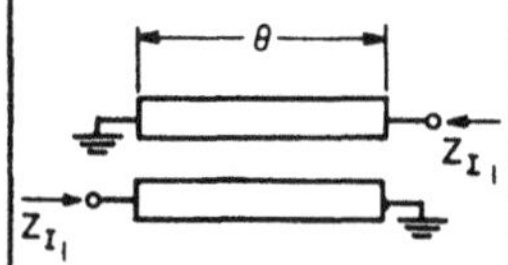 2-BAND PASS	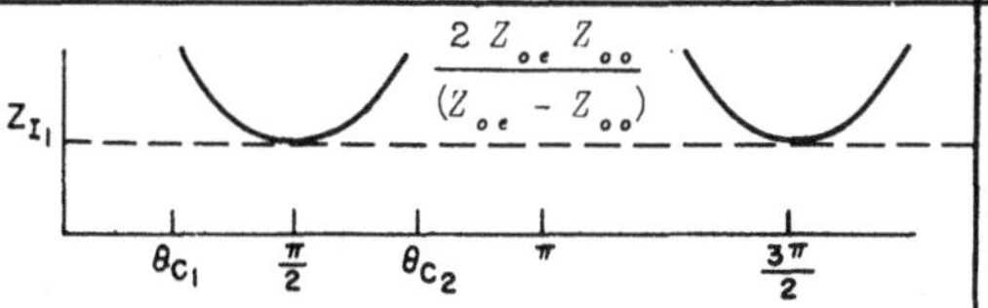 $Z_{I_1} \quad \dfrac{2 Z_{oe} Z_{oo} \sin\theta}{[(Z_{oe}-Z_{oo})^2 - (Z_{oe}+Z_{oo})^2 \cos^2\theta]^{1/2}}$	$\cosh(\alpha + j\beta) = \left[\dfrac{\frac{Z_{oe}}{Z_{oo}} + 1}{\frac{Z_{oe}}{Z_{oo}} - 1}\right] \cos\theta$ $\cos\theta_{c_1} = -\cos\theta_{c_2} = \left[\dfrac{\frac{Z_{oe}}{Z_{oo}} - 1}{\frac{Z_{oe}}{Z_{oo}} + 1}\right]$
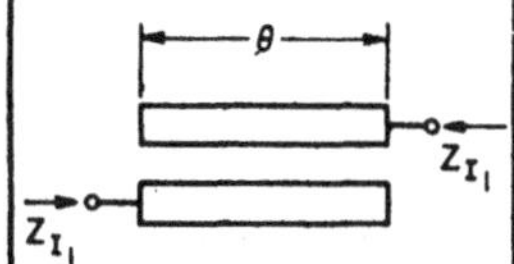 3-BAND PASS	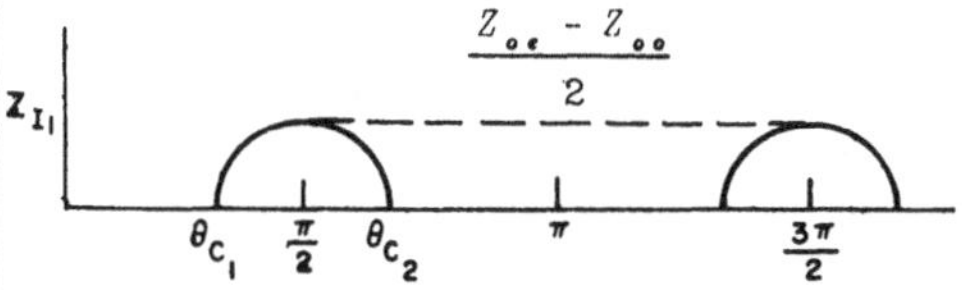 $Z_{I_1} = \dfrac{[(Z_{oe}-Z_{oo})^2 \quad (Z_{oe}+Z_{oo})^2 \cos^2\theta]^{1/2}}{2 \sin\theta}$	$\cos\theta_{c_1} = -\cos\theta_{c_2} = \left[\dfrac{\frac{Z_{oe}}{Z_{oo}} - 1}{\frac{Z_{oe}}{Z_{oo}} + 1}\right]$ $\cosh(\alpha + j\beta) = \left[\dfrac{\frac{Z_{oe}}{Z_{oo}} + 1}{\frac{Z_{oe}}{Z_{oo}} - 1}\right] \cos\theta$

Fig. 2(a)

CASE	Z_I	α, β and θ_c
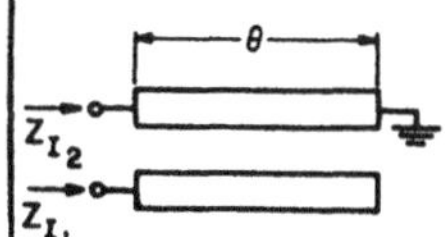 4-BAND PASS	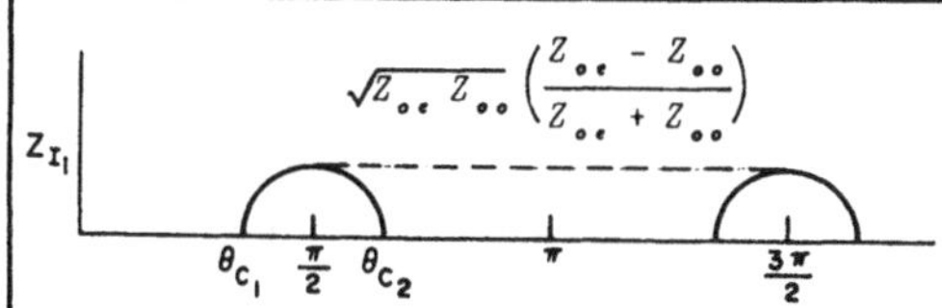 $Z_{I_1} = \dfrac{\sqrt{Z_{oe} Z_{oo}}\, [(Z_{oe}-Z_{oo})^2 - (Z_{oe}+Z_{oo})^2 \cos^2\theta]^{1/2}}{\sin\theta\, (Z_{oe}+Z_{oo})}$ $Z_{I_2} = \dfrac{Z_{oe} Z_{oo}}{Z_{I_1}}$	$\cosh(\alpha + j\beta) = \dfrac{1}{\sin\theta}\left[1 - \left(\dfrac{Z_{oe}+Z_{oo}}{Z_{oe}-Z_{oo}}\right)^2 \cos^2\theta\right]^{1/2}$ $\cos\theta_{c_1} = -\cos\theta_{c_2} = \left[\dfrac{\frac{Z_{oe}}{Z_{oo}} - 1}{\frac{Z_{oe}}{Z_{oo}} + 1}\right]$
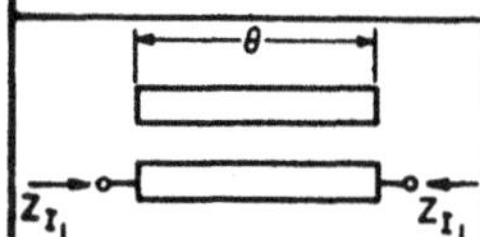 5- ALL PASS	$Z_{I_1} = \dfrac{Z_{oe} + Z_{oo}}{2}$	$\beta = \theta$
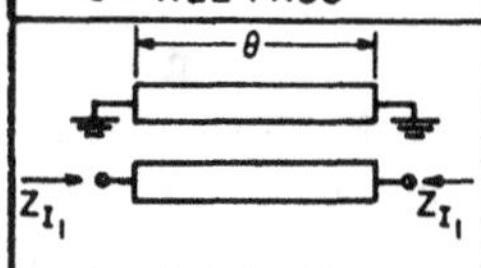 6- ALL PASS	$Z_{I_1} = \dfrac{2 Z_{oe} Z_{oo}}{Z_{oe} + Z_{oo}}$	$\beta = \theta$
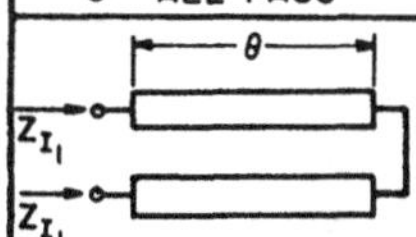 7- ALL PASS	$Z_{I_1} = \sqrt{Z_{oe} Z_{oo}}$	$\cos\beta = \dfrac{\frac{Z_{oe}}{Z_{oo}} - \tan^2\theta}{\frac{Z_{oe}}{Z_{oo}} + \tan^2\theta}$

Fig. 2(b)

CASE	Z_I	α, β, θ_c
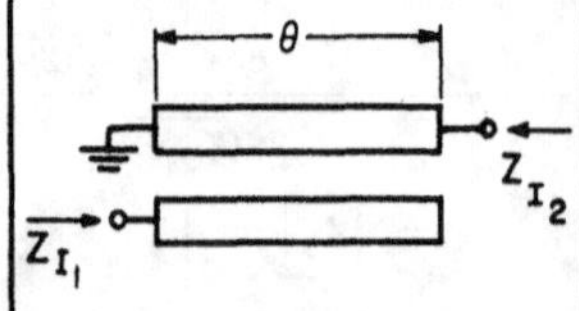 8 - ALL STOP	$Z_{I_1} = -j \frac{2 Z_{oo} Z_{oe}}{Z_{oo} + Z_{oe}} \cot\theta$ $Z_{I_2} = \frac{Z_{oe} Z_{oo}}{Z_{I_1}}$	$\alpha = \infty$
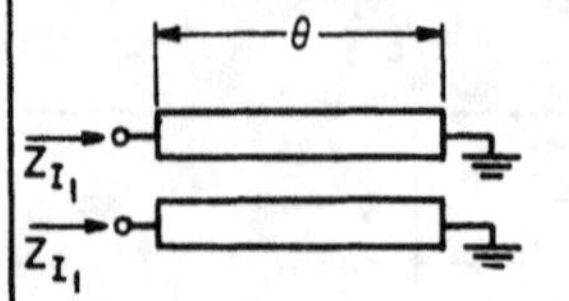 9 - ALL STOP	$Z_{I_1} = j\sqrt{Z_{oe} Z_{oo}} \tan\theta$	$\cosh\alpha = \frac{Z_{oe} + Z_{oo}}{Z_{oe} - Z_{oo}}$
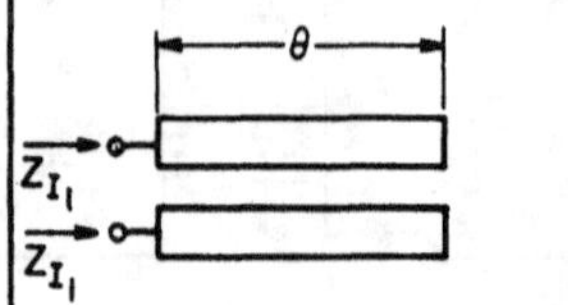 10 - ALL STOP	$Z_{I_1} = -j\sqrt{Z_{oe} Z_{oo}} \cot\theta$	$\cosh\alpha = \frac{Z_{oe} + Z_{oo}}{Z_{oe} - Z_{oo}}$

Fig. 2(c).—Image parameters for coupled transmission line filters.

When these equations are solved for the mode currents in terms of the terminal currents, it is found that

$$\begin{aligned} i_1 &= \tfrac{1}{2}(I_1 + I_2) \\ i_2 &= \tfrac{1}{2}(I_1 - I_2) \\ i_3 &= \tfrac{1}{2}(I_4 + I_3) \\ i_4 &= \tfrac{1}{2}(I_4 - I_3). \end{aligned} \qquad (6)$$

Since constant-current generators having infinite internal impedance are used in this derivation, each terminal voltage is the sum of all the mode voltages at that terminal. Thus

$$\begin{aligned} V_1 &= (v_{a_1} + v_{a_2} + v_{a_3} + v_{a_4})\big|_{z=o} \\ V_2 &= (v_{b_1} + v_{b_2} + v_{b_3} + v_{b_4})\big|_{z=o} \\ V_3 &= (v_{b_1} + v_{b_2} + v_{b_3} + v_{b_4})\big|_{z=l} \\ V_4 &= (v_{a_1} + v_{a_2} + v_{a_3} + v_{a_4})\big|_{z=l}. \end{aligned} \qquad (7)$$

When (1)–(4) and (6) are substituted in (7), the coefficients of I_1, I_2, I_3, and I_4 are easily identified as the elements of the impedance matrix. Performing this operation, it is found that the matrix elements are

$$Z_{11} = Z_{22} = Z_{33} = Z_{44} = -j(Z_{oe} + Z_{oo})\frac{\cot\theta}{2}$$

$$Z_{12} = Z_{21} = Z_{34} = Z_{43} = -j(Z_{oe} - Z_{oo})\frac{\cot\theta}{2}$$

$$Z_{13} = Z_{31} = Z_{24} = Z_{42} = -j(Z_{oe} - Z_{oo})\frac{\csc\theta}{2}$$

$$Z_{14} = Z_{41} = Z_{23} = Z_{32} = -j(Z_{oe} + Z_{oo})\frac{\csc\theta}{2} \qquad (8)$$

where θ is the electrical length of the coupled wires.

The admittance matrix may also be derived by substituting a double-T configuration of voltage generators at each end for the double-π configuration of constant current generators used above. The results of this analysis show that the elements of the admittance matrix are

$$Y_{11} = Y_{22} = Y_{33} = Y_{44} = -j(Y_{oo} + Y_{oe})\frac{\cot\theta}{2}$$

$$Y_{12} = Y_{21} = Y_{34} = Y_{43} = -j(Y_{oo} - Y_{oe})\frac{\cot\theta}{2}$$

$$Y_{13} = Y_{31} = Y_{24} = Y_{42} = -j(Y_{oo} - Y_{oe})\frac{\csc\theta}{2}$$

$$Y_{14} = Y_{41} = Y_{23} = Y_{32} = -j(Y_{oo} + Y_{oe})\frac{\csc\theta}{2}. \qquad (9)$$

Coupled Transmission-Line Filters

There are ten filters that can be obtained from a pair of coupled strip lines by placing open or short circuits on various terminal pairs, or by connecting ends of the lines together. Fig 2(a), (b), and (c), on this page and the previous page, show single sections of the ten possible coupled transmission-line filters together with their image parameters. In the schematic diagrams of the coupled line filters, the input and output terminal pairs are designated by small open circles. The image impedance, or admittance, seen looking into each of these terminal pairs is also shown near the terminal pair. Open-circuited terminal pairs of the coupled lines are shown with no connection in the filter schematic diagram, while short-circuited terminal pairs are designated with the standard grounding symbol.

Reference to Fig. 2 shows that it is possible to obtain low-pass, band-pass, and all-stop filters from a pair of coupled transmission lines. The low-pass filter 1 has an infinite number of pass bands centered about multiples of π, while the band-pass filters 2, 3, and 4 have an infinite number of pass bands centered about odd multiples of $\pi/2$. It is interesting to note that the bandwidth of the three band-pass filters is the same, while the bandwidth of the low-pass filter is just half that of the band-pass filters. Inspection of Fig. 2 also shows that there are close relationships among the image impedances of these filters. In particular, the product of the image impedances Z_{I_1} and Z_{I_2} of the unsymmetrical filters 1, 4, and 8 are each equal to $Z_{oo}Z_{oe}$. Furthermore, the product of the image impedances of the dual symmetrical filters 2 and 3, 5 and 6, and 9 and 10 are also equal to $Z_{oe}Z_{oo}$. Finally, the square of the image impedance of the symmetrical all-pass filter 7 is equal to $Z_{oe}Z_{oo}$.

Filter 8 does not have much practical utility; however, it does have interesting property of possessing an infinite image attenuation at all frequencies. Filters 9 and 10 are not very useful by themselves, but they can be made into band-pass filters by adding suitable reactive elements to their terminals. For example, the equivalent circuit of filter 9 is a symmetrical π composed of inductances. The susceptance of the inductances in each of the shunt arms is $-Y_{oe}\cot\theta$, while the susceptance of the series arm is

$$\frac{-(Y_{oo}-Y_{oe})\cot\theta}{2}.$$

The addition of series capacitances converts filter 9 into a well-known type of band-pass filter. The equivalent circuit of filter 10 is a symmetrical T network of capacitances. Each of the series capacitances has a reactance of $-Z_{oo}\cot\theta$, while the shunt capacitance has a reactance of

$$\frac{-(Z_{oe}-Z_{oo})\cot\theta}{2}.$$

In this case, series inductances must be added to this filter to convert it into a band-pass filter.

The image-parameter design equations presented in Fig. 2 are obtained in the following manner. First, the four-element impedance or admittance matrix of the filters is obtained from the four network equations

$$[V] = [Z][I] \tag{10}$$

or

$$[I] = [Y][V] \tag{11}$$

by applying the appropriate boundary conditions. Here $[V]$ and $[I]$ are four-element column matrices, while $[Z]$ and $[Y]$ are sixteen-element square matrices. For example, in filter 7 if it is assumed that terminal 3 and 4 are connected, the boundary conditions are $V_3 = V_4$ and $I_3 = -I_4$. In all the other filters the boundary conditions are $V=0$ or $I=0$ at terminals that are short-circuited or open-circuited, respectively. Second, the image impedance, Z_I, or image admittance, Y_I, and image transfer constant $(\alpha+j\beta)$ are calculated from the elements of the resulting four-element impedance matrix Z' or admittance matrix Y' by the following relations

$$Z_{I_1} = \left(Z_{11}'^2 - \frac{Z_{11}'Z_{12}'^2}{Z_{22}'}\right)^{1/2}, \quad Z_{I_2} = \left(Z_{22}'^2 - \frac{Z_{22}'Z_{12}'^2}{Z_{11}'}\right)^{1/2},$$

$$\cosh(\alpha+j\beta) = \frac{(Z_{11}'Z_{22}')^{1/2}}{Z_{12}'} \tag{12}$$

or

$$Y_{I_1} = \left(Y_{11}'^2 - \frac{Y_{11}'Y_{12}'^2}{Y_{22}'}\right)^{1/2}, \quad Y_{I_2} = \left(Y_{22}'^2 - \frac{Y_{22}'Y_{12}'^2}{Y_{11}'}\right)^{1/2},$$

$$\cosh(\alpha+j\beta) = \frac{Y(_{11}'Y_{22}')^{1/2}}{Y_{12}'}. \tag{13}$$

Physical Layout of Coupled-Strip-Line Filters

In most applications it is necessary to cascade several of the basic filter sections shown in Fig. 2 in order to achieve the desired performance from the filter. Fig. 3 illustrates the way these basic filter sections might be cascaded using strip line techniques. It is seen that where the input and output of basic sections occur at opposite ends of the strips (*i.e.*, as in filters 1, 2, 3, 5, 6, and 8) any number of sections may be cascaded. However, when the input and output of a single section occur at the same end, (*i.e.*, as in filters 4, 7, and 10) only two sections may be cascaded.

The image impedance of the coupled strip line filters is either higher or lower than the characteristic impedance of an isolated strip. Therefore, it is necessary to connect the filters to strips having different widths than the coupled strips, in order to reduce the mismatch loss at the terminals. For example, when the image impedance of the filter is less than the characteristic impedance of an isolated strip, the connecting strip is made wider than the coupled strips and, when the image impedance of the filter is greater than the characteristic impedance of an isolated strip, the connecting line is made narrower. The proper width of the connecting lines is illustrated for the filters in Fig. 3. At the bottom of Fig. 3 is shown the way in which the all-stop filters 9 and 10 may be converted to band-pass filters by adding series capacitors and series inductors, respectively.

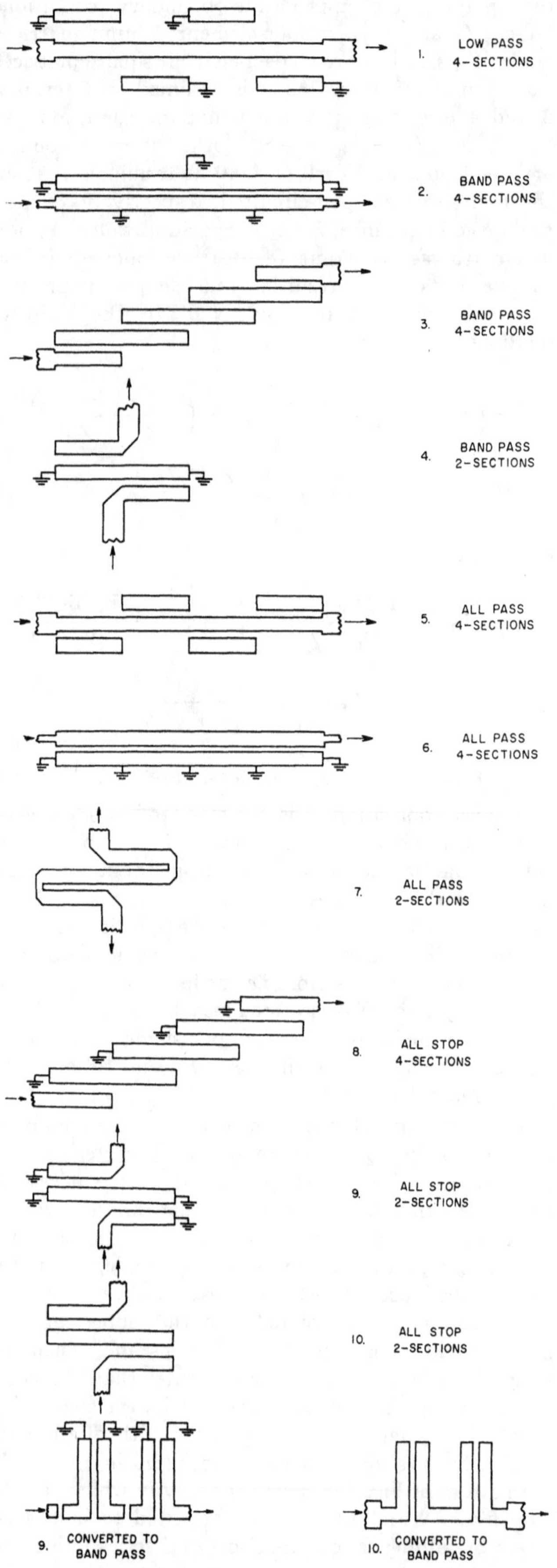

Fig. 3.—Physical layout of coupled strip line filters.

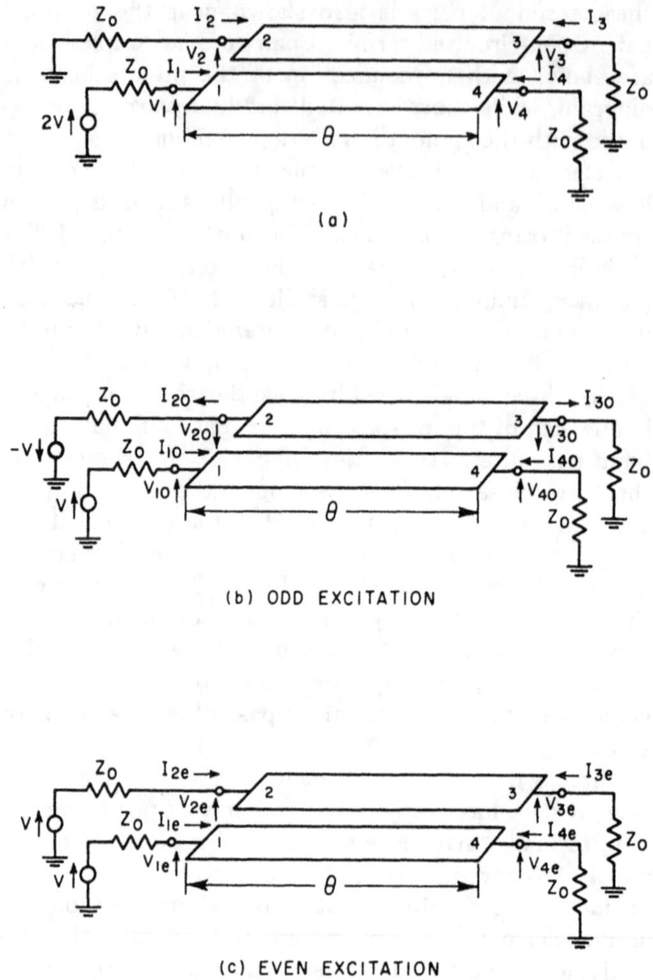

Fig. 4.—Strip line directional coupler with even and odd excitation.

Coupled-Strip-Line Directional Couplers

When the four terminal pairs of the coupled strip transmission lines illustrated in Fig. 1 are terminated in the proper constant resistance, the device behaves as a directional coupler with constant input impedance and infinite directivity at all frequencies, and for all degrees of coupling.[6] In this coupler energy is coupled backward instead of forward. Hence, if a signal is fed into one terminal pair, the coupled signal emerges from the adjacent terminal pair and no signal emerges from the diagonally opposite terminal pair.

The characteristics of this directional coupler can be determined rigorously from the impedance matrix. However, its behavior will be derived here from certain symmetry arguments, because it is believed that such a presentation gives the best physical picture of the directional coupler operation.

A simplified schematic diagram of the directional coupler is shown at the top of Fig. 4 with resistance terminations Z_o at each terminal. The coupler is ex-

[6] Oliver and Knechtli, *loc. cit.*, have also recently analyzed coupled transmission-line directional couplers and shown that they can be designed to have infinite directivity and constant input impedance for all degrees of coupling.

cited with a voltage $2V$ in series with terminal 1. At the bottom of Fig. 2, the same directional coupler is shown in two different states of excitation. In the odd excitation, out-of-phase voltages are applied in series with terminals 1 and 2, while the even excitation applies in-phase voltages in series with these terminals. Through the principle of superposition it may be seen that the behavior of the directional coupler with voltage $2V$ applied in series with terminal 1 can be obtained from its behavior with even and odd voltage excitations.

It is seen that the characteristic impedance of one of these strips to ground in the odd excitation is Z_{oo}, while the characteristic impedance of a strip to ground in the even excitation is Z_{oe}. In order for the directional coupler to have a perfect match at all frequencies it is necessary that the input impedance Z_{in} be equal to Z_o. Applying the principal of superposition, it is seen that the input impedance of the directional coupler terminated in Z_o can be written as

$$Z_{\text{in}} = \frac{V_{1o} + V_{1e}}{I_{1o} + I_{1e}} = \frac{\dfrac{Z_{1o}}{Z_o + Z_{1o}} + \dfrac{Z_{1e}}{Z_o + Z_{1e}}}{\dfrac{1}{Z_o + Z_{1o}} + \dfrac{1}{Z_o + Z_{1e}}}, \tag{14}$$

where

$$Z_{1o} = Z_{oo} \frac{(Z_o + jZ_{oo} \tan\theta)}{Z_{oo} + jZ_o \tan\theta} \tag{15}$$

and

$$Z_{1e} = Z_{oe} \frac{(Z_o + jZ_{oe} \tan\theta)}{Z_{oe} + jZ_o \tan\theta}. \tag{16}$$

When (15) and (16) are substituted in (14), it is found that $Z_{in} = Z_o$ when

$$Z_o = (Z_{oo} Z_{oe})^{1/2}. \tag{17}$$

This characteristic impedance of the coupler, $(Z_{oe} Z_{oo})^{1/2}$, is always less than the characteristic impedance of a single strip line of the same width. Therefore, the width of the lines connecting to the coupler is required to be slightly wider than the coupled strips. Although for couplings weaker than -15 db, the change in width is negligible.

Under the condition that $Z_o = (Z_{oe} Z_{oo})^{1/2}$, the voltage appearing at terminal 1 of the coupler is $V_1 = V$. The voltages $V_{2e} = V_{1e}$ and $V_{3e} = V_{4e}$ may be determined from the straightforward analysis of a transmission line of length θ, and characteristic impedance Z_{oe} terminated at either end by an impedance $(Z_{oe} Z_{oo})^{1/2}$ and fed by a voltage V. Likewise the voltages $V_{2o} = -V_{1o}$ and $V_{3o} = -V_{4o}$ may be determined from a similar analysis of a transmission line of length θ and characteristic impedance Z_{oo} terminated by $(Z_{oe} Z_{oo})^{1/2}$. The results of this analysis show that

$$V_2 = V_{2e} - V_{2o}$$

$$= j \frac{V \sin\theta \left[\sqrt{\dfrac{Z_{oe}}{Z_{oo}}} - \sqrt{\dfrac{Z_{oo}}{Z_{oe}}} \right]}{2\cos\theta + j \sin\theta \left[\sqrt{\dfrac{Z_{oe}}{Z_{oo}}} + \sqrt{\dfrac{Z_{oo}}{Z_{oe}}} \right]}, \tag{18}$$

$$V_3 = V_{3e} - V_{3o} = 0 \tag{19}$$

$$V_4 = V_{4e} - V_{4o}$$

$$= \frac{2V}{2\cos\theta + j \sin\theta \left[\sqrt{\dfrac{Z_{oe}}{Z_{oo}}} - \sqrt{\dfrac{Z_{oo}}{Z_{oe}}} \right]}. \tag{20}$$

The maximum coupled voltage, V_2, occurs when the coupler is a quarter wavelength long (*i.e.*, $\theta = 90°$). Therefore, from (18) a maximum coupling coefficient k may be defined as

$$\left| \frac{V_2}{V} \right| = k = \frac{\dfrac{Z_{oe}}{Z_{oo}} - 1}{\dfrac{Z_{oe}}{Z_{oo}} + 1}. \tag{21}$$

With this substitution, the voltage at terminal 2 can be written as

$$V_2 = V \frac{jk \sin\theta}{\sqrt{1-k^2}\cos\theta + j\sin\theta} \tag{22}$$

while the voltage at terminal 4 becomes

$$V_4 = \frac{V\sqrt{1-k^2}}{\sqrt{1-k^2}\cos\theta + j\sin\theta}. \tag{23}$$

Eqs. (22) and (23) are the same as those obtained by Oliver.[4] Reference to his Fig. 6 or to (22) above shows that for small values of k the coupler will operate over a 3 to 1 frequency band between 3 db points, while for stronger couplings even wider bandwidths of operation are obtained.

Coupled Transmission Line Impedances Z_{oe} and Z_{oo}

The characteristic impedances Z_{oe} and Z_{oo} of infinitesimally thin coupled strip lines have been calculated rigorously by Cohn[7] who presents the information in the form of a nomogram. The reader is also referred to Cohn's paper for approximate values of Z_{oe} and Z_{oe} for thick strips and strips printed on dielectric sheets.

In applications where these components must be operated at high rf potentials, the possibility of breakdown is minimized by using round conductors instead of strips. Approximate formulas for Z_{oe} and Z_{oo} for the round conductors have been derived by Honey.[8]

[7] S. B. Cohn, "Shielded coupled-strip line," Trans. IRE, vol. MTT-3, pp. 29–38; October, 1955.

[8] R. C. Honey, Stanford Research Institute, private communications.

He finds that

$$Z_{oe} - Z_{oo} = \frac{120}{\sqrt{\epsilon_r}} \ln \coth \frac{\pi s}{2b} \tag{24}$$

$$Z_{oe} + Z_{oo} = \frac{120}{\sqrt{\epsilon_r}} \ln \coth \frac{\pi d}{4b} \tag{25}$$

where s = center to center spacing of the wires, d = wire diameter, b = ground plane spacing, and ϵ_r = relative dielectric constant of medium surrounding the wires.

For configurations of coupled lines that are not amenable to analysis it is always possible to determine Z_{oo} and Z_{oe} from measurements of the static capacity of the strips. Thus

$$Z_{oo} = \frac{100\sqrt{\epsilon_r}}{3C_{oo}} = \frac{100\sqrt{\epsilon_r}}{3[C_{22} - C_{23}]} \tag{26}$$

while

$$Z_{oe} = \frac{100\sqrt{\epsilon_r}}{3C_{oe}} = \frac{100\sqrt{\epsilon_r}}{3[C_{22} + C_{23}]} \tag{27}$$

where

C_{oo} = one half the capacity between the two center conductors ($\mu\mu$f/cm),

C_{oe} = one half the capacity between the parallel combination of the two center conductors and ground ($\mu\mu$f/cm),

C_{22} = coefficient self capacity ($\mu\mu$f/cm) of either of the center conductors,

C_{22} = coefficient of induction ($\mu\mu$f/cm) between the two center conductors (a negative quantity).

The capacity C_{22} is easily measured as the capacity between either strip and ground when the other strip is grounded, while the capacity $C_{22} - C_{23}$ is just one half the capacity between the parallel combination of the two center conductors and ground.

Acknowledgement

The work reported in this paper was conducted under Contract No. AF 33(038)-7850 sponsored by the Air Force Cambridge Research Center, and Contract No. DA 36-039-sc-63232 and DA 36-039-sc-64625 sponsored by the Signal Corps Engineering Laboratories.

Reprinted from IRE TRANSACTIONS
ON MICROWAVE THEORY AND TECHNIQUES
Volume MTT-4, Number 2, April, 1956

PRINTED IN THE U.S.A.

The analytical equivalence of TEM-mode directional couplers and transmission-line stepped-impedance filters

Leo Young, M.A., M.S., Dr.Eng., Member

Summary

The coupling of two parallel transmission lines is utilized in directional couplers and filters which are simple to construct and have excellent electrical characteristics from high frequencies to microwave frequencies. The paper establishes the mathematical equivalence of such couplers to transmission-line stepped-impedance filters.

It also places new demands on that theory in so far as new insertion-loss functions, having, for instance, equi-ripple characteristics in the stop-band, will be required.

A new directional filter circuit, or channel-dropping filter, is suggested.

1 Introduction

When two transmission lines are coupled uniformly over a common length, it is possible to obtain a backward (contra-directional) coupler with theoretically perfect matching and perfect directivity at all frequencies, provided that all modes are propagated with the same phase velocity and the coupler impedances are properly chosen. Moreover, the coupling ratio can be designed to change only slightly over wide frequency bands. Such couplers have been described extensively,[1-9] and are suitable for frequencies ranging from television to microwaves.

In the paper it will be shown that every such coupler (4-port) can be analysed simply in terms of a length of transmission line (2-port), and indeed is mathematically equivalent to it. For instance, the 3-section stepped-impedance transmission-line directional coupler[4, 5] is completely solved as a 3-section stepped-impedance quarter-wave filter.*

2 Equivalence relations and the single-section coupler

Two parallel transmission lines coupled uniformly over an electrical length θ are shown in Fig. 1. The four ports

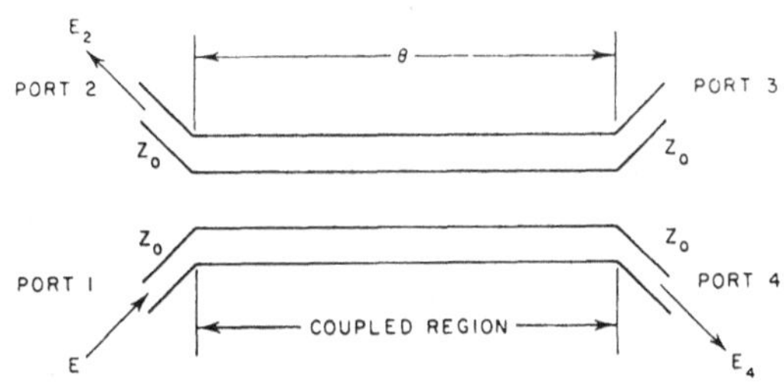

Fig. 1

TEM-mode quarter-wave directional coupler of one section

k = coupling factor

$|E|^2 = |E_2|^2 + |E_4|^2$

marked 1–4 are otherwise supposed to be completely isolated. All four ports are connected to transmission lines of characteristic impedances Z_0. The even and odd modes travelling along the coupled lines are both assumed to be TEM-waves, and therefore have the same phase velocity. It can be shown that if there is only one input, a wave of amplitude E (Fig. 1), and if the odd- and even-mode impedances, Z_{oo} and Z_{oe}, satisfy the condition

$$Z_{oo}Z_{oe} = Z_0^2 \quad . \quad . \quad . \quad . \quad . \quad . \quad (1)$$

then the only non-zero outputs are waves of amplitude E_2 and E_4, shown in Fig. 1 and given by [4, 5]

$$\frac{E_2}{E} = \frac{jk \sin\theta}{\sqrt{(1-k^2)}\cos\theta + j\sin\theta} \quad . \quad . \quad . \quad (2)$$

$$\frac{E_4}{E} = \frac{\sqrt{(1-k^2)}}{\sqrt{(1-k^2)}\cos\theta + j\sin\theta} \quad . \quad . \quad . \quad (3)$$

where k is the mid-band amplitude coupling factor between the lines, given by

$$k = \frac{(Z_{oe}/Z_{oo}) - 1}{(Z_{oe}/Z_{oo}) + 1} \quad . \quad . \quad . \quad . \quad . \quad (4)$$

Consider now a transmission line with two impedance steps, one up and one down, as shown in Fig. 2, separated by an

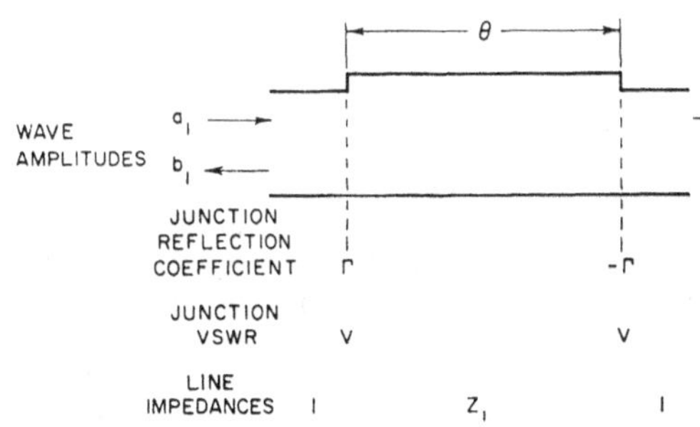

Fig. 2

Quarter-wave filter prototype for TEM-mode coupler of Fig. 1

$V = \frac{1+\Gamma}{1-\Gamma}$

electrical distance θ. With the notation of Fig. 2 it can be shown[10] that

$$\frac{b_1}{a_1} = \frac{\Gamma(1-\varepsilon^{-2j\theta})}{1-\Gamma^2\varepsilon^{-2j\theta}} \quad . \quad . \quad . \quad . \quad . \quad (5)$$

* The term 'quarter-wave filter' is introduced to distinguish the resulting physically symmetrical structure from a quarter-wave impedance transformer with a monotone sequence of steps. The analysis is the same in either case

Paper 4064E, first received 9th July and in final form 8th October 1962

Dr. Young is at the Stanford Research Institute, California, U.S.A.

$$= \frac{\left(\frac{2\Gamma}{1+\Gamma^2}\right) j \sin\theta}{\left(\frac{1-\Gamma^2}{1+\Gamma^2}\right)\cos\theta + j\sin\theta} \quad . \quad . \quad . \quad . \quad (6)$$

$$\frac{a_2}{a_1} = \frac{(1-\Gamma^2)\varepsilon^{-j\theta}}{1-\Gamma^2\varepsilon^{-2j\theta}} \quad . \quad . \quad . \quad . \quad (7)$$

$$= \frac{\left(\frac{1-\Gamma^2}{1+\Gamma^2}\right)}{\left(\frac{1-\Gamma^2}{1+\Gamma^2}\right)\cos\theta + j\sin\theta} \quad . \quad . \quad . \quad . \quad (8)$$

A comparison of eqns. 2 and 3 with eqns. 6 and 8 shows that b_1/a_1 corresponds to E_2/E, that a_2/a_1 corresponds to E_4/E, and that these equations are mathematically equivalent when we set

$$k = \frac{2\Gamma}{1+\Gamma^2} = \frac{Z_1^2 - 1}{Z_1^2 + 1} \quad . \quad . \quad . \quad . \quad (9)$$

when Z_1 is chosen greater than unity. Conversely

$$\Gamma = \frac{1 - \sqrt{(1-k^2)}}{k} \quad . \quad . \quad . \quad . \quad (10)$$

It follows that quarter-wave-coupled TEM-mode directional couplers can be solved as quarter-wave transformers,[11, 12] or rather quarter-wave filters. It can be shown from eqn. 3 that, for the single-section coupler of Fig. 1,

$$\left|\frac{E}{E_4}\right|^2 = 1 + \frac{(R-1)^2}{4R}\sin^2\theta \quad . \quad . \quad . \quad (11)$$

where $$R = Z_1^2 = \frac{1+k}{1-k} \quad . \quad . \quad . \quad . \quad . \quad (12)$$

is the v.s.w.r. of the prototype transformer (Fig. 2) at centre frequency ($\theta = \pi/2$). Compare eqn. 11 with Reference 12. The overall power-coupling ratio is

$$\left.\begin{aligned} \left|\frac{E_2}{E}\right|^2 &= 1 - \left|\frac{E_4}{E}\right|^2 \\ &= \frac{\frac{(R-1)^2}{4R}\sin^2\theta}{1 + \frac{(R-1)^2}{4R}\sin^2\theta} \\ &= \frac{k^2\sin^2\theta}{1 - k^2\cos^2\theta} \end{aligned}\right\} \quad . \quad . \quad . \quad (13)$$

The equivalence of the TEM-mode coupler and the quarter-wave filter can also be seen directly from the duality on an impedance/admittance basis of the even and odd modes, based on eqn. 1. The above derivation helps bring out explicitly the principal corresponding parameters.

The results pertaining to the analysis of quarter-wave transformers[12] can be taken over directly; however, the desired performance characteristics are essentially different (see below), and so the numerical data of Reference 12 are not applicable to TEM-mode couplers.

From eqns. 1 and 4, the even- and odd-mode impedances are

$$Z_{oe} = Z_0\sqrt{\frac{1+k}{1-k}} \quad . \quad . \quad . \quad . \quad (14)$$

and $$Z_{oo} = Z_0\sqrt{\frac{1-k}{1+k}} \quad . \quad . \quad . \quad . \quad (15)$$

The physical dimensions necessary to give the required even- and odd-mode impedances Z_{oe} and Z_{oo} of various strip transmission lines can be obtained from References 7 and 13–15.

3 The symmetrical 3-section coupler

The analysis of any coupler consisting of several cascaded sections of equal length θ can be carried out by multiplying transmission-line transformation matrices as for a quarter-wave transformer. This procedure is very tedious for a large number of sections. The single-section coupler has already been treated. The symmetrical 3-section coupler[4, 5] is shown in Fig. 3. The quantity $|E/E_4|^2$ corresponds to the

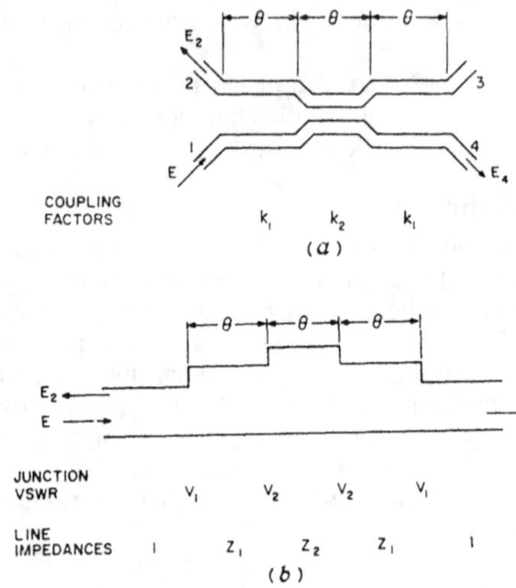

Fig. 3

Symmetrical TEM-mode quarter-wave directional coupler of three sections and its quarter-wave filter prototype

(a) TEM-mode coupler
$|E|^2 = |E_2|^2 + |E_4|^2$

(b) Prototype quarter-wave filter

'transducer loss ratio' or 'insertion loss function' of the prototype. It is given by a polynomial in $\cos^2\theta$ which should be compared with Reference 12 and which, for a 3-section coupler, is

$$\left|\frac{E}{E_4}\right|^2 = A_0 + A_1\cos^2\theta + A_2\cos^4\theta + A_3\cos^6\theta \quad . \quad (16)$$

where A_0, A_1, A_2, and A_4 are functions of the end- and middle-section coupling factors k_1 and k_2. The As are given by

$$\left.\begin{aligned} A_0 &= L_0^2 \\ A_1 &= (L_2 - 1)^2 - L_0^2 - L_0 L_1 L_2 \\ A_2 &= \left(\frac{L_1 L_2}{2}\right)^2 + L_0 L_1 L_2 - 2L_2(L_2 - 1) \\ A_3 &= L_2^2 - \left(\frac{L_1 L_2}{2}\right)^2 \end{aligned}\right\} \quad . \quad . \quad (17)$$

in which (see Fig. 3)

$$\left.\begin{aligned} L_0 &= \tfrac{1}{2}\left(\frac{V_1}{V_2} + \frac{V_2}{V_1}\right) = \frac{1 + k_1^2 - 2k_1k_2}{(1-k_1^2)\sqrt{(1-k_2^2)}} \\ L_1 &= V_1 + \frac{1}{V_1} = \frac{2}{\sqrt{(1-k_1^2)}} \\ L_2 &= V_2 + 2 + \frac{1}{V_2} = 2 + \frac{2(1-k_1k_2)}{\sqrt{[(1-k_1^2)(1-k_2^2)]}} \end{aligned}\right\} \quad . \quad (18)$$

The coupling voltage E_2 is found from

$$\left|\frac{E_2}{E}\right|^2 = 1 - \left|\frac{E_4}{E}\right|^2 \quad . \quad . \quad . \quad . \quad . \quad (19)$$

The mid-band v.s.w.r. of the prototype [Fig. 3(*b*)] is

$$R = \left(\frac{V_2}{V_1}\right)^2 \quad . \quad . \quad . \quad . \quad . \quad . \quad (20)$$

and the mid-band insertion loss of the prototype is

$$A_0 \equiv \left|\frac{E}{E_4}\right|^2_{\theta=\pi/2} = \frac{(R+1)^2}{4R} \quad . \quad . \quad . \quad . \quad (21)$$

The mid-band coupling k_0, corresponding to the mid-band reflection coefficient of the prototype, is

$$k_0 \equiv \left|\frac{E_2}{E}\right|_{\theta=\pi/2} = \left(\frac{R-1}{R+1}\right) \quad . \quad . \quad . \quad (22)$$

The coupling factors k_1 and k_2 of the individual sections are related to the corresponding prototype impedances Z_1 and Z_2 shown in Fig. 3 by an extension of the relations for the single-section coupler,

$$k_1 = \frac{Z_1^2 - 1}{Z_1^2 + 1} \text{ and } k_2 = \frac{Z_2^2 - 1}{Z_2^2 + 1} \quad . \quad . \quad . \quad (23)$$

The even- and odd-mode impedances Z_{oe1}, Z_{oo1}, Z_{oe2}, and Z_{oo2} are related to the coupling factors k_1 and k_2 by

$$Z_{oe_i} = Z_0\sqrt{\left(\frac{1+k_i}{1-k_i}\right)} \qquad i = 1, 2 \quad . \quad . \quad (24)$$

and

$$Z_{oo_i} = Z_0\sqrt{\left(\frac{1-k_i}{1+k_i}\right)} \qquad i = 1, 2 \quad . \quad . \quad (25)$$

since each Z_{oe_i} and Z_{oo_i} separately must satisfy eqn. 1.

It will be convenient to use the first junction v.s.w.r. V_1, shown in Fig. 3, as the independent parameter. Then V_2, Z_1, and Z_2 are determined from

$$V_2 = V_1\sqrt{R} \quad . \quad . \quad . \quad . \quad . \quad (26)$$

$$Z_1 = V_1 \qquad Z_2 = V_1V_2 = V_1^2\sqrt{R} \quad . \quad . \quad (27)$$

3.1 Maximally flat coupling

To obtain a maximally flat response at $\theta = \pi/2$,

$$A_1 = 0 \quad . \quad . \quad . \quad . \quad . \quad . \quad . \quad (28)$$

in eqn. 16. Eliminating V_2 leaves a quartic equation in V_1:

$$\left(V_1\sqrt{R} + 1 + \frac{1}{V_1\sqrt{R}}\right)^2 - \frac{R+1}{2\sqrt{R}}\left(2V_1 + \frac{2}{V_1} + V_1^2\sqrt{R} + \frac{1}{V_1^2\sqrt{R}}\right) = \frac{3(R+1)^2}{4R} \quad . \quad . \quad . \quad . \quad . \quad (29)$$

The solution of this equation is shown in Fig. 4, where $V_1 = V_{1MF}$ is plotted against the mid-band coupling in decibels. Selecting $V_1 = V_{1MF}$ from Fig. 4 or eqn. 29 will give a maximally flat response. It turns out that, when A_1 is thus made to vanish, A_2 is not zero. To obtain a flatter response with both A_1 and A_2 zero, the coupler or filter must not be symmetrical between ends.

When the mid-band coupling is maintained constant while V_1 is increased above the curve in Fig. 4, the coupling/frequency characteristic turns from maximally flat to equal-ripple. The ripple region for V_1 is indicated in Fig. 4.

3.2 10 dB couplers

A maximally flat 10 dB coupler has the quarter-wave filter parameters $V_1 = Z_1 = 1{\cdot}041$, $V_2 = Z_2/Z_1 = 1{\cdot}446$ (so $Z_2 = 1{\cdot}505$), and hence $k_1 = 0{\cdot}0403$, $k_2 = 0{\cdot}388$. As V_1

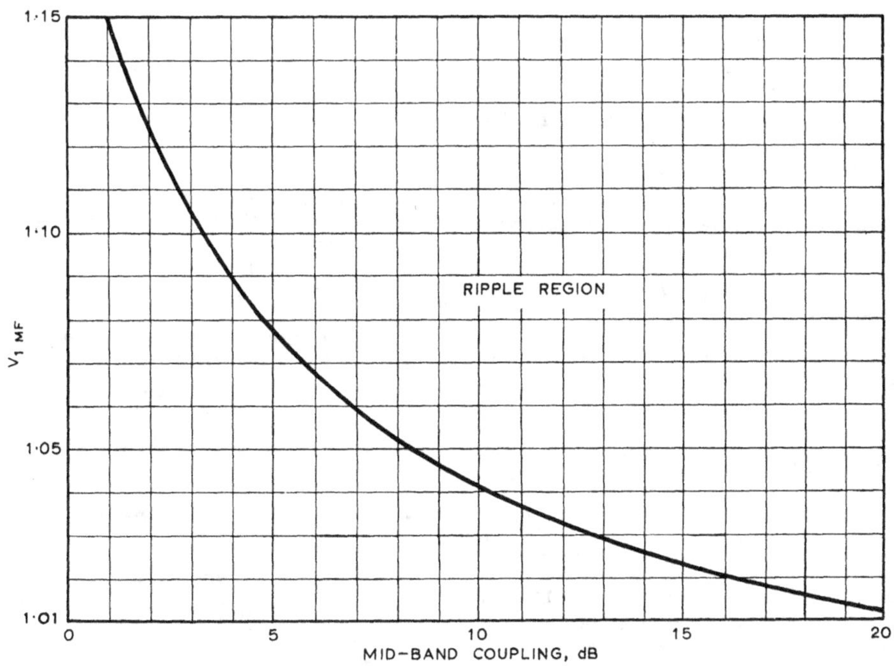

Fig. 4
End-junction v.s.w.r. (V_{1MF}) of maximally flat symmetrical 3-section coupler as a function of mid-band coupling

is increased, the coupling response develops ripples, as can be seen from Fig. 5, which shows six curves of coupling against frequency for $V_1 = 1{\cdot}04$, $1{\cdot}06$, $1{\cdot}08$, $1{\cdot}10$, $1{\cdot}12$, and $1{\cdot}14$. The coupling is symmetrical about the centre frequency in Fig. 5, and is therefore plotted only for frequencies above the mid-band. The mid-band coupling has been maintained at 10 dB by satisfying eqn. 22 with $R = 1{\cdot}926$. This R is obtained from eqn. 22 with $k_0 = 0{\cdot}3162$, corresponding to 10 dB coupling. For any given V_1, the parameters k_1 and k_2, and the even- and odd-mode impedances, Z_{oe_i} and Z_{oo_i} can be derived from eqns. 23–27.

3.3 3 dB couplers

Similar curves for 3 dB couplers are given in References 4 and 5, except that there it is the average coupling that is

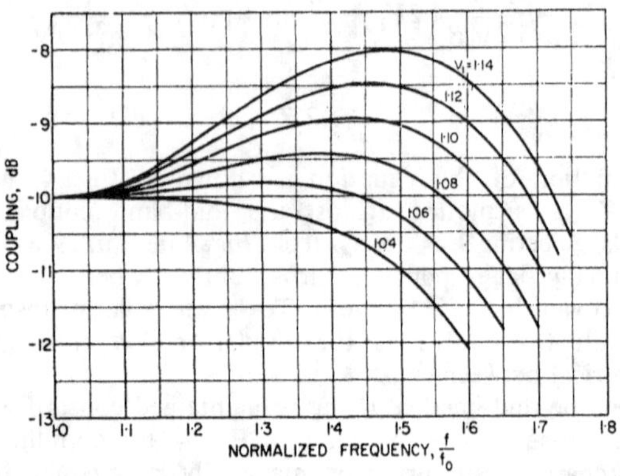

Fig. 5
Coupling as a function of frequency for six different symmetrical 3-section couplers having 10 dB mid-band coupling

maintained constant at 3 dB. In Fig. 5, the mid-band coupling was maintained constant at 10 dB.

134

3.4 0 dB coupling

It can be shown for the maximally flat 3-section coupler that in the limit as R tends to infinity, and as the coupling ratio therefore tends to unity (0 dB), V_1 tends ultimately to $\sqrt{(3/2)}$ (V_2 tends to infinity). Thus V_1 for the maximally flat 3-section coupler never exceeds $\sqrt{(3/2)} = 1{\cdot}224745$, or equivalently, k_1 never exceeds $0{\cdot}2$. Of course, it is not possible ever to attain 0 dB coupling, just as it is impossible ever to attain complete reflection in a stepped-impedance filter.

3.5 Equal-ripple couplers

Let f_1 and f_2 be the band-edge frequencies in the equal-ripple pass-band. Let the fractional bandwidth w be defined, as usual, by

$$w = \frac{f_2 - f_1}{f_0} = 2\left(\frac{f_2 - f_1}{f_2 + f_1}\right) = \frac{2}{\pi}(\theta_2 - \theta_1) \quad . \quad (30)$$

since

$$f/f_0 = 2\theta/\pi \; . \quad . \quad . \quad . \quad . \quad . \quad (31)$$

The ripple, or maximum deviation from average coupling, in the pass-band, may be defined as half the difference between the maximum coupling and the minimum coupling in the pass-band; for instance, for the $V_1 = 1{\cdot}14$ curve in Fig. 5 the ripple is nearly $1{\cdot}0$ dB. The fractional bandwidth w is plotted against the ripple in decibels in Fig. 6. The middle curve,

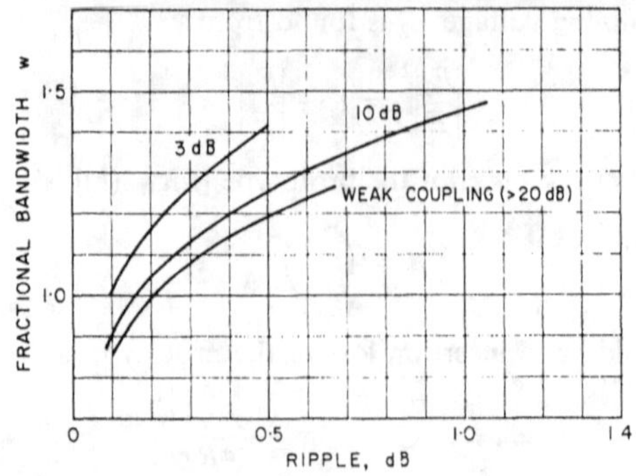

Fig. 6
Fractional bandwidth as a function of ripple for symmetrical 3-section couplers with 3 dB, 10 dB, and weak mid-band coupling

10 dB, is taken off Fig. 5. The 3 dB curve* is adapted from References 4 and 5. The first-order curve also comes from this source.

The fractional bandwidth w and the ripple are plotted against $(V_1 - 1)/(V_{1MF} - 1)$ in Fig. 7, where V_{1MF} is the

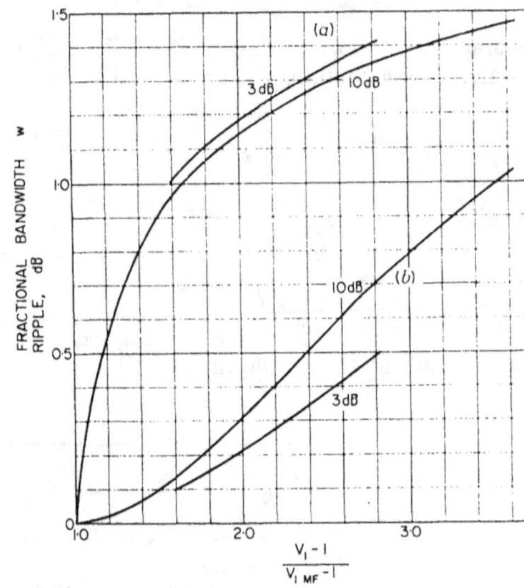

Fig. 7
Fractional bandwidth and ripple as V_1 is increased for symmetrical 3-section couplers with 3 dB, 10 dB, ~~and weak~~ mid-band coupling
(*a*) Fractional bandwidth
(*b*) Ripple

value of V_1 needed to achieve maximally flat coupling response. This is plotted for 10 dB and 3 dB couplers, and it is seen that, plotted in this way, the curves are fairly independent of the coupling strength, and can be used in conjunction with Fig. 4 to design for any coupling and ripple with fair accuracy.

4 Symmetrical couplers of five and more sections

4.1 First-order design theory

It is not difficult to show that for weak couplings the amplitude coupling ratio, $k(\theta)$, of an n-section symmetrical coupler is, as a function of θ,

* There is a slight inconsistency in these curves owing to a lack of more complete numerical data: the 10 dB curve refers to 10 dB at centre frequency; the 3 dB curve refers to 3 dB average over the band

$$\left|\frac{E_2}{E}\right| = k_1 \sin n\theta + (k_2 - k_1) \sin (n-2)\theta + \ldots$$
$$+ (k_i - k_{i-1}) \sin (n - 2i + 2)\theta + \ldots$$
$$+ [k_{(n+1)/2} - k_{(n-1)/2}] \sin \theta \quad . \quad . \quad . \quad . \qquad (32)$$

in the notation of Fig. 8. If, for example, it is required to

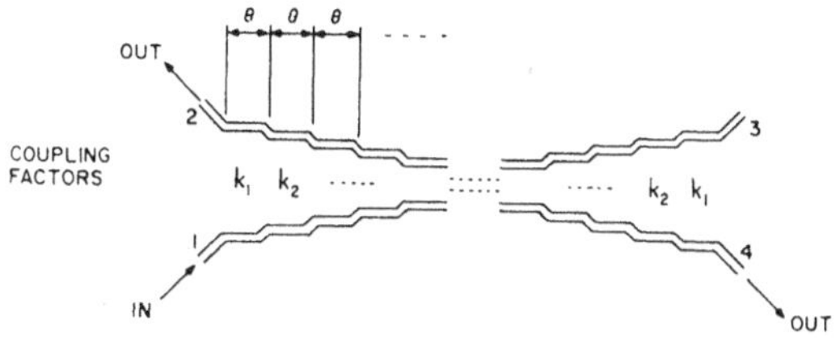

Fig. 8
Symmetrical TEM-mode multi-section directional coupler

obtain maximally flat coupling, the k_i must satisfy a set of $(n-1)/2$ linear equations

$$\left(\frac{d^p k(\theta)}{d\theta^p}\right)_{\theta=\pi/2} = 0 \quad . \quad . \quad . \quad . \quad . \qquad (33)$$

for $\qquad p = 2, 4, 6, \ldots n-1$

in addition to one equation, determining the centre-frequency coupling. The solutions of the system of eqns. 33 for $n = 3$ and $n = 5$ can be shown to be,

$$\text{for } n = 3 \qquad \left.\begin{aligned} k_2 &= 10k_1 \\ k_0 &= k_2 - 2k_1 = 8k_1 \end{aligned}\right\} \quad . \quad . \quad . \qquad (34)$$

$$\text{and for } n = 5 \qquad \left.\begin{aligned} &\frac{k_3}{536} = \frac{k_2}{84} = \frac{k_1}{9} \\ &k_0 = k_3 - 2k_2 + 2k_1 = 386k_1/9 \end{aligned}\right\} . \qquad (35)$$

It can also be shown that eqn. 29 for the maximally flat 3-section coupler reduces in the first-order approximation to

$$V_1 = \frac{7 + \sqrt{R}}{8} \quad . \quad . \quad . \quad . \quad . \quad . \qquad (36)$$

Eqn. 32 is a Fourier series with $(n+1)/2$ terms, and can be fitted closely to, for instance, a rectangular coupling response, thus giving nearly constant coupling over a prescribed frequency band. The coupling coefficients k_i can be found by equating coefficients in the truncated Fourier expansion of the ideal rectangular response. This method has been used by K. I. Khoury.[16]

4.2 Solutions for 5-section couplers

As for the 3-section coupler, the solution of the 5-section coupler is most easily expressed in terms of the normalized quarter-wave filter impedances Z_1, Z_2, and Z_3, which are again related to the coupling factors k_i, (Fig. 8) by

$$\left.\begin{aligned} Z_i^2 &= \frac{1 + k_i}{1 - k_i} \\ \text{or} \qquad k_i &= \frac{Z_i^2 - 1}{Z_i^2 + 1} \\ i &= 1, 2, 3 \end{aligned}\right\} \quad . \quad . \quad . \qquad (37)$$

corresponding to eqn. 23 for the 3-section coupler. Again the $Z_{oe\,i}$ and $Z_{oo\,i}$ are obtained as in eqns. 24 and 25, but with $i = 1, 2, 3$.

The first-order solution in eqn. 35 for maximally flat coupling must become less accurate as the k_i approach unity, since they can never exceed unity. Eqn. 35 becomes more accurate for appreciable couplings if we substitute $\log Z_i$ for k_i. Compare this with the similarly expedient substitution used in Reference 12 in the first-order theory of quarter-wave transformers. Then for maximally flat couplers, to a good approximation,

$$\frac{\log Z_3}{536} = \frac{\log Z_2}{84} = \frac{\log Z_1}{9} \quad . \quad . \quad . \quad . \qquad (38)$$

The Z_1, Z_2 and Z_3 are completely determined when the mid-band coupling factor k_0 is also specified, from the formula

$$\log Z_3 - 2\log Z_2 + 2\log Z_1 = \tfrac{1}{2}\log\left(\frac{1+k_0}{1-k_0}\right) \quad . \qquad (39)$$

The first-order equations, 38 and 39, for maximally flat coupling, have been found to hold very well for coupling as tight as 10 dB ($k_0^2 = 0{\cdot}1$). The coupling becomes weaker as the frequency deviates from mid-band. The following fractional bandwidths are obtained for various coupling deviations from the mid-band k_0, when k_0 corresponds to 10 dB or weaker coupling: $w = 0{\cdot}82$ for 0·1 dB deviation; $w = 1{\cdot}01$ for 0·25 dB deviation; $w = 1{\cdot}13$ for 0·5 dB deviation; $w = 1{\cdot}32$ for 1·0 dB deviation; and $w = 1{\cdot}51$ for 2·0 dB deviation.

The first-order relations in eqns. 38 and 39 for maximally flat coupling were even solved for a 3 dB coupler ($k_0^2 = 0{\cdot}5$) with the following results. The solution is easily found to be

$$\left.\begin{aligned} Z_1 &= 1{\cdot}0207 \\ Z_2 &= 1{\cdot}2114 \\ Z_3 &= 3{\cdot}4004 \end{aligned}\right\} \quad . \quad . \quad . \quad . \quad . \qquad (40)$$

As the frequency departs from mid-band, the coupling at first becomes stronger, going from 3·01 dB to 2·97 dB, and then becomes weaker. It is 3·11 dB at the edges of a 103% band ($w = 1{\cdot}03$), and 3·51 dB at the edges of a 124% ($w = 1{\cdot}24$). 135

An exact solution for a particular 3 dB coupler has been given by Cohn and Koontz.[17] They found by trial and error that with

$$\left.\begin{aligned} Z_1 &= 1{\cdot}098 \\ Z_2 &= 1{\cdot}417 \\ Z_3 &= 4{\cdot}130 \end{aligned}\right\} \quad . \quad . \quad . \quad . \quad . \qquad (41)$$

an equal-ripple characteristic with three ripples was obtained, the coupling being $3{\cdot}01 \pm 0{\cdot}163$ dB over a fractional bandwidth of $w =$ ~~1·70~~ The mid-band coupling is 2·847 dB. 1.41

The coupler is always determined by the Z_i, which give the k_i from eqn. 37 and the $Z_{oe\,i}$ and $Z_{oo\,i}$ from eqns. 24 and 25. The physical dimensions can then be obtained from References 7 or 13–15.

5 Characteristics of symmetrical and asymmetrical couplers with *n* sections

5.1 Insertion-loss formulae for directional couplers

The analysis of a coupler or filter is straightforward in principle, but for more than a few sections would, in practice, have to be carried out by a digital computer. On the other hand, the synthesis of such a coupler depends on finding the appropriate insertion-loss function and then applying Riblet's procedure,[11] which is based on Richards's theorem.[18, 19] Such a function must be a polynomial in $\mu^2 = \cos^2\theta$ and

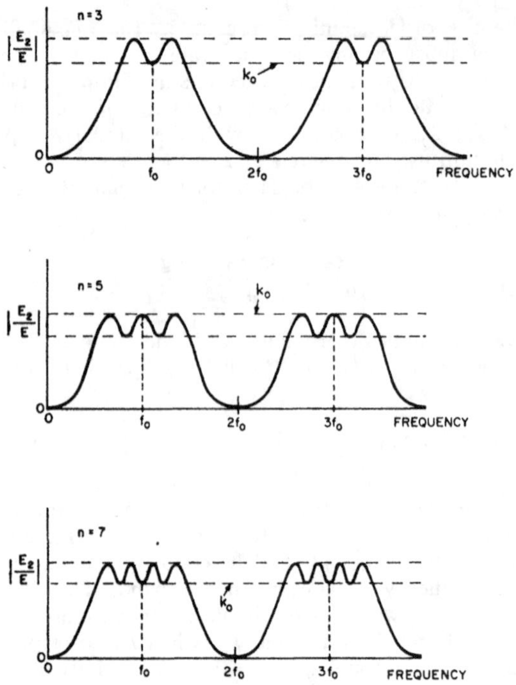

Fig. 9
Typical coupling characteristics of TEM-mode directional couplers. (The values of n *are for symmetrical couplers)*

should appear like the curves shown in Fig. 9. For instance the function

$$\left|\frac{E}{E_4}\right|^2 = 1 + \epsilon^2[T_n(1/\mu_0^2) - T_n(\mu^2/\mu_0^2)] \quad . \quad . \quad (42)$$

where T_n is a Chebÿshev polynomial and ϵ, μ_0 are constants, has for $n = 3$ the right appearance and number of ripples to fit Fig. 5. Unfortunately, it cannot be realized, at least for $n = 3$, by a symmetrical coupler. Another candidate is the function

$$\left|\frac{E}{E_4}\right|^2 = 1 + \epsilon^2[T_n^2(1/\mu_0) - T_n^2(\mu/\mu_0)] \quad . \quad . \quad (43)$$

proposed by Fel'dshtein.[9] It yields one more ripple for the same n than does the function in eqn. 42 but could not be fitted to Fig. 5; again, it cannot be realized by a symmetrical coupler, at least for $n = 3$. No insertion-loss function has as yet been discovered by the author which involves the mid-band coupling and the ripple as explicit parameters, and which is physically realizable with a symmetrical coupler. R. Levy[16] has recently been the first to construct asymmetrical couplers, which were exhibited in London.[16] He has arrived independently at a method of synthesis which is also based on a quarter-wave filter prototype circuit.

To ensure physical realizability, the desired insertion-loss function should yield impedances Z_i which are all greater than, or all less than, Z_0, since the ratio Z_{oe}/Z_{oo} in all known practical structures is greater than unity.

5.2 Phase relation between outputs

For a symmetrical coupler, the two outputs E_2 and E_4, shown for a 3-section coupler in Fig. 3, are in quadrature at all frequencies, i.e. differ by 90° when measured at the extremities of the coupler. This result follows at once from the quarter-wave-filter analogy, since for any symmetrical dissipation-free 2-port circuit, the reflected and transmitted wave amplitudes are orthogonal.[20]

5.3 Summary of basic equations

The TEM-mode coupler of n sections satisfies eqn. 1 for each section, and eqns. 22, 24, 25 and 37, with $i = 1, 2, \ldots n$. The mid-band v.s.w.r. R of the quarter-wave filter is

$$R = \left(\frac{Z_1 Z_3 Z_5 \ldots}{Z_2 Z_4 Z_6 \ldots}\right)^{\pm 2} > 1 \quad . \quad . \quad (44)$$

which for symmetrical couplers reduces to

$$\left.\begin{aligned} R &= (V_1 V_3 V_5 \ldots V_n)^2 \\ &= (V_2 V_4 V_6 \ldots V_{n+1})^2 \end{aligned}\right\} \quad . \quad . \quad . \quad (45)$$

6 Directional (channel-dropping) filters

The TEM-mode directional coupler is rooted in the impedance/admittance duality of the even and odd modes, as explained in Section 2. In the quarter-wave coupler this is expressed by eqn. 1. The duality concept can be generalized by substituting the even- and odd-mode input impedances Z_{ine} and Z_{ino} for the several characteristic impedances $Z_{oe\,i}$ and $Z_{oo\,i}$. The condition for a reflectionless directional coupler then generalizes to

$$Z_{ine} Z_{ino} = Z_0^2 \quad . \quad . \quad . \quad . \quad . \quad (46)$$

It is difficult to obtain good selectivity with a stepped-impedance filter or coupler because it is difficult to realize high impedance ratios with practical dimensions. However, ideal stepped-impedance filters have characteristics similar to those of practical reactance-coupled filters,[21] and indeed serve as prototypes[12, 22, 23] for them. In particular, a series-capacitance coupled filter is the dual of a shunt-inductance coupled filter; if the even mode can be made to operate as a series-capacitance coupled filter and the odd mode as a shunt-inductance coupled filter, or vice versa, both filters having the same discontinuity v.s.w.r., then instead of a wide-band directional coupler we can obtain a narrow-band directional filter. A possible circuit which could be made to function in this manner is suggested in Fig. 10. In the even mode the two parallel lines are at the same potential, and the principal discontinuities are due to the series-capacitive gaps; in the odd mode the two parallel lines carry balanced currents in opposite directions and the principal discontinuities are due to the shunt-inductive strips which join them. Since the discontinuity v.s.w.r.s have to be the same in each mode, and the two filters are dual on an impedance-admittance basis, the spacing between discontinuities is the same.[21, 22]

It is of interest to note that the directional filter of Fig. 10 differs from the directional filter of Coale[24] only in the distribution of lengths along a resonator rectangle, which is here long instead of square, but is again about one wavelength in circumference.

7 Conclusion

The analytical equivalence of TEM-mode coupled transmission lines, when $Z_{oe}Z_{oo} = Z_0^2$, and single transmission lines, has been established. This makes available an exact quantitative theory of stepped-impedance filters for the design of TEM-mode directional couplers.

It also makes new demands on that theory. For TEM-mode directional couplers, the prototype response is to be optimized in its stop-band, which becomes the pass-band of interest for the coupler. Thus, new insertion-loss functions have to be found, unlike those previously considered, having, for instance, equal-ripple characteristics in the transformer stop-band instead of in its pass-band (see Fig. 9).

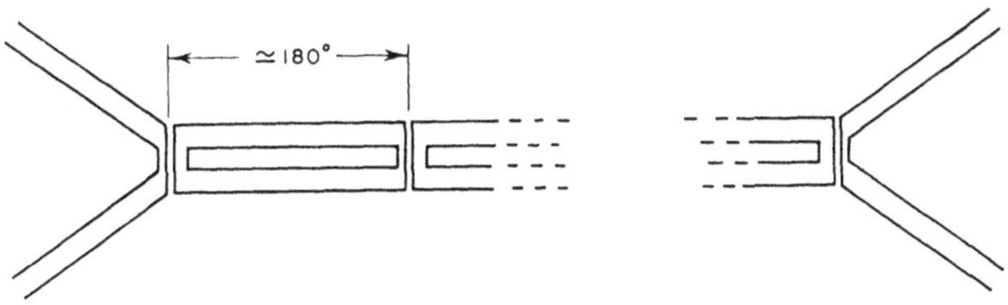

Fig. 10
A suggested directiona. filter circuit

A new directional filter has been suggested, and is shown in Fig. 10.

8 Acknowledgments

The author wishes to thank W. J. Getsinger and Dr. G. L. Matthaei for several helpful discussions. The work reported herein was supported by the U.S. Army Signal Research and Development Laboratory.

9 References

1 FIRESTONE, W. L.: 'Analysis of transmission line directional couplers,' *Proc. Inst. Radio Engrs*, 1954, **42**, p. 1529
2 OLIVER, B. M.: 'Directional electromagnetic couplers,' *ibid.*, 1954, **42**, p. 1686
3 MONTEATH, G. D.: 'Coupled transmission lines as symmetrical directional couplers', *Proceedings I.E.E.*, 1955, **102** B, p. 383
4 SHIMIZU, J. K.: 'Strip-line 3 dB directional couplers', *Institute of Radio Engineers, Wescon Convention Record*, 1957, **1**, Part 1, p. 4
5 SHIMIZU, J. K., and JONES, E. M. T.: 'Coupled-transmission-line directional couplers', *Trans Inst. Radio Engrs*, 1958, **MTT-6**, p. 403
6 JONES, E. M. T., and BOLLJAHN, J. T.: 'Coupled-strip-transmission-line filters and directional couplers', *ibid.*, 1956, **MTT-4**, p. 75
7 GETSINGER, W. J.: 'A coupled strip-line configuration using printed-circuit construction that allows very close coupling,' *ibid.*, 1961, **MTT-9**, p. 535
8 KOIKE, R.: 'Recent advances in the use of coupled transmission lines as directional couplers', *Proceedings I.E.E.*, 1961, **108 B**, p. 120
9 FEL'DSHTEIN, A. L.: 'Synthesis of stepped directional couplers,' *Radiotekhnika i Electronika*, 1961, **6**, No. 2, p. 234. English translation, Pergamon Press, 1961, p. 75
10 YOUNG, L.: 'Analysis of a transmission cavity wavemeter,' *Trans Inst. Radio Engrs*, 1960, **MTT-8**, p. 436 (The exponent in the numerator of eqn. 20 should be $-J\theta$ and not $-2J\theta$.)
11 RIBLET, H. J.: 'General synthesis of quarter-wave impedance transformers,' *ibid.*, 1957, **MTT-5**, p. 36
12 YOUNG, L.: 'Stepped-impedance transformers and filter prototypes,' *ibid.*, 1962, **MTT-10**, p. 339
13 COHN, S. B.: 'Shielded coupled-strip transmission lines,' *ibid.*, 1955, **MTT-3**, p. 29
14 COHN, S. B.: 'Characteristic impedances of broadside-coupled strip transmission lines,' *ibid.*, 1960, **MTT-8**, p. 633
15 COHN, S. B.: 'Thickness corrections for capacitive obstacles and strip conductors', *ibid.*, 1960, **MTT-8**, p. 638
16 YOUNG, LEO: Conference on components for microwave circuits, *Microwave*, 1962, **5**, No. 12, p. 28
17 COHN, S. B., and KOONTZ, R. H.: 'Microwave coupler study program,' 1962, Third Quarterly Progress Report, Rantec Corporation, Calabasas, California
18 RICHARDS, P. I.: 'Resistor-transmission-line circuits,' *Proc. Inst. Radio Engrs*, 1948, **36**, p. 217
19 RICHARDS, P. I.: 'A special class of functions with positive real part in a half-plane,' *Duke math. J.*, 1947, **14**, p. 777
20 YOUNG, LEO: 'A theorem on lossless symmetrical networks,' *Trans Inst. Radio Engrs*, 1960, **CT-7**, p. 75
21 COHN, S. B.: 'Direct-coupled-resonator filters', *Proc. Inst. Radio Engrs*, 1957, **45**, p. 187
22 YOUNG, LEO: 'The quarter-wave transformer prototype circuit', *Trans Inst. Radio Engrs*, 1960, **MTT-8**, p. 483
23 MATTHAEI, G. L., and YOUNG, L.: 'Microwave filters and coupling structures', 1962, Fifth Quarterly Progress Report, Stanford Research Institute, Menlo Park, California
24 COALE, F. S.: 'A travelling-wave directional filter,' *Trans Inst. Radio Engrs*, 1956, **MTT-4**, p. 256
25 SWEET, L. O.: 'Analysis of a two-section coupler,' *ibid.*, 1962, **MTT-10**, p. 295

138

Reprinted from IEEE TRANSACTIONS
ON MICROWAVE THEORY AND TECHNIQUES
Volume MTT-14, Number 7, July, 1966
Pp. 337-346

Coupled-Transmission-Line Directional Couplers with Coupled Lines of Unequal Characteristic Impedances

EDWARD G. CRISTAL, SENIOR MEMBER, IEEE

***Abstract*—A new class of coupled-transmission-line directional couplers, called "nonsymmetrical directional couplers," is described. Unlike conventional directional couplers, nonsymmetrical directional couplers use coupled lines of unequal characteristic impedances. The principal difference between the performance of nonsymmetrical directional couplers and that of conventional designs is the impedance level of the coupled waves, which may be changed to higher or lower impedance levels than that of the incident wave. These directional couplers may be designed to have infinite directivity and to be matched at all frequencies, or they may be designed to have infinite directivity at all frequencies and a specified maximum VSWR. Coupling relationships and design equations for both cases are presented, and the relative properties of both cases are discussed. The theoretical limitation on the maximum coupling and the maximum impedance transformation that can be obtained simultaneously are derived. Techniques for broadbanding by cascading additional sections of coupled lines are described. Experimental results of a trial −10-dB coupler with coupled lines of 50 and 75 ohms are presented.**

Manuscript received March 1, 1966; revised April 3, 1966. The work reported in this paper was supported in part by the U. S. Army Electronics Command Laboratories, Fort Monmouth, N. J., under Contract DA 28-043 AMC-01271(E).

The author is with Stanford Research Institute, Menlo Park, Calif.

Introduction

In the past, coupled-transmission-line directional couplers have been designed with coupled lines of equal characteristic impedances [1]–[6]. These couplers are used in many applications: power samplers, reflectometers, directional detectors, directional filters, and multiplexers are several. In this paper a new class of coupled-transmission-line directional couplers, called nonsymmetrical directional couplers, is described. In contrast to conventional directional couplers, nonsymmetrical directional couplers use coupled lines of unequal characteristic impedances. The principal difference between the performance of nonsymmetrical directional couplers and that of conventional directional couplers is the impedance level of the coupled waves, which may be changed to higher or lower impedance levels than that of the incident wave. Nonsymmetrical directional couplers, therefore, act as conventional directional couplers combined with impedance transformers at the two ports of one of the transmission lines. Thus, nonsymmetrical directional couplers should prove useful in applications calling for directional coupling and impedance transforming in combination, since these functions can be accomplished in a single device. For example, a practical application for which the new device may prove useful is in directional detectors [7], wherein the diode detector might be better matched in a line of different impedance from that of the main line.

The nomenclature *nonsymmetrical* directional coupler pertains to the side-by-side asymmetry of the directional coupler. It should not be confused with *cascaded* asymmetrical [4] or *cascaded* symmetrical [5], [6] directional couplers, which use coupled lines of equal characteristic impedances and have end-to-end asymmetry or symmetry, respectively.

Figure 1 shows the nonsymmetrical coupled-transmission-line directional coupler in diagrammatical form, and it also specifies voltage and current coordinates used throughout this paper for such couplers. The structure in the figure is to be regarded as two uniformly coupled transmission lines of unequal characteristic impedances that have equal propagation constants in balanced (odd-mode) and unbalanced (even-mode) excitation [1]. Although the structure shown has end-to-end geometrical symmetry, the conclusions and design equations presented herein only require that the structure have end-to-end electrical symmetry. While nonsymmetrical directional couplers may be realized in many configurations that satisfy these conditions, the electrical characteristics are independent of the physical configuration. The admittance matrix corresponding to the voltage and current coordinates of Fig. 1 is [8][1]

$$\begin{bmatrix} \frac{Y_{oo}^a+Y_{oe}^a}{2}\cdot\frac{1}{s} & -\frac{Y_{oo}^a-Y_{oe}^a}{2}\cdot\frac{1}{s} & \frac{Y_{oo}^a-Y_{oe}^a}{2}\cdot\frac{\sqrt{1-s^2}}{s} & -\frac{Y_{oo}^a+Y_{oe}^a}{2}\cdot\frac{\sqrt{1-s^2}}{s} \\ -\frac{Y_{oo}^b-Y_{oe}^b}{2}\cdot\frac{1}{s} & \frac{Y_{oo}^b+Y_{oe}^b}{2}\cdot\frac{1}{s} & -\frac{Y_{oo}^b+Y_{oe}^b}{2}\cdot\frac{\sqrt{1-s^2}}{s} & \frac{Y_{oo}^b-Y_{oe}^b}{2}\cdot\frac{\sqrt{1-s^2}}{s} \\ \frac{Y_{oo}^b-Y_{oe}^b}{2}\cdot\frac{\sqrt{1-s^2}}{s} & -\frac{Y_{oo}^b+Y_{oe}^b}{2}\cdot\frac{\sqrt{1-s^2}}{s} & \frac{Y_{oo}^b+Y_{oe}^b}{2}\cdot\frac{1}{s} & -\frac{Y_{oo}^b-Y_{oe}^b}{2}\cdot\frac{1}{s} \\ -\frac{Y_{oo}^a+Y_{oe}^a}{2}\cdot\frac{\sqrt{1-s^2}}{s} & \frac{Y_{oo}^a-Y_{oe}^a}{2}\cdot\frac{\sqrt{1-s^2}}{s} & -\frac{Y_{oo}^a-Y_{oe}^a}{2}\cdot\frac{1}{s} & \frac{Y_{oo}^a+Y_{oe}^a}{2}\cdot\frac{1}{s} \end{bmatrix} \tag{1}$$

where $s=\tanh \gamma L$, γ being the complex propagation constant and L the length of the coupled section. In the important case of lossless lines, $s=j\tan\theta$, where θ is the electrical length of the coupled sections, and $j=\sqrt{-1}$.

- Y_{oe}^a is the admittance measured at Port 1 for an infinite section of coupled lines excited in the even mode with voltage sources at Ports 1 and 2.
- Y_{oo}^a is the admittance measured at Port 1 for an infinite section of coupled lines excited in the odd mode with voltage sources at Ports 1 and 2.
- Y_{oe}^b is the admittance measured at Port 2 for an infinite section of coupled lines excited in the even mode with voltage sources at Ports 1 and 2.
- Y_{oo}^b is the admittance measured at Port 2 for an infinite section of coupled lines excited in the odd mode with voltage sources at Ports 1 and 2.

The following notation simplifies the design equations presented later and is therefore adopted:

$$A=\frac{Y_{oo}^a+Y_{oe}^a}{2} \tag{2}$$

$$B=\frac{Y_{oo}^b+Y_{oe}^b}{2} \tag{3}$$

$$D=\frac{Y_{oo}^a-Y_{oe}^a}{2}\equiv\frac{Y_{oo}^b-Y_{oe}^b}{2}. \tag{4}$$

G_a and G_b are the terminating admittances of lines a and b, respectively.

In the derivations and equations that follow, excitation at Port 1 has been assumed. However, had excitation at Port 2 been assumed, the results could be made

[1] B. M. Schiffman pointed out that the Y-matrix given in Ozaki and Ishii [8] is incorrect with respect to the signs of its entries; the correct matrix is given by (1).

139

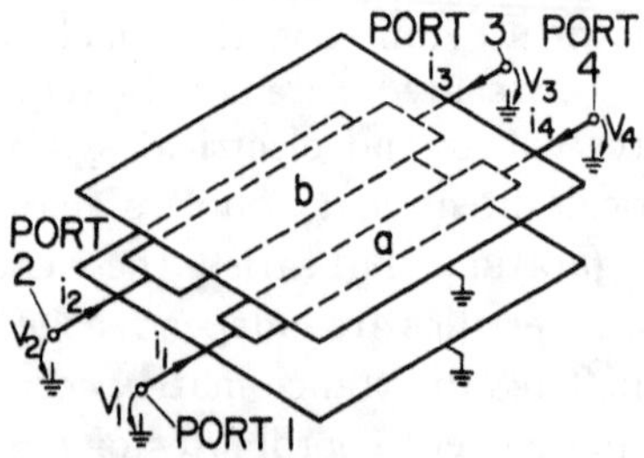

Fig. 1. Nonsymmetrical coupled - transmission - line directional coupler with coupled lines of unequal characteristic impedances.

to carry over directly by an interchange of G_a and G_b and an interchange of A and B. All formulas presented in the following sections were obtained by operating on the admittance matrix of (1) directly. The resulting formulas are thus exact for all values of coupling.

Condition for Infinite Directivity

The directivity of a directional coupler may be defined as follows. Let each transmission line of the coupler be terminated in its respective load admittance at Ports 2, 3, and 4. Let a signal be incident at Port 1. The directivity in decibels is given by

$$20 \log_{10} \left| \frac{v_2}{v_3} \right| . \tag{5}$$

Thus the condition for infinite directivity is that v_3 be equal to zero.

A method of solving for the directivity is given in Appendix I. In the interest of shortening the presenta-
140 tion, only the final result is given here. The condition for infinite directivity is found to be[2]

$$G_a G_b = AB - D^2. \tag{6}$$

There is no other constraint. Note that G_a and G_b need not be real to satisfy (6), although in the remainder of this section they are so assumed.

Conditions for Impedance Matching

A method of determining the conditions for impedance matching is given in Appendix II; again, only the final results are presented here. There are two conditions for impedance matching. They are

$$\frac{G_a}{G_b} = \frac{A}{B} \tag{7}$$

$$G_a G_b = AB - D^2. \tag{8}$$

Notice that the condition for infinite directivity is included in the conditions for impedance matching.

Thus a matched directional coupler must have infinite directivity. The converse, however, need not be so. A nonsymmetrical directional coupler may have infinite directivity without being matched. (This conclusion may also be established using the scattering matrix for a directional coupler [9].)

Additional consequences of these conditions are described later in the section in which design equations for mismatched directional couplers having infinite directivity are presented.

Equations (7) and (8) may be solved for G_a and G_b, giving

$$G_a = \sqrt{\frac{A}{B}} \sqrt{AB - D^2} \tag{9}$$

$$G_b = \sqrt{\frac{B}{A}} \sqrt{AB - D^2}. \tag{10}$$

Note that the right sides of (9) and (10) are real positive numbers for all physically realizable values of A, B, and D. Therefore, the right side of (9) and (10) may be interpreted as the characteristic admittances of lines a and b in the presence of each other.

Voltage and Power Coupling Relationships

By procedures similar to those discussed in Appendixes I and II, the transfer ratio v_2/i_1 in the lossless case is found to be

$$\frac{v_2}{i_1} = \frac{jk \tan\theta}{\sqrt{1-k^2} + j \tan\theta} \frac{1}{\sqrt{G_a G_b}}, \tag{11}$$

and the coupled power ratio is found to be

$$\frac{P_2}{P_a} = \frac{k^2 \tan^2\theta}{(1-k^2) + \tan^2\theta}, \tag{12}$$

where

$$k = \frac{D}{\sqrt{AB}} \tag{13}$$

is the square root of the power coupling coefficient, and P_a is the available power at Port 1 from a current source of strength I with internal conductance G_a.

Similarly, it is found that

$$\frac{v_4}{i_1} = \frac{1}{G_a} \frac{\sqrt{1-k^2}\sqrt{1+\tan^2\theta}}{\sqrt{1-k^2} + j\tan\theta}, \tag{14}$$

$$\frac{P_4}{P_a} = \frac{(1-k^2)(1+\tan^2\theta)}{(1-k^2) + \tan^2\theta}. \tag{15}$$

From (11) and (14) it is seen that v_2 and v_4 are 90 degrees out of phase at all frequencies. Comparisons of (12) and (15) with the results given in Jones and Bolljahn [1] show that the coupled power and bandwidth of nonsymmetrical directional couplers are identical to those of conventional directional couplers, provided that the coefficients of coupling in the two cases are the same.

[2] Using a low frequency approximation, Firestone [9] showed that the condition for infinite directivity for coupled open-wire lines is (in our notation) $G_a G_b = C_m / L_m$, where C_m and L_m are the mutual capacitance and inductance per unit length of the coupled lines. It can be shown that this equation and (6) are equivalent, so that in fact the expression originally derived by Firestone for electrically short couplers is true in general.

Design Equations

Recall that the coefficient of coupling and the characteristic admittance of the coupled lines are given by

$$k = D/\sqrt{AB},$$

$$G_a = \sqrt{\frac{A}{B}}\sqrt{AB - D^2},$$

$$G_b = \sqrt{\frac{B}{A}}\sqrt{AB - D^2}. \quad (16)$$

Equation (16) may be solved in conjunction with (2)–(4) to yield

$$Y_{oe}{}^a = \frac{G_a - k\sqrt{G_aG_b}}{\sqrt{1 - k^2}} \quad (17)$$

$$Y_{oo}{}^a = \frac{G_a + k\sqrt{G_aG_b}}{\sqrt{1 - k^2}} \quad (18)$$

$$Y_{oe}{}^b = \frac{G_b - k\sqrt{G_aG_b}}{\sqrt{1 - k^2}} \quad (19)$$

$$Y_{oo}{}^b = \frac{G_b + k\sqrt{G_aG_b}}{\sqrt{1 - k^2}} \cdot \quad (20)$$

Equations (17)–(20) are the fundamental equations from which nonsymmetrical (or conventional) directional couplers may be designed from specifications that give the coupling and terminating admittances.

The theoretical limitation on the amount of impedance transformation and coupling that can be accomplished simultaneously is readily derived from (17) and (19). In order for $Y_{oe}{}^a$ and $Y_{oe}{}^b$ to be positive, it is necessary that

$$\frac{1}{k^2} \geq \frac{G_a}{G_b} \text{ and } \frac{G_b}{G_a} \cdot \quad (21)$$

Equation (21) may be stated in words: *Let the impedance (or admittance) transformation ratio of the directional coupler be taken as larger than 1. Then, the reciprocal of the power coupling coefficient must be greater than or equal to the impedance (or admittance) transformation ratio.*

Thus, for example, a single-section −3.01-dB coupler has a theoretical limit of impedance transformation ratio of 2 to 1; a single-section −10-dB coupler has a theoretical limit of impedance transformation ratio of 10 to 1.

Figure 2 shows in diagrammatical form a general type of geometry frequently used in constructing shielded coupled-transmission-line directional couplers. Also shown are the several capacitances associated with the coupled lines when operated in the TEM mode.

The capacitance matrix associated with Fig. 2 is

$$C = \begin{pmatrix} C_a + C_{ab} & -C_{ab} \\ -C_{ab} & C_b + C_{ab} \end{pmatrix}. \quad (22)$$

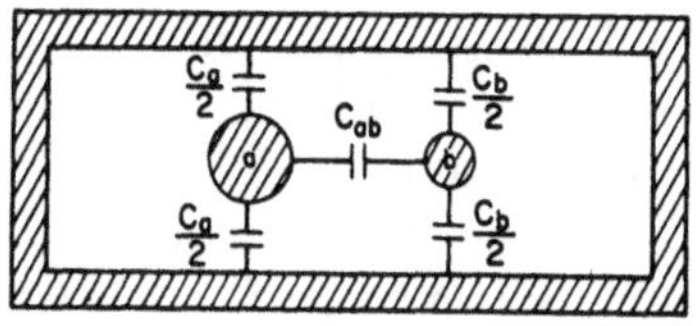

Fig. 2. A type of geometry frequently used in constructing shielded coupled-transmission-line directional couplers. (The lines a and b may be of any useful geometrical form.)

It is evident from Fig. 2 and (22) that the electrical characteristics of a coupled-transmission-line directional coupler of the type shown in Fig. 2 are completely specified when the capacitances C_a, C_b, and C_{ab} are known. The capacitances are related to the even- and odd-mode admittances by the following relationships:

$$Y_{oe}{}^a = vC_a \quad (23)$$

$$Y_{oe}{}^b = vC_b \quad (24)$$

$$Y_{oo}{}^a = v(C_a + 2C_{ab}) \quad (25)$$

$$Y_{oo}{}^b = v(C_b + 2C_{ab}), \quad (26)$$

where v is the velocity of propagation in the medium in which the waves travel.

Equations (17)–(20) may be solved in conjunction with (23)–(26) to yield the following convenient dimensionless expressions:

$$\frac{C_a}{\epsilon} = \frac{376.7}{\sqrt{\epsilon_r}}\left\{\frac{G_a - k\sqrt{G_aG_b}}{\sqrt{1 - k^2}}\right\} \quad (27)$$

$$\frac{C_b}{\epsilon} = \frac{376.7}{\sqrt{\epsilon_r}}\left\{\frac{G_b - k\sqrt{G_aG_b}}{\sqrt{1 - k^2}}\right\} \quad (28)$$

$$\frac{C_{ab}}{\epsilon} = \frac{376.7}{\sqrt{\epsilon_r}} \frac{k\sqrt{G_aG_b}}{\sqrt{1 - k^2}}, \quad (29)$$

where ϵ is the dielectric constant of the medium in which the waves propagate, ϵ_r is the relative dielectric constant of the medium, and 376.7 is the impedance of free space, calculated from $\sqrt{\mu_0/\epsilon_0}$.

By means of various available data [10]–[12], the normalized capacitances given in (27)–(29) can be related to physical dimensions of structures, thus permitting several possible realizations of the couplers to be made.

Broadbanding Matched Nonsymmetrical Directional Couplers

Matched nonsymmetrical directional couplers may be broadbanded, just as conventional directional couplers, by cascading two or more single-section couplers. Moreover, the existing tables of designs for asymmetrical and symmetrical multisection couplers [4]–[6] may be applied directly to designs for multisection nonsymmetrical directional couplers by interpreting the tables of designs in the proper way.

It is readily proved that nonsymmetrical directional couplers are mathematically equivalent to conventional directional couplers with transformers connected at both ports of one of the coupled lines. To show this, first replace B by A in (1) so that the matrix now represents a conventional directional coupler. Next, multiply rows 2 and 3 and columns 2 and 3 by $N=\sqrt{B/A}$. This operation returns the matrix to its original form with D replaced by ND. Since the preceding mathematical operation is physically equivalent to connecting transformers [14] of turns ratio $1:\sqrt{B/A}$ to Ports 2 and 3 of the conventional directional coupler, the equivalence $A=A'$, $N=\sqrt{B'/A'}$, and $D=D'/N$ is established, and the proof is complete. In the immediately preceding equations the primed symbols refer to the parameters of the nonsymmetrical directional coupler, and the unprimed symbols to those of the conventional directional coupler. Substitution of the primed and unprimed parameters into (13) shows that the coupling is invariant under these operations.

On the basis of this result, it can be concluded that the mathematical equivalence between stepped impedance lines and conventional directional couplers [15] carries over to nonsymmetrical directional couplers as well. By the same argument, the tables of designs for multisection conventional directional couplers [4]–[6] may be taken over for designs of multisection nonsymmetrical directional couplers.

For multisection nonsymmetrical directional couplers the design method is to cascade nonsymmetrical directional couplers which, when considered individually, have coupling coefficients corresponding to those in the tables. Because the data of Levy [4], Toulios and Todd [5], and Cristal and Young [6] are given in terms of normalized even-mode impedances, the coefficients of coupling may be obtained from the tables by the relationship

$$k = \frac{Z_{oe}^2 - 1}{Z_{oe}^2 + 1}. \tag{30}$$

In practice, the design procedure is as follows: 1) Choose a suitable design from one of the tables for multisection couplers; 2) calculate the coupling coefficient for each section by means of (30); and 3) obtain the even- and odd-mode admittances by (17)–(20), or the capacitances by means of (27)–(29). An example will be given to illustrate this procedure.

It is desired to design a −10-dB directional coupler to have less than 0.55-dB ripple and to have a 4 to 1 equal-ripple bandwidth. Let it also be assumed that an impedance transformation of 50 to 75 ohms is required. From Levy's design tables [4] it is found that a two-section asymmetrical directional coupler meets the requirements. The design obtained from the tables calls for one section to have $Z_{oe}^{(1)}/Z_0=1.5729$, and the second section to have $Z_{oe}^{(2)}/Z_0=1.1573$, where the superscript indicates the section number. The coefficient of coupling of each section is then computed from (30), giving

$$k^{(1)} = 0.4243 \tag{31}$$

$$k^{(2)} = 0.1450. \tag{32}$$

Finally, from (27)–(29), it is determined that the normalized capacitances for the two sections are

$$\begin{aligned} \text{Section 1}\quad & C_a^{(1)}/\epsilon = 5.438/\sqrt{\epsilon_r} \\ & C_b^{(1)}/\epsilon = 2.664/\sqrt{\epsilon_r} \\ & C_{ab}^{(1)}/\epsilon = 2.882/\sqrt{\epsilon_r} \end{aligned} \tag{33}$$

$$\begin{aligned} \text{Section 2}\quad & C_a^{(2)}/\epsilon = 6.713/\sqrt{\epsilon_r} \\ & C_b^{(2)}/\epsilon = 4.174/\sqrt{\epsilon_r} \\ & C_{ab}^{(2)}/\epsilon = 0.9018/\sqrt{\epsilon_r}. \end{aligned} \tag{34}$$

Mismatched Directional Couplers Having Infinite Directivity

The equations presented in previous sections showed that nonsymmetrical directional couplers may be mismatched and still have infinite directivity. It has been found that there are certain advantages to be obtained by purposely designing nonsymmetrical directional couplers to be mismatched: for a given coupling coefficient, both greater bandwidth and larger maximum impedance transformation ratio are obtained. These advantages must, of course, be weighed against the disadvantages of the mismatch. The exact relationships are presented in this section, along with coupling equations, design equations, and other data.

The condition for infinite directivity as given by (6) was

$$G_aG_b = AB - D^2.$$

If only this constraint is used, it is found that

$$\frac{v_4}{i_1} = \frac{G_b\sqrt{1+\tan^2\theta}}{G_aG_b + jAG_b\tan\theta} \tag{35}$$

$$\frac{v_2}{i_1} = \frac{jD\tan\theta}{G_aG_b + jAG_b\tan\theta}. \tag{36}$$

By applying the principle of conservation of energy, we obtain

$$|i_1|^2 \operatorname{Re}(Z_{in}) = |v_4|^2G_b + |v_2|^2G_a. \tag{37}$$

Substituting (35) and (36) and solving for the Re (Z_{in}) gives

$$G_a \operatorname{Re}(Z_{in}) = \frac{(1-k^2)+\tan^2\theta}{(1-k^2)+R\tan^2\theta}, \tag{38}$$

where $R=AG_b/BG_a$ is the normalized input admittance at Port 1 at $\theta=90$ degrees.

From (38), we can readily reconstruct [13] $Z_{in}(s)$ as follows:

$$G_aZ_{in}(s) = \frac{\sqrt{1-k^2}+(1/\sqrt{R})s}{\sqrt{1-k^2}+\sqrt{R}\,s}. \tag{39}$$

Lastly, from (35), (36), and (39), the power coupling ratios and reflected power ratio are obtained:

$$\frac{P_2}{P_a} = \frac{k^2 \tan^2 \theta}{(1 - k^2) + \dfrac{(R+1)^2}{4R} \tan^2 \theta} \tag{40}$$

$$\frac{P_4}{P_a} = \frac{(1 - k^2)(1 + \tan^2 \theta)}{(1 - k^2) + \dfrac{(R+1)^2}{4R} \tan^2 \theta} \tag{41}$$

$$\frac{P_r}{P_a} = \frac{\left(\dfrac{R-1}{2\sqrt{R}}\right)^2 \tan^2 \theta}{(1 - k^2) + \dfrac{(R+1)^2}{4R} \tan^2 \theta}. \tag{42}$$

From (42) it is found that

$$r_{max} = (\text{vswr})_{max}$$

$$= R \quad \text{or} \quad \frac{1}{R} \ \text{(whichever is greater than 1)} \tag{43}$$

and that r_{max} occurs at $\theta = 90$ degrees.

It is worth pointing out several conclusions that may be drawn from the previous expressions. First, dividing (36) by (35) gives

$$\frac{v_2}{v_4} = \frac{jD \sin \theta}{G_b}. \tag{44}$$

Thus v_2 and v_4 are 90 degrees out of phase for all frequencies.

Second, the ratio of P_2 to P_4 is

$$\frac{P_2}{P_4} = \frac{k^2}{1 - k^2} \sin^2 \theta, \tag{45}$$

which shows that the ratio of the power split between Ports 2 and 4 is independent of the amount of mismatch.

Third, from (39) it can be shown that

$$\{G_a Z_{in}(s)\}_{\text{Port 1}} = \left\{\frac{Y_{in}(s)}{G_b}\right\}_{\text{Port 2}}. \tag{46}$$

Thus the reflection coefficients at Ports 1 and 2 are negatives of each other.

Fourth, the ratio of P_2/P_a to $(P_2/P_a)_{max}$ is given by

$$\frac{(P_2/P_a)}{(P_2/P_a)_{max}} = \frac{\mathcal{R} u^2}{1 + \mathcal{R} u^2} \tag{47}$$

where

$$u = \frac{\tan \theta}{\sqrt{1 - k^2}} \tag{48}$$

$$\mathcal{R} = \frac{(r_{max} + 1)^2}{4 r_{max}}. \tag{49}$$

P_2/P_a will be 3 dB down from its maximum value when

$$\frac{\mathcal{R} u^2}{1 + \mathcal{R} u^2} = \frac{1}{2}. \tag{50}$$

Solving (50) gives

$$u = \mathcal{R}^{-1/2}. \tag{51}$$

Therefore,

$$\theta_{3\,dB} = \tan^{-1}\left\{\sqrt{1 - k^2}\, \frac{2\sqrt{r_{max}}}{r_{max} + 1}\right\}, \tag{52}$$

where $\theta_{3\,dB}$ is the angle satisfying (50).

Note that for a given k, $\theta_{3\,dB}$ is a monotonic decreasing function of r_{max}, and is a maximum for $r_{max} = 1$. Since the 3-dB bandwidths are given by formula

$$BW = \frac{180^\circ}{\theta^\circ_{3\,dB}} - 1, \tag{53}$$

it is readily demonstrated that the 3-dB bandwidths of nonsymmetrical directional couplers are always increased by designing the couplers to be mismatched. However, for weak coupling ($k^2 < 0.1$) and $r_{max} < 2$, the increase in 3-dB bandwidth is less than 3.8 percent; and for larger values of k, the percentage increase in bandwidth is less. For $k = 0.707$ (−3.01-dB coupling for matched couplers) and $r_{max} = 2$, the increase in bandwidth is only 2.9 percent. The fractional bandwidth[3] versus r_{max} for several values of coupling is shown in Fig. 3.

143

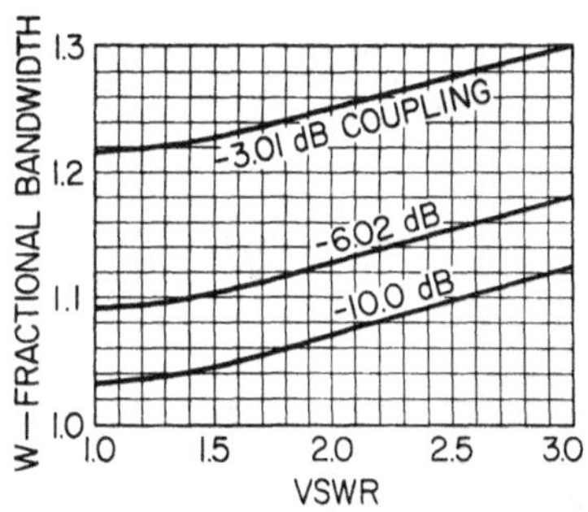

Fig. 3. Fractional bandwidth vs. maximum VSWR for nonsymmetrical directional couplers.

Design Equations for Mismatched Nonsymmetrical Directional Couplers

Recall that R as defined by (38) is the normalized input admittance at Port 1 when $\theta = 90°$, and is also the maximum VSWR or its reciprocal of the directional coupler. Therefore, in the design equations below, R

[3] The fractional-bandwidth w, is defined as

$$w = 2\left\{\frac{BW - 1}{BW + 1}\right\}.$$

should be chosen equal to the maximum VSWR (or its reciprocal) permitted by the design specifications. Either of the choices for the value of R will result in the same power relationships for the directional coupler. However, it generally will be easier to realize the coupler physically for one of the two choices. The design equations for mismatched nonsymmetrical directional couplers are as follows:

$$Y_{oe}{}^{a} = \frac{\sqrt{R}\,G_a - k\sqrt{G_aG_b}}{\sqrt{1-k^2}} \tag{54}$$

$$Y_{oo}{}^{a} = \frac{\sqrt{R}\,G_a + k\sqrt{G_aG_b}}{\sqrt{1-k^2}} \tag{55}$$

$$Y_{oe}{}^{b} = \frac{G_b/\sqrt{R} - k\sqrt{G_aG_b}}{\sqrt{1-k^2}} \tag{56}$$

$$Y_{oo}{}^{b} = \frac{G_b/\sqrt{R} + k\sqrt{G_aG_b}}{\sqrt{1-k^2}}. \tag{57}$$

For directional coupler geometries such as that shown in Fig. 2, the following equations apply:

$$\frac{C_a}{\epsilon} = \frac{376.7}{\sqrt{\epsilon_r}}\,\frac{\sqrt{R}\,G_a - k\sqrt{G_aG_b}}{\sqrt{1-k^2}} \tag{58}$$

$$\frac{C_b}{\epsilon} = \frac{376.7}{\sqrt{\epsilon_r}}\,\frac{G_b/\sqrt{R} - k\sqrt{G_aG_b}}{\sqrt{1-k^2}} \tag{59}$$

$$\frac{C_{ab}}{\epsilon} = \frac{376.7}{\sqrt{\epsilon_r}}\,\frac{k\sqrt{G_aG_b}}{\sqrt{1-k^2}}. \tag{60}$$

The theoretical limit on the amount of impedance transformation and coupling is now determined by (54) and (56). For $Y_{oe}{}^{a}$ and $Y_{oe}{}^{b}$ to be positive, it is necessary that

$$\frac{1}{k^2} \geq \frac{RG_a}{G_b} \quad \text{and} \quad \frac{G_b}{RG_a}, \tag{61}$$

whichever is greater.

If R is chosen advantageously (61) may be stated in words: *Let the impedance (or admittance) transformation ratio of the directional coupler be taken as greater than one. Then, the reciprocal of the power coupling coefficient must be greater than or equal to the impedance (or admittance) transformation ratio divided by the maximum VSWR of the coupler.* Thus another result of designing nonsymmetrical directional couplers to be mismatched is that, for a given impedance transformation ratio, a larger coupling coefficient is theoretically possible; similarly, for a given coupling coefficient, a larger impedance transformation ratio is theoretically possible.

Experimental Results

To substantiate the analytical results for nonsymmetrical matched directional couplers, a trial -10-dB coupler having coupled lines of 50 and 75 ohms, respectively, was constructed and tested. From (27)–(29) it was determined that

$$C_1/\epsilon = 5.891 \tag{62}$$

$$C_2/\epsilon = 3.244 \tag{63}$$

$$C_{12}/\epsilon = 2.050, \tag{64}$$

where the subscripts 1 and 2 refer to the 50- and 75-ohm lines, respectively.

It was decided to construct the coupler using shielded coupled rectangular bars. Using the data of Getsinger [10], the following dimensions were determined.

$$t/b = 0.400 \text{ (chosen arbitrarily)} \tag{65}$$

$$(w/b)_1 = 0.508 \tag{66}$$

$$(w/b)_2 = 0.111 \tag{67}$$

$$s/b = 0.233. \tag{68}$$

Here b is the ground plane spacing, t is the thickness of the rectangular bars, w is the width of a coupled rectangular bar, and s is the spacing between coupled bars. The trial coupler was constructed with $\frac{5}{8}$-inch ground plane spacing in air dielectric. The nominal center frequency was 1.5 Gc/s. A drawing of the coupler, showing important dimensions, is given in Fig. 4, and a photograph of the constructed coupler is given in Fig. 5.

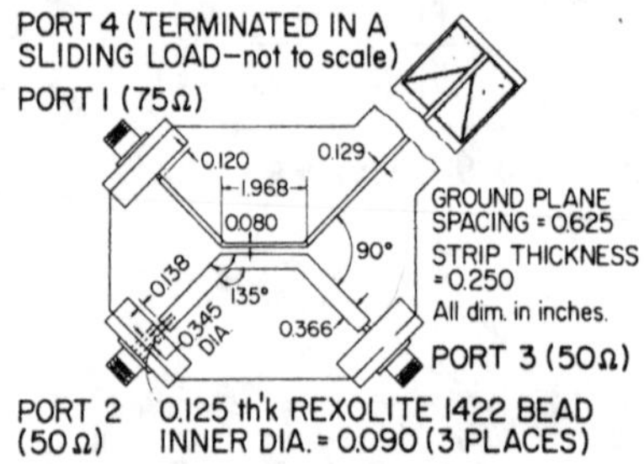

Fig. 4. Drawing of a trial nonsymmetrical -10-dB directional coupler with coupled lines of 50 and 75 ohms.

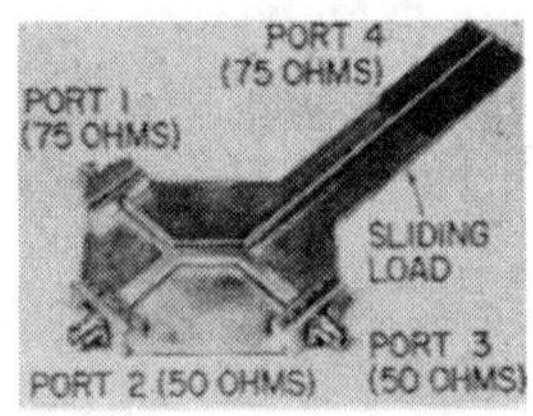

Fig. 5. Photograph of a trial nonsymmetrical -10-dB directional coupler with coupled lines of 50 and 75 ohms.

Type N connectors of 50 ohms were used at Ports 1, 2, and 3, while Port 4 was terminated internally in a sliding load. The 50-ohm connector at Port 1 (the 75-ohm port) was chosen for convenience. It was anticipated beforehand that the required measurements could be performed without being substantially affected by the mismatch.

The VSWR at Port 3 was measured with Port 2

terminated in a well-matched load. Since Port 1 is decoupled from Port 3, the discontinuity at Port 1 was not seen. The load at Port 4 had a sufficiently low VSWR when seen at Port 1 so that it also did not affect the measurements.

The VSWR at Port 1 was measured with Port 2 terminated in a well-matched load. The sliding load at Port 4 was adjusted to give first a maximum VSWR and next a minimum VSWR. During these measurements it was also noted whether the standing-wave minimum shifted 90 degrees or not. If there was a 90-degree phase shift, the VSWR at Port 1 was determined by

$$r = \sqrt{r_{max}/r_{min}}.$$

If there was no phase shift, the VSWR was determined by

$$r = \sqrt{r_{max} r_{min}}.$$

In Fig. 6 the results of these measurements are shown. The VSWR at Port 3 is seen to be quite low. It is less than 1.08 over most of the frequency band, 0.8 to 2.3 Gc/s. At Port 1, the VSWR varies about 1.5 (it should be exactly 1.5 in theory, as measured on a 50-ohm slotted line) without departing much from this value except at 0.8 and 0.9 Gc/s, where an unexplained resonance appears to occur. It should be pointed out that only a small effort was made to perfect the transition from the coaxial connector to the 75-ohm strip line, and that, undoubtedly, a portion of the VSWR at Port 1 is due to the transition itself. The power coupling from Port 2 to Port 1 was measured without compensating the discontinuity at Port 1. The VSWR of 1.5 at Port 1 corresponds to slightly less than 0.2-dB mismatch loss. The results of the measurements are given in Table I.

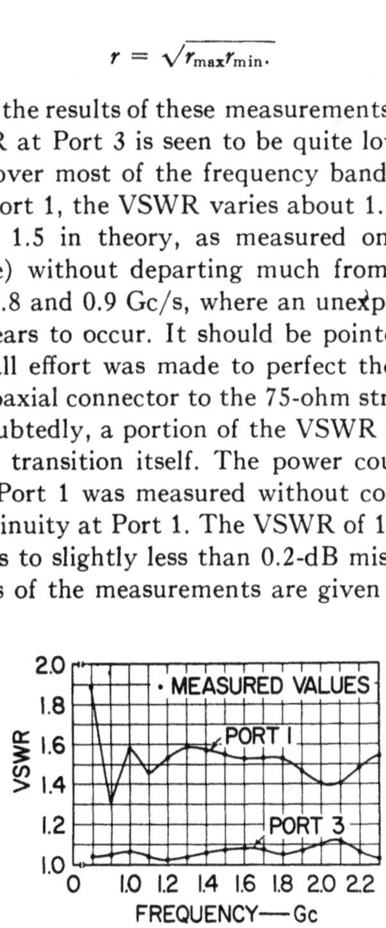

Fig. 6. VSWR at Ports 1 and 3 of trial nonsymmetrical −10-dB directional coupler. (The VSWR at Port 1 is with respect to a 50-ohm line.)

TABLE I

POWER-COUPLING FROM PORT 2 TO PORT 1

Frequency (Gc/s)	Calculated (dB)	Measured (dB)
0.800	−12.4	−12.5
1.000	−11.1	−11.7
1.200	−10.4	−11.2
1.500	−10.0	−10.3
1.800	−10.4	−10.9
2.250	−12.8	−12.5
2.700	−19.8	−19.2

Originally it was intended to make directivity measurements using a double-sliding-load method. However, during the measurements it was found that the VSWR of the connector of the sliding load itself was too large to make the measurements meaningful. It was then decided to use a single sliding load (the load at Port 4) and to terminate Port 2 in a matched load.

Basically, the measurement procedure using a single sliding load is as follows. Power is made incident to Port 1. A reference level (denoted as R-dB) is established by measuring the power available at Port 2. Next, Port 2 is terminated in a matched load (by using a good load and a double stub tuner to give unity VSWR), and the power available at Port 3 is monitored. The sliding load at Port 4 is adjusted to give maximum power out of Port 3; let this value be Max-dB. Next, the sliding load is adjusted to give a minimum of power at Port 3; let this value be Min-dB. Also, let Δ equal the difference of Max-dB and Min-dB. Then it can be shown that the intrinsic directivity is as follows:

$$\text{Directivity} = 10 \log_{10} \left| \frac{S_{12}}{S_{13}} \right|^2$$

$$= (R - \text{Max}) - 10 \log_{10} \left\{ \frac{1 \pm 10^{-|\Delta/20|}}{2} \right\}^2 \text{dB}.$$

The ambiguity in the above formula can be resolved by making an appropriate phase measurement. However, this was not done in the present case; instead, the minimum value of the above formula was taken as the directivity.

Initially, directivity measurements were made without compensating the junctions of the coupled and uncoupled lines. The measured directivities at 1 and 1.5 Gc/s were approximately 18 dB. Next, compensating blocks similar to those used in the 3-dB coupler of Shimizu-Jones (see Fig. 5 of their paper) were located to maximize the directivity at 2.3 Gc/s and were fixed in that location for the remaining measurements. The results of the directivity measurements are given in Table II.

TABLE II

DIRECTIVITY MEASUREMENTS

Frequency (Gc/s)	Directivity (dB)
1.0	27
1.25	24
1.50	26
1.75	26.5
2.00	19
2.30	28.5

CONCLUSIONS

Nonsymmetrical directional couplers are equivalent to conventional directional couplers with impedance transformers at the two ports of one of the transmission lines. They may be designed to have infinite directivity and to be matched at all frequencies; or they may be designed to have infinite directivity at all frequencies, and a specified maximum mismatch. Mismatched nonsymmetrical directional couplers were shown to have greater bandwidth than matched nonsymmetrical directional couplers, although in most practical cases the increased bandwidth will be less than four percent. Both mismatched and matched nonsymmetrical directional couplers have theoretical limits on the amount of coupling and impedance transformation ratio that can be obtained simultaneously. The rule is that the reciprocal of the power coupling coefficient must be greater than or equal to the impedance transformation ratio, taken greater than one, divided by the maximum VSWR of the coupler. Matched nonsymmetrical directional couplers may be broadbanded by cascading additional nonsymmetrical couplers. Existing tables of designs for multisection asymmetrical and symmetrical couplers may be taken over directly by interpreting the design data as described in this paper.

Several possible applications for nonsymmetrical directional couplers are as directional detectors, power samplers, and reflectometers; they may be applied advantageously to the design of directional filters by permitting a change of impedance level of the resonant lines. A novel application is the construction of impedance transformers having dc isolation by properly cascading several nonsymmetrical directional couplers to make a 0-dB coupler.

146

Appendix I

Derivation of the Condition for Infinite Directivity

The Y-matrix of the nonsymmetrical coupled-transmission-line directional coupler was given in (1). For the following derivation let us write the defining equations as

$$\begin{pmatrix} I_1 \\ I_2 \end{pmatrix} = \begin{pmatrix} y_{11} & Y_{12} \\ Y_{12}^T & Y_{22} \end{pmatrix} \begin{pmatrix} V_1 \\ V_2 \end{pmatrix} \tag{69}$$

where

$$\begin{aligned} y_{ij} &= \text{the } i,j \text{ entry of the matrix of (1)} \\ I_1 &= i_1 \\ I_2 &= \text{col } (i_2, i_3, i_4) \\ V_1 &= v_1 \\ V_2 &= \text{col } (v_2, v_3, v_4) \\ Y_{12} &= \text{row } (y_{12}, y_{13}, y_{14}) \\ Y_{12}^T &= \text{transpose of } Y_{12} \\ Y_{22} &= \begin{bmatrix} y_{22} & y_{23} & y_{24} \\ y_{23} & y_{22} & y_{12} \\ y_{13} & y_{12} & y_{11} \end{bmatrix}. \end{aligned} \tag{70}$$

By terminating Ports 2, 3, and 4 in their respective loads, the following constraint is applied:

$$I_2 = -GV_2, \tag{71}$$

where

$$G = \begin{bmatrix} G_b & 0 & 0 \\ 0 & G_b & 0 \\ 0 & 0 & G_a \end{bmatrix}. \tag{72}$$

Equations (69) and (71) are easily combined, and upon the elimination of I_2 and V_1 yield

$$\frac{Y_{12}^T}{y_{11}} I_1 = \left\{ \frac{Y_{12}^T Y_{12}}{y_{11}} - (G + Y_{22}) \right\} V_2. \tag{73}$$

Solving for V_2 gives

$$V_2 = H Y_{12}^T I_1, \tag{74}$$

where H is a 3×3 matrix having entries h_{ij} given by

$$H = \{ Y_{12}^T Y_{12} - y_{11}(G + Y_{22}) \}^{-1}. \tag{75}$$

The voltages at Ports 2, 3, and 4 are given by

$$\begin{bmatrix} v_2 \\ v_3 \\ v_4 \end{bmatrix} = \begin{bmatrix} 1 & 0 & 0 \\ 0 & 1 & 0 \\ 0 & 0 & 1 \end{bmatrix} H Y_{12}^T I_1. \tag{76}$$

Thus the directivity is given by

$$20 \log_{10} \left| \frac{v_2}{v_3} \right| = 20 \log_{10} \left| \frac{h_{11}y_{12} + h_{12}y_{13} + h_{13}y_{14}}{h_{12}y_{12} + h_{22}y_{13} + h_{23}y_{14}} \right|, \tag{77}$$

and the condition for infinite directivity is

$$h_{12}y_{12} + h_{22}y_{13} + h_{23}y_{14} \equiv 0. \tag{78}$$

Substituting in the values for h_{ij} and y_{ij} results in

$$AB - D^2 = G_a G_b. \tag{79}$$

Appendix II

Derivation of the Condition for an Impedance Match

The Y-matrix of the nonsymmetrical coupled-transmission-line directional coupler was given in (1). For the following derivation let us write the defining equations as

$$\begin{pmatrix} I_1 \\ I_2 \end{pmatrix} = \begin{pmatrix} Y_{11} & -Y_{11}(tC) \\ -(tC)Y_{11} & CY_{11}C \end{pmatrix} \begin{pmatrix} V_1 \\ V_2 \end{pmatrix}, \tag{80}$$

where

$$\left. \begin{aligned} I_1 &= \text{col } (i_1, i_2) \\ I_2 &= \text{col } (i_3, i_4) \\ V_1 &= \text{col } (v_1, v_2) \\ V_2 &= \text{col } (v_3, v_4) \\ t &= \sqrt{1 - s^2} \\ C &= \begin{pmatrix} 0 & 1 \\ 1 & 0 \end{pmatrix} \\ Y_{11} &= \frac{1}{s} \begin{pmatrix} A & -D \\ -D & B \end{pmatrix} \end{aligned} \right\}. \tag{81}$$

By terminating Ports 3 and 4 in their respective loads, the following constraint is applied:

$$I_2 = -(CGC)V_2, \tag{82}$$

where

$$G = \begin{pmatrix} G_a & 0 \\ 0 & G_b \end{pmatrix}. \tag{83}$$

Equations (82) and (80) may be combined, and upon elimination of I_2 and V_2, yield

$$I_1 = Y_{11}\{\mathcal{I}_2 - t^2(G + Y_{11})^{-1}Y_{11}\}V_1, \tag{84}$$

where $\mathcal{I}_2$ is a 2×2 identity matrix. Equation (84) may be simplified by noting that

$$-(G + Y_{11})^{-1} = \frac{T(G + Y_{11})T}{|(G + Y_{11})|} \tag{85}$$

where

$$T = \begin{pmatrix} 0 & 1 \\ -1 & 0 \end{pmatrix}, \tag{86}$$

and $|(G+Y_{11})|$ is the determinant of the matrix $G+Y_{11}$. Thus,

$$I_1 = Y_{11}\left\{\mathcal{I}_2 + t^2\frac{T(G + Y_{11})TY_{11}}{|G + Y_{11}|}\right\}V_1. \tag{87}$$

Let the entries of an M matrix be M_{ij}, and let

$$M = Y_{11}\left\{\mathcal{I}_2 + t^2\frac{T(G + Y_{11})TY_{11}}{|G + Y_{11}|}\right\}. \tag{88}$$

Then

$$\frac{i_1}{v_1} = (Y_{in})_a = \frac{\begin{vmatrix} M_{11} & M_{12} \\ M_{12} & G_b + M_{22} \end{vmatrix}}{G_b + M_{22}}$$

$$= M_{11} - \frac{M_{12}^2}{G_b + M_{22}}. \tag{89}$$

The condition for a match at Port 2 is

$$G_a - Y_{in} \equiv 0, \tag{90}$$

which when combined with (89) yields the following determining equation:

$$\text{Numerator of } \{G_aG_b + G_aM_{22} - G_bM_{11} - (M_{11}M_{22} - M_{12}^2)\} \equiv 0. \tag{91}$$

Substituting the appropriate values for M_{ij} into (91), and equating coefficients of the variables s_i equal to zero, yields

$$\left.\begin{aligned} \frac{G_a}{G_b} &= \frac{A}{B} \\ G_aG_b &= AB - D^2 \end{aligned}\right\}. \tag{92}$$

Acknowledgment

The author wishes to thank R. Pierce who constructed the trial nonsymmetrical directional coupler, and E. Fernandes who made the laboratory tests. Acknowledgments are due also to L. Robinson who suggested several possible techniques for measuring the directivity of the trial directional coupler, and to Dr. L. Young and B. M. Schiffman with whom the author had several discussions which helped to clarify several points in the paper.

References

[1] E. M. T. Jones and J. T. Bolljahn, "Coupled-strip-transmission-line filters and directional couplers," *IRE Trans. on Microwave Theory and Techniques*, vol. MTT-4, pp. 75–81, April 1956.
[2] J. K. Shimizu and E. M. T. Jones, "Coupled-transmission-line directional couplers," *IRE Trans. on Microwave Theory and Techniques*, vol. MTT-6, pp. 403–410, October 1958.
[3] H. Seidel and J. Rosen, "Multiplicity in cascade transmission-line synthesis—part 2," *IEEE Trans. on Microwave Theory and Techniques*, vol. MTT-13, pp. 398–407, July 1965.
[4] R. Levy, "Tables for asymmetric multi-element coupled-transmission-line directional couplers," *IEEE Trans. on Microwave Theory and Techniques*, vol. MTT-12, pp. 275–279, May 1964.
[5] P. P. Toulios and A. C. Todd, "Synthesis of symmetrical TEM-mode directional couplers," *IEEE Trans. on Microwave Theory and Techniques*, vol. MTT-13, pp. 536–544, September 1965.
[6] E. G. Cristal and L. Young, "Theory and tables of optimum symmetrical TEM-mode coupled-transmission-line directional couplers," *IEEE Trans. on Microwave Theory and Techniques*, vol. MTT-13, pp. 544–558, September 1965.
[7] "New coaxial couplers for reflectometers, detection, and monitoring," *Hewlett-Packard J.*, vol. 16, pp. 2–5, February 1965.
[8] H. Ozaki and J. Ishii, "Synthesis of a class of strip-line filters," *IRE Trans. on Circuit Theory*, vol. CT-5, pp. 104–109, June 1958.
[9] W. L. Firestone, "Analysis of transmission line directional couplers," *Proc. IRE*, vol. 42, pp. 1529–1538, October 1954. (See also W. L. Firestone, "Transmission-line directional couplers," Ph.D. dissertation, Northwestern University, Evanston, Ill., May 1952.)
[10] W. J. Getsinger, "Coupled rectangular bars between parallel plates," *IEEE Trans. on Microwave Theory and Techniques*, vol. MTT-10, pp. 65–72, January 1962.
[11] E. G. Cristal, "Coupled circular cylindrical rods between parallel ground plates," *IEEE Trans. on Microwave Theory and Techniques*, vol. MTT-12, pp. 428–439, July 1964.
[12] W. J. Getsinger, "Fringing capacitances for offset coupled strips in shielded strip lines," Lincoln Lab. Rept., Cambridge, Mass., Tech. Note 1965-33, Contract AF 19(628)-5167, Air Force Systems Command, July 1, 1965.
[13] M. E. VanValkenberg, *Introduction to Modern Network Synthesis*. New York: Wiley, 1960, ch. 8.
[14] E. A. Guillemin, *Communication Networks*, Vol. II, New York: Wiley, 1935, ch. 6.
[15] L. Young, "The analytical equivalence of TEM-mode directional couplers and transmission-line stepped-impedance filters," *Proc. IEE (London)* vol. 110, pp. 275–281, February 1963.

An Equivalence Principle for Nonuniform Transmission-Line Directional Couplers

CHARLES B. SHARPE, SENIOR MEMBER, IEEE

Abstract—**The analysis of transmission-line directional couplers is formulated in terms of a pair of first-order matrix differential equations. It is shown that for every nonuniform directional coupler that is electrically symmetric, there exists an equivalent pair of dual nonuniform transmission lines. It is also shown that a matched, transmission-line directional coupler having an absolutely continuous characteristic impedance matrix must be symmetric. Restrictions on the terminating impedances and the implications of these restrictions on the realizability of transmission-line couplers are investigated. Finally, the tapered-line magic *T* is treated as an example.**

INTRODUCTION

THE EQUIVALENCE between stepped, transmission-line directional couplers and stepped, transmission-line filters is well known and has been established analytically by several authors [1], [2]. These couplers, which consist of sections of equal-length uniform line, are usually assumed to be matched at each port and to have perfect directivity at all frequencies. The equivalence states that for such structures a two-port transmission-line network can be found which yields as a solution the desired parameters or characteristics of the four-port coupler. The principal purpose of this paper is to show that this equivalence holds for a general class of symmetric, nonuniform directional couplers which includes the stepped coupler as a special case.

It should be said that the existence of such an equivalence is implicit in the work of previous authors [3]–[5]. However, a rigorous justification for its application has not been given. While this may appear unnecessary from a practical point of view, it will be shown that such an approach does lead to some explicit results regarding the realizability of transmission-line directional couplers. Moreover, an analytical treatment of equivalence will provide a basis for the introduction to the coupled-line problem of synthesis techniques recently developed for single, nonuniform lines.

The fact that such an equivalence is possible is suggested by considering the scattering matrix of the general contra-directional coupler depicted in Fig. 1,

$$S(s) = \begin{bmatrix} 0 & S_{12}(s) & S_{13}(s) & 0 \\ S_{12}(s) & 0 & 0 & S_{24}(s) \\ S_{13}(s) & 0 & 0 & S_{34}(s) \\ 0 & S_{24}(s) & S_{34}(s) & 0 \end{bmatrix}, \tag{1}$$

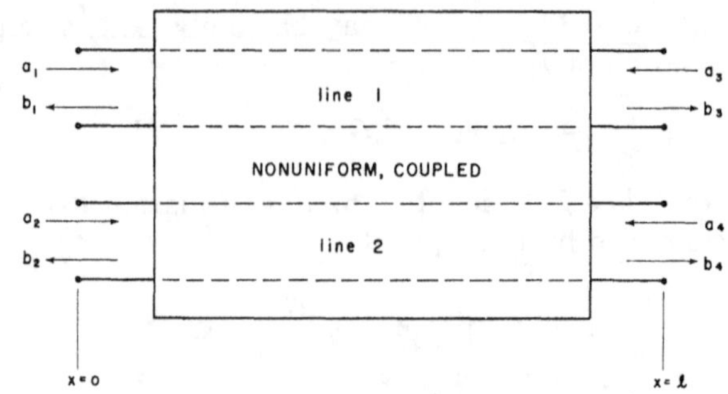

Fig. 1. A four-port transmission-line coupler.

where $s=\sigma+j\omega$ denotes the complex frequency variable. It is well known that for any lossless linear network, $S(s)$ must be unitary. That is, on the real frequency axis, $\bar{S}'(j\omega)S(j\omega)=I$, where the prime denotes the transpose, the bar the complex conjugate, and I the 4×4 unit matrix. This relation is a special case of the para-unitary condition,

$$S'(-s)S(s) = I, \tag{2}$$

which can be derived directly from the Wronskian of the matrix differential equation describing the system.

In the following it will be assumed that the directional coupler possesses side-by-side electrical symmetry, that is, $S_{13}(s)=S_{24}(s)$.[1] In the Appendix it is shown that one has very little freedom in choosing S_{13} and S_{24} independently. In fact, for the class of absolutely continuous coupled lines of equal length S_{13} must equal S_{24}. For directional couplers which are symmetric in this sense (2) reduces to three independent equations that can be combined into

$$[S^{\pm}(-s)]'S^{\pm}(s) = I, \tag{3}$$

where the reduced 2×2 matrices are defined by

$$S^{\pm}(s) = \begin{bmatrix} \pm S_{12}(s) & S_{13}(s) \\ S_{13}(s) & \pm S_{34}(s) \end{bmatrix}. \tag{4}$$

In (3) and hereafter I will be used to denote the 2×2 unit matrix. Equation (4) suggests that there exist two single transmission lines having as their reflection coefficients S_{12} and $-S_{12}$, respectively, and having S_{13} as their transmission coefficient. It will be shown that this conjecture is true for the class of nonuniform couplers considered.

Manuscript received November 14, 1966; revised March 15, 1967. This work was supported by the National Science Foundation under Grant GK-601.

The author is with the Radar and Optics Lab., Willow Run Labs., Institute of Science and Technology, University of Michigan, Ann Arbor, Mich.

[1] This definition of symmetry does not necessarily imply physical symmetry of any kind.

The Matrix Line Equations

It will be obvious that the discussion of this section applies equally well to n coupled lines. However, nothing is gained by the added generality, and in the interest of simplicity the discussion will be limited to the case where $n=2$. Denote by the column matrices,

$$e(z) = \begin{bmatrix} e_1(z) \\ e_2(z) \end{bmatrix}$$

$$i(z) = \begin{bmatrix} i_1(z) \\ i_2(z) \end{bmatrix} \tag{5}$$

the voltages and currents, respectively, on the two lines depicted in Fig. 1. The coupled, nonuniform line equations for the lossless case can then be written

$$\frac{de(z)}{dz} = -sL(z)i(z) \tag{6a}$$

$$\frac{di(z)}{dz} = -sC(z)e(z), \tag{6b}$$

where $L(z)$ and $C(z)$ are real, symmetric 2×2 matrices. It will be assumed that the elements of L and C are bounded and integrable functions of z and at each point on the line satisfy conditions consistent with TEM-mode propagation. Thus, from energy considerations both L and C must be positive semidefinite for every z. For reasons which will be apparent shortly it will be assumed that L and C are actually positive definite.

In the case of uniformly coupled, parallel conductors in a homogeneous media the inductance and capacitance matrices in (6a) and (6b) are, of course, independent of z. Moreover, it is known [6] that $LC=CL=\mu\epsilon I$. It is easily shown from Maxwell's equations that this relation as well as (6a) and (6b) still hold when the media is inhomogeneous in the z direction, that is, when $\mu=\mu(z)$ and $\epsilon=\epsilon(z)$. It makes ense to assume, therefore, that in the general nonuniform case in which the cross section of the structure may also vary with z that there still exists a positive function $v(z)$ such that

$$L(z)C(z) = C(z)L(z) = \frac{1}{v^2(z)} I. \tag{7}$$

This condition is consistent with the TEM assumption since it implies that at each point z the propagation on both lines is characterized by the same "local" velocity, $v(z)$. While (7) is difficult to justify in the case where the propagating field is not strictly TEM, it will now be shown that this assumption does lead to the definition of a unique "local" characteristic impedance matrix which reduces to the conventional result for the case of uniform coupled lines. Thus, at each point on the nonuniform structure it is possible to associate a system of coupled, uniform lines having the same inductance and capacitance matrices at that point. As in the case of single lines it would be expected that the error made in this type of approximation would be acceptable for coupled lines having only a moderate degree of nonuniformity.

Introduce the new variable

$$x(z) = \int_0^z \frac{d\xi}{v(\xi)}. \tag{8}$$

Noting that z is determined uniquely in terms of x by this relation, we will express the spatial dependence of the various functions as follows: $e[z(x)]=e(x)$, $L[z(x)]=L(x)$, etc. Then,

$$\frac{de(z)}{dz} = \frac{de[z(x)]}{dx}\frac{dx}{dz} = \frac{de(x)}{dx}\frac{1}{v(z)}. \tag{9}$$

Since every positive definite matrix has an inverse and a unique positive definite square root, (7) and (9) and a similar equation in $i(x)$ can be used to put (6a) and (6b) in the symmetric form,

$$\frac{de(x)}{dx} = -sZ_0(x)i(x) \tag{10a}$$

$$\frac{di(x)}{dx} = -sZ_0^{-1}(x)e(z), \tag{10b}$$

where the characteristic impedance matrix is defined by

$$Z_0(x) = C(x)^{-1/2}L(x)^{1/2}. \tag{11}$$

From the assumed properties of $L(x)$ and $C(x)$ it can be shown that $Z_0(x)$ is symmetric and positive definite. The proof follows from the well-known result that if two symmetric matrices commute they can be diagonalized by the same orthogonal matrix [7]. It is evident from (7) that (11) can also be written

$$Z_0(x) = \frac{1}{v(x)} C^{-1}(x), \tag{12}$$

so that, given $v(x)$, there is a one-to-one correspondence between $Z_0(x)$ and $C(x)$. The question of whether or not a particular $C(x)$ can be physically realized in the quasi TEM sense previously discussed will be deferred until later.

Properties of Transmission-Line Directional Couplers

The scattering matrix formulation is often the most convenient one to use when discussing directional couplers. This is particularly true of transmission-line couplers if the scattering matrix is partitioned so as to distinguish between input and output ports. No assumption will be made in this section regarding symmetry. It will be assumed initially that the ith port of the coupler is terminated in a uniform, lossless line of characteristic impedance $r_i>0$. Therefore, the scattering matrix is appropriately normalized with respect to the real characteristic impedance matrix,

$$R = \left[\begin{array}{cc|cc} r_1 & 0 & 0 & 0 \\ 0 & r_2 & 0 & 0 \\ \hline 0 & 0 & r_3 & 0 \\ 0 & 0 & 0 & r_4 \end{array}\right] = \left[\begin{array}{c|c} R_1 & 0 \\ \hline 0 & R_2 \end{array}\right]. \tag{13}$$

If we define the column matrices,

$$\boldsymbol{E}(s) = \begin{bmatrix} e_1(0, s) \\ e_2(0, s) \\ e_1(l, s) \\ e_2(l, s) \end{bmatrix}$$

$$\boldsymbol{I}(s) = \begin{bmatrix} i_1(0, s) \\ i_2(0, s) \\ -i_1(l, s) \\ -i_2(l, s) \end{bmatrix}, \tag{14}$$

then the scattering matrix $S(s)$ satisfies

$$\boldsymbol{b} = S\boldsymbol{a}, \tag{15}$$

where

$$2\boldsymbol{a}(s) = R^{-1/2}\boldsymbol{E}(s) + R^{1/2}\boldsymbol{I}(s)$$

$$2\boldsymbol{b}(s) = R^{-1/2}\boldsymbol{E}(s) - R^{1/2}\boldsymbol{I}(s) \tag{16}$$

and, referring to Fig. 1, $\boldsymbol{a}=(a_1a_2a_3a_4)'$ and $\boldsymbol{b}=(b_1b_2b_3b_4)'$.

The fundamental matrix solution to the system is now defined as the 2×2 matrices,

$$E(x, s) = \begin{bmatrix} e_1^{(1)}(x, s) & e_1^{(2)}(x, s) \\ e_2^{(1)}(x, s) & e_2^{(2)}(x, s) \end{bmatrix}$$

$$I(x, s) = \begin{bmatrix} i_1^{(1)}(x, s) & i_1^{(2)}(x, s) \\ i_2^{(1)}(x, s) & i_2^{(2)}(x, s) \end{bmatrix} \tag{17}$$

in which the corresponding columns are independent vector solutions of (10a) and (10b) satisfying the boundary conditions, $E(l, s)=R_2^{1/2}$ and $I(l, s)=R_2^{-1/2}$. The dependence of the solution on the frequency variable s is now explicitly indicated. Clearly we can write

$$\frac{dE(x, s)}{dx} = -sZ_0(x)I(x, s) \tag{18a}$$

$$\frac{dI(x, s)}{dx} = -sZ_0^{-1}(x)E(x, s). \tag{18b}$$

Associated with the two vector solutions are the 4×2 parameter matrices,

$$\begin{bmatrix} A_\alpha \\ \hline A_\beta \end{bmatrix} = \begin{bmatrix} a_1^{(1)} & a_1^{(2)} \\ a_2^{(1)} & a_2^{(2)} \\ \hline a_3^{(1)} & a_3^{(2)} \\ a_4^{(1)} & a_4^{(2)} \end{bmatrix}$$

$$\begin{bmatrix} B_\alpha \\ \hline B_\beta \end{bmatrix} = \begin{bmatrix} b_1^{(1)} & b_1^{(2)} \\ b_2^{(1)} & b_2^{(2)} \\ \hline b_3^{(1)} & b_3^{(2)} \\ b_4^{(1)} & b_4^{(2)} \end{bmatrix}. \tag{19}$$

Evidently,

$$\begin{bmatrix} B_\alpha \\ \hline B_\beta \end{bmatrix} = S \begin{bmatrix} A_\alpha \\ \hline A_\beta \end{bmatrix}, \tag{20}$$

where, for a matched directional coupler,

$$S = \left[\begin{array}{cc|cc} 0 & S_{12} & S_{13} & 0 \\ S_{12} & 0 & 0 & S_{24} \\ \hline S_{13} & 0 & 0 & S_{34} \\ 0 & S_{24} & S_{34} & 0 \end{array}\right] = \left[\begin{array}{c|c} S_{\alpha\alpha} & S_{\alpha\beta} \\ \hline S_{\beta\alpha} & S_{\beta\beta} \end{array}\right]. \tag{21}$$

The partitioning indicated in (21) allows us to treat the four-port coupler as a two-port network. This expedient is useful as long as the partitioned matrices are nonsingular, which is guaranteed by our choice of the fundamental matrix solution. From these equations it follows that

$$2A_\alpha = R_1^{-1/2}E(0, s) + R_1^{1/2}I(0, s) \tag{22a}$$

$$2B_\alpha = R_1^{-1/2}E(0, s) - R_1^{1/2}I(0, s) \tag{22b}$$

$$2A_\beta = R_2^{-1/2}E(l, s) - R_2^{1/2}I(l, s) \tag{22c}$$

$$2B_\beta = R_2^{-1/2}E(l, s) + R_2^{1/2}I(l, s). \tag{22d}$$

Inserting the boundary conditions in (22c) and (22d) we find that $A_\beta=0$ and $B_\beta=I$. It follows from (20) through (22b) that

$$E(0, s) = R_1^{1/2}(A_\alpha + B_\alpha) = R_1^{1/2}(I + S_{\alpha\alpha})S_{\beta\alpha}^{-1} \tag{23}$$

and

$$I(0, s) = R_1^{-1/2}(A_\alpha - B_\alpha) = R_1^{-1/2}(I - S_{\alpha\alpha})S_{\beta\alpha}^{-1}. \tag{24}$$

A basic premise in transmission-line theory is the uniqueness of voltage and current. Mathematically, this property can be derived by expressing (18a) and (18b) in the form of a matrix Volterra integral equation. In this way it can be shown that $E(x, s)$ and $I(x, s)$ are continuous functions of x for all s and that either function is uniquely determined by the values of E and I at one point, say $x=l$. This property is subject only to the condition that $Z_0(x)$ be integrable. In the practical situation where the elements of $Z_0(x)$ are sectionally continuous this requirement is certainly satisfied. A second property which can be deduced from the integral representation is that $E(x, s)$ and $I(x, s)$ are entire functions of s for all x, $0 \leq x \leq l$. We now wish to examine the implications of uniqueness on the terminal impedances of the coupler. Consider (18a) when $s=0$. The solution is $E(x, 0)=C$, where C is a constant 2×2 matrix.

But from the boundary conditions for the fundamental solution, $C=R_2^{1/2}$, and it is concluded that $E(0, 0)=R_2^{1/2}$ and $I(0, 0)=R_2^{-1/2}$. Subtracting (24) from (23) and employing (21) we obtain for $s=0$,

$$R_1^{-1/2}R_2^{1/2} - R_1^{1/2}R_2^{-1/2} = 2S_{\alpha\alpha}(0)S_{\beta\alpha}^{-1}(0) = \begin{bmatrix} 0 & \dfrac{2S_{12}(0)}{S_{24}(0)} \\ \dfrac{2S_{12}(0)}{S_{13}(0)} & 0 \end{bmatrix}. \tag{25}$$

Since R_1 and R_2 are diagonal, it follows that $R_1=R_2$. In other words, to be a directional coupler the lines must be loaded in a symmetric fashion so that $r_1=r_3$ and $r_2=r_4$. It also follows from (25) that $S_{12}(0)=0$ and from (23), $S_{13}(0)=S_{24}(0)=1$.

Analysis

In view of the above constraint on the terminal impedances, the following transformation is suggested:

$$\hat{E}(x, s) = R_1^{-1/2}E(x, s)$$
$$\hat{I}(x, s) = R_1^{1/2}I(x, s) \cdot \tag{26}$$

Substituting (26) in (18a) and (18b), the matrix line equations become

$$\frac{d\hat{E}(x, s)}{dx} = -sR_0(x)\hat{I}(x, s) \tag{27a}$$

$$\frac{d\hat{I}(x, s)}{dx} = -sR_0^{-1}(x)\hat{E}(x, s), \tag{27b}$$

where

$$R_0 = R_1^{-1/2}Z_0(x)R_1^{-1/2} \cdot \tag{28}$$

It will now be assumed that the coupler possesses electrical symmetry in the sense that $S_{13}=S_{24}$. The boundary conditions for the normalized voltage and current at $x=0$ and $x=l$ can then be written,

$$\hat{E}(0, s) = \begin{bmatrix} 1/S_{13} & S_{12}/S_{13} \\ S_{12}/S_{13} & 1/S_{13} \end{bmatrix}$$

$$\hat{I}(0, s) = \begin{bmatrix} 1/S_{13} & -S_{12}/S_{13} \\ -S_{12}/S_{13} & 1/S_{13} \end{bmatrix} \tag{29}$$

and $\hat{E}(l, s)=\hat{I}(l, s)=I$. The "double symmetry" displayed in the boundary conditions plays a key role in the following analysis. We shall consider the class of solutions of (27a) and (27b) which are doubly symmetric for all x in the interval $0\leq x\leq l$; that is, solutions for which $\hat{e}_1^{(1)}(x, s)=\hat{e}_2^{(2)}(x, s)$ and $\hat{e}_2^{(1)}(x, s)=\hat{e}_1^{(2)}(x, s)$, and similarly for the components of the current matrix. The existence of such solutions will be shown to follow from the existence of the solutions for two single-line problems. Under this assumption it is concluded from (27a) that $R_0(x)$ must also be doubly symmetric for all x. That is, $R_0(x)$ has the form,

$$R_0(x) = \begin{bmatrix} R_{11}(x) & R_{12}(x) \\ R_{12}(x) & R_{11}(x) \end{bmatrix}. \tag{30}$$

Adding and subtracting columns of (27a) and (27b) gives

$$\frac{d}{dx}[\hat{e}_1^{(1)}(x, s) \pm \hat{e}_1^{(2)}(x, s)] = -s[R_{11}(x) \pm R_{12}(x)][\hat{\imath}_1^{(1)}(x, s) \pm \hat{\imath}_1^{(2)}(x, s)] \tag{31a}$$

$$\frac{d}{dx}[\hat{\imath}_1^{(1)}(x, s) \pm \hat{\imath}_1^{(2)}(x, s)] = -s[G_{11}(x) \pm G_{12}(x)][\hat{e}_1^{(1)}(x, s) \pm \hat{e}_1^{(2)}(x, s)], \tag{31b}$$

where $G_{11}(x)$ and $G_{12}(x)$ are the corresponding elements of $G_0(x)=R_0^{-1}(x)$. Thus this procedure, which is analogous to the familiar even- and odd-mode analysis of physically symmetric microwave structures, has resulted in the separation of the matrix transmission-line equations into two scalar equations. Stated in other terms, the condition of double symmetry makes it possible to diagonalize simultaneously both (27a) and (27b) with a constant, orthogonal matrix. After making the definitions, $E^{\pm}(x, s)=\hat{e}_1^{(1)}(x, s)\pm\hat{e}_1^{(2)}(x, s)$, $I^{\pm}(x, s)=\hat{\imath}_1^{(1)}(x, s)\pm\hat{\imath}_1^{(2)}(x, s)$, and $R^{\pm}(x)=R_{11}(x)\pm R_{12}(x)$ and noting that

$$G^{\pm}(x) \equiv G_{11}(x) \pm G_{12}(x) = 1/R^{\pm}(x), \tag{32}$$

(31a) and (31b) become

$$\frac{dE^{\pm}(x, s)}{dx} = -sR^{\pm}(x)I^{\pm}(x, s) \tag{33a}$$

$$\frac{dI^{\pm}(x, s)}{dx} = -s1/R^{\pm}(x)E^{\pm}(x, s). \tag{33b}$$

The boundary conditions for these normalized equations are

$$E^{\pm}(0, s) = \frac{1\pm S_{12}(s)}{S_{13}(s)}$$

$$I^{\pm}(0, s) = \frac{1 \mp S_{12}(s)}{S_{13}(s)} \tag{34a}$$

$$E^{\pm}(l, s) = I^{\pm}(l, s) = 1. \tag{34b}$$

It is evident that (33) and (34) describe two nonuniform lines each terminated in one ohm and possessing the scattering parameters $S_{11}^{\pm}(s)=\pm S_{12}(s)$ and $S_{12}^{\pm}(s)=S_{13}(s)$ normalized to one ohm. From (28) and the fact that $Z_0(x)$ is positive definite it follows that $R_{11}>|R_{12}|$. Consequently, $R^{\pm}(x)>0$, $0\leq x\leq l$, and the sum and difference lines are always realizable. This completes the proof of the conjecture made in the Introduction.

It will now be shown that the sum and difference transmission-line problems are not independent; in fact, one line is the dual of the other. The proof of this statement follows

151

from the uniqueness of the synthesis problem for nonuniform lines. The question of uniqueness and general realizability for continuously nonuniform lines has only recently begun to receive attention. Wohlers [8] has shown that for lossless lines possessing a local characteristic impedance $R(x)$ that is twice continuously differentiable, the reflection coefficient $S_{11}(s)$ uniquely determines $R(x)$. Heim and Sharpe [9] have obtained the same result under the more general condition that $R(x)$ be only absolutely continuous. The uniqueness of the synthesis problem for general discontinuous lines has yet to be proved, although there are strong heuristic reasons for assuming that the realizability of $S_{11}(s)$ always guarantees the uniqueness of $R(x)$. In the familiar case of stepped, uniform lines of equal length it is evident from the method of extracting sections that the synthesis problem has a unique solution [10].[2] It is also known that $S_{11}(s)$ uniquely determines the other elements of the scattering matrix if R is absolutely, sectionally continuous[3] and l is known. This implies that either $E(0, s)$ or $I(0, s)$ is sufficient to prescribe the synthesis of a lossless line terminated in a real impedance.

Assume that the synthesis problem for the sum and difference lines associated with (33) has a unique solution. That is, given a realizable $E^+(0, s)$, $R^+(x)$ is uniquely determined and similarly for $E^-(0, s)$ and $R^-(x)$. But from (34a) $E^\pm(0, s) = I^\mp(0, s)$. It follows that[4]

$$R^+(x) = 1/R^-(x), \qquad 0 \le x \le l. \tag{35}$$

Furthermore, the voltage on the sum line is equal to the current on the difference line, and vice versa, for every x. Thus, the two lines are complete duals. Since $R^+(x)$ corresponds to the "even-mode impedance" Z_{0e}, and $R^-(x)$ to the "odd-mode impedance" Z_{0o}, (35) is a generalization of the well-known condition, $Z_{0e}Z_{0o} = 1$, which is necessary for the equivalence of stepped lines and stepped directional couplers.

It has been shown that the realizability of a symmetric coupler guarantees the realizability of the sum and difference lines. Unfortunately, the converse is not true. This state of affairs results from the fact that the positive definite character of $C(x)$ is not sufficient to guarantee at each value of x the existence of a uniform multiconductor line having the capacitance matrix $C(x)$. It is known that a necessary condition for realizability is that $C(x)$ be hyperdominant [6]. That is, the elements of $C(x)$ must satisfy for all x, $0 \le x \le l$,

$$\sum_{j=1}^{2} C_{ij}(x) \ge 0, \qquad i = 1, 2 \tag{36}$$

and

$$C_{12}(x) = C_{21}(x) \le 0. \tag{37}$$

These conditions cannot be inferred from the positive definite character of $C(x)$. Since (36) and (37) imply that $C(x)$ is positive semidefinite, however, it is possible that the hyperdominant restriction is sufficient as well as necessary, although to the author's knowledge this has not been proved. In any event (36) and (37) do impose restrictions on the characteristic impedance of the sum and difference lines which may complicate the synthesis problem when unequal terminating impedances are specified.

In order to determine these restrictions it will be convenient to deal with the characteristic admittance $Y_0(x) = Z_0^{-1}(x)$ and terminating conductances, $g_1 = 1/r_1$, and $g_2 = 1/r_2$. From (12) and (28),

$$C(x) = \frac{1}{v(x)}\begin{bmatrix}\sqrt{g_1} & 0\\ 0 & \sqrt{g_2}\end{bmatrix}\begin{bmatrix}G_{11} & G_{12}\\ G_{12} & G_{11}\end{bmatrix}\begin{bmatrix}\sqrt{g_1} & 0\\ 0 & \sqrt{g_2}\end{bmatrix}$$

$$= \frac{1}{v(x)}\begin{bmatrix}g_1G_{11} & \sqrt{g_1g_2}\,G_{12}\\ \sqrt{g_1g_2}\,G_{12} & g_2G_{11}\end{bmatrix}. \tag{38}$$

Condition (36) in conjunction with (32) and (35) leads to the right-hand inequality of

$$1 \ge G^+(x) \ge \sqrt{\left|\frac{\sqrt{g_1} - \sqrt{g_2}}{\sqrt{g_1} + \sqrt{g_2}}\right|} \tag{39}$$

while (37) leads to the left-hand inequality. In summary, a symmetric transmission-line directional coupler is realizable only if the corresponding characteristic admittance $G^+(x)$ of the equivalent line satisfies (39) for all x.

Unequal Terminating Impedances

In order to investigate the role of the impedance transformation given in (26) consider the familiar single-section stepped coupler with line (1) terminated at ports 1 and 3 in a conductance g_1, and line (2) terminated at ports 2 and 4 in a conductance g_2. This example will also bring out the importance of the uniqueness property. The appropriate sum and difference problems are depicted in Fig. 2. The reflection coefficients $S_{11}^\pm$ are readily found from the solutions to (33a) and (33b). The result is

$$S_{11}^\pm = \pm S_{12} = \frac{jk^\pm \sin \omega l}{\sqrt{1 - (k^\pm)^2}\cos \omega l + j \sin \omega l}, \tag{40}$$

where

$$k^\pm = \frac{(R^\pm)^2 - 1}{(R^\pm)^2 + 1}. \tag{41}$$

A unique solution for S_{12} will be obtained only if $k^- = -k^+$. It follows that

$$R^+R^- = G^+G^- = 1 \tag{42}$$

and

$$k^+ = \frac{R^+ - R^-}{R^+ + R^-} = \frac{G^- - G^+}{G^- + G^+}. \tag{43}$$

[2] This statement of uniqueness does not conflict with the multiplicity of solutions to the synthesis problem when the insertion loss rather than $S_{11}(s)$ is specified [11].

[3] $R(x)$ is absolutely continuous except for a finite number of discontinuities.

[4] *Added in proof:* The sufficiency of this condition for physically symmetric directional couplers has recently been established by S. Yamamoto, T. Azakami, and K. Itakura, ("Coupled nonuniform transmission line and its applications," *IEEE Trans. Microwave Theory and Techniques*, vol. MTT-15, pp. 220–231, April 1967).

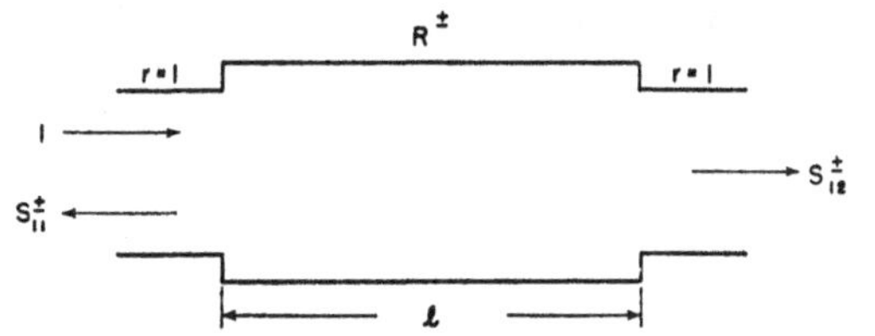

Fig. 2. Equivalent lines for a single-section directional coupler.

It remains to relate these parameters to the familiar even- and odd-mode characteristic admittances. Using Cristal's notation [12] the characteristic admittance matrix will be written

$$Y_0 = \begin{bmatrix} \dfrac{Y_{0e_1} + Y_{00_1}}{2} & \dfrac{Y_{0e_1} - Y_{00_1}}{2} \\ \dfrac{Y_{0e_1} - Y_{00_1}}{2} & \dfrac{Y_{0e_2} + Y_{00_2}}{2} \end{bmatrix} \equiv \begin{bmatrix} A & -D \\ -D & B \end{bmatrix}. \tag{44}$$

$$\frac{S_{12}(j\omega)}{S_{13}(j\omega)} \cong e^{j\omega l} \frac{K(\Gamma_2 e^{-2j\omega l} - \Gamma_1) + \int_0^l [e^{-2j\omega\xi} - \Gamma_1\Gamma_2 e^{-2j\omega(l-\xi)}]P(\xi)d\xi}{(1+\Gamma_2)(1-\Gamma_1)}, \tag{47}$$

From (28)

$$G_0 = R_0^{-1} = \begin{bmatrix} A/g_1 & -D/\sqrt{g_1g_2} \\ -D/\sqrt{g_1g_2} & B/g_2 \end{bmatrix}. \tag{45}$$

G_0, and hence R_0, will be doubly symmetric only if $A/g_1 = B/g_2$, which is Cristal's matching condition. Equations (32) and (42) lead to $AB - D^2 = g_1g_2$, which is Cristal's condition for infinite directivity. Also, from (43), the coupling coefficient is $k^+ = D/\sqrt{AB}$. Other formulas giving g_1 and g_2 in terms of the components of the L and C matrices have been given elsewhere [13] and will not be repeated here. It is clear from the example that the so-called nonsymmetric coupler is not basically different from the more common coupler for which $g_1 = g_2$. What is important is that the coupler possess electrical symmetry as defined previously in terms of the scattering parameters. Finally, it should be noted that if $S_{12}(s)$ or $G^+(x)$ is given, the terminating conductances g_1 and g_2 cannot be chosen arbitrarily.

Analysis of Nonuniform Directional Couplers

The principle advantage offered by continuously tapered couplers lies in the possibility of obtaining extremely wide coupling bandwidths. This contrasts with the periodic coupling characteristic of stepped directional couplers having equal-length sections. Theoretically, the coupling bandwidth will be infinite if $R^+(x)$ can be made discontinuous. However, the practical difficulty of realizing such a step in impedance leaves this ultimate bandwidth open to question. A more recognizable advantage seems to lie in the use of continuous nonuniform couplers to eliminate (in a quasi-TEM sense) reactive discontinuities altogether, thus removing a major deterrent to the achievement of high directivity.

It is often convenient to specify the properties of directional couplers in terms of the power division factor

$$\eta = \frac{|S_{12}|^2}{|S_{13}|^2}. \tag{46}$$

It can be shown that in the case of stepped couplers η is an even polynomial in $\cos \omega l$, where l is the normalized length of each section [3]. In the continuous case this parameter can be readily approximated using techniques developed for the analysis of single nonuniform lines. In order to accommodate possible discontinuities in $R^+(x)$ at $x=0$ and $x=l$, we will employ the formulation given by Youla [14]. Noting the boundary conditions given in (34) for the equivalent transmission-line problem, we obtain for the first-order approximation,

where

$$P(\xi) = 1/2 \frac{d}{d\xi} \ln R^+(\xi) \tag{48a}$$

$$\Gamma_1 = \frac{1 - R^+(0)}{1 + R^+(0)}$$

$$\Gamma_2 = \frac{1 - R^+(l)}{1 + R^+(l)} \tag{48b}$$

and

$$K = 1 - \ln \sqrt{R^+(l)/R^+(0)}. \tag{48c}$$

As an illustration in the application of this expression it will be instructive to consider the tapered-line magic T [15]. For reasons of simplicity we will assume that $R^+(x)$ and $R^-(x)$ have an exponential taper in the nonuniform region rather than the Chebychev taper originally employed. The variation of $R^+(x)$ is depicted by the solid line in Fig. 3. Even though the output terminals are located at $x=2l$, better accuracy will be obtained if (47) is applied only to the nonuniform section, $0 \leq x \leq l$, and a correction for the phase shift introduced by the uniform section is made afterward. Setting $\kappa = 1/R^+(l) = e^{-2\alpha l}$ and noting that $\Gamma_1 = 0$ and $\Gamma_2 = (\kappa-1)/(\kappa+1)$, we find that

$$\frac{S_{12}(j\omega)}{S_{13}(j\omega)} \cong \frac{\kappa+1}{2\kappa}\left[(1-\alpha l)\left(\frac{\kappa-1}{\kappa+1}\right) + \alpha l e^{j\omega l}\frac{\sin \omega l}{\omega l}\right]. \tag{49}$$

153

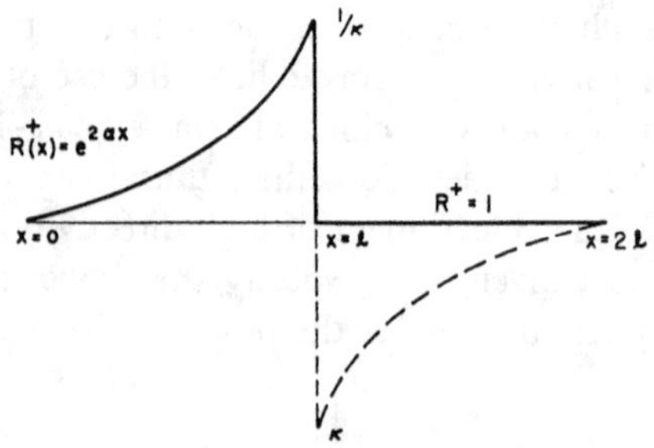

Fig. 3. Impedance variation of the equivalent line for a tapered-line magic T.

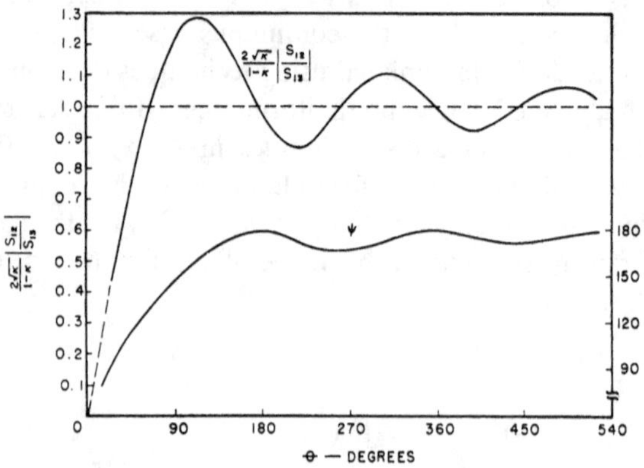

Fig. 4. Amplitude and phase variation of the power division factor for a tapered-line magic T.

This result, while reasonably accurate at intermediate frequencies, is not correct at either $\omega=0$ or $\omega=\infty$. The correct
154 asymptotic behavior at infinity, as given by DuHamel and Armstrong, [15] can be obtained by making the substitutions, $\alpha l \cong (1-\kappa)/(1+\kappa)$ and $1-\alpha l \cong \sqrt{\kappa}$. This can be justified when $\alpha l \ll 1$, since these approximations are of the same order as the approximations inherent in the derivation of (47). The final result is

$$\frac{S_{12}(j\omega)}{S_{13}(j\omega)} = \sqrt{\eta}\, e^{j\psi} = \frac{\kappa-1}{2\sqrt{\kappa}}\left[1 - \frac{e^{j\omega l}}{\sqrt{\kappa}} \frac{\sin \omega l}{\omega l}\right]. \tag{50}$$

The magnitude and phase of this expression are plotted in Fig. 4 versus $\theta=\omega l$ for the case where $\kappa=0.818$, corresponding to a 20 dB asymptotic coupling coefficient. Equation (47) provides a convenient formula for evaluating other tapers for directional couplers. For example, it can be shown by employing a double taper such as that suggested by the dotted line in Fig. 3, that the variation of the phase $\psi(\omega)$ can be substantially reduced. Unfortunately, because this characteristic violates (39), it does not lead to a realizable transmission-line coupler.[5] As in the case of stepped couplers, ψ can be made to approximate 90 degrees by employing an $R^+(x)$ characteristic which is symmetric about $x=l$ instead of antisymmetric as in the previous case. However, infinite coupling bandwidth is then no longer possible since, in the absence of a discontinuity in $R^+(x)$, η approaches zero for large frequencies. In spite of this restriction Tresselt [5] has recently reported obtaining a 7 to 1 bandwidth with such a design.

[5] This type of coupler was first discussed by Oliver [4]. He observed that it could be realized by transposing conductors, an operation which is ruled out in the present theory.

Conclusion

The important implication of the equivalence principle presented here is that whenever an exact solution to the analysis or synthesis problem for nonuniform transmission lines can be found, an exact solution to the corresponding directional coupler problem is also available, subject to the realizability conditions stated earlier. Thus, nonuniform directional couplers can be designed without necessarily making the usual assumption that the coupling is small or that the lines are only slightly nonuniform. Although there is little in the literature on exact methods of nonuniform transmission-line synthesis, there is increasing interest in the subject and hopeful signs that important advances will soon be forthcoming. In what may be the first practical solution to this problem, Heim [16] has recently devised an exact technique for constructing the characteristic impedance of a class of absolutely continuous lines. However, much work remains to be done, particularly on the approximation problem and on the realizability problem encountered here where the characteristic impedance is bounded from above and below.

Appendix

In this section a proof will be given for the following theorem: A matched, transmission-line directional coupler having an absolutely continuous characteristic impedance matrix must be symmetric; that is, $S_{13}(s)=S_{24}(s)$. It is understood that the two lines making up the coupler are of equal length.

It can be shown from the properties of the solution that $S_{13}(s)$ and $S_{24}(s)$ are meromorphic functions of s and analytic in the half plane, $\sigma \geq 0$. Therefore, S_{13}/S_{24} is also a meromorphic function, and we can write in general,

$$\frac{S_{13}(s)}{S_{24}(s)} = \frac{F(s)}{G(s)}, \tag{51}$$

where $F(s)$ and $G(s)$ are entire functions of s. The phase law,

$$S_{13}(s)S_{13}(-s) = S_{24}(s)S_{24}(-s), \tag{52}$$

which can be deduced from (2), implies that $G(s)$ and $F(-s)$ have the same zeros. It follows from Hadamard's factorization theorem [17] that

$$S_{24}(s) = S_{13}(s)\, \frac{F(-s)}{F(s)}\, e^{f(s)}, \tag{53}$$

where $f(s)$ is a polynomial of finite degree. We have seen that in order for a matched directional coupler to exhibit perfect directivity at all frequencies it is necessary that $S_{13}(0) = S_{24}(0) = 1$. From this condition we conclude that $f(0)=0$. It will now be shown that $F(s)$ can have no zeros. From (23) and (24),

$$S_{\beta\alpha}^{-1} = \begin{bmatrix} 1/S_{13} & 0 \\ 0 & 1/S_{24} \end{bmatrix}$$
$$= \tfrac{1}{2}[R_1^{-1/2}E(0, s) + R_1^{1/2}I(0, s)]$$
$$= \tfrac{1}{2}[\hat{E}(0, s) + \hat{I}(0, s)]. \quad (54)$$

Since the right side is an entire function of s, neither $S_{13}(s)$ or $S_{24}(s)$ can be zero for any finite value of s. Referring to (53) any zeros of $F(-s)$ which are canceled by poles of $S_{13}(s)$ produce poles of $S_{24}(s)$ in the right half plane. Consequently, $F(s)$ can have no zeros, and S_{13} and S_{24} differ at most by an exponential factor.

It remains to evaluate the asymptotic behavior of $S_{13}(s)$ and $S_{24}(s)$ as $s\to\infty$. In general the existence of a fundamental solution satisfying $\hat{E}(l, s)=\hat{I}(l, s)=I$ does not permit us to infer anything about the asymptotic behavior of $\hat{E}(x, s)$ and $\hat{I}(x, s)$ at $x=0$. However, it can be shown by extending previous results for the single line [9] to the matrix case that if the elements of $Z_0(x)$ are absolutely continuous, then there exists a fundamental solution to (27) such that for all x, $0 \le x \le l$,

$$\hat{E}(x, s) \to \begin{bmatrix} e^{s(l-x)} & 0 \\ 0 & e^{s(l-x)} \end{bmatrix} \quad (55)$$

and

$$\hat{I}(x, s) \to \begin{bmatrix} e^{s(l-x)} & 0 \\ 0 & e^{s(l-x)} \end{bmatrix} \quad (56)$$

uniformly as $s\to\infty$ in the half plane, $\sigma \ge 0$. Since the solution is unique, it follows from (54) that $S_{13}(s)$ and $S_{24}(s)$ have the same asymptotic behavior and are, therefore, equal.

References

[1] L. Young, "The analytical equivalence of TEM-mode directional couplers and transmission-line stepped-impedance filters," *Proc. IEE* (London), vol. 110, pp. 275–281, February 1963.

[2] R. Levy, "General synthesis of asymmetric multi-element coupled-transmission-line directional couplers," *IEEE Trans. Microwave Theory and Techniques*, vol. MTT-11, pp. 226–237, July 1963.

[3] A. L. Fel'dshtein, "Synthesis of stepped directional couplers," *Radiotekhn. i Electron.*, vol. 6, pp. 234–240, February 1961.

[4] B. M. Oliver, "Directional electromagnetic couplers," *Proc. IRE*, vol. 42, pp. 1686–1692, November 1954.

[5] C. P. Tresselt, "The design and construction of broadband, high-directivity, 90-degree couplers using nonuniform line techniques," *IEEE Trans. Microwave Theory and Techniques*, vol. MTT-14, pp. 647–656, December 1966.

[6] D. C. Youla, "An introduction to coupled-line network theory—part I," Polytechnic Institute of Brooklyn, Brooklyn, N. Y., Memo, PIBM-960-61, October 1961.

[7] R. Bellman, *Introduction to Matrix Analysis.* New York: McGraw-Hill 1960, p. 56.

[8] M. R. Wohlers, "A realizability theory for smooth lossless transmission lines," *IEEE Trans. Circuit Theory* vol. CT-13, pp. 356–363, December 1966.

[9] D. S. Heim and C. B. Sharpe, "The synthesis of nonuniform lines of finite length—part I," (submitted for publication to *IEEE Trans. Circuit Theory*).

[10] H. J. Riblet, "General synthesis of quarter-wave impedance transformers," *IRE Trans. Microwave Theory and Techniques*, vol. MTT-5, pp. 36–43, January 1957.

[11] H. Seidel and J. Rosen, "Multiplicity in cascade transmission line synthesis—part I," *IEEE Trans. Microwave Theory and Techniques*, vol. MTT-13, pp. 275–283, May 1965.

[12] E. G. Cristal, "Coupled-transmission-line directional couplers with coupled lines of unequal characteristic impedances," *IEEE Trans. Microwave Theory and Techniques*, vol. MTT-14, pp. 337–346, July 1966.

[13] E. Ott, "A network approach to the design of multiline 2N-port directional couplers," Polytechnic Institute of Brooklyn, Brooklyn, N. Y., Tech. Rept. RADC-TR-65-41, April 1965.

[14] D. C. Youla, "Analysis and synthesis of arbitrarily terminated lossless nonuniform lines," *IEEE Trans. Circuit Theory*, vol. CT-11, pp. 363–372, September 1964.

[15] R. H. DuHamel and M. E. Armstrong, "A wide-band monopulse antenna utilizing the tapered-line magic-*T*," Fifteenth Annual Symposium, Air Force Avionics Lab., Wright-Patterson AFB, in cooperation with the University of Illinois, Monticello, October 1965.

[16] D. S. Heim, "The synthesis of nonuniform lines of finite length—part II" (in preparation).

[17] E. C. Titchmarsh, *The Theory of Functions*, 2nd ed. London: Oxford University Press, 1939, p. 250.

Reprinted from IEEE TRANSACTIONS
ON MICROWAVE THEORY AND TECHNIQUES
Volume MTT-15, Number 7, July, 1967
pp. 398-405

The paper by Cohn (p. 159) provides an ingenious way to couple two lines very tightly. Essentially each line is tightly coupled to a common floating block. The block has two holes in it, through which the two lines pass with only a slight clearance. Since both lines pass through the same (floating) block, and each is coupled tightly to it, the block assumes a midway potential and merely serves to couple the two lines tightly to each other. The idea has taken root and floating strips are now commonly used as a "re-entrant" structure for tightly coupled strip lines or microstrip lines.

The paper by Cristal and Young (p. 164) was motivated by a desire to obtain an optimum quadrature coupler. (It can be shown that the two outputs from a symmetrical coupler are always 90 degrees out of phase.) It is not possible to synthesize such a coupler using Chebyshev polynomials. The coefficients of odd polynomials with equal-ripple characteristics were found by computer and this led to the synthesis of symmetrical couplers. Numerical solutions up to and including nine sections are tabulated. The most useful tables are probably not the ones for 3-dB coupling, but for 8.34-dB coupling, since a 3-dB coupler requires extremely tight tight coupling in the center section (such as is provided by Cohn's re-entrant coupler), whereas two 8.34-dB coupler (with far less tight couplings) can be cascaded to also give a 3-dB coupler.

Two matched quadrature couplers can always be cascaded to form a new directional coupler with generally tighter coupling. This useful property can be explained most simply in terms of **forward** couplers. If one visualizes a backward coupler with crossed-over lines, then it becomes identical to a forward coupler. The explanation rests on the fact that in a forward coupler the "even" and "odd" modes (no longer the same even and odd modes as in our backward coupler!) are phase shifted different amounts through each coupler. When two forward couplers are cascaded, the phase shifts of the respective modes add, and the coupling thus generally increases. (If the total phase shift goes over 180 degrees, then

it decreases again.) The explanation for forward couplers is given by Young (Ref. 6, pp. 812-814). See also Monteath, (pp. 118-119).

Another way to avoid tight coupling in the center is to arbitrarily limit the maximum permissible coupling and then to add the balance of the required coupling in a cascaded coupler (or couplers), as demonstrated by Shelton and Mosko.[21]

Another piece of history may be instructive. At the time we worked on our symmetrical coupler (p. 164), and had barely completed the theory and numerical analysis and tabulation, E.G. Cristal happened to visit P.P. Toulios in the fall of 1964 at Illinois Institute of Technology, and discovered that Toulios under the direction of A.C. Todd had tackled the same problem for his Ph.D. Dissertation. Toulios had used a different technique and had obtained solutions up to five sections. They had fortuitously even picked almost the same couplings and tolerances for design! The two papers were received by the editor of the Transactions on the same day and published side by side in the same issue. Only the paper by Cristal and Young has been reprinted because it goes up to nine sections, compared to five sections worked out by Toulios and Todd.[22]

The Re-Entrant Cross Section and Wide-Band 3-db Hybrid Couplers*

SEYMOUR B. COHN†, FELLOW, IEEE

Summary—**A new type of parallel-coupled TEM-mode cross section is described and named the re-entrant cross section. An analysis of the even- and odd-mode characteristic impedances of the re-entrant cross section shows it to have advantages in the case of tight coupling when compared to previously used parallel-strip cross sections. Close tolerances are easily held, and considerable misalignment is permissible. Two single-section 3-db couplers were tested with coupling curves very close to theoretical, and with good directivity. Then a three-section coupler having a re-entrant center section was designed for the 400- to 2000-Mc band, yielding a coupling variation less than 0.4 db and a minimum directivity of 29 db. Next a three-section coupler was designed for the 1- to 5-Gc band. A series of modifications resulted in a final model having a coupling variation within 0.5 db, and a minimum directivity of about 22 db.**

INTRODUCTION

BACKWARD-COUPLING parallel strip line couplers can theoretically be made to have nearly constant coupling over very large bandwidths by properly connecting coupled quarter-wave sections in cascade.[1-3] For example, one section yields 2:1 bandwidth with -3 ± 0.3-db coupling, while three sections in a symmetrical configuration can be designed to yield 3:1 bandwidth with -3 ± 0.1-db coupling, or 5.1:1 bandwidth with -3 ± 0.4-db coupling.[2] However, when using more than one section, discontinuity effects have previously reduced the coupler's directivity and increased its VSWR. Furthermore, it has been difficult to maintain sufficiently tight dimensional tolerances in the very closely spaced strips of the center section. Because of these problems, prior work on three section 3-db couplers for 3:1 bands has resulted in low directivity.[2] The situation is even more severe for designs of greater bandwidth.

In order to alleviate the problems of discontinuity effects and tolerances, different configurations have been evolved that include a new TEM-mode coupling cross section (patent pending). This cross section, which was named the *re-entrant* cross section, offers a number of advantages: 1) its design equations are simple; 2) by its nature tight tolerances are easily held; 3) it can tolerate considerable misalignment of its parts; and 4) experience has shown that it may be joined with conventional strip line end sections without excessive discontinuity effects.

DESIGN FORMULAS FOR RE-ENTRANT LINE COUPLERS

Fig. 1 shows the new re-entrant line coupling cross section. Conductors A and B are coaxial line center con-

* Received March 14, 1963. This development program was supported by the U. S. Army Signal Supply Agency, Contract DA36-039-SC-87435.

† Rantec Corporation, Calabasas, Calif.

[1] B. M. Oliver, "Directional electromagnetic couplers," PROC. IRE, vol. 42, pp. 1686–1962; November, 1954.

[2] J. K. Shimizu, "Strip-line 3-db directional couplers," 1957 IRE WESCON CONVENTION RECORD, Pt. 1, pp. 4–15.

[3] J. K. Shimizu and E. M. T. Jones, "Coupled-transmission-line directional couplers," IRE TRANS. ON MICROWAVE THEORY AND TECHNIQUES, vol. MTT-6, pp. 403–410; October, 1958.

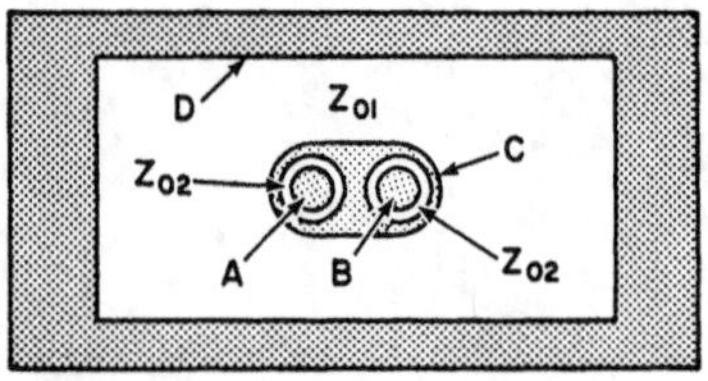

Fig. 1—Re-entrant coupled cross section.

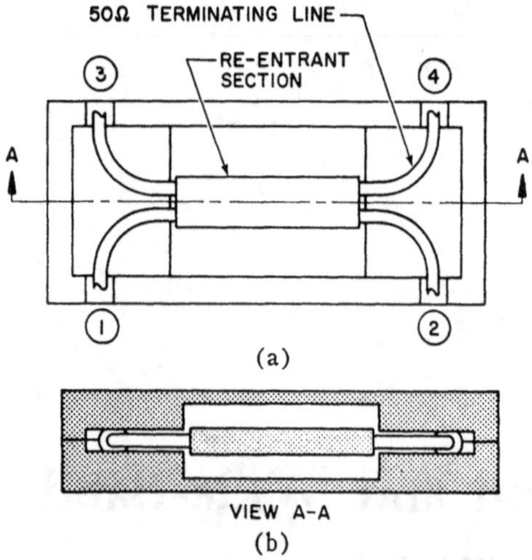

Fig. 2—Views of re-entrant quarter-wave directional coupler. (a) Top view, cover removed. (b) Longitudinal section.

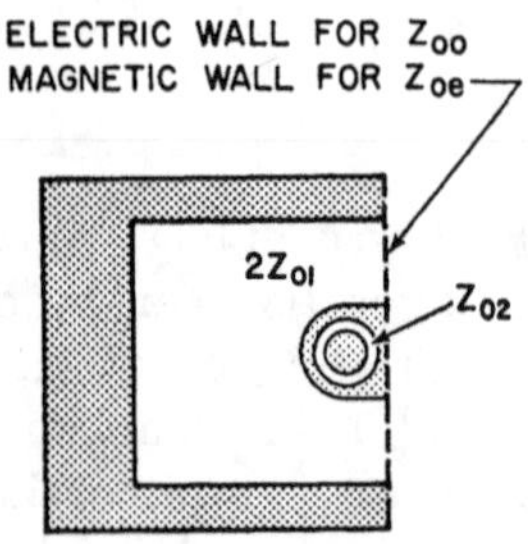

Fig. 3—Boundary conditions for even- and odd-modes in bisected re-entrant coupled cross section.

160

ductors of characteristic impedance Z_{o2} within the intermediate conductor, C, while C is the center conductor of a transmission line of characteristic impedance Z_{o1} within D. One method of connecting conductors A and B to the output ports is shown in Fig. 2. Note that the intermediate conductor is floating electrically. At first glance it appears that A and B in Fig. 1 are completely shielded from each other by C. However, the transmission line between C and D is in series with the two coaxial lines contained in C, and acts as the mutual coupling medium.

As in the case with conventional directional couplers, the re-entrant line coupler is best analyzed in terms of its even- and odd-mode characteristic impedances.[4] The even-mode characteristic impedance occurs between A and D when the structure is bisected through its vertical plane of symmetry by a magnetic wall, while the odd-mode characteristic impedance occurs when the bisecting plane is an electric wall.

Fig. 3 shows these two cases. For the even mode, Z_{oe} is equal to Z_{o2} and $2Z_{o1}$ in series

$$Z_{oe} = Z_{o2} + 2Z_{o1}. \tag{1}$$

For the odd mode, Z_{o1} is short-circuited by the electric wall, so that

$$Z_{oo} = Z_{o2}. \tag{2}$$

As with other forms of parallel transmission line directional couplers, this new structure is a backward-directed coupler. That is, if Port 1 in Fig. 2 is the input port, Port 3 will be the coupled port and Port 4 the isolated port. If discontinuity reactances are neglected, the isolation and input match will be perfect when the following condition is satisfied.

$$Z_o = \sqrt{Z_{oe}Z_{oo}} \tag{3}$$

where Z_o is the characteristic impedance of the terminating lines. The coupling factor is as follows at center frequency:

$$k = \frac{Z_{oe} - Z_{oo}}{Z_{oe} + Z_{oo}} = \frac{Z_{o1}}{Z_{o1} + Z_{o2}}. \tag{4}$$

In the design of a directional coupler, the desired value of k will first be determined. Than Z_{oe} and Z_{oo} may be calculated as follows:

$$Z_{oe} = Z_o\sqrt{\frac{1+k}{1-k}} \tag{5}$$

$$Z_{oo} = Z_o\sqrt{\frac{1-k}{1+k}}. \tag{6}$$

The necessary values of Z_{o1} and Z_{o2} are

$$Z_{o1} = \frac{1}{2}(Z_{oe} - Z_{oo}) = Z_o\left(\frac{k}{\sqrt{1-k^2}}\right) \tag{7}$$

$$Z_{o2} = Z_{oo} = Z_o\sqrt{\frac{1-k}{1+k}}. \tag{8}$$

Despite their simplicity, all of the above equations are exact for any value of k.

Single-Section Design

A quarter-wave section directional coupler will couple within -3 ± 0.3 db over an octave bandwidth if it is designed for a coupling value of -2.7 db (-2.69 db to be exact) at center frequency. In that case let $k=0.734$ and $Z_o=50$ ohms. By means of (7) and (8), the characteristic impedances in the re-entrant line cross section are found to be $Z_{o2}=19.6$ ohms and $Z_{o1}=\frac{1}{2}(127.8-19.6)=54.1$ ohms.

In two quarter-wave models that were constructed,

[4] E. M. T. Jones and J. T. Bolljahn, "Coupled strip-transmission line filters and directional couplers," IRE Trans. on Microwave Theory and Techniques, vol. MTT-4, pp. 75–81; Apirl, 1956.

the common outer conductor of the two coaxial lines consists of two pieces of 0.250-inch O.D. tubing soldered together with solder filling the spaces between the tubes. Thus, the outer perimeter is 0.250 inch thick and 0.500 inch wide. The wall thickness of the tubing is 0.022 inch, so that the inner diameter of the outer conductor is 0.206 inch. The ratio of conductor diameters to yield 19.6 ohms is 1.386, giving an inner conductor diameter of 0.149 inch. The gap between the conductors is thus about 0.028 inch. The inner conductor rods are supported by teflon rings one-sixteenth inch long.

Design data was not available for the Z_{o1} geometry. Therefore, the assumption was made that the Z_{o1} center conductor cross section is equivalent to a rectangular strip cross section having the same thickness and the same area. Thus if d_o is the outer diameter of the tubing, $t=d_o$ and $w=1.786d_o$. The accuracy of this equivalence is believed to be sufficiently good for this purpose. Then with the aid of thick strip characteristic impedance data,[5,6] it was determined that $Z_{o1}=54.1$ ohms for $d_o/b=0.335$. Thus for $d_o=0.25$ inch, $b=0.746$ inch.

Fig. 4 is a photograph of the internal arrangement of parts of the first experimental single section 3-db coupler, which was designed for $f_o=500$ Mc. Fifty-ohm slab line joins the re-entrant section to the connectors. The measured response in Fig. 5 shows that very good coupling accuracy and high directivity were achieved. The measurements were extended upwards in frequency in order to illustrate the periodic response of this type of coupler. The coupling and directivity in the band centered at $3f_o$ are essentially the same as that centered at f_o.

After the results of Fig. 5 were obtained, a second re-entrant coupler was designed for the 1000- to 2000-Mc band with $f_o=1500$ Mc. This coupler was made identical to the first coupler, except that the coupling region was made about one-third as long. Its mid-band coupling turned out to be -2.6 db, which is 0.1 db greater than the design value. The minimum directivity in the 1000- to 2000-Mc band is 29 db. The response curves are plotted in Fig. 6.

Fig. 4—Photograph of single section re-entrant line coupler for 340–680 Mc band.

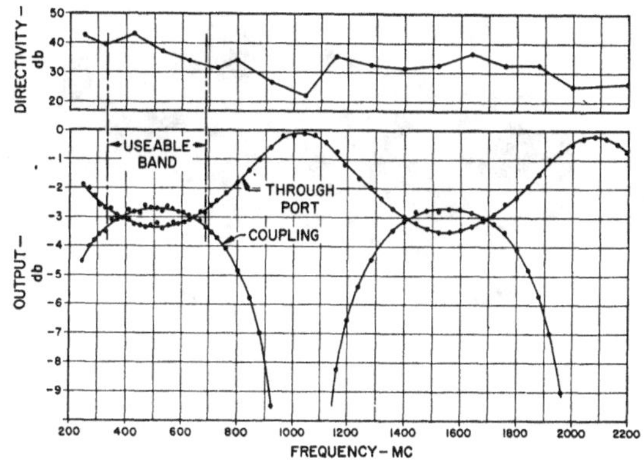

Fig. 5—Coupling and directivity response of quarter-wave re-entrant coupler after modification ($f_o=500$ Mc).

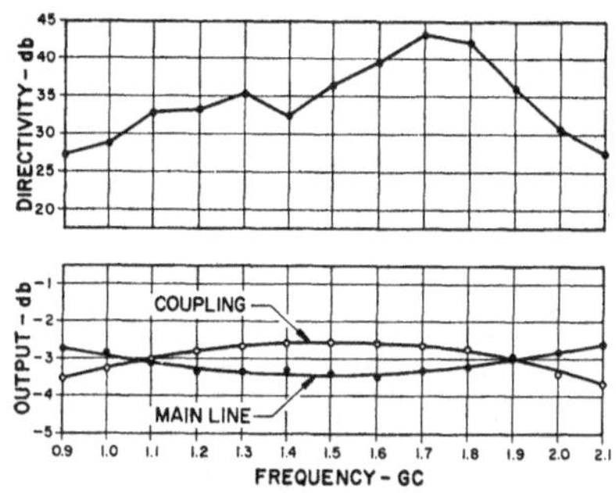

Fig. 6—Response of a 1- to 2-Gc single section re-entrant directional coupler.

Three-Section Design

The section couplings necessary to achieve a 5.1:1 bandwidth with -3 ± 0.4 db over-all coupling in a symmetrical three-section design are $k=0.8612$ (-1.28 db) and $k'=0.2337$ (-12.64 db), where k is the coupling factor of the center section and k' that of the end sections. With $Z_o=50$ ohms, the even- and odd-mode characteristic impedances of these sections are, respectively, $Z_{oe}=183$ ohms, $Z_{oe}'=63.5$ ohms, $Z_{oo}=13.66$ ohms, and $Z_{oo}'=39.4$ ohms. The re-entrant structure has been found suitable for the center section, but conventional parallel-coupled strip line is better for the end sections. The center-section characteristic impedances are obtained from (7) and (8) as follows: $Z_{o1}=84.7$ ohms and $Z_{o2}=13.66$ ohms. The dimensions were computed in the same way as in the case of the quarter-wave coupler. Based on a tubing outer diameter of 0.1875 inch and inner diameter 0.143 inch, the coaxial center conductor diameters are 0.114 inch, and the plate spacing is 0.956 inch.

The end sections were designed for copper-clad Tellite 3A dielectric material, which has a dielectric constant of 2.32. The thickness of each board is 0.125 inch so that the ground plane spacing is 0.250 inch. The copper thickness is 0.0014 inch. By means of published for-

[5] S. B. Cohn, "Problems in strip transmission lines," IRE Trans. on Microwave Theory and Techniques, vol. MTT-3, pp. 119–126; March, 1955.

[6] R. H. T. Bates, "The characteristic impedance of the shielded slab line," IRE Trans. on Microwave Theory and Techniques, vol. MTT-4, pp. 28–33; January, 1956.

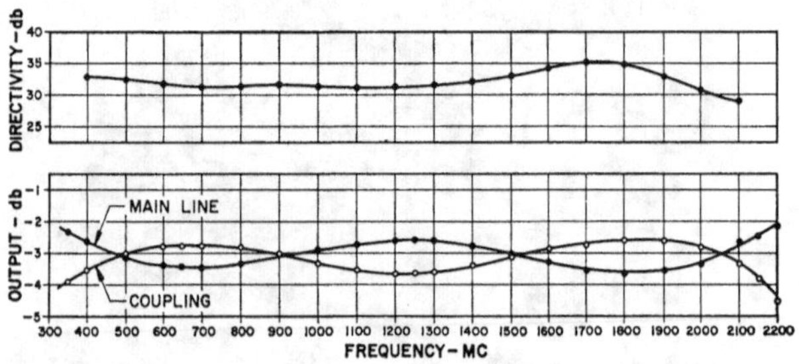

Fig. 7—Response curves of 400–2000 Mc three section directional coupler with re-entrant center section, after modifying end sections.

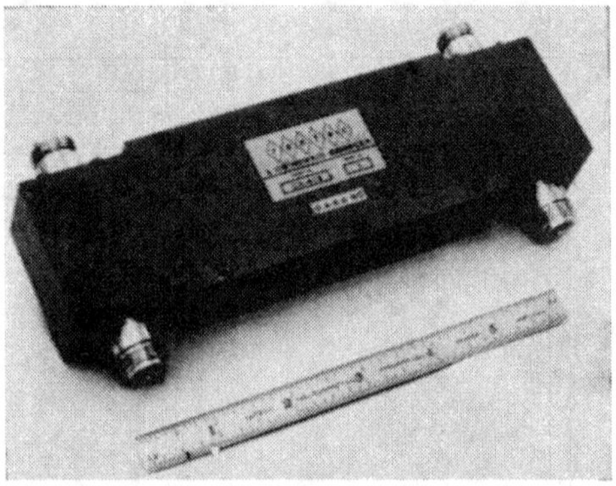

Fig. 8—Photograph of completed 400–2000-Mc hybrid coupler.

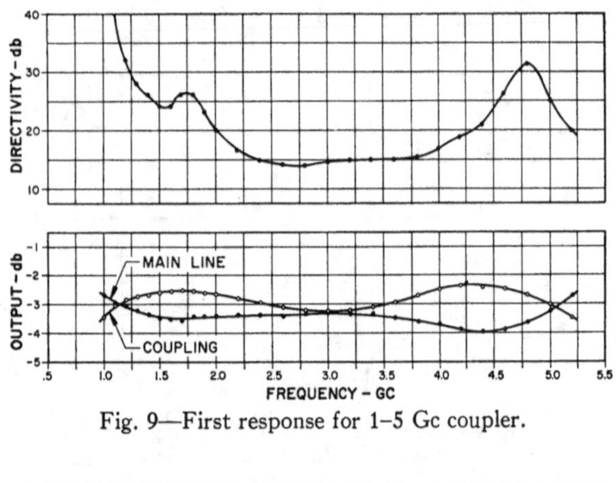

Fig. 9—First response for 1–5 Gc coupler.

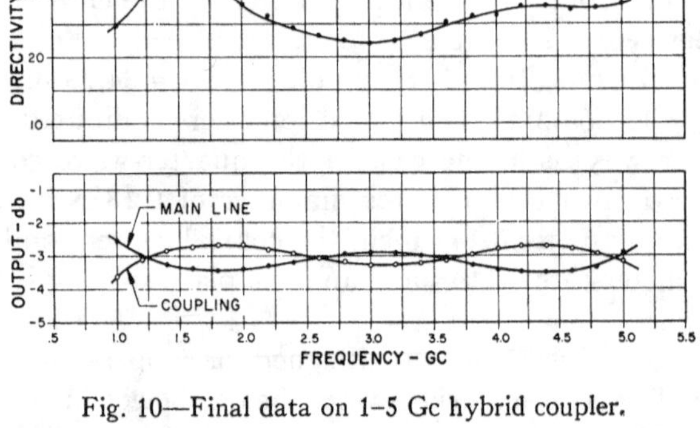

Fig. 10—Final data on 1–5 Gc hybrid coupler.

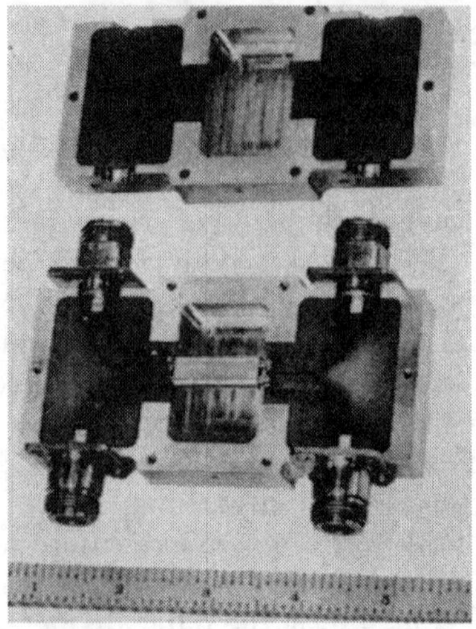

Fig. 11—Internal view of final 1–5 Gc hybrid coupler.

Fig. 12—View of completed 1–5 Gc hybrid coupler.

mulas,[7] the essential dimensions were computed in terms of the desired Z_{oe} and Z_{oo} values, taking thickness into account. In the coupled region, the strip width is $w=0.173$ inch and the gap is $s=0.0253$ inch. The 50-ohm terminating strips are 0.194 inch wide.

A coupler designed for the 400–2000 Mc band was built and tested. The first test revealed a fair conformance to the desired coupling response, but poor directivity. After some minor configuration changes near the ends of the end section strips to cancel discontinuity effects, the good results of Fig. 7 were achieved. A photograph of the completed model is shown in Fig. 8.

It was next decided to construct a model for 1 to 5 Gc having the same cross section dimensions as the 0.4- to 2.0-Gc model, but with lengths reduced in a 2.5-to-1 ratio. The measured curves for the first 1–5 Gc three-section coupler are shown in Fig. 9. It is evident that the coupling is too great, and that it increases gradually

[7] S. B. Cohn, "Shielded couple-strip transmission lines," IRE Trans. on Microwave Theory and Techniques, vol. MTT-3, pp. 29–38; October, 1955.

with frequency. This effect may be attributed to discontinuity effects at the junctions between the re-entrant and strip line sections. In view of the large plate spacing compared to wavelength ($b/\lambda = 0.405$ at 5.0 Gc in the re-entrant section), it is not surprising that discontinuity effects are severe. The poor directivity is also attributable to the discontinuities.

In order to decrease the coupling of the re-entrant section, the outer characteristic impedance, Z_{o1}, was reduced from 84.7 to about 76.3 ohms by reducing the plate spacing from 0.956 inch to 0.832 inch. This produced the desired triple loop appearance of the coupling and main line response curves. The center loop was 0.75 db, the lower loop 0.3 db, and the upper loop 0.7 db. The directivity was almost unaffected, however.

A number of changes were then made in the end sections to cancel discontinuity effects and increase the directivity. The final data for the 1–5-Gc hybrid coupler is shown in Fig. 10. Photographs of the coupler disassembled and assembled are given in Figs. 11 and 12.

Conclusion

The advantages of the re-entrant cross section have been verified experimentally in both single-section and three-section 3-db hybrid-coupler models. For cross-section heights in the range of 0 to 0.16λ, the simple design formulas have proved to yield very good accuracy. Beyond 0.16λ the coupling appears to increase gradually, but nevertheless good performance was achieved even with a cross section height as large as 0.4λ, after a minor adjustment of dimensions.

In the case of a three-section design for 5:1 bandwidth, a center section coupling of -1.28 db is required. This very strong coupling is easily achieved by the re-entrant cross section, and may be maintained accurately in production.

Theory and Tables of Optimum Symmetrical TEM-Mode Coupled-Transmission-Line Directional Couplers

164

E. G. CRISTAL, MEMBER, IEEE, AND L. YOUNG, SENIOR MEMBER, IEEE

***Abstract*—New equal-ripple polynomials were determined and applied to the synthesis of symmetrical TEM-mode coupled-transmission-line directional couplers (using exact methods). Tables of designs for symmetrical couplers of three, five, seven, and nine sections having mean couplings of −3.01, −6, −8.34, −10, and −20 dB, and having several equal-ripple tolerances in the coupling band are presented. Symmetrical maximally-flat directional-coupler designs having three, five, seven, and nine sections are also presented to complete the tables.**

Manuscript received November 23, 1964; revised June 1, 1965. This work was supported by the U. S. Army Electronics Laboratories, Fort Monmouth, N. J. under Contract DA-36-039-AMC-00084(E). At the time of submission, it was called to the attention of the authors that P. Toulios of the Illinois Institute of Technology Research Institute, Chicago, in an Internal Report dated September 8, 1964, had also independently compiled tables of designs for the three- and five-section symmetrical couplers utilizing a similar synthesis approach.

The authors are with Stanford Research Institute, Menlo Park, Calif.

I. Introduction

A. General Properties of the Couplers

A SYMMETRICAL TEM-mode coupled-transmission-line directional coupler is shown schematically in Fig. 1. Note that the symmetrical directional coupler has symmetry with respect to two planes: Ports 1 and 2 have end-to-end symmetry with respect to Ports 3 and 4; Ports 2 and 3 have side-to-side symmetry with respect to Ports 1 and 4.

A TEM-mode coupled-transmission-line directional coupler, whether symmetrical or not, has the following properties, when a signal generator is connected to Port 1:

1) There is transfer of power from Port 1 to Port 2.
2) There is transfer of power from Port 1 to Port 4.
3) There is no transfer of power from Port 1 to Port 3.
4) There is no reflected wave out of Port 1.

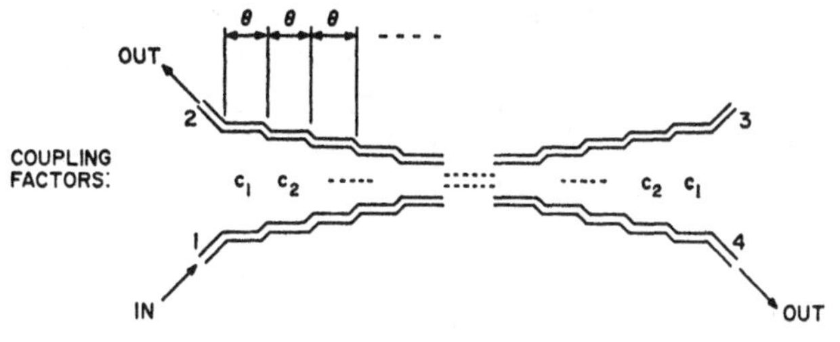

Fig. 1. Symmetrical TEM-mode coupled-transmission-line directional coupler.

The *symmetrical* directional coupler has, in addition, the unique and valuable property that

5) The two outputs at Ports 2 and 4 differ in phase by 90 degrees, at all frequencies.

It is this additional property that makes symmetrical couplers of particular importance.

Designs of optimum asymmetrical directional couplers of two to six sections were recently published by Levy [1]. These coupler designs have an equal-ripple approximation to the mean coupling. They are optimum in the sense that they provide a maximum bandwidth for a given number of sections, a given mean coupling, and a given coupling tolerance. To date, however, the exact design of optimum symmetrical couplers, i.e., symmetrical couplers having an equal-ripple approximation to the mean coupling, has been limited to couplers of at most three sections [2], [3]. Although symmetrical directional couplers of more than three sections can be synthesized on the basis of a first-order theory [4] these designs do not maximize the bandwidth since they do not necessarily provide an equal-ripple approximation to the mean coupling. Furthermore, for strong coupling, such as is required for 3-dB couplers, the first-order theory does not guarantee physically realizable results.

The first hurdle in the synthesis of optimum symmetrical couplers is to determine the appropriate equal-ripple polynomials for the insertion-loss function of the coupler. (These polynomials cannot be expressed in terms of known Chebyshev polynomials, as is the case for asymmetrical couplers.) Next, having obtained the insertion-loss function, extract the parameters of the coupler using exact synthesis procedures. Curves are plotted in Figs. 2, 3, and 4, showing typical responses of some of the five-, seven-, and nine-section 3-dB couplers that were obtained in this work.[1] These curves were calculated using the analytic expression for the coupling response rather than from an analysis of the synthesized couplers themselves.

[1] Whenever 3-db couplers are referred to, we shall mean equal power division, i.e., 3.0103-dB coupling.

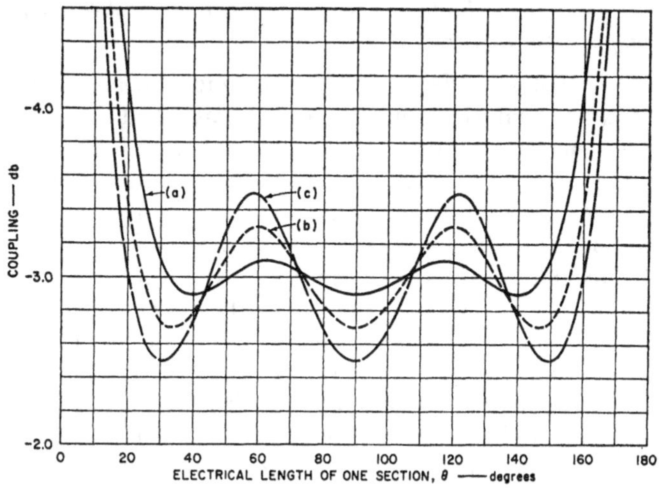

Fig. 2. Theoretical responses of symmetrical five-section couplers. (a) 0.1 dB ripple. (b) 0.3 dB ripple. (c) 0.5 dB ripple.

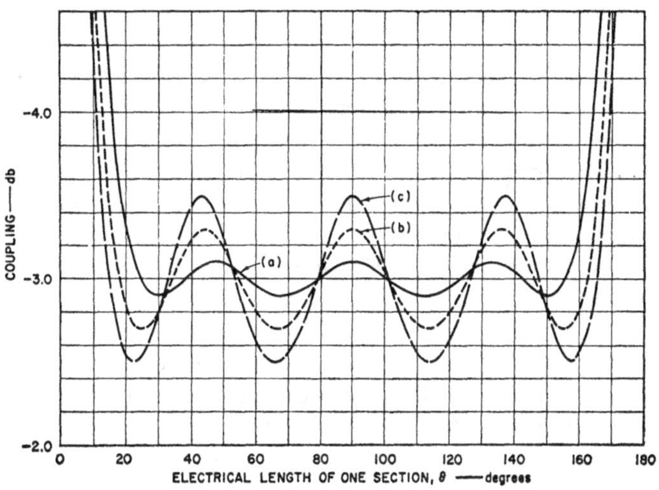

Fig. 3. Theoretical responses of symmetrical seven-section couplers, (a) 0.1 dB ripple. (b) 0.3 dB ripple. (c) 0.5 dB ripple.

165

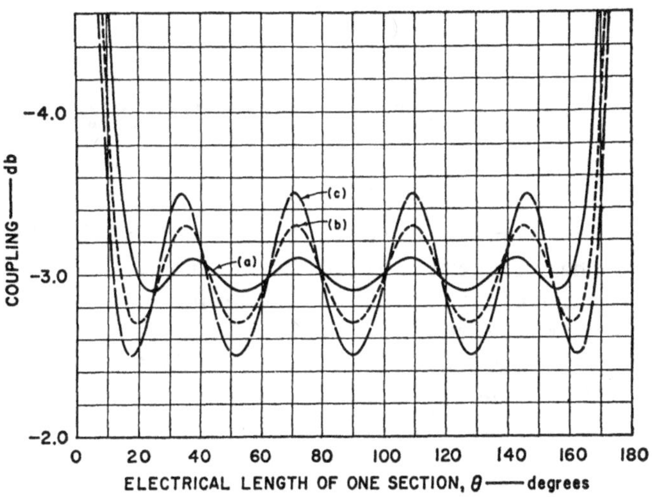

Fig. 4. Theoretical responses of symmetrical nine-section couplers. (a) 0.1 dB ripple. (b) 0.3 dB ripple. (c) 0.7 dB ripple.

B. Analytical Equivalence of Directional Couplers to Stepped-Impedance Filters

In papers by Feldshtein [5], Young [3], and Levy [6], an analytic equivalence is established between TEM-mode directional couplers and cascaded transmission lines. Briefly, the equivalence is that the reflected wave of the cascaded transmission lines corresponds to the backward-coupled wave of the directional coupler, and the transmitted wave of the cascaded transmission lines corresponds to the forward-coupled wave of the directional coupler. The use of this equivalence reduces the synthesis of TEM-mode coupled-transmission-line directional couplers having a prescribed coupling coefficient to the synthesis of cascaded transmission lines having a prescribed reflection coefficient.

C. The Insertion-Loss Function Theorem

Riblet [7] proved a theorem concerning the conditions that were necessary and sufficient if a given impedance function was to be realized as a cascade of equal-length transmission line sections. He also stated the most general insertion-loss function for a *quarter-wave transformer* (giving optimum match between two impedance levels). Levy [6] gave the most general insertion-loss function for the optimum asymmetrical quarter-wave filter (used as a prototype for the optimum asymmetrical TEM-mode coupler). Seidel and Rosen [8] have stated the necessary and sufficient conditions for making a prescribed insertion-loss function realizable as a cascade of equal-length transmission-line sections. They did for the insertion-loss function what Riblet did for the impedance function, and they arrived at a more concise statement. We shall state Seidel and Rosen's theorem as follows:

The necessary and sufficient conditions that an insertion-loss function L represents a homogeneous[2] stepped-impedance filter is that it be a polynomial of the form

$$L = L(\sin^2 \theta) \tag{1}$$

and that L be a polynomial greater than or equal to unity for all real values of θ.

D. The Symmetry Condition

The symmetry of a two-port filter is closely related to its phase properties. It can readily be shown (by an extension of the argument in [9], for instance) that the necessary and sufficient condition for a two-port filter to be symmetrical is that the phase of the transmission coefficient and the phase of the reflection coefficient differ by 90 degrees. This is the property that makes the symmetrical coupler so interesting and useful.

It is shown in textbooks on network synthesis [10], [11] that the necessary and sufficient condition for having an insertion-loss function represent a (lumped-constant) symmetrical network is that the function have the form: Unity plus the square of an odd function of frequency. By Richards' transformation [12], [13] we can now extend this result to resistor transmission-line circuits, including the stepped-impedance filters under present consideration. When this is done, it is found that the insertion-loss function is of the form: Unity plus the square of an odd function of θ. Combining this result with the insertion-loss theorem, we conclude that:

The necessary and sufficient condition that an insertion-loss function L represent[3] a *symmetical* homogeneous stepped-impedance filter, of n equal-line-length sections, is that it be of the form [21]

$$L = 1 + [P_n(\sin \theta)]^2 \tag{2}$$

where P_n is an odd polynomial in $\sin \theta$, of degree n.

Thus, the synthesis of a symmetrical directional coupler reduces to:

1) Finding the optimum odd polynomials P_n $(\sin \theta)$ and
3) Extracting the transmission-line impedances from the resulting insertion-loss function.

The synthesis procedure just outlined can produce many coupler designs, the number depending on the degree n of $P_n(\sin \theta)$. To obtain a symmetrical design, it is necessary to select the complex zeros of the reflection coefficient (obtained analytically from the insertion-loss function) such that symmetry with respect to the j-axis in the complex plane is achieved. This will be discussed further in Section IV.

II. Tables of Symmetrical Couplers

Tables of equal-ripple and maximally-flat symmetrical coupler designs are presented in the Appendix. TEM-mode coupled-transmission-line directional couplers are conveniently analyzed on the basis of the even- and odd-mode impedances of the individual sections of the coupler [14]. The tables of coupler designs presented here are, therefore, given on this basis. In addition, the signs have been normalized so that the product of even- and odd-mode impedances is unity; that is,

$$Z_{0e} Z_{00} = 1 \tag{3}$$

where Z_{0e} and Z_{00} are the even- and odd-mode impedances, respectively. For any particular application, the even- and odd-mode impedances are scaled by multiplying each normalized impedance of the tables by the value of the impedance of the terminating line. Since Z_{00} may be obtained from (3), only Z_{0e} is tabu-

[2] By *homogeneous* we mean that the ratios of the impedances of the cascaded lines, and the ratios of the wavelengths in them, be independent of frequency, *cf.* [20].

[3] At the same time, it should be pointed out that an insertion-loss function of the stated form can be represented by a number of networks, and this theorem states that at least one of them is symmetrical.

lated. To keep the tables as compact as possible, only one-half of the even-mode impedances for each design are presented, since the couplers are symmetrical.

A parameter of frequent interest to the designer is c_i, the coefficient of coupling of the ith section of the coupler. Because (3) is normalized, the coefficient of coupling can be obtained from the values of Z_{oe} in the tables by the formula

$$c_i = \frac{(Z_{oe})_i^2 - 1}{(Z_{oe})_i^2 + 1}. \tag{4}$$

Two bandwidth definitions for directional couplers are in common use, and both are presented in the tables for the convenience of the reader. The *fractional bandwidth*, denoted as w is given by

$$w = \frac{f_2 - f_1}{f_0} \tag{5}$$

where f_1 and f_2 are the lower and upper frequencies at the equal-ripple band edge (see Fig. 6), and

$$f_0 = \frac{f_2 + f_1}{2} \tag{6}$$

is the arithmetic mean of f_1 and f_2.

The second definition of bandwidth is the *bandwidth ratio*, denoted here by B; it is given by

$$B = \frac{f_2}{f_1} \tag{7}$$

where f_2 and f_1 are defined as before.

In the cases of maximally-flat coupler designs (Tables A-21 to A-24) the frequencies f_2 and f_1 used in the formulas for bandwidth refer to the frequencies that are 3 dB lower than the mean coupling.

The parameters M and δ that appear in Tables A-1 through A-20 in the Appendix, pp. 554–557, are explained by the representative coupling curve shown in Fig. 5. The symbol M represents the mean coupling in decibels. The symbol δ represents the maximum deviation (ripple value) from the mean in decibels. All designs presented in the tables have each coupling section one-quarter wavelength long at the mean frequency f_0.

Symmetrical coupler designs of three sections are included in the tables, although these designs may be obtained from prior work by Shimizu and Jones [2] or Young [3]. The design values in the tables were arrived at independently by the methods given in Section III. They are included only to make the tables as complete and useful as possible. Maximally-flat coupler designs are also presented in the tables. Exact designs of the three-section maximally-flat coupler have been published previously [3], but the design is included here. The designs for maximally flat couplers of 5, 7, and 9 sections are believed to be new.

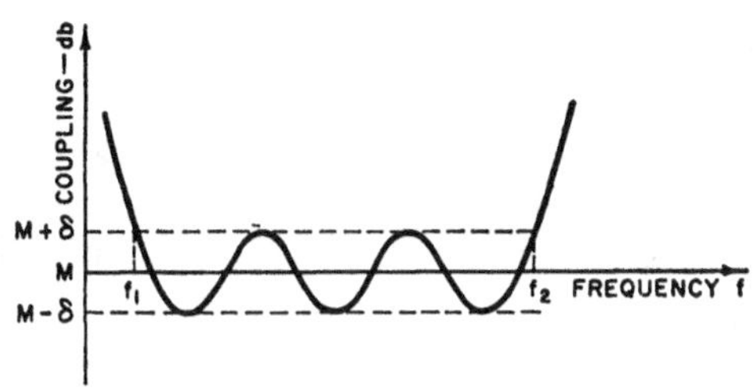

Fig. 5. Typical directional coupler characteristics (the particular curve shown represents a five-section coupler).

III. Deviation of the Tables

The derivation of the coupler parameters from the given insertion-loss function is presented here. (The derivation of the insertion-loss functions themselves will be presented in Section IV.) The synthesis of stepped-impedance lines from a prescribed insertion-loss function has been adequately described in previous papers [6], [7]. For this reason the synthesis description here is presented as a step-by-step procedure only, and is stated without proof.

Let $P_n(x)$ be an odd polynomial in x of degree n that makes the function

$$L(x) = 1 + P_n^2(x) \tag{8}$$

an equal-ripple function on the interval 0 to 1.

Letting $x = \sin\theta$, the function $L(x)$ may be identified as an equal-ripple insertion-loss function for a symmetrical cascade of transmission lines.

From (8), the magnitude squared of the reflection coefficient is obtained:

$$|\Gamma|^2 = \frac{P_n^2(x)}{1 + P_n^2(x)} \tag{9}$$

where Γ is the reflection coefficient, and again $x = \sin\theta$.

Next, using Richards' transformation in the form [12]

$$\tan\theta = s/j, \tag{10}$$

where s is a complex variable, and $j = \sqrt{-1}$, the substitution

$$x = \frac{s/j}{\sqrt{1 + (s/j)^2}} = \sin\theta \tag{11}$$

may be made for the variable x. To facilitate the computations which follow, however, the transformation (11) is accomplished in two parts:

Part 1) Replace x by $\hat{s}/j$, where in this part $\hat{s}$ is an intermediate variable and not equal to $j\tan\theta$ (12)

Part 2) Replace $\hat{s}/j$ by $(s/j)/\sqrt{1+(s/j)^2}$ (13)

Equation (13) completes the transformation of (11). Using the previous two-part transformation process, the synthesis procedure is as follows:

Step A: Transformation (12) is applied to (9) and the right side of (9) is factored into the form

$$\hat{\Gamma}(\hat{s})\hat{\Gamma}(-\hat{s}) = \Gamma(\hat{s}/j)\Gamma(-\hat{s}/j) \propto \frac{\prod_{i=1}^{n}(\hat{s}-z_i)(-\hat{s}-z_i)}{\prod_{i=1}^{n}(\hat{s}-p_i)(-\hat{s}-p_i)} \tag{14}$$

where $\hat{\Gamma}(\hat{s})$ is defined to be $\Gamma(\hat{s}/j)$, z_i is a zero of the numerator, and p_i is a zero of the denominator. Expression (14) is not stated as an equation because the necessary constant factors in the numerator and denominator are excluded at this point. The factoring of (9) is simplified by solving the lower-order equations

$$P_n(\hat{s}/j) = 0 \tag{15}$$

and

$$P_n(\hat{s}/j) = \pm j \tag{16}$$

rather than (9) itself. It is clear that the zeros of (15) are to be taken as double.

It can be demonstrated that where $P_n(x)$ is an odd polynomial, the values of s that are solutions to

$$P_n(\hat{s}/j) = j \tag{17}$$

are the negative of the values of $\hat{s}$ that satisfy

$$P_n(\hat{s}/j) = -j. \tag{18}$$

so that it is only necessary to solve one of the equations of (16).

Step B: Next, the zeros of the numerator and denominator are mapped into new zeros, denoted by primes, by the transformation of (13). This is equivalent to

$$z_i' = \frac{z_i}{\sqrt{1+(z_i)^2}} \tag{19}$$

$$p_i' = \frac{p_i}{\sqrt{1+(p_i)^2}}\,. \tag{20}$$

The reflection coefficient $\Gamma'(s)$ is then constructed in the following way: Zeros of the numerator of $\Gamma'(s)$ are chosen from the z' so that they are symmetrical with respect to the j axis. This is necessary to ensure that the network will be symmetrical. The zeros of the denominator of $\Gamma'(s)$ are chosen from the p' so they lie in the left half plane. This latter selection process is necessary to ensure that the reflection coefficient is analytic in the right half plane. (Network symmetry will be discussed in more detail subsequently.) After the zero and pole selection is completed, the complex reflection coefficient is constructed according to (21),

$$\Gamma'(s) = \frac{\gamma_N'(s)}{\gamma_D'(s)} = \frac{a\prod_{i=1}^{n}(s-z_i')}{b\prod_{i=1}^{n}(s-p_i')} \tag{21}$$

where $\gamma_N'(s)$ and $\gamma_D'(s)$ are equal to the numerator and denominator, respectively, of the right side of (21).

Step C: The constants a and b in (21) are evaluated as follows: An examination of (11) reveals that the point $\sqrt{2}$ goes into $\sqrt{2}$. Furthermore, the transformation (11) also requires that

$$P_n^2(\sqrt{2}) = a^2 \left.\frac{\prod_{i=1}^{n}(s-z_i')\prod_{i=1}^{n}(-s-z_i')}{(-1)^n}\right|_{s=\sqrt{2}}. \tag{22}$$

Solving gives

$$a = \left\{\frac{(-1)^n P_n^2(\sqrt{2})}{\prod_{i=1}^{n}(s-z_i')\prod_{i=1}^{n}(-s-z_i')}\right\}^{1/2}_{s=\sqrt{2}}. \tag{23}$$

The sign of a is chosen to that $\gamma_N'(s)\to+0$ as $s\to 0$. This choice of sign is justified by noting that the normalized even-mode impedances are always greater than one, which requires the complex reflection coefficient to be positive as $s\to 0$. Similarly, the constant b may be found from:

$$b = \left\{\frac{(-1)^n[1+P_n^2(\sqrt{2})]}{\prod_{i=1}^{n}(s-p_i')\prod_{i=1}^{n}(-s-p_i')}\right\}^{1/2}_{s=\sqrt{2}}. \tag{24}$$

Step D: When

$$\Gamma'(s) = \frac{\gamma_N'(s)}{\gamma_D'(s)}$$

has been obtained, the impedance function is determined from the relationship

$$Z(s) = \frac{1+\Gamma'}{1-\Gamma'} = \frac{\gamma_D'+\gamma_N}{\gamma_D'-\gamma_N'}\,. \tag{25}$$

From Z, the $ABCD$ transmission matrix is constructed[4] as follows:

A is identified with the even part of $\gamma_D'+\gamma_N'$,
B is identified with the odd part of $\gamma_D'+\gamma_N'$,
C is identified with the odd part of $\gamma_D'-\gamma_N'$,
D is identified with the even part of $\gamma_D'-\gamma_N'$. (26)

It can now be seen why the zeros of $\Gamma'(s)$ must be chosen symmetrical with respect to the j axis. For a network to be symmetrical it is necessary that its $ABCD$ matrix have $A=D$. From (26) it is seen that A is the even part of $\gamma_D'+\gamma_N'$ while D is the even part of $\gamma_D'-\gamma_N'$. The only way to ensure that $A=D$ is to make γ_N' an odd polynomial. Choosing the zeros of the numerator of $\Gamma'(s)$ to be symmetrical with respect to the j axis (and recalling that there is a zero in the origin) forces γ_N' to

[4] A constant premultiplier of the $ABCD$ matrix, $(1/\sqrt{1-s^2})^n$, is neglected in the construction process.

be an odd polynomial. Other selections of the zeros of the numerator $\Gamma'(s)$ will result in asymmetrical structures, although the same insertion-loss function will result.

Step E: Next, the transmission-line impedances are extracted from the *ABCD* matrix.

The first line impedance is given by the formula

$$Z_1 = \left.\frac{A(s)}{C(s)}\right|_{s=1} = \left.\frac{B(s)}{D(s)}\right|_{s=1}. \tag{27}$$

Next, the matrix multiplication

$$\frac{1}{1-s^2}\begin{bmatrix} 1 & -Z_1 s \\ \frac{s}{-Z_1} & 1 \end{bmatrix}\begin{bmatrix} A & B \\ C & D \end{bmatrix} = \begin{bmatrix} \overline{A} & \overline{B} \\ \overline{C} & \overline{D} \end{bmatrix} \tag{28}$$

is performed. The result is a new *ABCD* matrix, which has the same form as the preceding *ABCD* matrix, except that the new input impedance

$$\overline{Z}(s) = \frac{A + \overline{B}}{\overline{C} + \overline{D}} \tag{29}$$

is of one degree less than the preceding $Z(s)$. The next line impedance is given by (27),

$$Z_2 = \left.\frac{A(s)}{\overline{C}(s)}\right|_{s=1} \tag{30}$$

and the reduction process of (28) is repeated.

In this way all the line impedances may be obtained. Since the structure is symmetrical, however, it is only necessary to perform the cycle a total of $(n+1)/2$ times where n is the (odd) number of sections in the line.

Example: A numerical example will serve to illustrate this synthesis procedure. The insertion-loss function of a three-section maximally-flat symmetrical coupler with equal-power division at $x = \sin\theta = 1$ can be shown to be[5]

$$L = 1 + (1.5x - 0.5x^3)^2$$
$$= 1 + P_3^2(x). \tag{31}$$

Using *Step A*, we find that

$$P_3(\hat{s}/j) = 0 \tag{32}$$

has double roots

$$z = 0, \quad j1.732, \quad -j1.732 \tag{33}$$

and that

$$P_3(\hat{s}/j) = j \tag{34}$$

has roots

$$p = -0.5961, \quad (0.2980 \pm j1.8073). \tag{35}$$

[5] The general form of maximally-flat odd polynomials is presented in Part IV.

Therefore, the roots of $P_3(\hat{s}/j) = -j$ are

$$p = +0.5961, \quad -(0.2980 \pm j1.8073). \tag{36}$$

Using *Step B* next, it is found that the roots are transformed into

$$z' = 0, \quad \pm(1.2247) \tag{37}$$

and

$$p' = \pm 0.5120, \quad \pm(1.1726 \pm j0.07782). \tag{38}$$

From *Step C*,

$$a \text{ is found to be } 1.0$$
$$b \text{ is found to be } 1.414. \tag{39}$$

From *Step D*,

$$A(s) = 1.0 + 4.0407s^2$$
$$B(s) = 5.1512s + 0.41421s^3$$
$$C(s) = 2.1512s + 2.4142s^3$$
$$D(s) = 1.0 + 4.0407s^2. \tag{40}$$

Last, the impedances are extracted as explained in *Step E*. The results are

$$Z_1 = \frac{A(1)}{C(1)} = \frac{5.0407}{4.5657} = 1.104 \tag{41}$$

which is also equal to Z_3,

Next, Z_2 is found to be

$$Z_2 = \frac{A(1)}{\overline{C}(1)} = 2.943. \tag{42}$$

IV. Derivation of the Equal-Ripple and Maximally-Flat Polynomials

A. *Equal-Ripple Polynomials*

The determination of the equal-ripple polynomials used in the insertion-loss functions of the symmetrical couplers is described here. This problem may be viewed as the determination of odd polynomials of degree n that give an equal-ripple approximation to a constant on the interval of zero to one. The polynomials take their last "equal-ripple value" at $x = 1$.

Consider for an example the case of a fifth-order odd polynomial approximating unity on $0 < x \leq 1$. This is shown in Fig. 6. It is clear from the figure that there are two points at thich the first derivative of the polynomial is zero. Let these points be denoted as x_1 and x_2. Because the polynomial is odd, $P_5'(x)$ (where the prime denotes d/dx) may be written as

$$P_5' = C(x^2 - x_1^2)(x^2 - x_2^2). \tag{43}$$

From (43), $P_5(x)$ is determined by integration:

$$P_5(x) = C\int_0^x (u^2 - x_1^2)(u^2 - x_2^2)du. \tag{44}$$

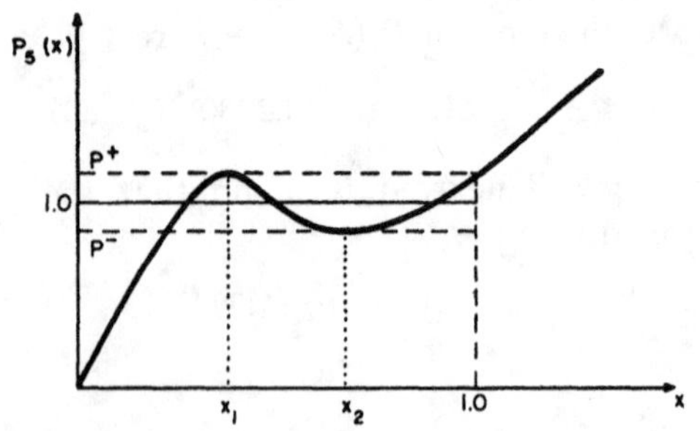

Fig. 6. Example of a fifth-order odd polynomial approximating unity on the interval zero to one.

The constant C is determined by the condition that

$$P_5(1) = P^+(\text{see Fig. 7 for definition of } P^+). \tag{45}$$

The result is

$$P_5(x) = c_1(x_1, x_2)x + c_3(x_1, x_2)x^3 + c_5(x_1, x_2)x^5. \tag{46}$$

The coefficients c_i are determined by the conditions that

$$\begin{aligned} P_5(x_1) &= P^+ \\ P_5(x_2) &= P^- \end{aligned} \tag{47}$$

where P^+ and P^- are the equal-ripple extremes (see Fig. 7).

Before describing the solution of (47), we wish to generalize the procedure just given. For an nth-order odd polynomial approximating a constant on the interval zero to one, we have

$$P_n'(x) = C \prod_{i=1}^{k} (x^2 - x_i^2) \tag{48}$$

170 where

$$k = \frac{n-1}{2} \tag{49}$$

$$P_n(x) = C \int_0^x \prod_{i=1}^{k} (u^2 - x_i^2)du \tag{50}$$

and

$$C = \frac{P_n(1)}{\int_0^1 \prod_{i=1}^{k} (u^2 - x_i^2)du} \tag{51}$$

where $P_n(1)$ equals P^+ or P^- depending on the value of n. The result of the preceding operations is an expression for $P_n(x)$ in the form

$$P_n(x) = \sum_{i=1}^{n} c_i x^i \tag{52}$$

where c_i are functions of the x_i and all coefficients with even i are zero. The c_i are determined by the condition that

$$\begin{aligned} P_n(x, x_1, x_2, \cdots, x_k)\big|_{x=x_1} &= P^+ \\ P_n(x, x_1, x_2, \cdots, x_k)\big|_{x=x_2} &= P^- \\ &\vdots \\ P_n(x, x_1, x_2, \cdots, x_k)\big|_{x=x_k} &= P^+ \text{ or } P^- \\ &\text{depending on } n. \end{aligned} \tag{53}$$

The following method was used to solve the set of nonlinear equations in (53). Considering P_n as a multivariable function of the x_i each equation of the set was linearized by taking the first two terms of the generalized Taylor expansion of the functions on the left in (53). Initial guesses for the values of $x_1, x_2, \cdots, x_k$ were substituted into the resultant linear set, and a solution was obtained by standard methods. The new solution was used as a second approximation to the initial guess and the process was repeated. (This method is equivalent to Newton's method for a single variable but is generalized here to the $(n-1)/2$ variable case [15].) The linear set of equations takes the form

$$\begin{pmatrix} a_{11} & a_{12} & \cdots & a_{1,k} \\ a_{21} & a_{22} & \cdots & a_{2,k} \\ \vdots & & & \\ a_{k,1} & & \cdots & a_{k,k} \end{pmatrix} \begin{pmatrix} \Delta x_1 \\ \Delta x_2 \\ \vdots \\ \Delta x_k \end{pmatrix} = \begin{pmatrix} b_1 \\ b_2 \\ \vdots \\ b_k \end{pmatrix} \tag{54}$$

where

$$k = \frac{n-1}{2},$$

Δx_i is the correction to previous x_i,
$b_i = P^+ - P(x_i)$ or $P^- - P(x_i)$, whichever applies.

Also, for $i \neq j$ and for $i, j = 1, 2, 3 \cdots k$

$$a_{ij} = \sum_{n=1}^{k} \frac{\partial c_n}{\partial x_j} x_i^n, \qquad (n \text{ odd}) \tag{55}$$

and for $i = 1, 2 \cdots k$

$$a_{ii} = \sum_{n=1}^{k} \left\{ \frac{\partial c_n}{\partial x_i} x_i^n + n c_n x_i^{n-1} \right\}, \qquad (n \text{ odd}). \tag{56}$$

This iterative method of solution was found to be a very rapidly convergent process. In a few cases where the initial guesses were not close to the solutions, the iterative process still converged but gave some x_i outside the interval zero to one.[6] However, these solutions were scaled to the interval zero to one to obtain the sought-for answers.

For cases where the polynomials are to approximate a constant value M, P^+ may be defined as $M+\delta$ where δ is the maximum equal-ripple deviation; P^- may then be defined as $M-\delta$ which gives M as the arithmetic

[6] A consideration of the set (53) of equations shows that there exist multiple solutions outside the interval zero to one, depending on the value of n.

Table 1

COEFFICIENTS FOR EQUAL-RIPPLE POLYNOMIALS APPROXIMATING UNITY ON THE INTERVAL ZERO TO ONE

δ	C_1	C_3	C_5	C_7	C_9
0.1	2.1952143	− 1.2952143			
0.1	3.4118700	− 5.7150991	3.4032291		
0.1	4.6700175	−15.3542579	22.6907481	− 11.106507	
0.1	5.94547862	−32.3085509	85.6573300	− 97.6641647	39.4699070
0.2	2.5923814	− 1.7923814			
0.2	4.1112287	− 8.2285005	5.3172718		
0.2	5.66731189	−22.4205945	36.0524481	− 18.4991655	
0.2	7.238206	−47.4794411	137.138390	−164.008299	68.3111441

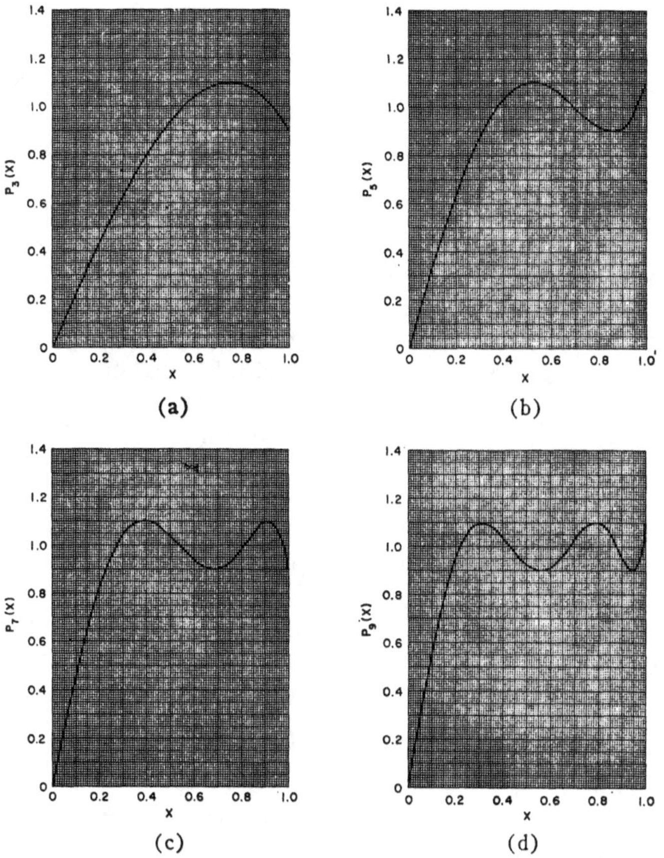

Fig. 7. Equal-ripple approximations to unity on the interval zero to one (equal-ripple tolerance 0.1). (a) Third-order polynomial. (b) Fifth-order polynomial. (c) Seventh-order polynomial. (d) Ninth-order polynomial.

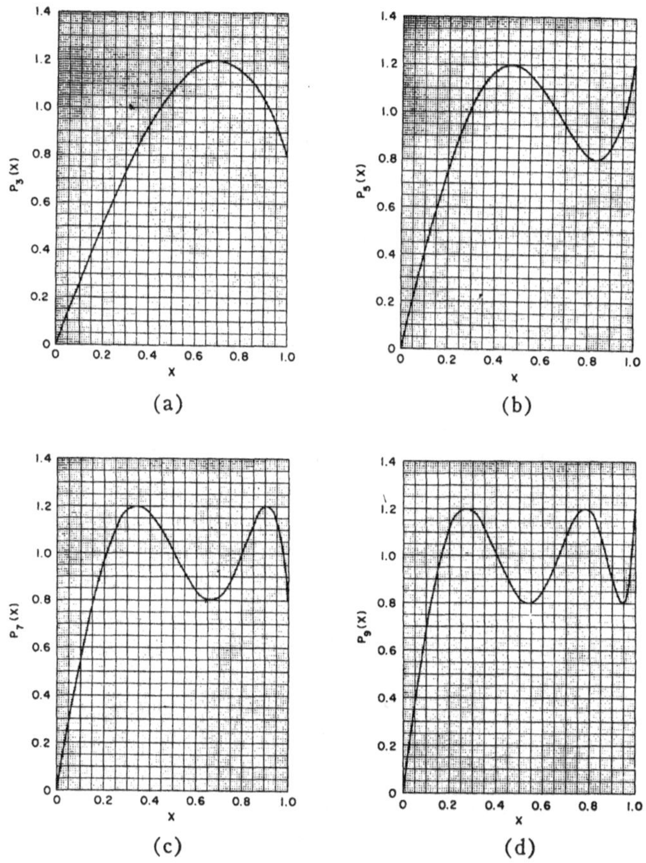

Fig. 8. Equal-ripple approximations to unity on the interval zero to one (equal-ripple tolerance 0.2). (a) Third-order polynomial. (b) Fifth-order polynomial. (c) Seventh-order polynomial. (d) Ninth-order polynomial.

mean. Another possibility is to define P^- as $1/(M+\delta)$ which gives M as the geometric mean. For the cases involving the synthesis of the symmetric couplers, it was required that the coupling in decibels be an equal-ripple function. For these cases the values of P^+ and P^- were defined as

$$P^+ = \left\{ \frac{1}{\dfrac{A(M)}{A(\delta)} - 1} \right\}^{1/2}$$

$$P^- = \left\{ \frac{1}{A(M)A(\delta) - 1} \right\}^{1/2} \tag{57}$$

where the function $A(x)$ is defined as

$A(x) = \text{antilog}_{10}(x/10)$.

M is the mean coupling value in decibels.

δ is the maximum deviation in decibels from the mean coupling.

To illustrate a result of the previously described method of obtaining equal-ripple polynomials, several representative polynomials which approximate unity on the interval zero to one are tabulated in Table 1, and shown in Figs. 7 and 8. The polynomials in Table 1 approximate unity in an arithmetic-mean sense: that is, $P^+ = 1+\delta$ and $P^- = 1-\delta$. The case for $\delta = 0.1$ and 0.2 are shown in Figs. 7 and 8, respectively. [Table 1 was not used to synthesize couplers, since another equal-ripple criterion was used, as previously explained, namely (57) and (58).]

B. Maximally-Flat Polynomials

The derivation of the maximally-flat polynomials is given in this part. A suitable starting point is (48) where all s_i are set equal to unity. That is,

$$P_n'(x) = C(x^2 - 1)^k. \tag{62}$$

Integrating (62) gives

$$P_n(x) = C\int_0^x (u^2 - 1)^k du \tag{63}$$

and C is evaluated by

$$C = \frac{P_n(1)}{\int_0^1 (u^2 - 1)^k du}\cdot \tag{64}$$

For the maximally-flat polynomials, $P_n(1)$ equals either (57) or (58) with δ set equal to zero.

V. Discussion

A. On 3.01-dB Directional Couplers as Equal Power Splitters

In Section V the derivation of the equal-ripple polynomials was described, and several possible definitions of P^+ and P^- (the equal-ripple extremes) were stated. The particular definition that is most appropriate will depend on the application. For the synthesis of symmetrical couplers, (57) and (58) were used for P^+ and P^-. These definitions give an equal-ripple approximation of the mean coupling in decibels, in an arithmetic sense. Since in applications involving couplers, specifications are usually stated in terms of decibels, these definitions are certainly appropriate. However, in many applications involving 3-dB couplers,[7] what is usually desired is an equal power splitter. A specification like 3 ± 0.3 dB (for example) is, therefore, not strictly appropriate since it cannot apply to both output ports simultaneously. Thus, 3 ± 0.3 dB corresponds to a coupling coefficient having squared magnitude between 0.4677 and 0.5370, which gives an equal-ripple approximation (in the arithmetic mean for coupled power) to an average of 0.5023 rather than 0.5000. The forward power varies from 0.5323 to 0.4630, having an arithmetic mean value of 0.4977. As the ripple deviation in decibels gets larger, the arithmetic mean of the coupling coefficient becomes even further from 0.5. In view of this problem, consideration was given to designing the "3-dB hybrids" on the basis that the squared magnitude of the coupling coefficient be $0.5000 \pm \delta$, so that the power division out of both ports would be equal-ripple in the same way. It was finally decided to stick to the conventional approach, and to synthesize on the basis of (57) and (58). The difference in any practical problem is small. In addition, the synthesis on the basis of equal-ripple coupling in decibels, i.e., on a log basis, is in keeping with previously published work, and is also more appropriate for other-than-3-dB couplings.

[7] The designation 3-dB coupler is intended to imply equal-power division. A more correct designation would be 3.0103-dB coupler.

B. Applications of Symmetrical 3-dB Couplers

Directional couplers with 3-dB coupling are power dividers in which the power is evenly divided between the two outputs. In general, there need not be any particular phase relationship between the outputs. If the two output arms are short-circuited or open-circuited in two positions where the emerging waves are 90 degrees out of phase, then the reflected waves add up in the remaining (fourth) arm, and no power is reflected into the input arm. This property is very useful for some applications, in many of which a circulator could also have been used. Much greater bandwidths can be realized with TEM-mode directional couplers than with circulators, and the couplers can operate in any frequency band. Other typical advantages may include higher isolation and better input VSWR. Applications for which symmetrical couplers are best suited include diplexers and multiplexers, directional filters, phase shifters, balanced mixers, duplexers, negative-resistance amplifiers, and others in which the 90-degree phase-difference property is essential. It is only in *symmetrical* couplers that two positions can be found in the output arms where the 90-degree phase difference is maintained independent of frequency. A further practical advantage of symmetrical couplers is that the strongest coupling region is in the center and not at one end [1], [6] so that it becomes less difficult to connect to all four ports.

C. Other Designs Realizing the Same Insertion-Loss Function

In the synthesis procedure described in Section III, the magnitude squared of the reflection coefficient for real frequencies is to be generalized (by analytic continuation) to

$$|\Gamma|^2 \to \Gamma(s)\Gamma(-s).$$

If the symmetrical coupler is to be realized, the numerator of $\Gamma(s)$ must have zeros that are symmetrical with respect to the j axis. However, unsymmetrical couplers can be obtained from the insertion-loss function by selecting other permissible choices for the zeros of $\Gamma(s)$. For example, Levy chose all zeros of his $\Gamma(s)$ to be in the left half plane, and obtained couplers having monotonically increasing Z_{oe} within the coupling region. By choosing other combinations of zeros, placing some zeros in the left half plane and others in the right half plane, realizations are possible that have Z_{oe} neither monotonically increasing nor symmetrical. The maximum number of possible realizations will depend on the degree n of the insertion-loss polynomial.

D. Physical Realization of the Couplers

This paper is concerned only with the design of the

circuit parameters of symmetrical couplers. For the realization of the final physical dimensions, the reader is referred to the extensive literature on the subject. He will find a good account with many references up to 1963 in the book by Matthaei Young, and Jones [4]. For weak to moderate coupling, coupler designs may be realized with coupled round rods. Design data of Honey [14] are appropriate for weak coupling, while the data of Cristal [16] are more accurate for stronger coupling, although they may be used for weak coupling also. For moderate to tight coupling, Getsinger [17] presents data on rectangular bars, and Cohn [18] describes an ingenious re-entrant coupling mechanism. More recently, Shelton [19] has reported some novel strip-line configurations.

It can be shown that the even-mode impedance of coupled TEM-mode transmission lines is always greater than the odd-mode impedance. Because of this physical requirement, the realizability of the stepped-impedance prototype filter does not guarantee the realizability of the coupler. To insure realizability of the coupler, the impedances of the stepped-impedance prototype filter must all be greater than (or all less than) the terminating impedance. The restrictions on the insertion-loss polynomial that would guarantee realizability of the coupler are, as yet, unknown. However, the requirement that the normalized even-mode impedances of the coupler be greater than unity is satisfied in all the numerical solutions presented in this paper.

E. On −8.34-dB Directional Couplers

Recently, Shelton, Wolfe, and Van Wagoner [19] presented design data and demonstrated practical techniques of constructing −3.01-dB directional couplers by connecting two −8.34-dB couplers in tandem in the appropriate fashion. A −8.34-dB coupler is considerably easier to construct than a −3.01-dB coupler, and, therefore, the method of Shelton, Wolfe, and Van Wagoner [19] is of practical value. It is for this reason that the −8.34-dB coupler designs are included in the design tables.

VI. Conclusions

Design tables of optimum symmetrical TEM-mode coupled-transmission-line directional couplers of three, five, seven, and nine sections were presented. The designs give the maximum bandwidth for symmetrical couplers of a given number of sections, given mean coupling, and given coupling tolerance. To complete the tables, designs of maximally-flat directional couplers were also presented. An iterative method for determining new equal-ripple polynomials was presented, and a few examples of equal-ripple polynomials that approximate unity on the interval zero to one were given.

VII. Acknowledgment

The computer programs for obtaining equal-ripple and maximally-flat polynomials, and for the synthesis of the symmetrical couplers, were written by W. Wiebenson, P. Omlor and J. Ulrich, and Miss Elizabeth Tessman.

References

[1] R. Levy, "Tables for asymmetric multi-element coupled transmission line directional couplers," *IRE Trans. on Microwave Theory and Techniques*, vol. MTT-12, pp. 275–279, May 1964.
[2] J. K. Shimizu and E. M. T. Jones, "Coupled transmission-line directional couplers," *IRE Trans. on Microwave Theory and Techniques*, vol. MTT-6, pp. 403–411, October 1958.
[3] L. Young, "The analytical equivalence of TEM-mode directional couplers and transmission-line stepped-impedance filters," *Proc. IEE (London)*, vol. 110, pp. 275–281, February 1963.
[4] G. L. Matthaei, L. Young, and E. M. T. Jones, *Microwave Filters, Impedance-Matching Networks, and Coupling Structures*, ch. 13, New York: McGraw-Hill, 1964.
[5] A. L. Feldshtein, "Synthesis of stepped directional couplers," *Radiotekn. i Electron.*, vol. 6, pp. 234–240, February 1961. English translation, New York: Pergamon, pp. 74–85.
[6] R. Levy, "General synthesis of asymmetric multi-element coupled-transmission-line directional couplers," *IRE Trans. on Microwave Theory and Techniques*, vol. MTT-11, pp. 226–237, July 1963.
[7] H. J. Riblet, "General synthesis of quarter-wave impedance transformers," *IRE Trans. on Microwave Theory and Techniques*, vol. MTT-5, pp. 36–43, January 1957.
[8] H. Seidel and J. Rosen, "Multiplicity in cascade transmission line synthesis—Part I," *IEEE Trans. on Microwave Theory and Techniques*, vol. MTT-13, pp. 275–283, May 1965; and Part II pp. 398–407, July 1965.
[9] L. Young, "A theorem in lossless symmetrical networks," *IRE Trans. on Circuit Theory (Correspondence)*, vol. CT-7, p. 75, March 1960.
[10] L. Weinberg, *Network Analysis and Synthesis*, New York: McGraw-Hill, 1962, pp. 296, 344, and 595.
[11] E. A. Guillemin, *Synthesis of Passive Networks*, New York: Wiley, 1957, p. 469.
[12] P. I. Richards, "Resistor-transmission-line circuits," *Proc. IRE*, vol. 36, pp. 217–220, February 1948.
[13] P. I. Richards, "A special class of functions with positive real parts in a half-plane," *Duke Math. J.*, vol. 14, pp. 777–786, September 1947.
[14] E. M. T. Jones and J. T. Bolljahn, "Coupled-strip-transmission line filters and directional couplers," *IRE Trans. on Microwave Theory and Techniques*, vol. MTT-4, pp. 75–81, April 1956.
[15] F. B. Hildebrand, *Introduction to Numerical Analysis*, New York: McGraw-Hill, 1956, p. 45.
[16] E. G. Cristal, "Coupled circular cylindrical rods between parallel ground planes," *IEEE Trans. on Microwave Theory and Techniques*, vol. MTT-12, pp. 428–439, July 1964.
[17] W. J. Getsinger, "Coupled rectangular bars between parallel plates," *IRE Trans. on Microwave Theory and Techniques*, vol. MTT-10, pp. 479–491, November 1962.
[18] S. B. Cohn, "The re-entrant cross-section and wide-band 3 dB hybrid coupler," *IRE Trans. on Microwave Theory and Techniques*, vol. MTT-11, pp. 254–258, July 1963.
[19] J. P. Shelton, R. Van Wagoner, and J. J. Wolfe, "Tandem couplers and phase shifters: a new class of unlimited bandwidth components," 1964 14th Ann. Symp., USAF Antenna Research and Development Program, sponsored by Air Force Avionics Lab., Wright-Patterson AFB, Dayton, Ohio, held at Monticello, Ill., or, see *Microwave*, vol. 4, pp. 14–19, April 1965.
[20] L. Young, "Stepped-impedance transformers and filter prototypes," *IRE Trans. on Microwave Theory and Techniques*, vol. MTT-10, pp. 339–359, September 1962. See definitions on p. 340.
[21] H. Seidel, "Synthesis of a class of microwave filters," *IRE Trans. on Microwave Theory and Techniques*, vol. MTT-5, pp. 107–114, April 1957.
[22] S. O. Rice, "Steady state solutions of transmission line equations," *Bell Syst. Tech. J.*, vol. 20, no. 2, pp. 131–178, April 1941.
[23] W. R. Firestone, "Analysis of transmission line directional couplers," *Proc. IRE*, vol. 42, pp. 1529–1538, October 1954.
[24] B. M. Oliver, "Directional electromagnetic couplers," *Proc. IRE*, vol. 42, pp. 1686–1692, November 1954.
[25] E. F. Barnett, P. D. Lacy, and B. M. Oliver, "Principle of directional coupling in reciprocal systems," *1954 Proc. Symp. Modern Advances in Microwave Techniques*, Polytechnic Inst. of Brooklyn, Microwave Research Inst., N. Y., vol. 14, pp. 251–268.
[26] A. L. Feldshtein and Y. S. Zhavoronkova, "Design of Chebyshev directional couplers with weak coupling," *Telecommunications and Radio Engineering (USSR)*, vol. 17, January 1962. English translation, New York: Pergamon, pp. 39–50.
[27] V. D. Kuznetsov and V. K. Paramonov, "Stepped directional couplers," *Telecommunications and Radio Engineering (USSR)*, vol. 19, no. 1, January 1964. English translation, New York: Pergamon, pp. 100–113.

Appendix

Tables of Parameters for Symmetrical TEM-Mode Coupled-Transmission-Line Directional Couplers

Table A-1

EVEN-MODE IMPEDANCES FOR EQUAL-RIPPLE SYMMETRICAL −3.01-db COUPLERS OF THREE SECTIONS

$(Z_{4-i} = Z_i)$

δ	Z_1	Z_2	w	B
0.05	1.14888	3.16095	0.86101	2.51187
0.10	1.17135	3.25984	1.00760	3.03063
0.15	1.19039	3.34049	1.10168	3.45275
0.20	1.20776	3.41242	1.17199	3.83085
0.25	1.22415	3.47932	1.22844	4.18429
0.30	1.23992	3.54311	1.27572	4.52271
0.35	1.25528	3.60495	1.31645	4.85178
0.40	1.27036	3.66560	1.35225	5.17521
0.45	1.28527	3.72563	1.38420	5.49559
0.50	1.30008	3.78546	1.41305	5.81489
0.60	1.32964	3.90585	1.46353	6.45616
0.70	1.35942	4.02894	1.50670	7.10860
0.80	1.38970	4.15648	1.54440	7.77966
0.90	1.42073	4.29005	1.57788	8.47591
1.00	1.45274	4.43120	1.60798	9.20361

Table A-2

EVEN-MODE IMPEDANCES FOR EQUAL-RIPPLE SYMMETRICAL −6-db COUPLERS OF THREE SECTIONS

$(Z_{4-i} = Z_i)$

δ	Z_1	Z_2	w	B
0.10	1.10298	2.09445	0.91996	2.70356
0.20	1.12090	2.14693	1.07404	3.31984
0.30	1.13625	2.18999	1.17223	3.83226
0.40	1.15038	2.22865	1.24518	4.29931
0.50	1.16381	2.26488	1.30345	4.74258
0.60	1.17680	2.29968	1.35201	5.17291
0.70	1.18952	2.33366	1.39364	5.59673
0.80	1.20208	2.36724	1.43006	6.01830
0.90	1.21454	2.40072	1.46241	6.44068
1.00	1.22698	2.43431	1.49150	6.86621

Table A-3

EVEN-MODE IMPEDANCES FOR EQUAL-RIPPLE SYMMETRICAL −8.34-db COUPLERS OF THREE SECTIONS

$(Z_{4-i} = Z_i)$

δ	Z_1	Z_2	w	B
0.05	1.06661	1.69824	0.76021	2.22636
0.10	1.07434	1.71858	0.89286	2.61290
0.15	1.08073	1.73468	0.97882	2.91703
0.20	1.08644	1.74864	1.04355	3.18211
0.25	1.09171	1.76127	1.09583	3.42397
0.30	1.09670	1.77299	1.13986	3.65041
0.35	1.10146	1.78405	1.17796	3.86595
0.40	1.10606	1.79461	1.21159	4.07347
0.45	1.11054	1.80478	1.24170	4.27495
0.50	1.11490	1.81463	1.26898	4.47178
0.55	1.11918	1.82424	1.29391	4.66502
0.60	1.12339	1.83365	1.31688	4.85550
0.65	1.12754	1.84289	1.33817	5.04386
0.70	1.13164	1.85200	1.35801	5.23063
0.75	1.13570	1.86101	1.37658	5.41623
0.80	1.13973	1.86993	1.39403	5.60103
0.85	1.14373	1.87878	1.41049	5.78534
0.90	1.14770	1.88759	1.42607	5.96943
0.95	1.15166	1.89636	1.44084	6.15354
1.00	1.15560	1.90510	1.45488	6.33787

Table A-4

EVEN-MODE IMPEDANCES FOR EQUAL-RIPPLE SYMMETRICAL −10-db COUPLERS OF THREE SECTIONS

$(Z_{4-i} = Z_i)$

δ	Z_1	Z_2	w	B
0.20	1.06945	1.57423	1.03140	3.12968
0.40	1.08475	1.60708	1.19816	3.98852
0.60	1.09817	1.63470	1.30282	4.73738
0.80	1.11075	1.66014	1.37959	5.44739
1.00	1.12290	1.68458	1.44020	6.14545

Table A-5

EVEN-MODE IMPEDANCES FOR EQUAL-RIPPLE SYMMETRICAL -20-db COUPLERS OF THREE SECTIONS

$(Z_{4-i} = Z_i)$

δ	Z_1	Z_2	w	B
0.20	1.02070	1.14914	1.00980	3.03958
0.40	1.02497	1.15617	1.17423	3.84396
0.60	1.02866	1.16197	1.27772	4.53804
0.80	1.03208	1.16720	1.35381	5.19011
1.00	1.03534	1.17213	1.41398	5.82570

Table A-6

EVEN-MODE IMPEDANCES FOR EQUAL-RIPPLE SYMMETRICAL -3.01-db COUPLERS OF FIVE SECTIONS

$(Z_{6-i} = Z_i)$

δ	Z_1	Z_2	Z_3	w	B
0.05	1.05972	1.32624	3.81243	1.20488	4.03071
0.10	1.07851	1.37268	3.97615	1.32559	4.93114
0.15	1.09451	1.40890	4.10191	1.39889	5.65437
0.20	1.10921	1.44029	4.21023	1.45184	6.29714
0.25	1.12314	1.46883	4.30864	1.49333	6.89474
0.30	1.13659	1.49551	4.40089	1.52744	7.46462
0.35	1.14973	1.52091	4.48917	1.55639	8.01698
0.40	1.16266	1.54541	4.57491	1.58152	8.55845
0.45	1.17547	1.56926	4.65912	1.60371	9.09367
0.50	1.18822	1.59265	4.74253	1.62357	9.62609
0.60	1.21370	1.63864	4.90924	1.65791	10.69292
0.70	1.23941	1.68425	5.07867	1.68691	11.77568
0.80	1.26555	1.73013	5.25363	1.71196	12.88720
0.90	1.29235	1.77678	5.43655	1.73402	14.03860
1.00	1.31998	1.82466	5.62978	1.75370	15.24047

Table A-7

EVEN-MODE IMPEDANCES FOR EQUAL-RIPPLE SYMMETRICAL -6-db COUPLERS OF FIVE SECTIONS

$(Z_{6-i} = Z_i)$

δ	Z_1	Z_2	Z_3	w	B
0.10	1.04501	1.21972	2.38181	1.25446	4.34522
0.20	1.06052	1.25302	2.46010	1.37766	5.42738
0.30	1.07392	1.27919	2.52068	1.45202	6.29953
0.40	1.08633	1.30203	2.57332	1.50548	7.08866
0.50	1.09818	1.32294	2.62159	1.54720	7.83386
0.60	1.10969	1.34262	2.66727	1.58135	8.55462
0.70	1.12099	1.36146	2.71142	1.61023	9.26242
0.80	1.13217	1.37976	2.75470	1.63520	9.94482
0.90	1.14328	1.39772	2.79760	1.65716	10.66721
1.00	1.15438	1.41542	2.84048	1.67673	11.37370

Table A-8

EVEN-MODE IMPEDANCES FOR EQUAL-RIPPLE SYMMETRICAL -8.34-db COUPLERS OF FIVE SECTIONS

$(Z_{6-i} = Z_i)$

δ	Z_1	Z_2	Z_3	w	B
0.05	1.02538	1.14102	1.85802	1.11764	3.53328
0.10	1.03211	1.15690	1.89019	1.23184	4.20727
0.15	1.03770	1.16899	1.91418	1.30256	4.73524
0.20	1.04271	1.17918	1.93414	1.35395	5.19150
0.25	1.04737	1.18822	1.95170	1.39442	5.60527
0.30	1.05179	1.19648	1.96764	1.42783	5.99090
0.35	1.05602	1.20417	1.98243	1.45627	6.35662
0.40	1.06012	1.21142	1.99635	1.48104	6.70767
0.45	1.06412	1.21833	2.00960	1.50296	7.04761
0.50	1.06803	1.22497	2.02232	1.52262	7.37898
0.55	1.07187	1.23138	2.03462	1.54043	7.70370
0.60	1.07565	1.23760	2.04658	1.55670	8.02323
0.65	1.07939	1.24367	2.05826	1.57168	8.33872
0.70	1.08309	1.24960	2.06971	1.58554	8.65112
0.75	1.08675	1.25542	2.08098	1.59844	8.96119
0.80	1.09039	1.26114	2.09210	1.61050	9.26959
0.85	1.09401	1.26678	2.10310	1.62182	9.57685
0.90	1.09761	1.27235	2.11401	1.63247	9.88347
0.95	1.10119	1.27785	2.12484	1.64253	10.18985
1.00	1.10476	1.28331	2.13562	1.65206	10.49635

Table A-9

EVEN-MODE IMPEDANCES FOR EQUAL-RIPPLE SYMMETRICAL -10-db COUPLERS OF FIVE SECTIONS

$(Z_{6-i} = Z_i)$

δ	Z_1	Z_2	Z_3	w	B
0.20	1.03418	1.14316	1.70922	1.34442	5.10148
0.40	1.04784	1.16808	1.75305	1.47118	6.56407
0.60	1.05996	1.18815	1.78805	1.54675	7.82513
0.80	1.07140	1.20606	1.81943	1.60053	9.01322
1.00	1.08249	1.22280	1.84912	1.64210	10.17639

Table A-10

EVEN-MODE IMPEDANCES FOR EQUAL-RIPPLE SYMMETRICAL −20-db COUPLERS OF FIVE SECTIONS

$(Z_{6-i} = Z_i)$

δ	Z_1	Z_2	Z_3	w	B
0.20	1.01016	1.04183	1.17873	1.32734	4.94656
0.40	1.01406	1.04855	1.18767	1.45350	6.31936
0.60	1.01747	1.05386	1.19463	1.52888	7.49038
0.80	1.02066	1.05851	1.20073	1.58261	8.58338
1.00	1.02371	1.06280	1.20638	1.62420	9.64410

Table A-11

EVEN-MODE IMPEDANCES FOR EQUAL-RIPPLE SYMMETRICAL −3.01-db COUPLERS OF SEVEN SECTIONS

$(Z_{8-i} = Z_i)$

δ	Z_1	Z_2	Z_3	Z_4	w	B
0.05	1.03635	1.14905	1.50280	4.39954	1.40024	5.6693
0.10	1.05240	1.18406	1.56753	4.61180	1.49705	6.9531
0.15	1.06643	1.21168	1.61640	4.77112	1.55447	7.9780
0.20	1.07950	1.23581	1.65795	4.90662	1.59539	8.8860
0.25	1.09201	1.25786	1.69523	5.02872	1.62715	9.7283
0.30	1.10419	1.27860	1.72975	5.14254	1.65308	10.5302
0.35	1.11615	1.29840	1.76238	5.25103	1.67497	11.3064
0.40	1.12798	1.31754	1.79367	5.35611	1.69388	12.0666
0.45	1.13975	1.33622	1.82400	5.45909	1.71051	12.8174
0.50	1.15149	1.35457	1.85365	5.56097	1.72534	13.5637
0.60	1.17505	1.39069	1.91172	5.76434	1.75090	15.0578
0.70	1.19890	1.42654	1.96915	5.97094	1.77238	16.5728
0.80	1.22323	1.46258	2.02682	6.18437	1.79087	18.1270
0.90	1.24820	1.49918	2.08545	6.40775	1.80710	19.7360
1.00	1.27399	1.53668	2.14566	6.64407	1.82155	21.4147

Table A-12

EVEN-MODE IMPEDANCES FOR EQUAL-RIPPLE SYMMETRICAL −6-db COUPLERS OF SEVEN SECTIONS

$(Z_{8-i} = Z_i)$

δ	Z_1	Z_2	Z_3	Z_4	w	B
0.10	1.02686	1.10756	1.32930	2.62516	1.44052	6.1494
0.20	1.04246	1.13419	1.37278	2.72038	1.53802	7.6583
0.30	1.05449	1.15540	1.40584	2.79246	1.59558	8.8908
0.40	1.06580	1.17408	1.43416	2.85438	1.63645	10.0026
0.50	1.07670	1.19128	1.45977	2.91078	1.66806	11.0505
0.60	1.08735	1.20755	1.48367	2.96391	1.69378	12.0626
0.70	1.09787	1.22318	1.50642	3.01511	1.71541	13.0553
0.80	1.10831	1.23839	1.52841	3.06523	1.73404	14.0396
0.90	1.11872	1.25331	1.54989	3.11486	1.75036	15.0232
1.00	1.12915	1.26805	1.57104	3.16446	1.76487	16.0119

Table A-13

EVEN-MODE IMPEDANCES FOR EQUAL-RIPPLE SYMMETRICAL −8.34-db COUPLERS OF SEVEN SECTIONS

$(Z_{8-i} = Z_i)$

δ	Z_1	Z_2	Z_3	Z_4	w	B
0.05	1.01460	1.06403	1.21141	1.99183	1.32568	4.9319
0.10	1.02033	1.07694	1.23301	2.03194	1.42127	5.9117
0.15	1.02519	1.08680	1.24872	2.06076	1.47818	6.6655
0.20	1.02963	1.09518	1.26167	2.08436	1.51889	7.3140
0.25	1.03379	1.10266	1.27297	2.10489	1.55059	7.9005
0.30	1.03778	1.10953	1.28316	2.12339	1.57653	8.4458
0.35	1.04163	1.11595	1.29256	2.14044	1.59848	8.9622
0.40	1.04538	1.12204	1.30136	2.15641	1.61749	9.4572
0.45	1.04905	1.12786	1.30969	2.17156	1.63423	9.9360
0.50	1.05265	1.13346	1.31764	2.18606	1.64919	10.4022
0.55	1.05621	1.13889	1.32528	2.20004	1.66270	10.8588
0.60	1.05972	1.14417	1.33267	2.21361	1.67500	11.3077
0.65	1.06320	1.14933	1.33984	2.22683	1.68629	11.7507
0.70	1.06666	1.15438	1.34684	2.23979	1.69672	12.1891
0.75	1.07009	1.15934	1.35368	2.25251	1.70640	12.6240
0.80	1.07350	1.16423	1.36040	2.26506	1.71543	13.0563
0.85	1.07689	1.16904	1.36700	2.27746	1.72389	13.4870
0.90	1.08028	1.17381	1.37351	2.28975	1.73184	13.9165
0.95	1.08365	1.17852	1.37993	2.30194	1.73934	14.3456
1.00	1.08702	1.18319	1.38629	2.31408	1.74643	14.7747

Table A-14

EVEN-MODE IMPEDANCES FOR EQUAL-RIPPLE SYMMETRICAL −10-db COUPLERS OF SEVEN SECTIONS

$(Z_{8-i} = Z_i)$

δ	Z_1	Z_2	Z_3	Z_4	w	B
0.20	1.02360	1.07622	1.20802	1.81699	1.51198	7.1965
0.40	1.03597	1.09725	1.23839	1.86715	1.61028	9.2638
0.60	1.04718	1.11444	1.26213	1.90649	1.66773	11.0383
0.80	1.05786	1.12991	1.28298	1.94149	1.70815	12.7059
1.00	1.06834	1.14444	1.30229	1.97446	1.73917	14.3359

Table A-15

EVEN-MODE IMPEDANCES FOR EQUAL-RIPPLE SYMMETRICAL −20-db COUPLERS OF SEVEN SECTIONS

$(Z_{8-i} = Z_i)$

δ	Z_1	Z_2	Z_3	Z_4	w	B
0.20	1.00697	1.02256	1.05976	1.20128	1.49853	6.9766
0.40	1.01052	1.02846	1.06767	1.21112	1.59672	8.9188
0.60	1.01369	1.03320	1.07372	1.21863	1.65421	10.5678
0.80	1.01669	1.03740	1.07894	1.22515	1.69472	12.1029
1.00	1.01958	1.04129	1.08368	1.23116	1.72584	13.5903

Table A-16

EVEN-MODE IMPEDANCES FOR EQUAL-RIPPLE SYMMETRICAL −3.01-db COUPLERS OF NINE SECTIONS

$(Z_{10-i} = Z_i)$

δ	Z_1	Z_2	Z_3	Z_4	Z_5	w	B
0.05	1.02680	1.09163	1.24706	1.66958	4.93133	1.5218	7.365
0.10	1.04112	1.12024	1.29488	1.74863	5.18240	1.6012	9.030
0.15	1.05391	1.14328	1.33137	1.80742	5.36886	1.6478	10.356
0.20	1.06598	1.16366	1.36260	1.85696	5.52654	1.6807	11.528
0.25	1.07763	1.18248	1.39075	1.90116	5.66814	1.7062	12.615
0.30	1.08904	1.20027	1.41691	1.94192	5.79985	1.7269	13.649
0.35	1.10030	1.21737	1.44168	1.98035	5.92523	1.7444	14.648
0.40	1.11149	1.23397	1.46548	2.01711	6.04655	1.7594	15.627
0.45	1.12264	1.25023	1.48856	2.05271	6.16540	1.7726	16.594
0.50	1.13379	1.26625	1.51114	2.08747	6.28296	1.7844	17.554
0.60	1.15624	1.29789	1.55536	2.15551	6.51769	1.8046	19.475
0.70	1.17904	1.32941	1.59902	2.22278	6.75634	1.8216	21.423
0.80	1.20234	1.36117	1.64277	2.29038	7.00316	1.8362	23.421
0.90	1.22630	1.39348	1.68712	2.35918	7.26188	1.8490	25.488
1.00	1.25107	1.42660	1.73250	2.42995	7.53602	1.8604	27.644

Table A-17

EVEN-MODE IMPEDANCES FOR EQUAL-RIPPLE SYMMETRICAL −6-db COUPLERS OF NINE SECTIONS

$(Z_{10-i} = Z_i)$

δ	Z_1	Z_2	Z_3	Z_4	Z_5	w	B
0.10	1.02201	1.06888	1.17282	1.42807	2.83542	1.5550	7.989
0.20	1.03437	1.09137	1.20736	1.47877	2.94305	1.6345	9.943
0.30	1.04554	1.10967	1.23393	1.51676	3.02373	1.6809	11.535
0.40	1.05615	1.12599	1.25686	1.54902	3.09269	1.7136	12.969
0.50	1.06645	1.14117	1.27768	1.57805	3.15533	1.7389	14.319
0.60	1.07658	1.15561	1.29716	1.60504	3.21427	1.7594	15.622
0.70	1.08662	1.16957	1.31574	1.63070	3.27103	1.7765	16.900
0.80	1.09661	1.18320	1.33370	1.65546	3.32658	1.7913	18.166
0.90	1.10661	1.19661	1.35125	1.67962	3.38161	1.8042	19.431
1.00	1.11663	1.20989	1.36852	1.70342	3.43663	1.8157	20.702

Table A-18

EVEN-MODE IMPEDANCES FOR EQUAL-RIPPLE SYMMETRICAL −8.34-db COUPLERS OF NINE SECTIONS

$(Z_{10-i} = Z_i)$

δ	Z_1	Z_2	Z_3	Z_4	Z_5	w	B
0.05	1.01032	1.03838	1.10598	1.27508	2.10668	1.4599	6.406
0.10	1.01536	1.04904	1.12341	1.30048	2.15200	1.5392	7.681
0.15	1.01974	1.05735	1.13622	1.31862	2.18413	1.5858	8.658
0.20	1.02379	1.06452	1.14687	1.33341	2.21025	1.6190	9.498
0.25	1.02764	1.07099	1.15622	1.34622	2.23285	1.6446	10.256
0.30	1.03134	1.07697	1.16469	1.35771	2.25315	1.6656	10.960
0.35	1.03494	1.08261	1.17253	1.36825	2.27180	1.6832	11.627
0.40	1.03846	1.08798	1.17989	1.37809	2.28925	1.6985	12.265
0.45	1.04193	1.09314	1.18687	1.38738	2.30577	1.7119	12.883
0.50	1.04534	1.09813	1.19356	1.39622	2.32156	1.7238	13.484
0.55	1.04872	1.10298	1.19999	1.40471	2.33677	1.7346	14.072
0.60	1.05206	1.10771	1.20622	1.41290	2.35152	1.7444	14.650
0.65	1.05538	1.11234	1.21228	1.42085	2.36589	1.7534	15.221
0.70	1.05868	1.11689	1.21819	1.42858	2.37996	1.7617	15.785
0.75	1.06196	1.12137	1.22398	1.43615	2.39378	1.7694	16.345
0.80	1.06523	1.12579	1.22966	1.44356	2.40740	1.7766	16.901
0.85	1.06849	1.13016	1.23525	1.45084	2.42087	1.7833	17.455
0.90	1.07174	1.13448	1.24076	1.45802	2.43421	1.7896	18.008
0.95	1.07498	1.13877	1.24620	1.46511	2.44745	1.7955	18.560
1.00	1.07823	1.14302	1.25158	1.47211	2.46063	1.8011	19.111

Table A-19

EVEN-MODE IMPEDANCES FOR EQUAL-RIPPLE SYMMETRICAL −10-db COUPLERS OF NINE SECTIONS

$(Z_{10-i} = Z_i)$

δ	Z_1	Z_2	Z_3	Z_4	Z_5	w	B
0.20	1.01889	1.05161	1.11743	1.26387	1.90628	1.6133	9.345
0.40	1.03041	1.07004	1.14313	1.29777	1.96074	1.6927	12.016
0.60	1.04103	1.08543	1.16344	1.32390	2.00313	1.7386	14.303
0.80	1.05127	1.09945	1.18139	1.34672	2.04073	1.7708	16.450
1.00	1.06133	1.11271	1.19805	1.36779	2.07614	1.7954	18.547

Table A-20

EVEN-MODE IMPEDANCES FOR EQUAL-RIPPLE SYMMETRICAL −20-db COUPLERS OF NINE SECTIONS

$(Z_{10-i} = Z_i)$

δ	Z_1	Z_2	Z_3	Z_4	Z_5	w	B
0.20	1.00555	1.01529	1.03447	1.07471	1.21931	1.6024	9.061
0.40	1.00886	1.02054	1.04153	1.08328	1.22965	1.6818	11.571
0.60	1.01187	1.02485	1.04700	1.08974	1.23748	1.7278	13.697
0.80	1.01474	1.02871	1.05175	1.09527	1.24426	1.7601	15.674
1.00	1.01753	1.03232	1.05608	1.10028	1.25049	1.7848	17.588

Table A-21

EVEN-MODE IMPEDANCES FOR MAXIMALLY FLAT SYMMETRICAL COUPLERS OF THREE SECTIONS

$(Z_{4-i} = Z_i)$

-3.01 db Coupling

Z_1	Z_2	w	B
1.10410	2.94302	1.46466	6.47184

-6 db Coupling

Z_1	Z_2	w	B
1.06810	1.97911	1.37435	5.39334

-10 db Coupling

Z_1	Z_2	w	B
1.04110	1.50382	1.33123	4.98113

-20 db Coupling

Z_1	Z_2	w	B
1.01260	1.13358	1.30852	4.78465

Table A-22

EVEN-MODE IMPEDANCES FOR MAXIMALLY FLAT SYMMETRICAL COUPLERS OF FIVE SECTIONS

$(Z_{6-i} = Z_i)$

-3.01 db Coupling

Z_1	Z_2	Z_3	w	B
1.01634	1.18837	3.28771	1.56911	8.28314

-6 db Coupling

Z_1	Z_2	Z_3	w	B
1.01231	1.12196	2.13095	1.49525	6.92473

-10 db Coupling

Z_1	Z_2	Z_3	w	B
1.00755	1.07298	1.57346	1.45978	6.40442

-20 db Coupling

Z_1	Z_2	Z_3	w	B
1.00235	1.02216	1.14966	1.44104	6.15614

Table A-23

EVEN-MODE IMPEDANCES FOR MAXIMALLY FLAT SYMMETRICAL COUPLERS OF SEVEN SECTIONS

$(Z_{8-i} = Z_i)$

-3.01 db Coupling

Z_1	Z_2	Z_3	Z_4	w	B
1.00375	1.04173	1.25855	3.55017	1.62962	9.7996

-6 db Coupling

Z_1	Z_2	Z_3	Z_4	w	B
1.00254	1.02791	1.16589	2.24311	1.56569	8.2100

-10 db Coupling

Z_1	Z_2	Z_3	Z_4	w	B
1.00157	1.01708	1.09855	1.62369	1.53490	7.6003

-20 db Coupling

Z_1	Z_2	Z_3	Z_4	w	B
1.00049	1.00530	1.02968	1.16096	1.51861	7.3092

Table A-24

EVEN-MODE IMPEDANCES FOR MAXIMALLY FLAT SYMMETRICAL COUPLERS OF NINE SECTIONS

$(Z_{10-i} = Z_i)$

-3.01 db Coupling

Z_1	Z_2	Z_3	Z_4	Z_5	w	B
1.00081	1.01035	1.06619	1.31891	3.76506	1.6703	11.131

-6 db Coupling

Z_1	Z_2	Z_3	Z_4	Z_5	w	B
1.00055	1.00700	1.04409	1.20303	2.33294	1.6131	9.338

-10 db Coupling

Z_1	Z_2	Z_3	Z_4	Z_5	w	B
1.00034	1.00431	1.02689	1.11989	1.66324	1.5855	8.651

-20 db Coupling

Z_1	Z_2	Z_3	Z_4	Z_5	w	B
1.00011	1.00135	1.00831	1.03587	1.16969	1.5709	8.322

STEPPED ASYMMETRICAL COUPLERS VII

The two papers by Levy (p. 182), (p. 194) give the complete solution for the stepped coupler of maximum bandwidth for a given coupling, ripple, and number of sections (or conversely, having the minimum number of sections for a specified coupling, ripple, and bandwidth), with tables for up to six sections. We could not resist reprinting the one-page account in Japanese by Kuroda and his co-workers (p. 181), which had been alluded to earlier. (See under Section V, GENERAL THEORY.) Dr. Kuroda kindly added a few words of English, which, combined with the two figures, makes it clear what the paper is about. Compare also Ref. 19, as discussed before.

A word about asymmetrical couplers generally is in order. While they assure optimization of amplitude response (coupling), they do not control phase. Without adding phase equalizing networks, one cannot cascade two such couplers (to boost the coupling of one, for example), or use them as quadrature hybrids or magic T's. Such equalizing networks have been added by Kraker[23] and by Rozzi.[24] The question may be asked: In what way do the characteristic impedances of an unsymmetrical coupler stepped-impedance prototype differ from the characteristic impedances of a quarter-wave transformer having the same number of sections? The answer was given by Young.[11] In brief, the quarter-wave transformer design is for constant reflection coefficient **vector** (amplitude and phase), while the coupler prototype design is for constant reflection coefficient **amplitude** (without regard to phase). The former yields a small tolerance circle on the Smith chart, while the latter results in a narrow annulus on the Smith chart, as depicted in Fig. 40 of Ref.8, and reproduced here as Fig. 2. It has been found that an asymmetrical design based on a quarter-wave transformer prototype falls far short of an optimum coupler in bandwidth, but (because the phase is controlled) it can be turned into a magic T. (See the next section.)

Bib.(13) Convention Record: Nat. Meeting of Inst. Elec. Commun. Engr.

163　広帯域分布結合型方向性結合器の試作

An Experiment on a type of Wide-Band Directional couplers.

黒田一之　深井成人　井上暉朗　野村浩己
藤村健男　村田武夫　（静岡大学工学部）

1，まえがき

さきに筆者等の一人は，ＴＥＭ線路の分布結合型方向性結合器を多段従續接續することによつて，あらかじめ与えられた結合度の周波数特性を実現する方法を発表した。[1]

今回この方法によつて，ストリップ線路の方向性結合器を試作実験したところ，結合特性は一応理論値と近いものが得られることが分った。

２．試作分布結合型方向性結合器の構造

今回試作したものは，$n=2$，Tschebyscheff 特性のもので，中心周波数は２，０００ＭＣとし，結合度の偏差は＋0.5 db，最大結合度は１６ dbとした。

公称インピーダンスを５０Ωとして，各線路の特性インピーダンスをもめると，

$$Z^e_{01}=61.4\,\Omega,\qquad Z^e_{02}=52.8\,\Omega$$
$$Z^o_{01}=40.7\,\Omega,\qquad Z^o_{02}=47.3\,\Omega$$

となる。これからストリップ線路の寸法をきめると，[2] 第２図に示すようになる。

３．実験結果

結合度の周波数特性を，第１図に示す。これは引出線を，第２図のようにすべて４５°の角度で引き出した場合であつて，この他９０°にまげて引き出した場合等についても試みたが，４５°の場合が一番よいようであった。

方向性は理論上は全周波数について完全な筈であるが，今回の実験ではあまり良好でなく平均十数db程度で，かつかなり不規則な変化を示した。これは結合器自体の特性のみならず，測定法にも問題があるので，今後さらに検討したい。

同軸型ではもう少しよい結果が期待出来るのではないかと思われる。

（正　誤）前論文[1]の S および α の式が間違っていたので，この機会をかりて次のように訂正する。

$$S=\sqrt{(1-\alpha^2)/(1-\alpha^2k^2)},\qquad 1/\alpha=\cos\left(n\cos^{-1}\sqrt{1+\omega_c^2}\right)$$

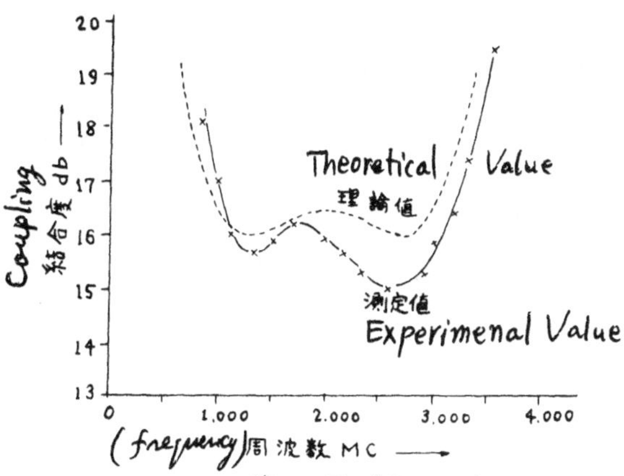

第１図（Fig.1）

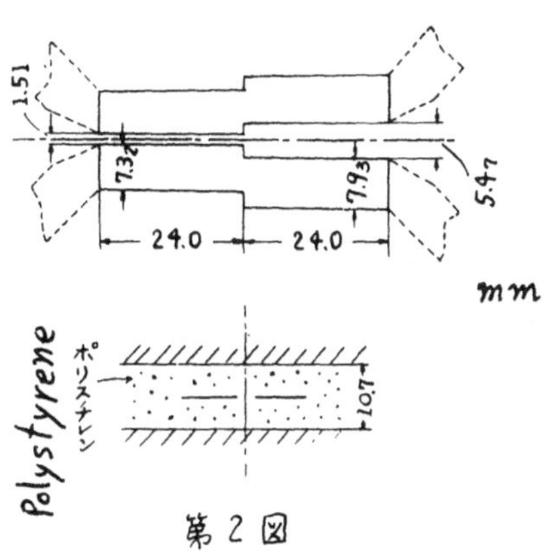

第２図（Fig.2）

（１）黒田一之：昭和３２年度電気通信学会全国大会，１３４．
（２）S. B. Cohn：Trans.I.R.E., MTT-3, Oct. 1955.

Reprinted from IEEE TRANSACTIONS
ON MICROWAVE THEORY AND TECHNIQUES
Volume MTT-11, Number 4, July, 1963

PRINTED IN THE U.S.A.

General Synthesis of Asymmetric Multi-Element Coupled-Transmission-Line Directional Couplers*

RALPH LEVY†

Summary—An exact synthesis procedure is derived for a class of asymmetric multi-element coupled-transmission-line directional couplers with any number of elements. It is based on the equivalence between the theory of the directional coupler and that of a stepped quarter-wavelength filter. This can be treated using Richards' theorem for the synthesis of transmission-line distributed networks, as described previously by Riblet. The method is extended to give a general expression for the input reflection coefficient of the stepped filter, which corresponds to the voltage coupling of the directional coupler. Explicit formulas for the parameters of two, three, four and five couplers are derived and the extension to larger number of elements is straightforward. Two and three element couplers have been designed on this theoretical basis, and show excellent agreement with theory, for example a three element coupler of 20 db ±0.5 db over a 6:1 bandwidth, and a two element coupler of 3.2 db ±0.85 db over a 6.7:1 bandwidth. It is possible to design a 3-db ±0.43 db-coupler for decade bandwidths using only four elements. The 3 db-couplers may be used as 90° hybrids by careful choice of reference planes in the output parts.

I. Introduction

COUPLED-TRANSMISSION-LINE directional couplers have been described by a number of authors (Oliver,[1] Jones and Bolljahn,[2] and Shimizu and Jones[3]). The simple quarter-wavelength directional coupler has perfect isolation and perfect input match, but the coupling varies according to

$$L = 1 + \tfrac{1}{4}(Z_{oe} - 1/Z_{oe})^2 \sin^2 \theta \tag{1}$$

where L is the power insertion loss from the input to arm 4 (Fig. 1). Z_{oe} and Z_{oo} are the impedances of the even and odd modes in the coupled line normalized to the impedance of the input lines, and are related by

$$Z_{oe}Z_{oo} = 1. \tag{2}$$

Eq. (1) indicates that a useful bandwidth of rather more than one octave is obtained from the simple quarter-wavelength section. Shimizu and Jones[3] have described how much greater bandwidths may be obtained by cascading three coupled-line sections to form a three-quarter wavelength coupler. Considerable simplification of their equation (10) for the coupling is possible (see Appendix I), leading to the formula

$$L = 1 + \frac{1}{4}\left[\left\{2\left(Z_{oe}' - \frac{1}{Z_{oe}'}\right) + \left(Z_{oe} - \frac{1}{Z_{oe}}\right)\right\} \cdot \sin\theta\cos^2\theta - \left(\frac{Z_{oe}'^2}{Z_{oe}} - \frac{Z_{oe}}{Z_{oe}'^2}\right)\sin^3\theta\right]^2 \tag{3}$$

where, as in (1), L is the insertion loss from the input

* Received October 24, 1962; revised manuscript received February 14, 1963. This paper will form part of a thesis for the Ph.D. degree at Queen Mary College, University of London, London, England.

† Systems Division, Mullard Research Laboratories, Redhill, Surrey, England.

[1] B. M. Oliver, "Directional electromagnetic couplers," PROC. IRE, vol. 42, pp. 1688–1692; November, 1954.

[2] E. M. T. Jones and J. T. Bolljahn, "Coupled-strip transmission line filters and directional couplers," IRE TRANS. ON MICROWAVE THEORY AND TECHNIQUES, vol. MTT-4, pp. 75–81; April, 1956.

[3] J. K. Shimizu and E. M. T. Jones, "Coupled-transmission-line directional couplers," IRE TRANS. ON MICROWAVE THEORY AND TECHNIQUES, vol. MTT-6, pp. 403–410; October, 1958.

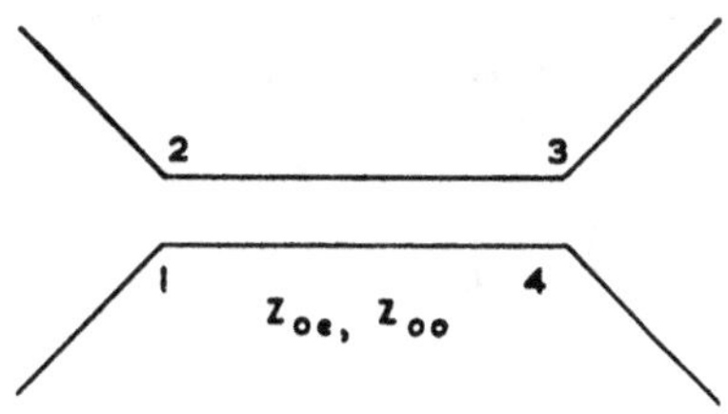

Fig. 1—Quarter-wavelength directional coupler.

to arm 4. This equation can be analyzed quite readily for any coupling and bandwidth required, *e.g.*, for a coupling of 3 db ±0.4 db the bandwidth is 5:1. In spite of this large bandwidth the three-quarter wavelength coupler is not optimum in the sense of having maximum bandwidth for a given coupling tolerance. Eq. (3) is a cubic in $\sin^2\theta$ or $\cos^2\theta$ and the coupling characteristic should have two equal ripples to be of optimum form. In fact it has only one ripple. The reason for this is the restriction placed on the form of coupling by making the two outer elements equal. When the three elements are allowed to be all unequal, then it is possible to obtain two ripples in the coupling characteristic. Fel'dshtein[4] has shown that for such an n-element asymmetric coupler the power division between arms 2 and 4 should take the form

$$\eta = \frac{|S_{12}|^2}{|S_{14}|^2} = \beta^2 - h^2 T_n{}^2\left(\frac{\cos\theta}{\cos\theta_0}\right) \qquad (4)$$

where

$$1/\cos\theta_0 = \cosh\left(\frac{1}{n}\cosh^{-1}\beta/h\right) \qquad (5)$$

and the pass band extends from θ_0 to $\pi-\theta_0$. The coupling or insertion loss to arm 4 is then

$$L = \frac{1}{|S_{14}|^2} = 1 + \beta^2 - h^2 T_n{}^2\left(\frac{\cos\theta}{\cos\theta_0}\right). \qquad (6)$$

T_n denotes the Chebyshev function of the first kind of degree n.

A key to the solution of the synthesis problem is found by noting that the couplings to arms 2 and 4 are respectively the reflection and transmission coefficients of a cascaded set of transmission lines of electrical length θ with the impedances $Z_1, Z_2, \cdots, Z_n$ (where as an abbreviation Z_r is written for Z_{oer}) terminated by lines of unit impedance. This equivalence, which is illustrated in Fig. 2, has been proved in current papers,[5,6] but it is

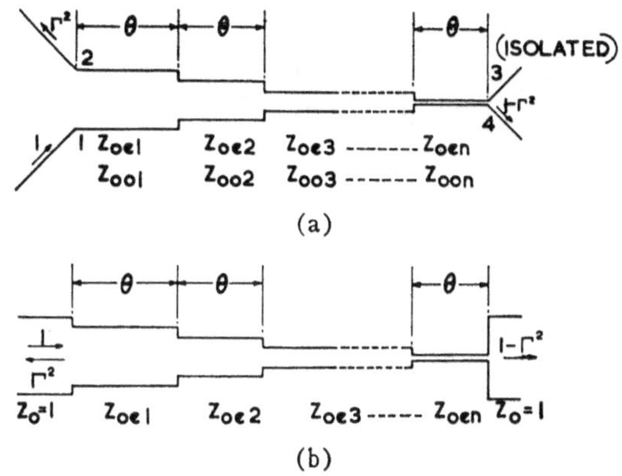

Fig. 2—(a) Equivalence between an n-element coupler (b) and n stepped quarter-wavelength lines.

presented in Appendix I for completeness. Its application leads immediately to the simple derivation of equations such as (1) and (3). In the general lossless case of n elements the transfer matrix is given by

$$\prod_{r=1}^{n}\begin{bmatrix}\cos\theta & jZ_r\sin\theta\\ j/Z_r\sin\theta & \cos\theta\end{bmatrix} = \begin{bmatrix}A_n & jB_n\\ jC_n & D_n\end{bmatrix} \qquad (7)$$

and the insertion loss to arm 4 of the coupler is the insertion loss of the two-port network, *i.e.*,

$$L = 1 + \tfrac{1}{4}(A_n - D_n)^2 + \tfrac{1}{4}(B_n - C_n)^2. \qquad (8)$$

Since (8) can be expressed as a polynomial in powers of $\cos^2\theta$ of degree n, the synthesis of the optimum coupling characteristic may be performed by equating coefficients of $\cos^{2r}\theta$ in (8) with corresponding coefficients in (6). This leads to n simultaneous equations in the n variables $Z_1, Z_2, \cdots, Z_n$, which may be solved for values of n up to $n=3$. Beyond this value of n the process rapidly becomes untractable. A solution of the problem is available, however, by recourse to modern synthesis techniques.

II. Synthesis Procedure

The synthesis is performed using a theorem due to Richards[7] and extended by Riblet[8] to the synthesis of quarter-wave impedance transformers. The present problem is similar to that treated by Riblet, but before applying the techniques, it is first necessary to extend his theorem to the synthesis of any transmission line stepped-impedance filter, as described in Appendix II. It follows that (6) can always be synthesized as a stepped impedance filter with real positive characteristic impedances. Introducing the propagation function γ, given by

$$\gamma = \sigma + j\theta \qquad (9)$$

[4] A. L. Fel'dshtein, "Synthesis of stepped directional couplers," *Radiotekh. i Electron.*, vol. 6, pp. 234–240; February, 1961.

[5] R. Levy, "Coupled Transmission Line Theory and the Design of Ultra-Broadband Microwave Components," presented at IEE Conference on Components for Microwave Circuits, London, England; September 1962.

[6] L. Young, "The analytical equivalence of T.E.M. mode directional couplers and transmission-line-stepped-impedance filters," Proc. I.E.E., vol. 110, pp. 275–281; February, 1963. Leo Young has informed the author that he has independently reached the conclusion that the analysis of T.E.M. mode couplers is equivalent to that of quarter-wave stepped transformers, and has heard of similar work by S. B. Cohn and H. J. Riblet.

[7] P. I. Richards, "Resistor-transmission-line circuits," Proc. IRE, vol. 34, pp. 217–220; September, 1948.

[8] H. J. Riblet, "General synthesis of quarter-wave impedance transformers," IRE Trans. on Microwave Theory and Techniques, vol. MTT-5, pp. 36–43; January, 1957.

where θ is the electrical length of the uniform line, and σ is the attenuation coefficient, then Richards showed that the frequency transformation

$$t = \tanh \gamma \tag{10}$$

maps the complex impedance plane $Z(\gamma)$ into the plane $Z(t)$, where t is a new complex variable equivalent to the complex frequency variable in ordinary lumped-element synthesis theory. Thus $Z(t)$ satisfies Brune's condition that it must be a positive real function if it is to represent a realizable impedance. For this condition to hold the reflection coefficient $\Gamma(t)$ must be a function which is regular (*i.e.*, with no poles) in the right-half plane. In the case of minimum phase networks there is the additional condition that $\Gamma(t)$ has no zeros in the right-half plane.

The required insertion loss function is given by (6), so that the square of the reflection coefficient is

$$\left|\Gamma\left(\frac{\cos\theta}{\cos\theta_0}\right)\right|^2 = \frac{L-1}{L} = \frac{\beta^2 - h^2 T_n{}^2\left(\dfrac{\cos\theta}{\cos\theta_0}\right)}{1+\beta^2 - h^2 T_n\left(\dfrac{\cos\theta}{\cos\theta_0}\right)}. \tag{11}$$

$|\Gamma(\cos\theta/\cos\theta_0)|^2$ represents the power reflection coefficient of a lossless network, and it is necessary to generalize (11) to give $|\Gamma(t)|^2$. The generalization will reduce to (11) for real frequencies (*i.e.*, for $t = \tanh j\theta = j\tan\theta$). The roots θ_r of the numerator in (11) are given by

$$\cosh\left(n\cosh^{-1}\frac{\cos\theta_r}{\cos\theta_0}\right) = \pm\frac{\beta}{h}$$

i.e.,

$$n\cosh^{-1}\frac{\cos\theta_r}{\cos\theta_0} = \cosh^{-1}\frac{\beta}{h} + jr\pi$$

$$r = 1, 2, \cdots, n).$$

Using (5) this becomes

$$\cos\theta_r = \frac{\cosh(1/n \cosh^{-1}\beta/h + jr\pi/n)}{\cosh(1/n\cosh^{-1}\beta/h)}. \tag{12}$$

$$r = 1, 2, \cdots, n)$$

This may now be written in terms of $\tan^2\theta_r$ and hence generalized to give the complex roots

$$t_r{}^2 = 1 - \frac{\cosh^2 J/n}{\cosh^2(J/n + jr\pi/n)} \tag{13}$$

where

$$J = \cosh^{-1}\beta/h. \tag{14}$$

After some trigonometrical manipulation (13) can be written in terms of its real and imaginary parts as

$$t_r{}^2 = 1 - \frac{\cosh^2 J/n(\cos^2 r\pi/n \cosh^2 J/n - \sin^2 r\pi/n \sinh^2 J/n)}{(\cosh^2 J/n - \sin^2 r\pi/n)^2} + j\frac{\cosh^2 J/n \sin 2r\pi/n \sinh 2J/n}{2(\cosh^2 J/n - \sin^2 r\pi/n)^2}. \tag{15}$$

Similarly the roots of the denominator in (11) are given by

$$t_r'^2 = 1 - \frac{\cosh^2 J/n(\cos^2 r\pi/n \cosh^2 H/n - \sin^2 r\pi/n \sinh^2 H/n)}{(\cosh^2 H/n - \sin^2 r\pi/n)^2} + j\frac{\cosh^2 J/n \sin 2r\pi/n \sinh 2H/n}{2(\cosh^2 H/n - \sin^2 r\pi/n)^2} \tag{16}$$

where

$$H = \cosh^{-1}\sqrt{1+\beta^2}/h. \tag{17}$$

The expression for $|\Gamma(t)|^2$ finally takes the form

$$|\Gamma(t)|^2 = K\frac{(t^2 - t_1{}^2)(t^2 - t_2{}^2)\cdots(t^2 - t_n{}^2)}{(t^2 - t_1'^2)(t^2 - t_2'^2)\cdots(t^2 - t_n'^2)} \tag{18}$$

$$K = \begin{cases} \dfrac{\beta^2}{1+\beta^2} = \left(\dfrac{\cosh J}{\cosh H}\right)^2 \quad (n \text{ odd}) & (19) \\[2ex] \dfrac{\beta^2 - h^2}{1+\beta^2-h^2} = \left(\dfrac{\sinh J}{\sinh H}\right)^2 \quad (n \text{ even}). & (20) \end{cases}$$

The values of the roots $t_r{}^2$, $t_r'^2$ given by (15) and (16) are either real, or occur in complex conjugate pairs. For example when n is odd there is a real root for $r = n$, and the others occur in complex conjugate pairs $(r, n-r)$ for $r = 1, 2, \cdots, \frac{1}{2}(n-1)$. Eq. (18) becomes, for n odd,

$$|\Gamma(t)|^2 = \frac{\beta^2}{(1+\beta^2)}\frac{(t^2 - t_n{}^2)}{(t^2 - t_n'^2)}\prod_{r=1}^{\frac{1}{2}(n-1)}\frac{(t^2 - t_r{}^2)(t^2 - t_r{}^{2*})}{(t^2 - t_r'^2)(t^2 - t_r'^{2*})}. \tag{21}$$

As previously stated, $\Gamma(t)$ must have no poles or zeros in the right half plane, leading to the formula

$$\Gamma(t) = \frac{\beta}{\sqrt{1+\beta^2}}\frac{(t+t_n)}{(t+t_n')}\prod_{r=1}^{\frac{1}{2}(n-1)}\frac{(t+t_r)(t+t_r{}^*)}{(t+t_r')(t+t_r'^*)} \tag{22}$$

where the real parts of the t_r, t_r' are all positive. Eq. (22) is further simplified by forming the product

$$(t+t_r')(t+t_r'^*) = t^2 + (t_r' + t_r'^*)t + |t_r'|^2$$

and deriving expressions for the coefficients of this quadratic in t from (16), *i.e.*,

$$T_r' = (t_r' + t_r'^*) = \sqrt{2\left[1 - \frac{2\cosh^2 J/n(\cosh 2H/n \cos 2r\pi/n + 1)}{(\cosh 2H/n + \cos 2r\pi/n)^2} + |t_r'|^2\right]} \tag{23}$$

$$|t_r'|^2 = \frac{\sqrt{(\cosh 2(H+J)/n - \cos 2r\pi/n)(\cosh 2(H-J)/n - \cos 2r\pi/n)}}{(\cosh 2H/n + \cos 2r\pi/n)}. \tag{24}$$

Similar expressions for $T_r = (t_r + t_r^*)$ and $|t_r|^2$ are obtained by substituting J for H in (23) and (24). All factors in (22) now have positive real coefficients, complex numbers having been eliminated.

The generalized reflection coefficient for n even is similarly

$$\Gamma(t) = \frac{\sqrt{\beta^2 - h^2}(t + t_{n/2})(t + t_n)}{\sqrt{1 + \beta^2 - h^2}(t + t_{n/2}')(t + t_n')} \cdot \prod_{r=1}^{\frac{1}{2}n-1} \frac{[t^2 + (t_r + t_r^*)t + |t_r|^2]}{[t^2 + (t_r' + t_r'^*)t + |t_r'|^2]}. \tag{25}$$

Eqs. (22) or (25) express $\Gamma(t)$ as the quotient of two positive real polynomials of the nth degree, and from them the input impedance $Z(t)$ is derived from the formula

$$Z(t) = \frac{1 + \Gamma(t)}{1 - \Gamma(t)} = \frac{P_n(t)}{Q_n(t)} \tag{26}$$

where $P_n(t)$ and $Q_n(t)$ are positive real polynomials of the nth degree.

The next step in the synthesis is to find the values of the impedances of the n lines of electrical length θ, terminated by lines of unit impedance (Fig. 2) which gives this value of $Z(t)$. In terms of the propagation coefficient γ [(9)] the transfer matrix of a single line element of impedance Z is

$$\begin{bmatrix} \cosh\gamma & Z\sinh\gamma \\ 1/Z\sinh\gamma & \cosh\gamma \end{bmatrix} = \frac{1}{\sqrt{1-t^2}}\begin{bmatrix} 1 & Zt \\ t/Z & 1 \end{bmatrix} \tag{27}$$

where

$$t = \tanh\gamma.$$

Hence, the over-all transfer matrix for n elements is

$$\frac{1}{(1-t^2)^{n/2}} \prod_{r=1}^{n} \begin{bmatrix} 1 & Z_r t \\ t/Z_r & r \end{bmatrix} = \frac{1}{(1-t^2)^{n/2}} \begin{bmatrix} A(t) & B(t) \\ C(t) & D(t) \end{bmatrix}. \tag{28}$$

The input impedance is given by

$$Z(t) = \frac{A(t) + B(t)}{D(t) + C(t)} \tag{29}$$

where $A(t)$, $D(t)$ are even and $B(t)$, $C(t)$ are odd polynomials in t.

Now if (26) is written

$$Z(t) = \frac{EP_n(t) + OP_n(t)}{EQ_n(t) + OQ_n(t)} \tag{30}$$

where E and O refer to the even and odd parts of the polynomials $P_n(t)$ and $Q_n(t)$, it is clear that the over-all transfer matrix is

$$\frac{1}{(1-t^2)^{n/2}} \begin{bmatrix} EP_n(t) & OP_n(t) \\ OQ_n(t) & EQ_n(t) \end{bmatrix}. \tag{31}$$

This is the product of n matrices similar to that of (27), *i.e.*,

$$\begin{bmatrix} EP_n(t) & OP_n(t) \\ OQ_n(t) & EQ_n(t) \end{bmatrix} = \prod_{r=1}^{n} \begin{bmatrix} 1 & Z_r t \\ t/Z_r & 1 \end{bmatrix}. \tag{32}$$

In order to express the over-all matrix (32) as the product of its constituent matrices, it is premultiplied by the inverse matrix

$$\frac{1}{1-t^2} = \begin{bmatrix} 1 & -Z_1 t \\ -t/Z_1 & 1 \end{bmatrix} \tag{33}$$

and the value of Z_1 is obtained from the condition that all elements of the resulting matrix must be divisible by $(1-t^2)$. This process is repeated to give all impedances Z_r as a function of the original variables β and h of (6).

Example n=2: As an example of the synthesis procedure it is useful and instructive to derive the impedances for a two-element coupler. This is a simple but important case; for example, one can achieve a bandwidth of 5:1 for a coupling of 3 db $\pm$0.5 db, or a bandwidth of 4:1 for a coupling of 10 db $\pm$0.5 db. 185

With $n=2$ the roots given by (15) and (16) are all real, reducing to

$$t_1^2 = \frac{\cosh^2 J/2 + \sinh^2 J/2}{\sinh^2 J/2} = \frac{2\beta}{\beta - h}$$

$$t_2^2 = 0$$

$$t_1'^2 = \frac{\sinh^2 H/2 + \cosh^2 J/2}{\sinh^2 H/2} = \frac{\sqrt{1+\beta^2} + \beta}{\sqrt{1+\beta^2} - h}$$

$$t_2'^2 = \frac{\cosh^2 H/2 - \cosh^2 J/2}{\cosh^2 H/2} = \frac{\sqrt{1+\beta^2} - \beta}{\sqrt{1+\beta^2} + h}. \tag{34}$$

Applying (25) the reflection coefficient is

$$\Gamma(t) = \frac{\sqrt{\beta^2+h^2}}{\sqrt{1+\beta^2-h^2}} \frac{t\left(t+\sqrt{\frac{2\beta}{\beta-h}}\right)}{\left(t+\frac{\sqrt{1+\beta^2}+\beta}{\sqrt{1+\beta^2}-h}\right)\left(t+\frac{\sqrt{1+\beta^2}-\beta}{\sqrt{1+\beta^2}+h}\right)}$$

$$= \frac{\sqrt{\beta^2-h^2}\,t^2+\sqrt{2\beta(\beta+h)}\,t}{\sqrt{1+\beta^2-h^2}t^2+(\sqrt{(1+\beta^2)+(\beta+h)\sqrt{1+\beta^2}+\beta h}+\sqrt{(1+\beta^2)-(\beta+h)\sqrt{1+\beta^2}+\beta h})t+1} \cdot \quad (35)$$

Eqs. (26) and (31) give the transfer matrix

$$\frac{1}{1-t^2}\begin{vmatrix} at^2+1 & bt \\ ct & dt^2+1 \end{vmatrix} \quad (36)$$

where

$$a = \sqrt{1+\beta^2-h^2}+\sqrt{\beta^2-h^2} \quad (37)$$

$$\begin{aligned} b = {} & \sqrt{(1+\beta^2)+(\beta+h)\sqrt{1+\beta^2}+\beta h} \\ & + \sqrt{(1+\beta^2)-(\beta+h)\sqrt{1+\beta^2}+\beta h} \\ & + \sqrt{2\beta(\beta+h)} \end{aligned} \quad (38)$$

$$\begin{aligned} c = {} & \sqrt{(1+\beta^2)+(\beta+h)\sqrt{1+\beta^2}+\beta h} \\ & + \sqrt{(1+\beta^2)-(\beta+h)\sqrt{1+\beta^2}+\beta h} \\ & - \sqrt{2\beta(\beta+h)} \end{aligned} \quad (39)$$

$$d = \sqrt{1+\beta^2-h^2}-\sqrt{\beta^2-h^2} \quad (40)$$

Premultiplying (36) by (33) [neglecting the $1/(1-t^2)$ factors] gives the matrix

$$\begin{bmatrix} 1 & -Z_1 t \\ -t/Z_1 & 1 \end{bmatrix}\begin{bmatrix} at^2+t & bt \\ ct & dt^2+1 \end{bmatrix} = \begin{bmatrix} (a-cZ_1)t^2+1 & -dZ_1t^3+(b-Z_1)t \\ -at^3/Z_1+(c-a/Z_1)t & (d-b/Z_1)t^2+1 \end{bmatrix}. \quad (41)$$

All elements of this matrix are divisible by $(1-t^2)$ giving the two conditions

$$Z_1 = \frac{a+1}{c} = \frac{b}{d+1} \quad (42)$$

which are equivalent, since the determinant of (36) is unity. Eq. (41) now reduces to

$$\begin{bmatrix} 1 & \frac{dbt}{d+1} \\ \frac{act}{a+1} & 1 \end{bmatrix} \equiv \begin{bmatrix} 1 & Z_2 t \\ t/Z_2 & 1 \end{bmatrix} \quad (43)$$

giving

$$Z_2 = Z_1/c = dZ_1. \quad (44)$$

Eqs. (43) and (44) are the expressions for the normalized even-mode impedances of the two sections of the coupler, and are functions of β and h only, *i.e.*, of the two parameters of the original insertion loss function [(4)–(6)]. In designing a coupler, the coupling, ripple and bandwidth are the quantities of interest, and these are perhaps more conveniently expressed in terms of J and H, themselves functions of β and h, as defined by (14) and (17). The bandwidth, in terms of the electrical length θ of one coupling section, extends from θ_0 to $\pi-\theta_0$, where θ_0 is defined by (15), which may be written in terms of J in the form

$$\frac{1}{\cos\theta_0} = \cosh J/n. \quad (5a)$$

Simple manipulation of the basic (4)–(6) give the following equations for the mean coupling C db and the coupling tolerance on ripple $\pm R$ db:

$$C = 10\log_{10}\frac{\sinh 2H}{\sinh 2J} \quad (45)$$

$$R = 10\log_{10}\frac{\tanh H}{\tanh J}\cdot \quad (46)$$

The bandwidth ratio $(\pi-\theta_0)/\theta_0$ as a function of the coupling tolerance of 3-db and 10-db couplers for a given number of elements is shown in Fig. 3. This graph was obtained from (5a), (45) and (46).

Eqs. (37)–(40), (42) and (44) have been applied to give several two-element couplers, design data for which are as follows:

a) Coupling 10 db$\pm$0.5 db
Bandwidth 0.5–2.0 Gc/s.

This data is satisfied by $\beta=0.3535$, $h=0.1672$. Direct substitution of these values into the equations gives

$$Z_1 = 1.562 \qquad Z_2 = 1.154.$$

Remembering that Z_1 and Z_2 are normalized even-mode impedances of the two-element coupler, being related to the odd-mode impedances by (2), and renormalizing to 50 ohms, gives the following ohmic values of the impedances in the two sections of the coupler:

Section	1	2
Z_{oe}	78.1	57.7
Z_{oo}	32.0	43.3

186

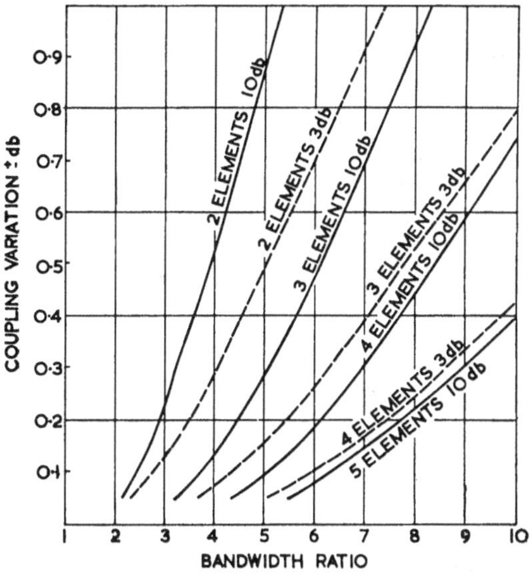

Fig. 3—Coupling tolerance of 3-db and 10-db stepped directional couplers as a function of bandwidth.

Fig. 4—Two-element coupler, 10 db ± 0.5 db, 0.8–2.0 Gc/s.

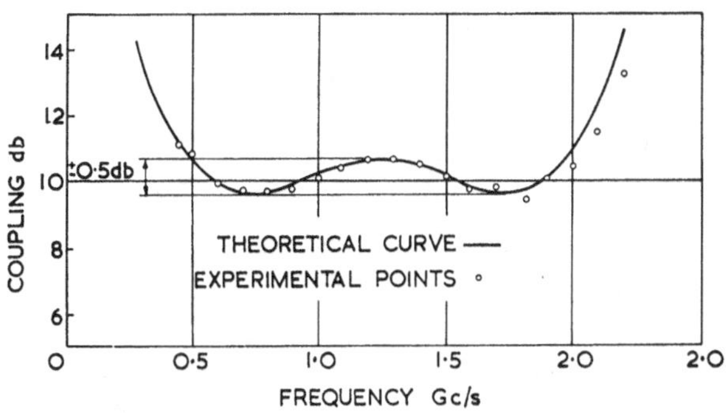

Fig. 5.—Two-element 10-db coupler.

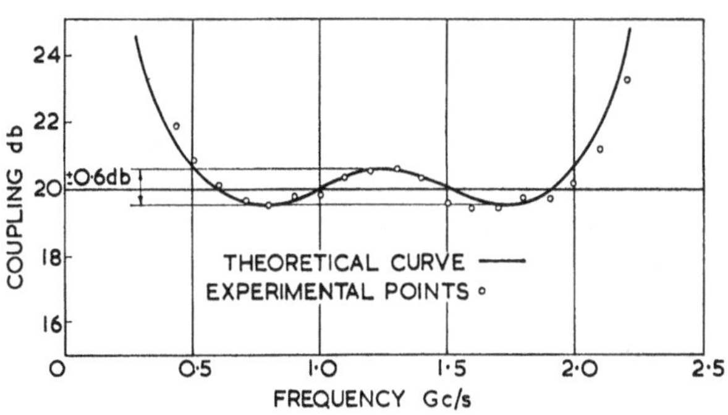

Fig. 6—Two-element 20-db coupler.

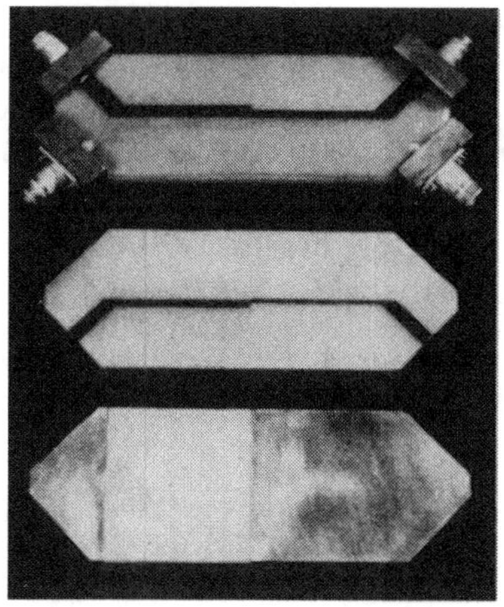

Fig. 7—Two-element coupler using interleaved-strip printed circuit construction.

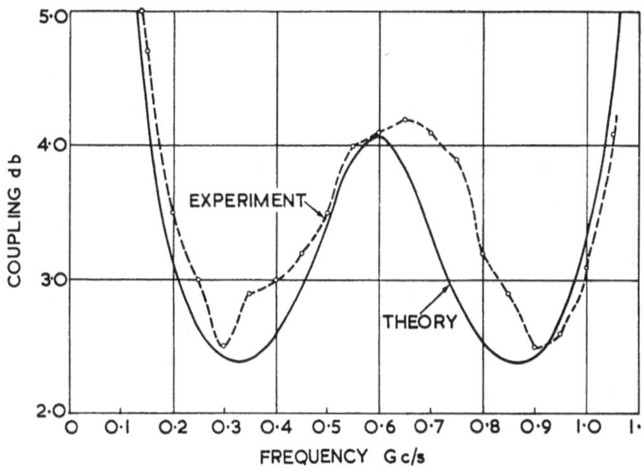

Fig. 8—Two-element hybrid.

This coupler was made in side-coupled stripline using a ground-plane spacing of 5/16 inch and a strip thickness of 1/16 inch. The dimensions were calculated using Getsinger's accurate method,[9] and a photograph of the coupler is shown in Fig. 4. The result for the coupling is given in Fig. 5, which indicates the very good agreement between the experimental results and the theory. The directivity is better than 18 db and the VSWR better than 1.13 over the design frequency band.

b) Coupling 20 db ± 0.6 db
Bandwidth 0.5–2.0 Gc/s
$\beta = 0.1070$, $h = 0.0519$.

The formulas give $Z_1 = 1.143$, $Z_2 = 1.046$, *i.e.*, the following values of the impedances in the coupler:

Section	1	2
Z_{oe}	57.4	52.3
Z_{oo}	43.6	47.8

[9] W. J. Getsinger, "Coupled bars between parallel plates," IRE Trans. on Microwave Theory and Techniques, vol. MTT-10, pp. 65–72; January, 1962.

This coupler is similar in construction to a) and the comparison between theory and experiment for the coupling is shown in Fig. 6. The directivity was better than 14 db and the VSWR better than 1.10 over the frequency band.

c) Coupling 2.4–4.07 db
Bandwidth 0.155–1.035 Gc/s.

This data gives $J=0.8423$, $H=1.200$, or $\beta=1.164$, $h=0.8487$, which upon substitution in the formulas give $Z_1=3.52$, $Z_2=1.692$, *i.e.*,

Section	1	2
Z_{oe}	176	84.6
Z_{oo}	14.2	29.5

The coupler was made in the interleaved printed circuit configuration, as described by Getsinger[10] and shown in Fig. 7. The ground-plane spacing in the more closely coupled section is greater than in the more loosely coupled section in order to give reasonable dimensions, in particular better matching between the dimensions of the inner strips. The comparison between the experimental results and theory is shown in Fig. 8. The directivity is greater than 20 db and the input VSWR better than 1.2 over the band.

The phase properties of these couplers are of interest in that, since they are asymmetric, the phase division of the output voltages is not 90° independent of frequency, but has a marked frequency dependence, as noted by Sweet.[11] The phase division for the two-element couplers is calculated in Appendix III, and the theoretical equations (82)–(85) have been verified by measurements on the two-element coupler c).

A. General Formulas: n = Z

It would be very useful to derive formulas for the Z_r in the general case similar to those found by Takahasi[12] and Green[13] for the low-pass filter. The present problem is more complicated, however, and although it is felt that general formulas do exist, no attempt has yet been made to find them. It is fortunate that only a small number of elements are ever required, for example a 10:1 bandwidth can be obtained for a coupling of 3 db ± 0.43 db with only four elements.

It is quite simple to write down general formulas for the Z_r in the case of a small number of elements, although the expressions are inevitably rather complicated. As an example, take $n=3$. Application of (22) gives

$$\Gamma(t) = \frac{\cosh J\, t(t^2 + T_1 t + |t_1|^2)}{\cosh H\, (t + t_3')(t^2 + T_1' t + |t_1'|^2)} \tag{47}$$

where T_1, T_1', $|t_1|^2$ and $|t_1'|^2$ are obtained from (23) and (24) with the unprimed terms given by substituting J for H. Application of (26) and (31) gives the transfer matrix

$$\begin{bmatrix} a_2 t^2 + 1 & b_3 t^3 + b_1 t \\ c_3 t^3 + c_1 t & d_2 t^2 + 1 \end{bmatrix} \tag{48}$$

where

$$b_3 = h(\cosh H + \cosh J)$$
$$c_3 = h(\cosh H - \cosh J)$$
$$a_2 = h(t_3' + T_1')\cosh H + hT_1 \cosh J$$
$$d_2 = h(t_3' + T_1')\cosh H - hT_1 \cosh J$$
$$b_1 = h(|t_1'|^2 + t_3' T_1') \cosh H + h|t_1|^2 \cosh J$$
$$c_1 = h(|t_1'|^2 + t_3' T_1') \cosh H - h|t_1|^2 \cosh J.$$

In deriving this matrix use has been made of the fact that

$$ht_3' |t_1'|^2 \cosh H = 1 \tag{49}$$

which enables the matrix to be normalized so that the coefficient of t^0 in the even polynomial is always unity. It would appear that an equation similar to (49) may be written down for any value of n. The matrix (48) may be broken down into its constituent matrices to give the general formulas

$$Z_1 = \frac{a_2 + 1}{c_3 + c_1} = \frac{b_3 + b_1}{d_2 + 1} \tag{50}$$

$$Z_2 = \frac{c_3 Z_1 + 1}{c_1 - 1/Z_1} = \frac{b_1 - Z_1}{b_3/Z_1 + 1} \tag{51}$$

$$Z_3 = \frac{b_3 Z_2}{Z_1} = \frac{Z_2}{c_3 Z_1}\cdot \tag{52}$$

It can be shown easily that the alternative forms given in (50)–(52) are equivalent by applying the condition that the determinant of matrix (48) is equal to $(1-t^2)^3$. Similar formulas for $n=4$ and $n=5$ are given in Appendix IV.

Eqs. (50) to (52) have been applied to design a coupler of 20 db ± 0.5 db over a 6:1 band, the relevant constants being $\beta=0.1063$, $h=0.0495$, leading to normalized even-mode impedances of $Z_1=1.164$, $Z_2=1.080$, $Z_3=1.030$, the actual impedance values in ohms in the three sections being as follows:

Section	1	2	3
Z_{oe}	58.2	54.0	51.5
Z_{oo}	42.95	46.3	48.55

[10] W. J. Getsinger, "A coupled stripline configuration using printed-circuit construction that allows very close coupling," IRE TRANS. ON MICROWAVE THEORY AND TECHNIQUES, vol. MTT-9, pp. 535–544; November, 1961.

[11] O. Sweet, "Analysis of a two-section coupler," IRE TRANS. ON MICROWAVE THEORY AND TECHNIQUES, vol. MTT-10, p. 295; July, 1962.

[12] L. Weinberg and P. Slepian, "Takahasi's results on Tchebycheff and Butterworth ladder networks," IRE TRANS. ON CIRCUIT THEORY, vol. CT-7, pp. 88–101; June, 1960.

[13] E. Green, "Synthesis of ladder networks to give Butterworth or Chebyshev response in the pass-band," *Proc. I.E.E.*, vol. 101 IV, pp. 192–203; January, 1954.

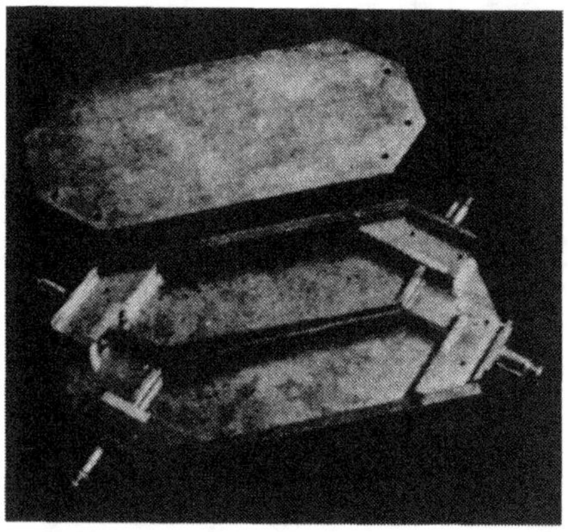

Fig. 9.—Three-element broadside-coupled stripline coupler, 20 db ±0.5 db, 0.5–3.0 Gd/s.

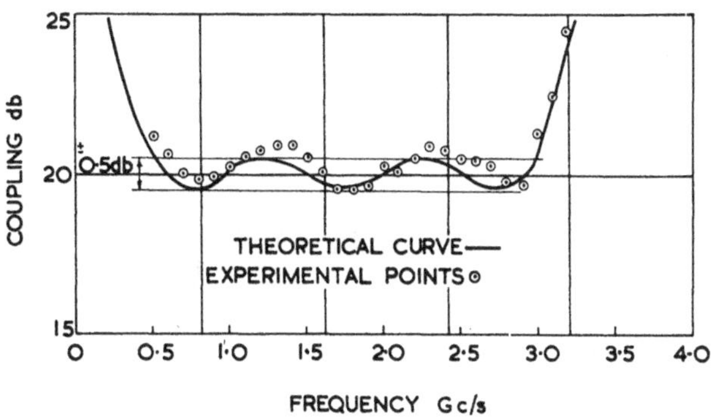

Fig. 10—Three-element 20-db coupler.

The coupler was constructed in broadside-coupled stripline, as shown in Fig. 9, using Cohn's formulas.[14,15] The remarkably good agreement obtained between theory and experiment is shown in Fig. 10. The directivity is better than 20 db and the VSWR better than 1.20 over the band.

III. Practical Disadvantages of Multi-Element Couplers

The main difficulty in applying the theory is the fact that the coupling of the tightest coupled element (considered as a single-element coupler) is always considerably tighter than the actual over-all coupling. This is not a serious drawback for loose couplers, *e.g.*, the coupling corresponding to Z_1 in the above example of a 20-db coupler is 16 db, which is easy to manufacture. In the case of 3-db couplers the tightest coupling may become extremely tight. Preliminary calculations on a four-element 3-db coupler for a bandwidth of 10:1 indicate that the tightest coupling is 1 db, which should be possible to manufacture, however.

Another difficulty is that it is necessary to maintain very close tolerances on all dimensions, and in particular the electrical lengths of the sections must be all equal. Multiple reflections within the coupler and at the ends, where it is joined to the input lines, may also be troublesome.

The problems of input VSWR and directivity are also associated with undesirable internal reflections within the coupler, and in differences in the effective electrical lengths between the even and odd modes. The introduction of dielectric which affects the velocity of one mode with respect to the other in such a way as to cancel the unwanted signal in the isolated arm would appear to be a useful nonempirical way of obtaining good performance.

IV. Conclusions

Having proved that a lumped-element ladder network forms a prototype for a stepped-impedance filter, it has been shown how to synthesize a class of asymmetric multi-element directional couplers. The synthesis leads to explicit formulas for the essential parameters, *i.e.*, the normalized even-mode impedance of each coupler element, as a function of the bandwidth, coupling and ripple. Examples of two- and three element couplers have been made and tested, giving excellent agreement with the theory. It now seems possible to design 3-db couplers for operation over bandwidths of one decade or more. The phase division is highly frequency dependent, but by careful choice of reference planes in the output ports, this can be made approximately 90° over the bandwidth of the coupler.

Appendix I

Equivalence between a Multi-Element Directional Coupler and a Stepped-Impedance Filter

The directional coupler shown in Fig. 1 consists of two identical lines 1–4 and 2–3 with uniform spacing over the electrical length θ. These lines are terminated by input and output ports of characteristic impedance unity. The coupler may be analyzed by the method of Reed and Wheeler,[16] which also gives a good physical picture of the directional coupler operation. The coupled lines may be analyzed in two modes known as the even and odd modes.[2,3] In the even mode, Fig. 11(a), in-phase signals are applied to ports 1 and 2 and a voltage maximum occurs on the line of symmetry. A magnetic wall may be located at the symmetry plane without affecting the field distribution in this mode. The odd mode is shown in Fig. 11(b), when out-of-phase signals are applied to ports 1 and 2 and a voltage zero occurs on the line of symmetry. Hence an electric wall (short circuit) may be located without affecting the operation. The electric field distribution for the two

[14] S. B. Cohn, "Characteristic impedances of broadside-coupled strip transmission lines," IRE Trans. on Microwave Theory and Techniques, vol. MTT-8, pp. 633–637; November, 1960.

[15] S. B. Cohn, "Thickness corrections for capacitive obstacles and strip conductors," IRE Trans. on Microwave Theory and Techniques, vol. MTT-8, pp. 638–644; November, 1960.

[16] J. Reed and G. J. Wheeler, "A method of analysis of symmetrical four port networks," IRE Trans. on Microwave Theory and Techniques, vol. MTT-4, pp. 246–252; October, 1956.

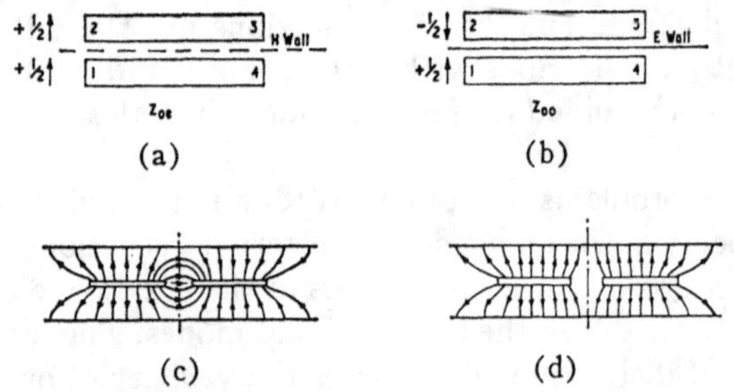

Fig. 11—Reed and Wheeler directional coupler theory.

modes is shown in Figs. 11(c) and 11(d) for the case of coupled striplines. In each case, the analysis reduces to that of a two-port network and, by superposition, the sum of the two cases is equivalent to a single signal of unit amplitude applied to port 1.

The amplitude and phase of the signals emerging from the four ports are[16]

$$\begin{aligned} A_1 &= \tfrac{1}{2}(\Gamma_{oe} + \Gamma_{oo}) \\ A_2 &= \tfrac{1}{2}(\Gamma_{oe} - \Gamma_{oo}) \\ A_3 &= \tfrac{1}{2}(T_{oe} - T_{oo}) \\ A_4 &= \tfrac{1}{2}(T_{oe} + T_{oo}) \end{aligned} \tag{53}$$

where Γ_{oe} and Γ_{oo} are the reflected waves, and T_{oe} and T_{oo} are the transmitted waves, for the even and odd two-port networks, respectively. The analysis of the two-port networks is most easily carried out by use of the transfer matrix of each, remembering the following results:

$$\Gamma = \frac{A + B - C - D}{A + B + C + D} \tag{54}$$

$$T = \frac{2}{A + B + C + D} \tag{55}$$

where Γ and T are the reflected and transmitted waves with a matched load on the output port, and where the matrix elements are normalized with respect to this load impedance. In the case of the coupled transmission lines, the transfer matrix for the even mode is

190

$$\begin{bmatrix} \cos\theta & jZ_{oe}\sin\theta \\ (j\sin\theta)/Z_{oe} & \cos\theta \end{bmatrix} \tag{56}$$

and for the odd mode it is

$$\begin{bmatrix} \cos\theta & jZ_{oo}\sin\theta \\ (j\sin\theta)/Z_{oo} & \cos\theta \end{bmatrix}. \tag{57}$$

Application of (54) and (55) to these transfer matrices gives

$$\Gamma_{oe} = \frac{j(Z_{oe} - 1/Z_{oe})\sin\theta}{2\cos\theta + j(Z_{oe} + 1/Z_{oe})\sin\theta}$$

$$\Gamma_{oo} = \frac{j(Z_{oo} - 1/Z_{oo})\sin\theta}{2\cos\theta + j(Z_{oo} + 1/Z_{oo})\sin\theta}$$

$$T_{oe} = \frac{2}{2\cos\theta + j(Z_{oe} + 1/Z_{oe})\sin\theta}$$

$$T_{oo} = \frac{2}{2\cos\theta + j(Z_{oo} + 1/Z_{oo})\sin\theta}. \tag{58}$$

When the condition

$$Z_{oe}Z_{oo} = 1 \tag{2}$$

is satisfied (58) reduces as follows:

$$\Gamma_{oe} = -\Gamma_{oo} = \frac{j(Z_{oe} - 1/Z_{oe})\sin\theta}{2\cos\theta + j(Z_{oe} + 1/Z_{oe})\sin\theta}$$

$$T_{oe} = T_{oo} = \frac{2}{2\cos\theta + j(Z_{oe} + 1/Z_{oe})\sin\theta}. \tag{59}$$

From (53) the signals emerging from the four ports of the directional coupler are then:

$$A_1 = 0 \tag{60}$$

$$A_2 = \Gamma_{oe} = \frac{j(Z_{oe} - 1/Z_{oe})\sin\theta}{2\cos\theta + j(Z_{oe} + 1/Z_{oe})\sin\theta} \tag{61}$$

$$A_3 = 0 \tag{62}$$

$$A_4 = T_{oe} = \frac{2}{2\cos\theta + j(Z_{oe} + 1/Z_{oe})\sin\theta}. \tag{63}$$

Eqs. (60) and (62) show that the coupler is perfectly matched and isolated at all frequencies, while the coupling to ports 2 and 4 vary with frequency.

The proof that the analysis of the directional coupler is equivalent to that of a stepped impedance filter [providing (2) holds] is now evident from (61) and (63). These are simply the reflection and transmission coefficients respectively of a uniform transmission line of electrical length θ and characteristic impedance Z_{oe}, normalized to its input and output terminations. It is a trivial extension to show that the multi-element coupler retains the perfect VSWR and isolation property if each element has the same effective characteristic impedance, *i.e.*, if

$$Z_{oe1}Z_{oo1} = Z_{oe2}Z_{oo2} = \cdots = Z_{oen}Z_{oon} = 1 \tag{64}$$

and hence the analysis of this stepped coupler is identical to that of a multi-element stepped filter, as shown in Fig. 2.

The simple formulas for the power insertion loss between arms 1 and 4 of the directional coupler, *e.g.*, (1) and (3), are given by application of the well-known formula

$$L = 1 + \tfrac{1}{4}(A - D)^2 - \tfrac{1}{4}(B - C)^2 \tag{65}$$

where A and D are the real and B, C the imaginary elements of the transfer matrix for a lossless networ

Appendix II

General Synthesis of Transmission-Line Stepped-Impedance Filters

Riblet[8] has proved that a rational function of $t=\tanh\theta$ (in Riblet's notation $t=1/p$) with real coefficients written in the form

$$Z(t) = \frac{A(t) + B(t)}{D(t) + C(t)} \tag{29}$$

with A and D even and B and C odd functions of t, is the input impedance of a stepped-impedance filter with equal line lengths θ if

1) Z is a positive real function of t, and

2) $A(t)D(t) - B(t)C(t) = C'(1-t^2)^n$ (66)

where C' is a constant.

Leo Young[17] has pointed out that a third condition is required, namely that

3) The degrees of the numerator and denominator in (29) are the same (otherwise a stub will be required).

Riblet's theorem may be extended by noting that any two-port lumped-element ladder network consisting of simple series reactances and shunt susceptances terminated by pure resistances has a transfer matrix of the form

$$\begin{bmatrix} A(p) & B(p) \\ C(p) & D(p) \end{bmatrix} \tag{67}$$

where p is the complex frequency variable, $A(p)$, $D(p)$ are even functions of p, and $B(p)$, $C(p)$ are odd functions of p. The insertion loss is therefore

$$L = 1 + \tfrac{1}{4}(A-D)^2 - \tfrac{1}{4}(B-C)^2 \tag{68}$$

and the square of the magnitude of the input reflection coefficient is

$$|\Gamma(p)|^2 = \frac{\tfrac{1}{4}(A-D)^2 - \tfrac{1}{4}(B-C)^2}{1 + \tfrac{1}{4}(A-D)^2 - \tfrac{1}{4}(B-C)^2}. \tag{69}$$

Now if the frequency variable is replaced by a linear function in $\cos\theta$ and $\sin\theta$, then it is easily shown, by multiplying the numerator and denominator of (69) by

$$\frac{1}{\cos^{2n}\theta} = (1-t^2)^n,$$

that

$$|\Gamma(t)|^2 = \frac{f_1(t^2) + f_2(t^2)}{(1-t^2)^n + f_1(t^2) + f_2(t^2)}. \tag{70}$$

[17] L. Young, "Concerning Riblet's theorem," IRE Trans. on Microwave Theory and Techniques, vol. MTT-7, pp. 477–478; October, 1959.

Richards[7] has shown that the right half of the p plane maps into the right half of the t plane, and since $Z(p)$ is positive real therefore $Z(t)$ must be positive real, proving that condition 1) holds. The impedance $Z(t)$ obtained from (70) will give (29). Since $Z(t)$ is positive real, all the coefficients appearing in Z are of the same sign, and the degrees of the numerator and denominator differ at most by unity. However, in the case of a prototype ladder network where the poles of the insertion loss function are all at infinity, this difference is zero. The reflection coefficient is

$$\Gamma(t) = \frac{(A-D) + (B-C)}{(A+D) + (B+C)} \tag{71}$$

giving

$$|\Gamma(t)|^2 = \frac{(A-D)^2 - (B-C)^n}{(A+D)^2 - (B+C)^2}. \tag{72}$$

Eqs. (70) and (72) are identical, and equating the differences between the numerator and denominator of each expression for $\Gamma(t)^2$ gives

$$A(t)\cdot D(t) - B(t)\cdot C(t) = \tfrac{1}{4}(1-t^2)^n$$

which is (66), *i.e.*, proving condition 2). The following theorem may now be stated:

An insertion loss function of the form

$$L = 1 + [f_1(a\cos\theta + b\sin\theta)]^2 + [f_2(a\cos\theta + b\sin\theta)]^2 \tag{73}$$

can be realized as a stepped impedance filter with real positive characteristic impedances if the function

$$L = 1 + [f_1(\omega)]^2 + [f_2(\omega)]^2 \tag{74}$$

having all its poles at infinity, is realizable as a two-port ladder network consisting of simple lossless series reactances and shunt susceptances terminated by resistances.

The restriction that the impedance function shall have a numerator and denominator of equal degree is implied by the form of (74), since this has all its poles at infinity, and therefore there is no need to state the restriction explicitly in the theorem.

The prototype function considered in the present context is given by

$$L = 1 + \beta^2 - h^2 T_n^2(\omega) \tag{75}$$

and it is quickly proved that this satisfies the requirements of the above theorem. The only difference between (75) and the usual Chebyshev insertion loss function encountered in ladder network filter theory is in the sign of the $h^2T_n^2(\omega)$ term. Eq. (75) can still be physically realized however, since the roots of $|\Gamma(j\omega)|^2$ formed from it are similar, giving

$$|\Gamma(j\omega)|^2 = \Gamma(j\omega)\cdot\Gamma(-j\omega) \tag{76}$$

with the roots of $\Gamma(j\omega)$ occurring as complex conjugate pairs. Hence $\Gamma(p)$ (where $p=\sigma+j\omega$) can be selected to be analytic in the right-half plane, *i.e.*, $Z(p)$ is positive real, a necessary and sufficient condition for the physical realization of the insertion loss function of (75). Application of the above theorem shows that the transformation $\omega \to \cos\theta/\cos\theta_0$ in (75), giving (6), enables the resulting insertion loss function to be realized as a stepped impedance filter.

An alternative and possibly more direct proof that (6) can be synthesized as a stepped impedance filter is obtained by applying the reasoning which proved the theorem of this appendix, and showing that (66) is true. The main text shows that $Z(t)$ is positive real, and (26) that the numerator and denominator of $Z(t)$ are of the same degree.

Appendix III

Phase Properties of the Asymmetric Coupler

The couplings to arms 2 and 4 of the asymmetric coupler are given by Γ and T (see Fig. 2) derived from the transfer matrix (7), *i.e.*,

$$\Gamma = \frac{(A_n - D_n) + j(B_n - C_n)}{(A_n + D_n) + j(B_n + C_n)} \tag{77}$$

$$T = \frac{2}{(A_n + D_n) + j(B_n + C_n)} \tag{78}$$

and the power division between arms 2 and 4 is

$$\eta = \frac{\Gamma}{T} = \frac{(A_n - D_n) + j(B_n - C_n)}{2}\cdot \tag{79}$$

In the case of a symmetric coupler, *e.g.*, the three-quarter wavelength coupler described previously,[3] the phase difference between Γ and T is 90° at all frequencies because $A_n = D_n$. This condition does not hold for an asymmetric coupler, and the phase difference is

$$\phi = \tan^{-1}\frac{B_n - C_n}{A_n - D_n}\cdot \tag{80}$$

Thus, while the symmetric 3-db coupler can be used
192 as a 90° hybrid, this is not immediately true of the asymmetric 3-db coupler, which has a phase shift ϕ varying in an approximately linear fashion over the operating frequency band of the coupler. However, by the addition of a length of line to one of the output arms, this phase variation can be approximately cancelled, and a usable 90° hybrid performance obtained. In other words, reference planes exist in arms 2 and 4 where the phases differ by approximately 90° over the band.

As an example take the case of the two element asymmetric coupler. This has a transfer matrix given by (36) to (40), but in this simple case it is preferable to form the matrix directly by analysis, giving

$$\begin{bmatrix} \cos^2\theta - (Z_1\sin^2\theta)/Z_2 & j(Z_1+Z_2)\sin\theta\cos\theta \\ j(1/Z_1+1/Z_2)\sin\theta\cos\theta & \cos^2\theta - (Z_2\sin^2\theta)/Z_1 \end{bmatrix}. \tag{81}$$

The phase difference is given by (54), *i.e.*,

$$\phi = \tan^{-1}(K_1\cot\theta) \tag{82}$$

where

$$K_1 = \frac{(Z_1+Z_2) - (1/Z_1 + 1/Z_2)}{(Z_2/Z_1 - Z_1/Z_2)}\cdot \tag{83}$$

At midband $\theta=\pi/2$ giving $\phi=0$. At zero frequency $\phi=\pi/2$, and at $\theta=\pi$, $\phi=-\pi/2$, so that by adding a line of electrical length $\pi/2$ at midband to ϕ the phase will be exactly $\pi/2$ at these three points. In order to decide on which arm to include the extra line it is necessary to find the exact meaning of the sign of the phase. Negative phase implies a forward traveling wave, *i.e.*, one emerging from an output arm, so by writing

$$\frac{\Gamma}{T} = \frac{|\Gamma|}{|T|}\frac{e^{-j\theta_2}}{e^{-j\theta_4}} = \frac{|\Gamma|}{|T|}e^{j\phi}$$

then

$$\phi = \theta_4 - \theta_2. \tag{84}$$

Hence, the extra line must be added to arm 4, which is the one leading into the closely coupled element. The phase difference between the new reference planes is now

$$\psi = \phi + \theta = \tan^{-1}(K_1\cot\theta) + \theta \tag{85}$$

where in the range $0<\theta<\pi$

$$-\frac{\pi}{2} < \tan^{-1}(K_1\cot\theta) < +\frac{\pi}{2}\cdot \tag{86}$$

Maximum and minimum values of ψ occur when

$$\frac{d\psi}{d\theta} = 0, \qquad i.e., \text{ for } \theta_m = \tan^{-1} \pm\sqrt{K_1} \tag{87}$$

and these values of θ substituted in (85) give the points of maximum deviation of ψ from 90°. For example, in the case of a two-element coupler designed to give a coupling of 3 db ± 0.9 db over a 7:1 band then $K_1 = 2.8$, giving

$$\theta_m = 59.3° \text{ and } 110.7°.$$

Substituting in (85) gives

$$= 118.6° \text{ and } 61.4° \text{ (respectively)}$$

i.e., a phase variation of $\pm 28.6°$ about 90°.

When the coupling ripple level is decreased, the phase linearity is improved, *e.g.*, for a 5:1 bandwidth giving ± 0.48 db ripple then the phase variation reduces to $\pm 22.2.°$

Variations of this order of magnitude are quite acceptable in many applications. In cases where the phase is required to be much more exactly equal to 90°, then the symmetrical form of the coupler should be used.

Appendix IV

General Formulas

A. $n=4$

Remembering (20), (25) becomes

$$\Gamma(t) = \frac{\sinh J}{\sinh H} \frac{t(t+t_2)(t^2+T_1t+|t_1|^2)}{(t+t_2')(t+t_4')(t^2+T_1't+|t_1'|^2)}$$

giving the transfer matrix

$$\begin{bmatrix} a_4t^4+a_2t^2+1 & b_3t^3+b_1t \\ c_3t^3+c_1t & d_4t^4+d_2t^2+1 \end{bmatrix} \tag{88}$$

where

$$\left.\begin{matrix} a_4 \\ d_4 \end{matrix}\right\} = h(\sinh H \pm \sinh J)$$

$$\left.\begin{matrix} b_3 \\ c_3 \end{matrix}\right\} = h(T_1'+t_2'+t_4')\sinh H \pm h(T_1+T_2)\sinh J$$

$$\left.\begin{matrix} a_2 \\ d_2 \end{matrix}\right\} = \begin{matrix} h[|t_1'|^2+T_1'(t_2'+t_4')+t_2't_4']\sinh H \\ \pm h[|t_1|^2+T_1t_2]\sinh J \end{matrix}$$

$$\left.\begin{matrix} b_1 \\ c_1 \end{matrix}\right\} = \begin{matrix} h[T_1't_2't_4'+|t_1'|^2(t_2'+t_4')]\sinh H \\ \pm h|t_1|^2 t_2 \sinh J. \end{matrix}$$

In the above equations the plus sign refers to the upper quantities on the left-hand side, and the negative sign to the lower quantities. The roots $|t_1'|^2$, T_1', etc., are obtained from (23), (24), *et seq.*, with $n=4$. In deriving the equations a relation similar to that of (49) has been used, namely that

$$h|t_1'|^2 t_2't_4' \sinh H = 1 \tag{89}$$

which is readily proved.

Expressions for the four normalized even-mode impedances are now obtained by breaking up matrix (88) into its constituent matrices, giving

$$Z_1 = \frac{a_4+a_2+1}{c_3+c_1} = \frac{b_3+b_1}{d_4+d_2+1} \tag{90}$$

$$Z_2 = \frac{c_3Z_1-a_4+1}{a_4/Z_1-1/Z_1+c_1} = \frac{d_4Z_1-Z_1+b_1}{b_3/Z_1-d_4+1} \tag{91}$$

$$Z_3 = \frac{a_4Z_2/Z_1+1}{c_1-1/Z_1-1/Z_2} = \frac{b_1-Z_1-Z_2}{d_4Z_1/Z_2+1} \tag{92}$$

$$Z_4 = \frac{Z_1Z_3}{a_4Z_2} = \frac{d_4Z_1Z_3}{Z_2}. \tag{93}$$

B. $n=5$

Eq. (22) gives

$$\Gamma(t) = \frac{\cosh J}{\cosh H} \cdot \frac{t(t^2+T_1t+|t_1|^2)(t^2+T_2t+|t_2|^2)}{(t+t_5')(t^2+T_1't+|t_1'|^2)(t^2+T_2't+|t_2'|^2)} \tag{94}$$

leading to the transfer matrix

$$\begin{bmatrix} a_4t^4+a_2t^2+1 & b_5t_5+b_3t^3 \\ c_5t^5+c_3t^3+c_1t & d_4t^4+d_2t^2+1 \end{bmatrix} \tag{95}$$

where

$$\left.\begin{matrix} b_5 \\ c_5 \end{matrix}\right\} = h(\cosh H \pm \cosh J)$$

$$\left.\begin{matrix} a_4 \\ d_4 \end{matrix}\right\} = h(T_1'+T_2'+t_5')\cosh H \pm h(T_1+T_2)\cosh J$$

$$\left.\begin{matrix} b_3 \\ c_3 \end{matrix}\right\} = \begin{matrix} h[|t_1'|^2+|t_2'|^2+T_1'T_2'+t_5'(T_1'+T_2')]\cosh H \\ \pm [|t_1|^2+|t_2|^2+T_1T_2]\cosh J \end{matrix}$$

$$\left.\begin{matrix} a_2 \\ d_2 \end{matrix}\right\} = \begin{matrix} h[T_1'|t_2'|^2+T_2'|t_1'|^2+t_5'(t_1'^2+|t_2'|^2 \\ +T_1'T_2')]\cosh H \pm h[T_1|t_2|^2+T_2|t_1|^2]\cosh J \end{matrix}$$

$$\left.\begin{matrix} b_1 \\ c_1 \end{matrix}\right\} = \begin{matrix} h[|t_1'|^2|t_2'|^2+t_5'(T_1'|t_2'|^2+T_2'|t_1'|^2)]\cosh H \\ \pm h|t_1|^2|t_2|^2\cosh J, \end{matrix}$$

and the normalized even mode impedances are

$$Z_1 = \frac{a_4+a_2+1}{c_5+c_3+1} = \frac{b_5+b_3+b_1}{d_4+d_2+1} \tag{96}$$

$$Z_2 = \frac{(c_5-c_1)Z_1+a_2}{(a_4-1)/Z_1-c_5+c_1} = \frac{(d_4-1)Z_1+b_5+b_1}{(b_5-b_1)/Z_1+d_2} \tag{97}$$

$$Z_3 = \frac{a_4Z_2/Z_1-c_5(Z_1+Z_2)+1}{c_5Z_1/Z_2+c_1-(1/Z_1+1/Z_2)} = \frac{b_5Z_2/Z_1+b_1-(Z_1+Z_2)}{d_4Z_1/Z_2-b_5(Z_1+Z_2)+1} \tag{98}$$

$$Z_4 = \frac{c_5Z_1Z_3/Z_2+1}{c_1-(1/Z_1+1/Z_2+1/Z_3)} = \frac{b_1-(Z_1+Z_2+Z_3)}{b_5Z_2/Z_1Z_3+1} \tag{99}$$

$$Z_5 = \frac{Z_2Z_4}{Z_1Z_3c_5} = \frac{Z_2Z_4b_5}{Z_1Z_3}. \tag{100}$$

Although at first glance the general formulas seem complicated, they are quite simple to apply and would be particularly appropriate for a computer program. They have a good deal of symmetry, which, as suggested in the main text, might imply the existence of general formulas for any value of n.

V. Acknowledgments

The numerical calculations on the two and three element couplers were performed by M. M. McDermott and Dr. P. T. Saaler, who also measured their characteristics. The author is grateful to Prof. M. W. Humphrey Davies for use of facilities in the Electrical Engineering Department, Queen Mary College, University of London, and to Dr. J. S. Seeley for discussions.

Tables for Asymmetric Multi-Element Coupled-Transmission-Line Directional Couplers

RALPH LEVY

Summary—**Tables are presented for the design of TEM-mode asymmetric multi-element coupled-transmission-line directional couplers, having a very broad-band Chebyshev equal-ripple coupling response. Coupler designs of 2 to 6 elements having bandwidths up to 20:1 and mean couplings of 3, 6, 10, 15 and 20 db are tabulated.**

Introduction

The MULTI-ELEMENT asymmetric coupled-transmission-line directional coupler is shown diagrammatically in Fig. 1. It consists of a cascaded set of coupled TEM-mode transmission lines, each of identical electrical length. The individual coupling of each section becomes progressively looser proceeding from left to right in Fig. 1, *i.e.*, the coupling is a stepped monotonic function of distance along the coupling region.

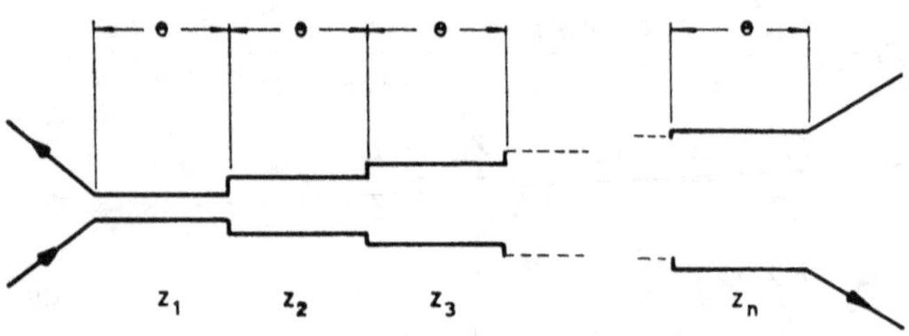

Fig. 1—Even-mode impedances for the *n*-element coupler.

The main features of this class of directional couplers, which it has in common with other TEM-mode couplers, are that the input VSWR and isolation are theoretically perfect at all frequencies. The coupling may be expressed as a simple function of frequency, and the structure is therefore amenable to synthesis by modern network techniques. In consequence, optimum Chebyshev equal-ripple coupling functions are realizable with extremely wide-band coupling characteristics, without necessarily requiring a very large number of coupled sections. On the other hand, since the coupler is asymmetric, the phase difference between the coupled waves at the ends of the coupler region is not 90° at all frequencies, as is the case with any symmetrical coupler. By choosing suitable reference planes, however, 90°-phase difference may be obtained approximately.[1]

The coupling of each section is defined by the even and odd mode impedances Z_{oe} and Z_{oo}, respectively.[2] In the present context these impedances are normalized with respect to the characteristic impedance of the input ports, and the relationship

$$Z_{oe}Z_{oo} = 1 \tag{1}$$

holds for each coupled element. Eq. (1) ensures the perfect input VSWR and isolation of the coupler. In consequence of (1), any individual coupled element is completely specified by its even-mode impedance Z_{oe}. Since $Z_{oe} > Z_{oo}$, $Z_{oe} > 1$. The coupling is determined by the ratio

$$Z_{oe}/Z_{oo} = Z_{oe}^2$$

which approaches unity for very loose coupling, so that the looser the coupling, the smaller is Z_{oe}. Suppressing the *oe* suffices, the several even-mode impedances of the *n* elements may be specified by the monotonically decreasing series $Z_1, Z_2, \cdots, Z_n$ as shown in Fig. 1, in which every value of Z_r is greater than unity.

The Computation

In a previous paper,[1] exact closed formulas were given for the even-mode impedances of the elements of asymmetric coupled-transmission-line directional couplers for 2 to 5 elements in terms of the bandwidth, coupling and ripple. The formulas are basically quite simple but require a considerable amount of computation to derive even one design. It was decided to prepare a set of tables from which any coupler may be designed rapidly. A computer program was written which derives the even-mode impedances directly by synthesis rather than by using the actual closed formulas. By supplying data giving the number of elements, the mean coupling and the bandwidth, the corresponding ripple is computed and printed with the derived even-mode impedances. Thus it is a general program and will give results for almost any number of elements, *n*, although the present computation gives results to only 6 elements. A limitation on *n* will occur because the accuracy of the

Received October 7, 1963; revised December 27, 1963. This paper will form part of a thesis for the Ph.D. degree at Queen Mary College, University of London, London, England.
The author is with the Systems Division, Mullard Research Laboratories, Redhill, Surrey, England.

[1] R. Levy, "General synthesis of asymmetric multi-element coupled-transmission-line directional couplers," IEEE Trans. on Microwave Theory and Techniques, vol. MTT-11, pp. 226–237; July, 1963.

[2] E. M. T. Jones and J. T. Bolljahn, "Coupled strip-transmission-line filters and directional couplers," IRE Trans. on Microwave Theory and Techniques, vol. MTT-4, pp. 75–81; April, 1956.

computer is limited by the effective number of decimal places to which numbers are held, 9 in the present case (*i.e.*, 30 binary digits). Synthesis for large values of n would demand better than 9-figure accuracy. A more accurate computer might then be used, but this was not justified in the present instance because generally only a very small number of elements is required. It is considered that the present program is capable of deriving couplers to orders higher than $n=6$ if greater bandwidths or other applications were to create a demand for this.

Use of Tables

The coupling characteristic is shown in Fig. 2, which defines the mean coupling C and the ripple R, both in units of decibels. The edges of the pass band are at frequencies f_1 and f_2, and the bandwidth is defined as the ratio

$$BW = f_2/f_1.$$

Tables I–V (pages 276–279) give the normalized even-mode impedances for 2- to 6-element couplers. Sufficient results are given for accurate interpolation.

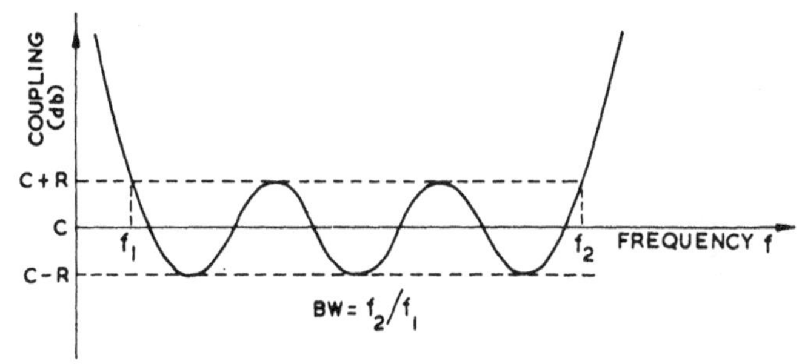

Fig. 2—Directional coupler characteristic.

It is interesting to compare the tabulated results with the examples given in the earlier paper,[1] which were calculated by logarithms and a desk calculator. The comparison is shown in Table VI (page 279), and indicates very close agreement. The inconsistencies in the values of R are due to errors in reading from a graph of R vs BW. Minor differences in the impedances are due to errors in the earlier calculations.

The impedances may be difficult to realize in practice for the case of tight coupling, *e.g.*, a 6-element 3-db coupler for a 20:1 bandwidth requires $Z_1=4.9$. With an input impedance of 50 ohms, this gives

$$Z_{oe} = 245 \text{ ohms}, \qquad Z_{oo} = 10.2 \text{ ohms}.$$

These values are impossible to realize by normal techniques, but the introduction of re-entrant cross-sectional coupled line[3] enables impedances of this order to be obtained with reasonable dimensions. In the above example, the re-entrant line impedances are given by

$$Z_{01} = \tfrac{1}{2}(Z_{oe} - Z_{oo}) = 117.4 \text{ ohms}$$

$$Z_{02} = Z_{oo} = 10.2 \text{ ohms}$$

i.e., this section would consist of a 117.4-ohm line containing two 10.2-ohm lines within its inner conductor. While this would be feasible at lower microwave frequencies, it would still present great difficulties at an upper working frequency of, say, 5 Gc per second, because of the limitation on the over-all size of the structures required by the necessity to prevent severe discontinuity effects[3] and propagation of waveguide type modes.

[3] S. B. Cohn, "The re-entrant cross section and wide-band 3-db hybrid couplers," IEEE Trans. on Microwave Theory and Techniques, vol. MTT-11, pp. 254–258; July, 1963.

TABLE I

Normalized Even-Mode Impedances of Two-Element Couplers

$N=2$, Coupling = 3.00 db

BW	2.0	3.0	4.0	5.0	6.0	7.0
R (db)	0.022	0.128	0.292	0.480	0.673	0.858
Z_1	2.9655	3.1292	3.2895	3.4455	3.6007	3.7574
Z_2	1.2308	1.3206	1.4235	1.5320	1.6434	1.7563

$N=2$, Coupling = 6.00 db

BW	2.0	3.0	4.0	5.0	6.0	7.0
R (db)	0.034	0.191	0.439	0.724	1.017	1.303
Z_1	1.9866	2.0596	2.1295	2.1958	2.2603	2.3240
Z_2	1.1481	1.2046	1.2680	1.3335	1.3994	1.4648

$N=2$, Coupling = 10.0 db

BW	2.0	3.0	4.0	5.0	6.0	7.0
R (db)	0.040	0.230	0.528	0.871	1.226	1.575
Z_1	1.5073	1.5412	1.5729	1.6025	1.6307	1.6581
Z_2	1.0882	1.1210	1.1573	1.1940	1.2302	1.2655

$N=2$, Coupling = 15.0 db

BW	2.0	3.0	4.0	5.0	6.0	7.0
R (db)	0.043	0.248	0.568	0.938	1.321	1.698
Z_1	1.2533	1.2688	1.2833	1.2966	1.3091	1.3211
Z_2	1.0480	1.0656	1.0847	1.1038	1.1232	1.1402

$N=2$, Coupling = 20.0 db

BW	2.0	3.0	4.0	5.0	6.0	7.0
R (db)	0.044	0.253	0.580	0.959	1.351	1.737
Z_1	1.1344	1.1423	1.1496	1.1562	1.1624	1.1684
Z_2	1.0266	1.0362	1.0466	1.0569	1.0668	1.0763

TABLE II

NORMALIZED EVEN-MODE IMPEDANCES OF THREE-ELEMENT COUPLERS

			$N=3$, Coupling = 3.00 db				
BW	4.0	5.0	6.0	7.0	8.0	9.0	10.0
R (db)	0.076	0.160	0.266	0.385	0.509	0.635	0.758
Z_1	3.5255	3.6516	3.7672	3.8765	3.9828	4.0881	4.1938
Z_2	1.6386	1.7528	1.8633	1.9072	2.0742	2.1759	2.2759
Z_3	1.1382	1.1925	1.2520	1.3149	1.3801	1.4469	1.5149
			$N=3$, Coupling = 6.00 db				
BW	4.0	5.0	6.0	7.0	8.0	9.0	10.0
R (db)	0.114	0.241	0.400	0.579	0.768	0.959	1.149
Z_1	2.2239	2.2752	2.3215	2.3646	2.4058	2.4462	2.4862
Z_2	1.3875	1.4510	1.5112	1.5683	1.6229	1.6754	1.7263
Z_3	1.0920	1.1276	1.1661	1.2063	1.2475	1.2891	1.3308
			$N=3$, Coupling = 10.0 db				
BW	4.0	5.0	6.0	7.0	8.0	9.0	10.0
R (db)	0.137	0.290	0.482	0.697	0.925	1.156	1.387
Z_1	1.6140	1.6364	1.6564	1.6747	1.6921	1.7089	1.7253
Z_2	1.2215	1.2555	1.2870	1.3166	1.3443	1.3707	1.3959
Z_3	1.0559	1.0772	1.1000	1.1235	1.1473	1.1710	1.1945
			$N=3$, Coupling = 15.0 db				
BW	4.0	5.0	6.0	7.0	8.0	9.0	10.0
R (db)	0.147	0.312	0.518	0.750	0.996	1.246	1.495
Z_1	1.3016	1.3115	1.3203	1.3283	1.3359	1.3431	1.3501
Z_2	1.1174	1.1346	1.1503	1.1649	1.1785	1.1912	1.2033
Z_3	1.0308	1.0424	1.0547	1.0672	1.0798	1.0922	1.1044
			$N=3$, Coupling = 20.0 db				
BW	4.0	5.0	6.0	7.0	8.0	9.0	10.0
R (db)	0.150	0.319	0.530	0.767	1.018	1.274	1.529
Z_1	1.1578	1.1636	1.1680	1.1719	1.1756	1.1792	1.1826
Z_2	1.0642	1.0733	1.0816	1.0892	1.0963	1.1030	1.1092
Z_3	1.0172	1.0236	1.0303	1.0372	1.0440	1.0507	1.0573

TABLE III

NORMALIZED EVEN-MODE IMPEDANCES OF FOUR-ELEMENT COUPLERS

$N=4$, Coupling = 3.00 db

BW	4.0	6.0	8.0	10.0	12.0	14.0
R (db)	0.020	0.105	0.250	0.427	0.613	0.796
Z_1	3.7032	3.9210	4.0974	4.2556	4.4088	4.5637
Z_2	1.8149	2.0471	2.2520	2.4379	2.6121	2.7792
Z_3	1.2407	1.3789	1.5184	1.6556	1.7904	1.9232
Z_4	1.0502	1.1108	1.1881	1.2753	1.3683	1.4648

$N=4$, Coupling = 6.00 db

BW	4.0	6.0	8.0	10.0	12.0	14.0
R (db)	0.030	0.158	0.376	0.642	0.925	1.208
Z_1	2.2934	2.3787	2.4459	2.5048	2.5608	2.6166
Z_2	1.4814	1.6028	1.7056	1.7961	1.8786	1.9560
Z_3	1.1561	1.2415	1.3249	1.4047	1.4810	1.5546
Z_4	1.0340	1.0749	1.1262	1.1831	1.2426	1.3031

$N=4$, Coupling = 10.0 db

BW	4.0	6.0	8.0	10.0	12.0	14.0
R (db)	0.036	0.190	0.452	0.773	1.115	1.459
Z_1	1.6438	1.6802	1.7083	1.7326	1.7553	1.7775
Z_2	1.2705	1.3328	1.3840	1.4279	1.4671	1.5032
Z_3	1.0933	1.1424	1.1891	1.2327	1.2735	1.3120
Z_4	1.0210	1.0459	1.0768	1.1105	1.1451	1.1796

$N=4$, Coupling = 15.0 db

BW	4.0	6.0	8.0	10.0	12.0	14.0
R (db)	0.038	0.204	0.487	0.832	1.201	1.572
Z_1	1.3147	1.3306	1.3428	1.3532	1.3629	1.3722
Z_2	1.1419	1.1726	1.1973	1.2181	1.2364	1.2531
Z_3	1.0509	1.0769	1.1012	1.1235	1.1441	1.1632
Z_4	1.0117	1.0254	1.0423	1.0604	1.0788	1.0968

$N=4$, Coupling = 20.0 db

BW	4.0	6.0	8.0	10.0	12.0	14.0
R (db)	0.039	0.209	0.497	0.851	1.229	1.608
Z_1	1.1652	1.1730	1.1790	1.1841	1.1888	1.1933
Z_2	1.0771	1.0932	1.1060	1.1168	1.1261	1.1346
Z_3	1.0282	1.0424	1.0555	1.0675	1.0784	1.0884
Z_4	1.0065	1.0142	1.0235	1.0335	1.0435	1.0533

TABLE IV

NORMALIZED EVEN-MODE IMPEDANCES OF FIVE-ELEMENT COUPLERS

			$N=5$, Coupling = 3.00 db				
BW	6.0	8.0	10.0	12.0	14.0	16.0	18.0
R (db)	0.042	0.123	0.239	0.377	0.524	0.673	0.818
Z_1	4.0463	4.2052	4.3384	4.4595	4.5767	4.6945	4.8155
Z_2	2.2021	2.4053	2.5831	2.7439	2.8936	3.0365	3.1751
Z_3	1.4917	1.6410	1.7815	1.9147	2.0424	2.1660	2.2869
Z_4	1.1831	1.2761	1.3726	1.4703	1.5682	1.6659	1.7633
Z_5	1.0510	1.0980	1.1557	1.2206	1.2903	1.3631	1.4381
			$N=5$, Coupling = 6.00 db				
BW	6.0	8.0	10.0	12.0	14.0	16.0	18.0
R (db)	0.062	0.184	0.360	0.567	0.790	1.017	1.241
Z_1	2.4247	2.4839	2.5326	2.5762	2.6178	2.6592	2.7012
Z_2	1.6779	1.7760	1.8592	1.9326	1.9995	2.0622	2.1221
Z_3	1.3064	1.3920	1.4702	1.5423	1.6099	1.6741	1.7356
Z_4	1.1208	1.1800	1.2400	1.2995	1.3580	1.4152	1.4713
Z_5	1.0349	1.0668	1.1055	1.1484	1.1939	1.2406	1.2880
			$N=5$, Coupling = 10.0 db				
BW	6.0	8.0	10.0	12.0	14.0	16.0	18.0
R (db)	0.075	0.222	0.432	0.682	0.952	1.226	1.499
Z_1	1.6992	1.7237	1.7436	1.7612	1.7778	1.7941	1.8104
Z_2	1.3696	1.4173	1.4568	1.4911	1.5218	1.5501	1.5767
Z_3	1.1781	1.2248	1.2664	1.3040	1.3386	1.3709	1.4013
Z_4	1.0730	1.1077	1.1422	1.1759	1.2083	1.2395	1.2696
Z_5	1.0216	1.0411	1.0646	1.0903	1.1171	1.1443	1.1715
			$N=5$, Coupling = 15.0 db				
BW	6.0	8.0	10.0	12.0	14.0	16.0	18.0
R (db)	0.081	0.238	0.465	0.735	1.025	1.321	1.616
Z_1	1.3388	1.3494	1.3579	1.3654	1.3723	1.3791	1.3859
Z_2	1.1902	1.2129	1.2315	1.2473	1.2614	1.2742	1.2862
Z_3	1.0954	1.1193	1.1402	1.1589	1.1758	1.1914	1.2060
Z_4	1.0401	1.0587	1.0770	1.0945	1.1113	1.1272	1.1424
Z_5	1.0120	1.0228	1.0357	1.0496	1.0640	1.0784	1.0927
			$N=5$, Coupling = 20.0 db				
BW	6.0	8.0	10.0	12.0	14.0	16.0	18.0
R (db)	0.082	0.244	0.476	0.751	1.048	1.351	1.653
Z_1	1.1771	1.1822	1.1864	1.1900	1.1934	1.1967	1.1999
Z_2	1.1024	1.1141	1.1236	1.1316	1.1387	1.1452	1.1512
Z_3	1.0524	1.0652	1.0763	1.0861	1.0950	1.1031	1.1106
Z_4	1.0223	1.0325	1.0425	1.0520	1.0610	1.0695	1.0775
Z_5	1.0067	1.0127	1.0199	1.0276	1.0355	1.0433	1.0511

TABLE V

NORMALIZED EVEN-MODE IMPEDANCES OF SIX-ELEMENT COUPLERS

$N=6$, Coupling = 3.00 db

BW	8.0	10.0	12.0	14.0	16.0	18.0	20.0
R (db)	0.060	0.134	0.232	0.344	0.465	0.588	0.711
Z_1	4.2976	4.4178	4.5218	4.6176	4.7105	4.8038	4.8992
Z_2	2.5378	2.7112	2.8640	3.0027	3.1320	3.2551	3.3743
Z_3	1.7504	1.8944	2.0274	2.1520	2.2704	2.3841	2.4944
Z_4	1.3572	1.4618	1.5639	1.6635	1.7606	1.8559	1.9496
Z_5	1.1540	1.2224	1.2946	1.3687	1.4441	1.5200	1.5962
Z_6	1.0523	1.0905	1.1363	1.1875	1.2426	1.3005	1.3603

$N=6$, Coupling = 6.00 db

BW	8.0	10.0	12.0	14.0	16.0	18.0	20.0
R (db)	0.090	0.201	0.348	0.518	0.700	0.888	1.077
Z_1	2.5164	2.5596	2.5965	2.6301	2.6623	2.6944	2.7270
Z_2	1.8357	1.9145	1.9821	2.0424	2.0976	2.1494	2.1988
Z_3	1.4506	1.5280	1.5976	1.6613	1.7206	1.7766	1.8301
Z_4	1.2284	1.2913	1.3512	1.4084	1.4632	1.5160	1.5672
Z_5	1.1026	1.1469	1.1928	1.2393	1.2857	1.3318	1.3773
Z_6	1.0360	1.0620	1.0929	1.1271	1.1635	1.2012	1.2397

$N=6$, Coupling = 10.0 db

BW	8.0	10.0	12.0	14.0	16.0	18.0	20.0
R (db)	0.109	0.242	0.418	0.623	0.843	1.070	1.299
Z_1	1.7368	1.7543	1.7691	1.7825	1.7952	1.8077	1.8203
Z_2	1.4452	1.4820	1.5130	1.5403	1.5650	1.5878	1.6094
Z_3	1.2554	1.2958	1.3313	1.3633	1.3926	1.4198	1.4455
Z_4	1.1350	1.1704	1.2035	1.2346	1.2639	1.2916	1.3182
Z_5	1.0624	1.0887	1.1155	1.1423	1.1686	1.1944	1.2196
Z_6	1.0223	1.0383	1.0571	1.0778	1.0994	1.1216	1.1440

$N=6$, Coupling = 15.0 db

BW	8.0	10.0	12.0	14.0	16.0	18.0	20.0
R (db)	0.117	0.261	0.450	0.670	0.907	1.153	1.400
Z_1	1.3550	1.3624	1.3687	1.3743	1.3796	1.3848	1.3900
Z_2	1.2259	1.2430	1.2573	1.2697	1.2808	1.2910	1.3005
Z_3	1.1346	1.1547	1.1721	1.1876	1.2016	1.2145	1.2266
Z_4	1.0731	1.0915	1.1086	1.1244	1.1392	1.1530	1.1661
Z_5	1.0343	1.0486	1.0629	1.0771	1.0909	1.1043	1.1172
Z_6	1.0124	1.0213	1.0316	1.0429	1.0546	1.0664	1.0783

$N=6$, Coupling = 20.0 db

BW	8.0	10.0	12.0	14.0	16.0	18.0	20.0
R (db)	0.119	0.266	0.460	0.685	0.928	1.179	1.431
Z_1	1.1850	1.1886	1.1916	1.1943	1.1969	1.1994	1.2019
Z_2	1.1207	1.1294	1.1366	1.1429	1.1484	1.1535	1.1583
Z_3	1.0733	1.0839	1.0930	1.1011	1.1083	1.1150	1.1211
Z_4	1.0403	1.0503	1.0595	1.0679	1.0758	1.0831	1.0899
Z_5	1.0191	1.0270	1.0348	1.0425	1.0500	1.0572	1.0642
Z_6	1.0070	1.0119	1.0176	1.0239	1.0303	1.0368	1.0433

TABLE VI

COMPARISON OF RESULTS OBTAINED BY PREVIOUS COMPUTATION[1] WITH THE COMPUTER RESULTS

N	2		2		2		3	
BW	4		4		6.677		6	
C (db)	10		20		3.235		20	
	Results[1]	Computer	Results[1]	Computer	Results[1]	Computer (interpolated)	Results[1]	Computer
R (db)	0.5	0.528	0.6	0.580	0.835	0.832	0.5	0.530
Z_1	1.562	1.5729	1.143	1.1496	3.52	3.59	1.164	1.1680
Z_2	1.154	1.1573	1.046	1.0466	1.69	1.70	1.080	1.0816
Z_3	—	—	—	—	—	—	1.030	1.0303

TAPERED LINES AND COUPLERS VIII

If the asymmetrical coupler is designed from a quarter-wave transformer prototype of very low VSWR, or better still, if its design is based on a long, gradual taper (instead of a quarter-wave transformer) then the coupler would look like a pair of coupled lines approaching very gradually and separating very suddenly. The prototype tapered transmission line then has essentially a single sharply defined mismatch, where the tapered end suddenly reverts to the input impedance Z_0 (see Fig. 28 of Ref. 11, reproduced above as part of Fig.2). The mismatch of the prototype circuit (corresponding to the coupling in the coupler) is then almost entirely due to one impedance step, and is nearly constant in amplitude and phase. Let us now add a length of line on the other side of the big impedance step, equal to the length of the tapered section (or equal to the length of the quarter-wave transformer). Furthermore, let us now measure the transmission coefficient at the end of the added line, and similarly let us measure the reflection coefficient at the beginning of the taper; these two positions are on opposite sides of the big impedance step, and equidistant from it (in opposite directions). Under those conditions (provided the taper or transformer is gradual enough) the reflection and transmission coefficients must always be in phase (or out of phase, depending on the definition of phase of reflection coefficient). For the coupler derived from such a prototype, the two coupled waves are either in phase or out of phase, thus resulting in a magic T. DuHamel and Armstrong (p. 207) appear to have been the first to point this out. Their often referenced paper is published here for the first time in an easily available publication. Note that when two 8.34-dB tapered-line couplers are cascaded, the two joining lengths of line can be shortened (so long as each is shortened by the same amount) to make a more compact 3-dB magic T.

Theoretically there may be little to choose between a taper and a stepped-impedance transformer, but in practice when two coupled lines come closer in steps, those steps cause "non-theoretical" discontinuities which have to be matched out rather carefully. One mostly uses tabs on the strips at the corners, but remember that those tabs must match **both** the even and the odd modes. Moreover, if there is any differential phase shift between the even and odd modes due to the tabs, then the coupler performance deteriorates, especially with regard to directivity (or isolation). For these reasons, a tapered design is usually preferred over a stepped design. Because of the importance of the tapered coupler or magic T, we have also reprinted two papers (p. 201), (p. 206) that deal with the design of the taper, although the former was written before these couplers were thought of.

The two papers by Tresselt (p. 234), (p. 244) also deal with couplers using tapered lines. Two related papers by Millican and Wales[25] and by Kammler[26] have not been reproduced.

The idea of cascading two couplers of lower coupling to obtain one coupler of greater coupling has already been explained in connection with Section VI. When designing 3-dB quadrature couplers, it is usual to cascade two couplers of 8.34 dB. It is even more necessary to cascade weaker couplers in the case of asymmetrical couplers giving in-phase couplers (or magic T's), than was the case for symmetrical couplers (giving quadrature hybrids). The reason is that in the case of an in-phase coupler the coupling is essentially due to a single step, requiring much tighter coupling there than in the case of a quadrature coupler (where the coupling is essentially distributed over an entire quarter-wave section).

A Transmission Line Taper of Improved Design*

R. W. KLOPFENSTEIN†

Summary—**The theory of the design of optimal cascaded transformer arrangements can be extended to the design of continuous transmission-line tapers. Convenient relationships have been obtained from which the characteristic impedance contour for an optimal transmission-line taper can be found.**

The performance of the Dolph-Tchebycheff transmission-line taper treated here is optimum in the sense that it has minimum reflection coefficient magnitude in the pass band for a specified length of taper, and, likewise, for a specified maximum magnitude reflection coefficient in the pass band, the Dolph-Tchebycheff taper has minimum length.

A sample design has been carried out for the purposes of illustration, and its performance has been compared with that of other tapers. In addition, a table of values of a transcendental function used in the design of these tapers is given.

Introduction

THE ANALYSIS of nonuniform transmission lines has been a subject of interest for a considerable period of time. One of the uses for such nonuniform lines is in the matching of unequal resistances over a broadband of frequencies. It has recently been shown that the theory of Fourier transforms is applicable to the design of transmission-line tapers.[1] It is the purpose of this paper to present a transmission-line taper design of improved characteristics. The performance of this taper is optimum in the sense that for a given taper length the input reflection coefficient has minimum magnitude throughout the pass band, and for a specified tolerance of the reflection coefficient magnitude the taper has minimum length.

For any transmission line system the applicable equations are

$$\frac{dV}{dx} = -ZI$$
$$\frac{dI}{dx} = -YV, \tag{1}$$

where

V = the voltage across the transmission line,
I = the current in the transmission line,
Z = the series impedance per unit length of line,
and
Y = the shunt admittance per unit length of line.

Fig. 1 illustrates the configuration to which the above equations are to be applied.

For nonuniform lines, the quantities Z and Y are known nonconstant functions of position along the line, and the properties of the system are determined through a solution of (1) along with the pertinent boundary conditions. Through use of the waveguide formalism[2] (1) is applicable to uniconductor waveguide as well as to transmission line. Strictly speaking, of course, (1) is not precisely applicable to any system since it accounts for the propagation of a single mode only. It furnishes an excellent description, however, as long as all modes but dominant mode are well below cutoff.

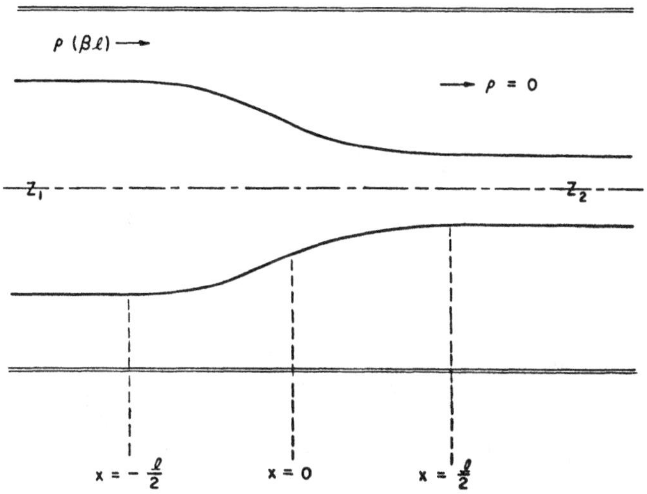

Fig. 1—Tapered transmission-line matching section.

Eq. (1) can be recast in a more directly useful form through the introduction of the quantities

$\gamma = \sqrt{ZY}$ = the propagation constant of the line,

$Z_0 = \sqrt{Z/Y}$ = the characteristic impedance of the line,

and

$$\rho = \frac{V/I - Z_0}{V/I + Z_0} = \text{the reflection coefficient at any point along the line.} \tag{2}$$

These lead to first order nonlinear differential equation[3]

$$\frac{d\rho}{dx} - 2\gamma\rho + \frac{1}{2}(1-\rho^2)\frac{d(\ln Z_0)}{dx} = 0. \tag{3}$$

This equation has the advantage that it is in terms of the quantity of direct interest in impedance matching problems. Likewise, a very natural approximation for impedance matching purposes can be made directly in this equation. If it is assumed that $\rho^2 \ll 1$, (3) becomes

$$\frac{d\rho}{dx} - 2\gamma\rho + F(x) = 0,$$

* Original manuscript received by the IRE, June 9, 1955.
† RCA Labs., Princeton, N. J.

[1] F. Bolinder, "Fourier transforms in the theory of inhomogeneous transmission lines," PROC. IRE, vol. 38, p. 1354; November, 1950.

[2] N. Marcuvitz, "Waveguide Handbook," McGraw-Hill Book Co., Inc., New York, N. Y., ch. 1, p. 7; 1951.

[3] L. R. Walker and N. Wax, "Nonuniform transmission lines and reflection coefficients," *Jour. Appl. Phys.*, vol. 17, pp. 1043–1045; December, 1946.

where

$$F(x) = \frac{1}{2}\,\frac{d(\ln Z_0)}{dx}, \tag{4}$$

which is a first-order linear differential equation in ρ.

Solution of the Differential Equation

A solution of the differential equation (4) is sought which satisfies the boundary condition $\rho = 0$ at $x = l/2$. An integrating factor for this equation is

$$G(x) = \exp\left[-2\int_a^x \gamma d\xi\right], \tag{5}$$

where the lower limit of the integral is arbitrary. Applying this to (4) it is found that the solution satisfying the boundary condition is given by[4]

$$\rho(x) = \int_x^{l/2} F(z)\exp\left[-2\int_x^z \gamma(\xi)d\xi\right]dz, \tag{6}$$

and, hence, the input reflection coefficient is

$$\rho = \int_{-l/2}^{l/2} F(z)\exp\left[-2\int_{-l/2}^z \gamma(\xi)d\xi\right]dz. \tag{7}$$

The solution given above is subject only to the restriction that the reflection coefficient is relatively small. It is equally applicable to lossless transmission line, lossy transmission line, and waveguide tapers. The physical interpretation of the solution as given is evident. The incremental reflection at each cross section is given by $F(z)$, and the exponential term expresses the total delay and attenuation of this reflected component at the input of the tapered section relative to incident input wave.

Optimal Design of Transmission-Line Tapers

An important special case of the general situation considered above is the lossless transmission-line taper as illustrated in Fig. 1. In this case, the characteristic impedance is a real number and is independent of frequency. The wave propagated in the line is essentially TEM in character, and the propagation constant is purely imaginary and proportional to the frequency.

Under these conditions the interior integration of (7) can be carried out, and the input reflection coefficient becomes

$$\rho \exp(j\beta l) = \int_{-l/2}^{l/2} F(z)\exp(-j2\beta z)dz. \tag{8}$$

This relationship can be inverted through the theory of Fourier transforms to obtain

$$F(x) = \frac{1}{\pi}\int_{-\infty}^{\infty}[\rho\exp(j\beta l)]\exp(j2\beta x)d\beta. \tag{9}$$

The analogy between the present problem and the synthesis of radiation patterns from line sources is evident. In each case, the quantities of interest are related through the Fourier transform.

In the use of (9) the reflection coefficient is specified so that its value at negative frequencies is equal to the complex conjugate of its value at the corresponding positive frequencies. This is necessary in order that the specified reflection coefficient shall correspond to a physically realizable structure.[5] This requirement then insures that the transform will be completely real as it must be in order that $F(x)$ have significance in terms of a transmission-line taper.

Collin has recently shown that optimum performance is obtained from a cascaded transformer structure when the power loss ratio is expressed in terms of the Tchebycheff polynomial of degree equal to the number of sections.[6] This is equivalent to having the input reflection coefficient proportional to the Tchebycheff polynomial of the same degree, when its square is small relative to one [an assumption already made in the derivation of transform pair (8) and (9)].

By allowing the number of sections to increase indefinitely for a fixed over-all length, the results of Collin can be extended to the case of a continuous transmission-line taper. In the case of a cascaded transformer arrangement, a secondary maximum in the reflection coefficient magnitude occurs at the first and at all succeeding frequencies where the individual section lengths become equal to a multiple of a half-wavelength. As the number of sections is allowed to increase without limit for a fixed over-all length, the frequency at which this first secondary maximum occurs also increases without limit so that the pass band consists of all frequencies beyond that for which the reflection coefficient first comes within the specified tolerance.

For maximum bandwidth with a fixed maximum magnitude of reflection coefficient then, input reflection coefficient for a continuous taper takes form

$$\rho\exp(j\beta l) = \rho_0 \frac{\cos\left[\sqrt{(\beta l)^2 - A^2}\right]}{\cosh(A)}, \tag{10}$$

which is the limiting form of the Tchebycheff polynomial as its degree increases without limit.[7] The specification of the parameter A determines the maximum magnitude of reflection coefficient in the pass band which consists of all frequencies such that $\beta l \geqq A$. The reflection coefficient magnitude takes on its maximum value $|\rho_0|$ at zero frequency, and it oscillates in the pass band with constant amplitude equal to $\rho_0/\cosh(A)$. A plot of the function given by (10) is shown in Fig. 2 for a number of different values of A.

The inversion of the above specified reflection coefficient through (9) yields[7]

[4] L. R. Ford, "Differential Equations," McGraw-Hill Book Co., Inc., New York, N. Y., ch. 2, pp. 36–39; 1933.

[5] H. W. Bode, "Network Analysis and Feedback Amplifier Design," D. Van Nostrand Co., Inc., New York, N. Y., ch. 7, p. 106; 1945.

[6] R. E. Collin, "Theory and design of wide-band multisection quarter-wave transformers," Proc. IRE, vol. 43, pp. 179–185; February, 1955.

[7] T. T. Taylor, "Dolph arrays of many elements," Tech. Memo. No. 320, Hughes Aircraft Co., Culver City, Calif.; August 18, 1953.

$$F(x) = \frac{\rho_0}{\cosh(A)}\left\{\frac{A^2}{l}\frac{I_1[A\sqrt{1-(2x/l)^2}]}{A\sqrt{1-(2x/l)^2}} + \frac{1}{2}\left[\delta\left(x-\frac{l}{2}\right)+\delta\left(x+\frac{l}{2}\right)\right]\right\}, \quad |x| \leq l/2,$$
$$= 0, \quad |x| > l/2, \tag{11}$$

where I_1 is the first kind of modified Bessel function of the first order, and δ is the unit impulse function.

The variation of characteristic impedance along the taper can be found by direct integration of $F(x)$, and it is given by

$$\ln(Z_0) = \frac{1}{2}\ln(Z_1Z_2) + \frac{\rho_0}{\cosh(A)}\left\{A^2\phi(2x/l, A) + U\left(x-\frac{l}{2}\right) - U\left(-x-\frac{l}{2}\right)\right\}, \quad |x| \leq l/2,$$
$$= \ln(Z_2), \quad x > l/2,$$
$$= \ln(Z_1), \quad x < -l/2. \tag{12}$$

U is the unit step function defined by,

$$U(z) = 0, \quad z < 0,$$
$$U(z) = 1, \quad z \geqq 0, \tag{13}$$

and ϕ is defined by

$$\phi(z, A) = -\phi(-z, A) = \int_0^z \frac{I_1(A\sqrt{1-y^2})}{A\sqrt{1-y^2}}\,dy, \quad |z| \leqq 1. \tag{14}$$

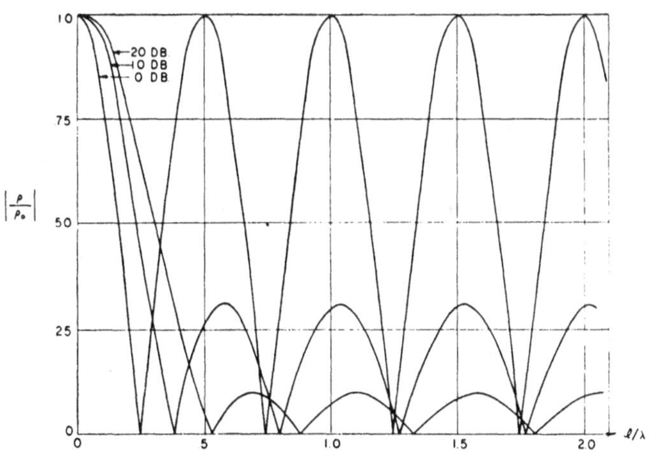

Fig. 2—Response of Dolph-Tchebycheff transmission-line tapers.

Eq. (12) furnishes the information required for the design of a Dolph-Tchebycheff tapered transition. The quantity ρ_0 is determined by the two impedances Z_2 and Z_1 which are to be matched, and A is selected on the basis of the allowed maximum reflection coefficient magnitude in the pass band. One of the interesting aspects of this design is that the taper has a discontinuous change of characteristic impedance at each end as well as a continuous change along the length of the taper. It is interesting to note that when the tolerated reflection coefficient approaches the initial reflection coefficient ρ_0, the parameter A approaches zero, and the bandwidth comprises all frequencies from zero to infinity. In this case, the Dolph-Tchebycheff taper design degenerates into the usual quarter-wavelength transformer design with a discontinuous change of characteristic impedance at each end and a constant characteristic impedance at intermediate points.

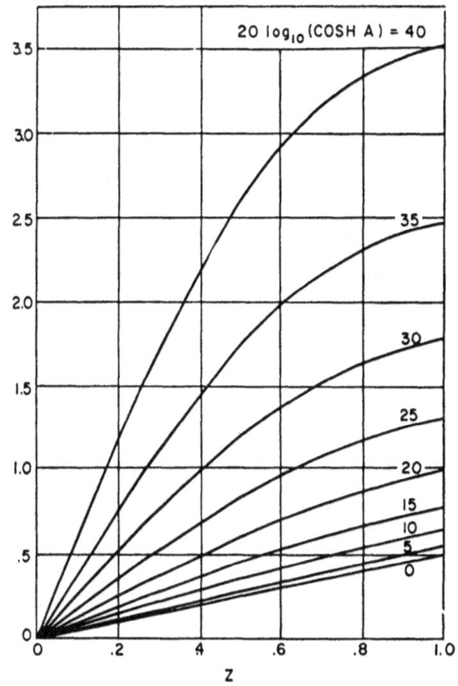

Fig. 3—Plot of the function $\phi(z, A)$.

The function $\phi(z, A)$ is not expressible in closed form except for special values of the parameters. Therefore, this function has been computed through standard integration formulas on an IBM CPC digital computer for a suitable range of values of the parameters. Tabulated values of $\phi(z, A)$ are given to six decimal places in Table I, p. 34, and the function is shown in Fig. 3. The special closed-form relationships

$$\phi(0, A) = 0,$$
$$\phi(z, 0) = z/2,$$

and

$$\phi(1, A) = \frac{\cosh(A) - 1}{A^2}, \tag{15}$$

are obtained for the end points of the parameter ranges.

One more comment should be made in regard to the application of the preceding design procedure. If one uses the natural value

$$\rho_0 = \frac{Z_2 - Z_1}{Z_2 + Z_1}, \tag{16}$$

in entering the equation (12), it will be found that the designed taper does not quite fit the final impedances

TABLE I

VALUES OF THE FUNCTION $\phi(z, A)$ FOR $z=0(0.05)1.00$ AND $20 \log_{10} (\cosh A)=0(5)40$

	$\phi(z, A)$								
	$20 \log_{10} (\cosh A)$								
z	0	5	10	15	20	25	30	35	40
0.00	0.000000	0.000000	0.000000	0.000000	0.000000	0.000000	0.000000	0.000000	0.000000
0.05	0.025000	0.029593	0.036848	0.048140	0.065590	0.092539	0.134313	0.199460	0.301772
0.10	0.050000	0.059161	0.073629	0.096137	0.130902	0.184564	0.267698	0.397268	0.600625
0.15	0.075000	0.088681	0.110276	0.143848	0.195661	0.275567	0.399242	0.591802	0.893707
0.20	0.100000	0.118128	0.146721	0.191132	0.259597	0.365055	0.528062	0.781511	1.178306
0.25	0.125000	0.147479	0.182899	0.237850	0.322448	0.452552	0.653321	0.964936	1.451913
0.30	0.150000	0.176708	0.218746	0.283869	0.383962	0.537610	0.774237	1.140744	1.712272
0.35	0.175000	0.205794	0.254197	0.329059	0.443899	0.619809	0.890101	1.307751	1.957437
0.40	0.200000	0.234711	0.289191	0.373296	0.502035	0.698767	1.000282	1.464942	2.185803
0.45	0.225000	0.263438	0.323667	0.416460	0.558163	0.774142	1.104238	1.611487	2.396134
0.50	0.250000	0.291950	0.357568	0.458441	0.612094	0.845635	1.201523	1.746753	2.587582
0.55	0.275000	0.320226	0.390837	0.499134	0.663658	0.912994	1.291792	1.870306	2.759684
0.60	0.300000	0.348244	0.423420	0.538444	0.712709	0.976019	1.374800	1.981918	2.912359
0.65	0.325000	0.375982	0.455266	0.576284	0.759120	1.034555	1.450409	2.081555	3.045886
0.70	0.350000	0.403418	0.486328	0.612574	0.802790	1.088504	1.518582	2.169376	3.160875
0.75	0.375000	0.430533	0.516559	0.647248	0.843641	1.137814	1.579377	2.245710	3.258228
0.80	0.400000	0.457305	0.545918	0.680245	0.881619	1.182484	1.632947	2.311049	3.339098
0.85	0.425000	0.483716	0.574365	0.711518	0.916692	1.222564	1.679531	2.366019	3.404835
0.90	0.450000	0.509746	0.601865	0.741027	0.948855	1.258145	1.719443	2.411361	3.456938
0.95	0.475000	0.535377	0.628386	0.768745	0.978123	1.289363	1.753065	2.447905	3.497000
1.00	0.500000	0.560591	0.653899	0.794653	1.004533	1.316391	1.780835	2.476547	3.526658

Z_2 and Z_1 at the end points. This fact is an evidence of the approximation $\rho^2 \ll 1$ which was made at the outset in differential equation (4). Discrepancy becomes larger as value of magnitude of ρ_0 increases. This design inconvenience can be eliminated, however, by taking

$$\rho_0 = \tfrac{1}{2} \ln (Z_2/Z_1), \tag{17}$$

as the initial value of the reflection coefficient instead of the true value of (16). The two expressions are identical to a second order of approximation for small differences between Z_2 and Z_1, and the use of the second expression will yield a taper design which exactly fits its end-point
204 impedances for all values of Z_2 and Z_1. The effect of the approximation $\rho^2 \ll 1$ will then be evidenced by a slight deviation from the performance given by (10) in the low-frequency range outside the pass band.

COAXIAL TRANSMISSION-LINE TAPER FROM 50 TO 75 OHMS

As an application of the preceding results, the design of an optimal 50–75 ohm coaxial transmission line taper will be indicated in detail. The taper is to be designed so that the input reflection coefficient magnitude does not exceed about one per cent in the pass band.

The initial value of the reflection coefficient in this case is equal to 0.2. The value of ρ_0 for use in the design of the taper is found from (17) to be

$$\rho_0 = \tfrac{1}{2} \ln(1.5) = 0.20274. \tag{18}$$

As observed previously, this value does not differ markedly from the zero frequency reflection coefficient.

It will be required that the maximum reflection coefficient magnitude in the pass band shall not exceed one-twentieth of ρ_0. Thus, from (10)

$$\cosh (A) = 20,$$

so that

$$A = 3.6887. \tag{19}$$

The characteristic impedance contour can now be obtained directly from (12). The resulting Z_0 curve is illustrated in Fig. 4, and the corresponding coaxial line-conductor contour is shown in Fig. 5.

Characteristic impedance has a discontinuous jump from 50 to 50.52 ohms at left-hand end and a corresponding jump from 74.24 to 75 ohms at right-hand end. Characteristic impedance at center of taper is equal to 61.24 ohms, geometric mean between 50 and 75 ohms.

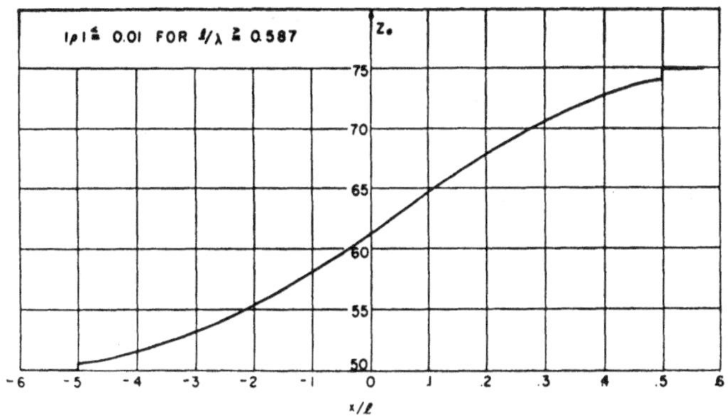

Fig. 4—50–75 ohm Dolph-Tchebycheff tapered transition.

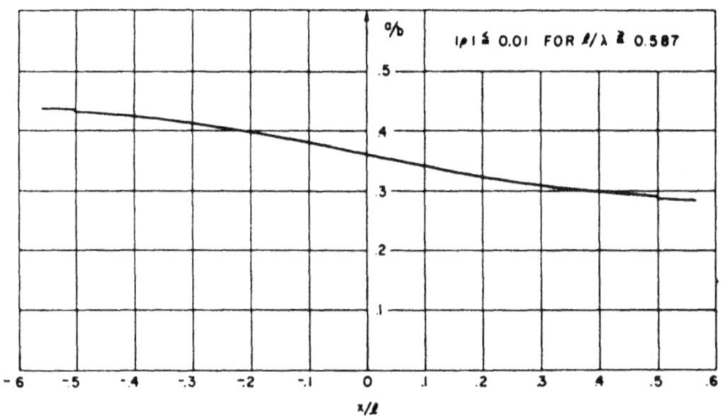

Fig. 5—Inner conductor contour for 50–75 ohm Dolph-Tchebycheff coaxial-line transition.

The performance of this taper is plotted in Fig. 6. The pass band consists of all frequencies greater than that for which $\ell/\lambda = 0.587$. For comparison, the performance of an exponential taper[8] and a hyperbolic taper[9] has been indicated on the same curve.

[8] C. R. Burrows, "The exponential transmission line," *Bell Sys. Tech. Jour.*, vol. 17, pp. 555–573; October, 1938.

[9] H. J. Scott, "The hyperbolic transmission line as a matching section," PROC. IRE, vol. 41, pp. 1654–1657; November, 1953.

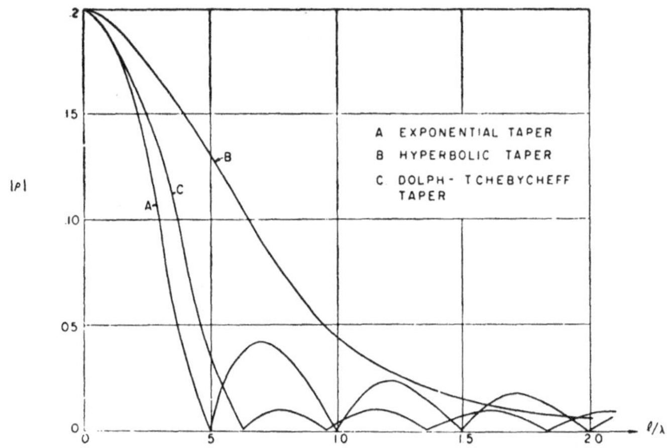

Fig. 6—Performance of 50–75 ohm Dolph-Tchebycheff tapered transition.

Conclusion

The theory of the design of optimal cascaded transformer arrangements can be extended to the design of continuous transmission-line tapers. Convenient relationships have been obtained from which the characteristic impedance contour for an optimal transmission-line taper can be found. Alternatively, this impedance contour can be thought of as the envelope of the pointwise specified characteristic impedance of a discrete cascaded transformer arrangement.

The performance of the Dolph-Tchebycheff transmission-line taper treated here is optimum in the sense that it has minimum reflection coefficient magnitude in the pass band for a specified length of taper, and, likewise, for specified maximum magnitude reflection coefficient in the pass band the Dolph-Tchebycheff taper has minimum length.

A sample design has been carried out for the purposes of illustration, and its performance has been compared with that of other tapers. In addition, a table of values of a transcendental function used in the design of these tapers is given in Table I.

Excerpted from Proc. IEEE, Vol. 56, No. 9, pp. 1629-1630 (September 1968).

Extremely Rapid Computation of the Klopfenstein Impedance Taper

Abstract—**The Klopfenstein impedance taper is widely used in the design of couplers, baluns, mixers, etc. However, its utility is limited by computational complexity. It is suggested that the application of some preliminary numerical analysis can remove all computational difficulties from the computer evaluation of the characteristic impedance.**

In a 1956 paper in the PROCEEDINGS, R. W. Klopfenstein[1] derived an expression for the variation of characteristic impedance with position along a tapered transmission line which is optimally designed in the "equal-ripple" sense.

In order to apply Klopfenstein's results, it is necessary to evaluate a function $\phi(z, A)$ which is defined as the integral of a modified Bessel function. Klopfenstein computes and tabulates this function by numerical integration techniques. It is the purpose of this letter to suggest an alternative to this time-consuming relatively complicated computational procedure. Indeed, by applying some preliminary numerical analysis, the computer evaluation of the function $\phi(z, A)$ can be made as simple and rapid as the evaluation of any elementary transcendental function sine, cosine, exponential, etc.

The function $\phi(z, A)$ is defined by

$$\phi(z, A) = \int_0^z \frac{I_1(A\sqrt{1 - y^2})dy}{A\sqrt{1 - y^2}}.$$

Now the incomplete Bessel function may be expanded in a power series[2]

$$\phi(z, A) = \int_0^z \frac{1}{2} \sum_{k=0}^{\infty} \frac{\left(\frac{A^2}{4}\right)^k (1 - y^2)^k}{k!(k + 1)!} dy.$$

Next, since the series convergence is uniform for y between -1 and $+1$, the series may be integrated term by term:

$$\phi(z, A) = \sum_{k=0}^{\infty} a_k b_k$$

where

$$a_k = \frac{A^{2k}}{4^k k!(k + 1)!}$$

and

$$b_k = \frac{1}{2}\int_0^z (1 - y^2)^k \, dy.$$

Next, recursion relations for a_k and b_k may be obtained, the b_k recursion being obtained through an integration by parts:

$$a_0 = 1; \quad a_k = \frac{A^2}{4k(k + 1)} a_{k-1}$$

$$b_0 = \frac{z}{2}; \quad b_k = \frac{\frac{z}{2}(1 - z^2)^k + 2kb_{k-1}}{2k + 1}.$$

Therefore, the evaluation of the ϕ-function may be performed by the following simple FORTRAN IV subprogram:

```
      FUNCTION PHI (Z,A)
      AK=1.0
      BK=0.5*Z
      CK=BK
      PHI=BK
      DO 100 K=1,20
      FK=K
      CK=CK*(1.0-Z**2)
      BK=(CK+2.0*FK*BK)/(2.0*FK+1.0)
      AK=A**2/(4.0*FK*(FK+1.0))*AK
      PHI=PHI+AK*BK
      IF (ABS(AK*BK) .LT. 1.0E-7) RETURN
  100 CONTINUE
      RETURN
      END
```

This routine has been used extensively. The results are in perfect agreement with Klopfenstein's tabulated results, and the computational time is some small fraction of a millisecond on either IBM 7094 or CDC 3600 computers.

MILTON A. GROSSBERG
Alto Systems Engineers, Inc.
Sunnyvale, Calif. 94086

Manuscript received June 10, 1968; revised June 27, 1968.

[1] R. W. Klopfenstein, "A transmission line taper of improved design," *Proc. IRE*, vol. 44, pp. 31–35, January 1956.

[2] M. Abramovitz and I. A. Stegun, Eds., *Handbook of Mathematical Functions*, Appl. Math. Ser. 55. Washington, D. C.: NBS, June 1964, p. 375, eq. 9.6.10.

THE TAPERED-LINE MAGIC-T*

R.H. DuHamel and M.E. Armstrong
Hughes Aircraft Company
Ground Systems Group
Fullerton, California

I INTRODUCTION

Since the introduction of log-spiral[1] and log-periodic[2] antennas, great progress has been made in the field of extremely broadband antennas and microwave components. This has allowed the development of many communication, search, ELINT, ECM, ECCM, homing, and tracking antenna systems with frequency bandwidths exceeding 10:1. Some of these antenna systems require the use of signal processing circuit components (such as hybrids and phase difference circuits) with comparable bandwidths. Due to the efforts of Shimizu, Jones, Cohn, Young, Shelton and many others (see References 3 and 4 for extensive bibliographies), multi-element symmetrical quadrature TEM hybrids are available for practically any desired bandwidth. Size requirements usually restrict the use of these devices to frequencies above UHF. The work of Schiffman[5] and Shelton et al.[6] has led to constant phase difference circuits with similar bandwidths and frequency ranges. Extremely wide-band 3-dB

---

*This paper is a consolidation of material presented in Hughes Internal Document ID64-14-111, "The Tapered-Line Magic-T, dated October 29, 1964; a paper entitled "A Wide-Band Monopulse Antenna Utilizing the Tapered-Line Magic-T," presented at the 15th Annual USAF Antenna Symposium at Allerton Park, University of Illinois, 1965; and a paper entitled "Dual Polarization Log-Periodic Monopulse Antennas and Frequency-Independent Feed Networks for Multi-Element Arrays, which was included in "Electromagnetics and Antennas," a series of lectures from the University of Illinois, College of Engineering, Urbana, Illinois, 1967.

magic-T's* utilizing ferrite materials are available for frequencies up to 2 GHz.

Even though great progress has been made in the circuit area there is still an urgent need for broadband magic-T's with arbitrary power split and equal power split above 2 GHz, and compact quadrature hybrids and phase difference circuits for frequencies below UHF. Although a quadrature hybrid and 90° phase difference circuit can be combined to perform as a magic-T, it is desirable to have one device perform as a magic-T.

The subject of this paper is a novel device, termed the tapered-line magic-T,† which may be designed for any power split and practically any desired bandwidth. Theoretically, it is a high pass device in the sense that it performs in the desired manner above a frequency determined by the length of the coupler. With the proper choice of reference planes the coupled outputs are either in or out of phase. The magnitude of the coupled output may be controlled in a simple manner. The tapered-line magic-T is essentially a single reflection device as opposed to the multiple reflection symmetrical and asymmetrical hybrids. At first it might be considered as the limiting case of the asymmetric multi-element directional coupler described by Levy[7] as the number of elements approaches infinity. However, Young[8] has pointed out that the phase characteristics of the Levy coupler are incorrect for magic-T's and that the tapered line magic-T is the limiting case of an asymmetric coupler based on quarter-wave transformer designs. After the development of the tapered-line magic-T, a review of the literature revealed that it is quite similar to a coupler proposed by Oliver[9] (see his Fig. 8) thirteen years ago. His coupler used exponential variation of the coupling with distance rather than Tchebycheff variation and apparently he failed to realize that with the proper choice of reference planes the coupled outputs would be in or out of phase independent of frequency. This latter property might have been overlooked since his theoretical analysis was approximate to the extent that it held only for lightly coupled lines. Even so, this is just another example of the lethargy of our profession in extending or using the published ideas of others.

208 In the following, the design procedure and theoretical analysis of the tapered-line magic-T are given. Experimental results for a 3-dB magic-T for the frequency range of 2 to 12 GHz are presented in a concluding section.

---

*In this paper, the term "magic-T" denotes hybrids for which the coupled outputs are either in or out of phase. The term "quadrature hybrid" denotes hybrids for which the coupled outputs are 90° out of phase.

†U.S. Patent 3,432,775.

II DESCRIPTION OF TAPERED-LINE MAGIC-T

A schematic diagram of the tapered-line magic-T is given in Figure 1. The device consists of two coupled transmission lines of length θ on the left and two uncoupled transmission lines of impedance Z_o and length θ on the right. The transmission line system is symmetrical about the center line. Because of this symmetry the circuit is most easily analyzed in terms of the even and odd modes of excitation. For example, the excitation of port one with a unit incident wave is equivalent (by superposition) to excitation of ports one and two with even and odd incident waves of magnitude 1/2. All of the ports of the four-port circuit are terminated in impedances Z_o. The even and odd mode characteristic impedances of the coupled transmission lines are a function of the distance x along the lines. The coupled transmission lines are designed such that the square root of the even and odd mode impedances is equal to Z_o along the complete length of the coupled lines. Figure 2 illustrates typical characteristic impedance functions for the magic-T. Notice that at x = 0 the even and odd mode impedances are both equal to Z_o, that is, the lines are uncoupled at this point. At x = 1 the even and odd mode characteristic impedances are Z_o/k and kZ_o respectively. The ratio of the coupled outputs is determined by the parameter k. Above a certain frequency the coupled lines act essentially as perfect transformers for both the even and odd modes of excitation.

The scattering coefficients of the four-port tapered-line Magic-T are simply related[10] to the reflection and transmission coefficients for the even and odd modes of excitation. The reflection and transmission coefficients for one mode of excitation are easily determined by the analysis of a cascade of two single transmission lines. The scattering coefficients are given by

$$\begin{aligned} S_{11} &= S_{22} = \frac{1}{2}(\Gamma_{1e} + \Gamma_{1o}) \\ S_{12} &= \frac{1}{2}(\Gamma_{1e} - \Gamma_{1o}) \\ S_{13} &= \frac{1}{2}(T_{1e} - T_{1o}) \\ S_{12} &= \frac{1}{2}(T_{1e} + T_{1o}) \end{aligned} \tag{1}$$

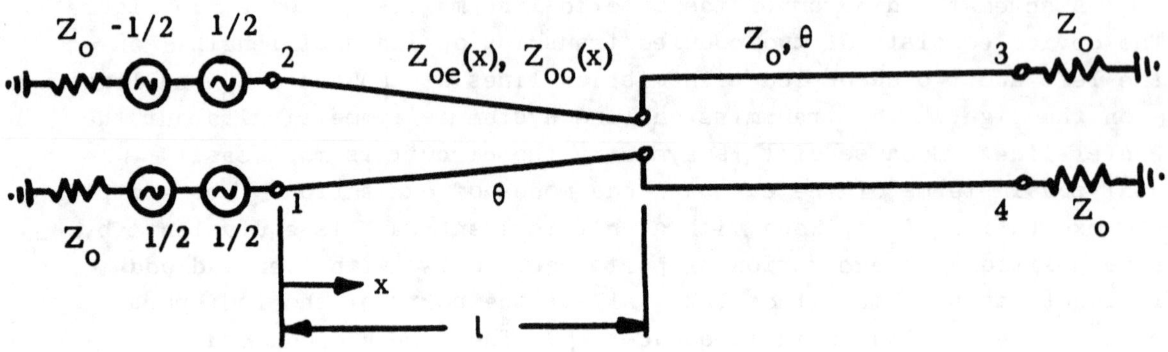

Figure 1. Schematic of Tapered-Line Magic-T

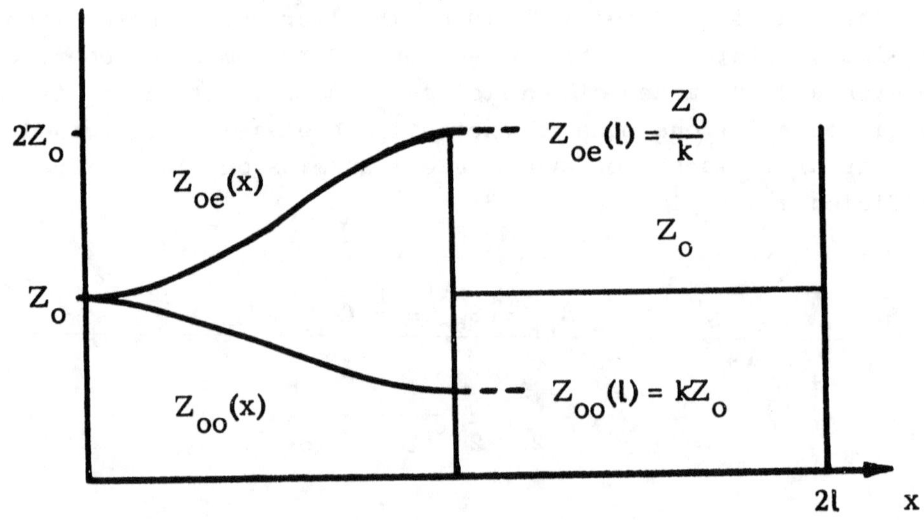

Figure 2. Impedance Variation Along Magic-T

where Γ_{1e} and Γ_{1o} are the input reflection coefficients and T_{1e} and T_{1o} are the transmission coefficients for the even and odd modes of excitation. Since the coupler is asymmetric, the scattering coefficients S_{44} and S_{43} are determined with even and odd generators connected to terminals three and four. The coefficients are given by

$$S_{33} = S_{44} = \frac{1}{2}(\Gamma_{2e} + \Gamma_{2o})$$
$$S_{43} = \frac{1}{2}(\Gamma_{2e} - \Gamma_{2o}) \qquad (2)$$

where the subscript 2 on Γ indicates that the generators are connected on the right side of the circuit. The remaining scattering coefficients are determined from symmetry conditions such as

$$S_{13} = S_{24}$$
$$S_{14} = S_{23} \quad , \text{ etc.} \qquad (3)$$

As will be shown the scattering matrix for the magic-T under ideal conditions is given by

$$S = \begin{vmatrix} 0 & -\alpha & 0 & \beta \\ -\alpha & 0 & \beta & 0 \\ 0 & \beta & 0 & \alpha \\ \beta & 0 & \alpha & 0 \end{vmatrix} \qquad (4)$$

where

$$\alpha = \frac{1-k}{1+k} e^{-j2\theta} \qquad \beta = \frac{2k^{1/2}}{1+k} e^{-j2\theta} \qquad (5)$$

Notice that this matrix exhibits the characteristic of a magic-T. That is, all of the ports are matched and there is isolation between diametrically opposed ports.

The performance of the magic-T is achieved by a reflection at the connection of the coupled and uncoupled lines. This is in contrast to the multi-element symmetric and asymmetric couplers wherein reflections occur at many impedance steps within the coupler. In the design of a tapered-line magic-T it is desired that the coupled lines act as reflectionless transformers for the even and odd modes.

III ANALYSIS OF THE MAGIC-T

In this section the magic-T will be analyzed on the assumption that the tapered lines act as perfect transformers. A simple explanation for the performance may be obtained as follows: with the circuit excited on the left in the even mode, a reflection will occur at the junction of the coupled and uncoupled lines. The reflection coefficient is given by

$$\Gamma_e = \frac{Z_o - \frac{Z_o}{k}}{Z_o + \frac{Z_o}{k}} = \frac{k - 1}{k + 1} \tag{6}$$

For the odd mode of excitation the reflection coefficient will be just the negative of Γ_e. Since the transformers are assumed to be perfect, the input reflection coefficient Γ_{1e} at terminal one is simply Γ_e delayed by 2θ degrees. Thus S_{12} in (4) follows immediately from (1). When the circuit is excited on the right, the reflection coefficients for the even and odd modes will be the negative of those when the circuit is excited from the left. Thus S_{34} of (4) follows immediately. The transmission coefficients and hence β may be obtained from the continuity of voltage and current at the reflection point. The reader may show that β is given as in (5).

For convenience the magic-T will be analyzed in terms of transmission matrices. The notations to be used are defined in Figure 3. Figure 3a illustrates the conventions for a two-port circuit for which the connecting transmission lines at ports one and two have characteristic impedances Z_1 and Z_2. The voltages and currents are related by the usual A B C D matrix.

$$\begin{bmatrix} V_1 \\ I_1 \end{bmatrix} = \begin{bmatrix} A & B \\ C & D \end{bmatrix} \begin{bmatrix} V_2 \\ -I_2 \end{bmatrix} \tag{7}$$

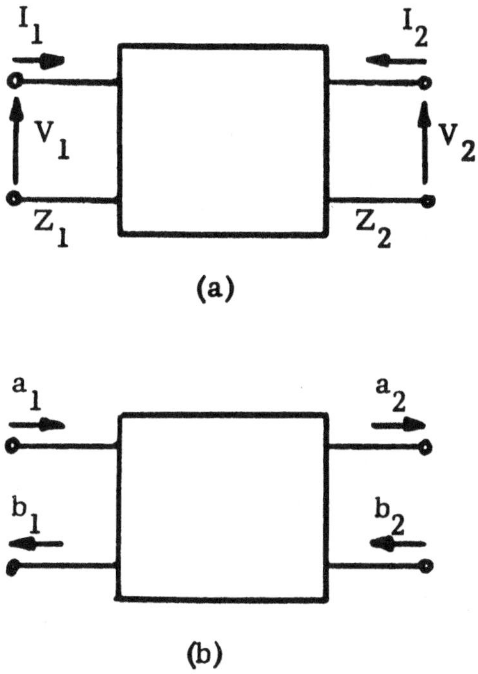

Figure 3. Two-Port Circuit Reference Directions

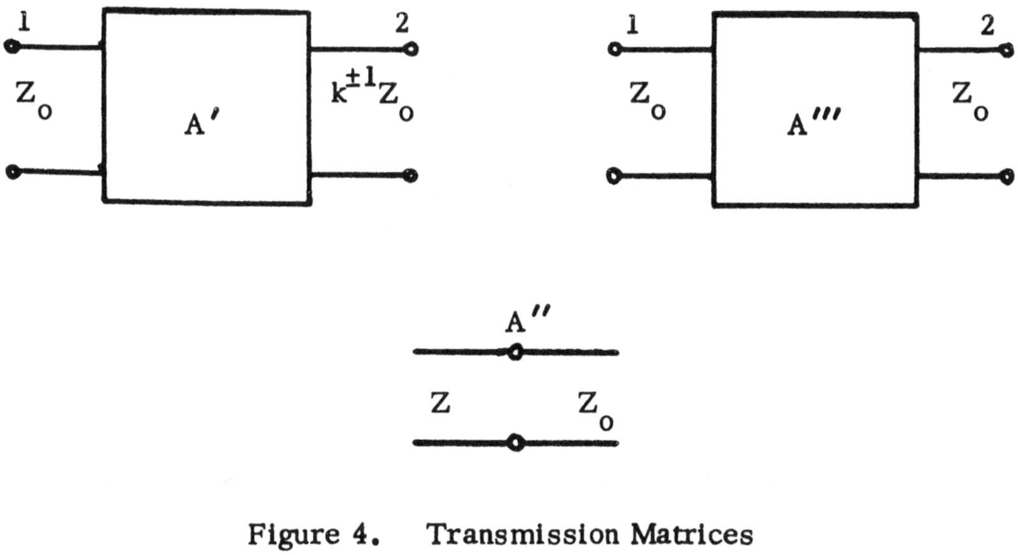

Figure 4. Transmission Matrices

Incident and reflected waves for the network are defined as shown in Figure 3b. These waves are related by the transmission matrix* as follows:

$$\begin{bmatrix} a_1 \\ b_1 \end{bmatrix} = \begin{bmatrix} A_{11} & A_{12} \\ A_{21} & A_{22} \end{bmatrix} \begin{bmatrix} a_2 \\ b_2 \end{bmatrix} \qquad (8)$$

The relations between the waves and the voltages and currents are defined as

$$\begin{aligned} V_1 &= Z_1^{1/2} (a_1 + b_1) , & V_2 &= Z_2^{1/2} (a_2 + b_2) \\ I_1 &= Z_1^{-1/2} (a_1 - b_1) & -I_2 &= Z_2^{-1/2} (a_2 - b_2) \end{aligned} \qquad (9)$$

Notice that the waves at the left and right of the two-port circuit are normalized to the characteristic impedances Z_1 and Z_2. This is slightly different than the more conventional transmission matrices. By substituting (9) into (7) it is easily shown that

$$A_{11} = \frac{1}{2} [A' + B' + C' + D']$$

$$A_{12} = \frac{1}{2} [A' - B' + C' - D']$$

$$A_{21} = \frac{1}{2} [A' + B' - C' - D']$$

$$A_{22} = \frac{1}{2} [A' - B' - C' + D']$$

*Actually this is the inverse of the usual transmission matrix. Since no name has been given for this matrix, the term transmission matrix will be used throughout.

where

$$A' = A\left[\frac{Z_2}{Z_1}\right]^{1/2} \quad , \quad B' = B[Z_1 Z_2]^{-1/2}$$

$$C' = C[Z_1 Z_2]^{1/2} \quad , \quad D' = D\left[\frac{Z_1}{Z_2}\right]^{1/2}$$

It is also easily shown that

$$A_{22}A_{11} - A_{12}A_{21} = 1 \tag{11}$$

It will be necessary to make use of the scattering matrix for two-port circuits. It may be written as

$$\begin{vmatrix} b_1 \\ \\ a_2 \end{vmatrix} = \begin{bmatrix} \Gamma_1 & T_2 \\ \\ T_1 & \Gamma_2 \end{bmatrix} \begin{vmatrix} a_1 \\ \\ b_2 \end{vmatrix} \tag{12}$$

The relations between the scattering elements and transmission elements are given by

$$\Gamma_1 = \frac{A_{21}}{A_{11}} \quad , \quad T_2 = \frac{1}{A_{11}} = T_1 \quad , \quad \Gamma_2 = -\frac{A_{12}}{A_{11}} \tag{13}$$

and

$$A_{11} = \frac{1}{T_1} \quad , \quad A_{12} = \frac{-\Gamma_2}{T_1} \quad , \quad A_{21} = \frac{\Gamma_1}{T_1} \quad , \quad A_{22} = T_1 - \frac{\Gamma_1\Gamma_2}{T_1} \tag{14}$$

Notice that the scattering matrix is symmetrical, that is $T_1 = T_2$. This would not be the case if the incident and reflected waves had not been normalized as in (9).

With the foregoing definitions of the basic relations for the transmission and scattering matrices, the analysis of the magic-T of Figure 1 will proceed. For the even and odd modes of excitation, magnetic and electric walls may be placed along the center line of the circuit. Thus, it is only necessary to analyze two transmission lines in cascade. Let the transmission matrices for the coupled and uncoupled lines be denoted by A' and A''' respectively. The characteristic impedances associated with the transmission matrices are illustrated in Figure 4. All the connecting transmission lines have a characteristic impedance of Z_o, except for the line on the right of A'. Now in using transmission matrices wherein the waves are normalized with respect to the characteristic impedance of the transmission lines, care must be taken when two lines of different impedances are connected. Let A'' represent the transmission matrix for the connection of two lines of impedance Z and Z_o as illustrated in Figure 4. It may be shown that the matrix is given by

$$A'' = \frac{Z_o + Z}{2(Z_o Z)^{1/2}} \begin{vmatrix} 1 & \Gamma \\ \Gamma & 1 \end{vmatrix} \tag{15}$$

where

$$\Gamma = \frac{Z_o - Z}{Z_o + Z} \tag{16}$$

This relation is established by making use of (9) and the continuity of voltage and current. Thus, the transmission matrix for the cascade of the coupled and uncoupled lines is given by

$$A_p = A' A''_p A''' \tag{17}$$

where

$$A' = A''' = \begin{vmatrix} e^{j\theta} & 0 \\ 0 & e^{-j\theta} \end{vmatrix} \tag{18}$$

and

$$A''_p = \frac{1 + k}{2k^{1/2}} \begin{vmatrix} 1 & \Gamma_p \\ \Gamma_p & 1 \end{vmatrix} \tag{19}$$

where the index p may be either e or o (denoting the even and odd modes). Γ_e and Γ_o are given by

$$\Gamma_o = - \Gamma_e = \frac{1 - k}{1 + k} \tag{20}$$

Notice that the transmission matrix for the coupled line for the even and odd modes is identical to that for the uncoupled line. Actually the line coupling is taken into account by the matrix A''. Carrying out the multiplication indicated by (17) the transmission matrix for the cascade of two lines is obtained as

$$A_p = \frac{1 + k}{2k^{1/2}} \begin{vmatrix} e^{j2\theta} & \Gamma_p \\ \Gamma_p & e^{-j2\theta} \end{vmatrix} \tag{21}$$

The scattering matrix elements for the two port circuit are obtained from (13) to give

$$\Gamma_{1p} = \frac{A_{21p}}{A_{11p}} = \Gamma_p\, e^{-j2\theta}$$

$$\Gamma_{2p} = - \frac{A_{12p}}{A_{11p}} = - \Gamma_p\, e^{-j2\theta} \tag{22}$$

$$T_{1p} = \frac{1}{A_{11}} = \frac{2k^{1/2}}{1 + k}\, e^{-j2\theta}$$

The scattering coefficient S_{11} for the four port magic-T may be obtained from (1) and (22) to give

$$S_{11} = \frac{1}{2}(\Gamma_{1e} + \Gamma_{1o}) = \frac{1}{2}(\Gamma_e + \Gamma_o)\, e^{-j2\theta} = 0 \tag{23}$$

In a similar manner the other scattering elements are determined to be

$$S_{12} = \frac{1}{2}(\Gamma_e - \Gamma_o)\, e^{-j2\theta} = -\frac{1-k}{1+k}\, e^{-j2\theta} = -\alpha$$

$$S_{13} = 0$$

$$S_{14} = \frac{2k^{1/2}}{1+k}\, e^{-j2\theta} = \beta \tag{24}$$

$$S_{44} = 0$$

$$S_{43} = \frac{1-k}{1+k}\, e^{-j2\theta} = \alpha$$

The ratio of the coupled to the direct outputs are given by

$$\frac{S_{12}}{S_{14}} = -\frac{1-k}{2k^{1/2}}$$

(25)

$$\frac{S_{43}}{S_{41}} = \frac{1-k}{2k^{1/2}}$$

For a 3 dB coupler it is found that k =0.1717 or Z_{oe} = 5.83 Z_o. This requires a coupling coefficient of 0.944 at the extreme right of the tapered lines. Of course, this is very difficult to achieve in practice. As will be discussed later two 8.36 dB magic-T's may be connected in cascade to form a 3 dB magic-T. In this case it is found that $k = 0.446$, $Z_{oe} = 2.24\ Z_o$ and the coupling coefficient is approximately 0.666 (slide rule accuracy). The latter conditions are more easily obtained in practice.

IV DESIGN OF COUPLED LINES

The objective is to design the coupled lines to act as perfect transformers for the even and odd modes of excitation. For the even mode it is desired to match impedance Z_o to Z_o/k. The design of continuously tapered transformers has been described by Klopfenstein.[10] He has derived a Fourier transform relation between the input reflection coefficient of the transformer and the logarithmic derivative of the characteristic impedance along the tapered line. Design information for transformers with an equal ripple reflection coefficient is presented. The input reflection coefficient of the transformer is given by

$$\rho_p = \rho_{op} \frac{\cos[(\beta\ell)^2 - A^2]^{1/2}}{\cosh A} e^{-j\beta\ell} \tag{26}$$

where ρ_{op} is the reflection coefficient at zero frequency, $\beta\ell$ is the electrical length of the transformer and $\rho_{op}/\cosh A$ is the maximum magnitude of the reflection coefficient in the pass band. The pass band occurs for $\beta\ell > A$. For the even mode of excitation the input reflection coefficient at zero frequency is given by

$$\rho_{oe} = \frac{1 - k}{1 + k} \tag{27}$$

Note that in his work the input reflection coefficient at zero frequency is actually calculated from

$$\rho_{oe} = \frac{1}{2} \ln \frac{1}{k} \tag{28}$$

This is necessary since the analysis is approximate in the sense that it assumes the reflection coefficient is always small. The solution for the characteristic impedance as a function of the distance x along the coupler is given by

$$\ln \frac{Z_{oe}(x)}{Z_o k^{-1/2}} = \frac{\rho_{oe}}{\cosh A} [A^2 \varphi (\frac{2x}{\ell}, A) + U(x - \frac{1}{2}) - U(- x - \frac{1}{2})] \tag{29}$$

where φ is a function tabulated in his paper and U is a step function. This formula has been written for the even mode characteristic impedance $Z_{oe}(x)$.

For the odd mode of excitation, it is desired to match Z_o to kZ_o. Hence $\rho_{oo} = - \rho_{oe}$ and the solution for $Z_{oo}(x)$ is given by

$$\ln \frac{Z_{oo}(x)}{Z_o k^{1/2}} = \frac{-\rho_{oe}}{\cosh A} [A^2 \varphi (\frac{2x}{\ell}, A) + U(x - \frac{1}{2}) - U(-x - \frac{1}{2})] \tag{30}$$

By adding (29) and (30) it may be shown that

$$Z_{oe}(x) Z_{oo}(x) = Z_o^2 \tag{31}$$

The parameter k along the length of the transformer is given by

$$k(x) = \sqrt{\frac{Z_{oo}(x)}{Z_{oe}(x)}} = \frac{Z_o}{Z_{oe}(x)} \tag{32}$$

Figure 5 illustrates the characteristic impedance variation along the length of the transformer. The resulting input reflection coefficient is shown in Figure 6. An important characteristic of the tapered-line transformer is that the reflection coefficient for the odd mode is the negative of that for the even mode when excited from the left. When the transformer is excited from the right the reflection coefficients are the negative of those when it is excited on the left.

220

The design procedure for the tapered-line transformer is as follows: determine A from the desired reflection coefficient magnitude in the pass band. Calculate k(x) from (29) and (32). The dimensions and spacing of the coupled lines may then be determined if impedance information on the particular configuration is available.

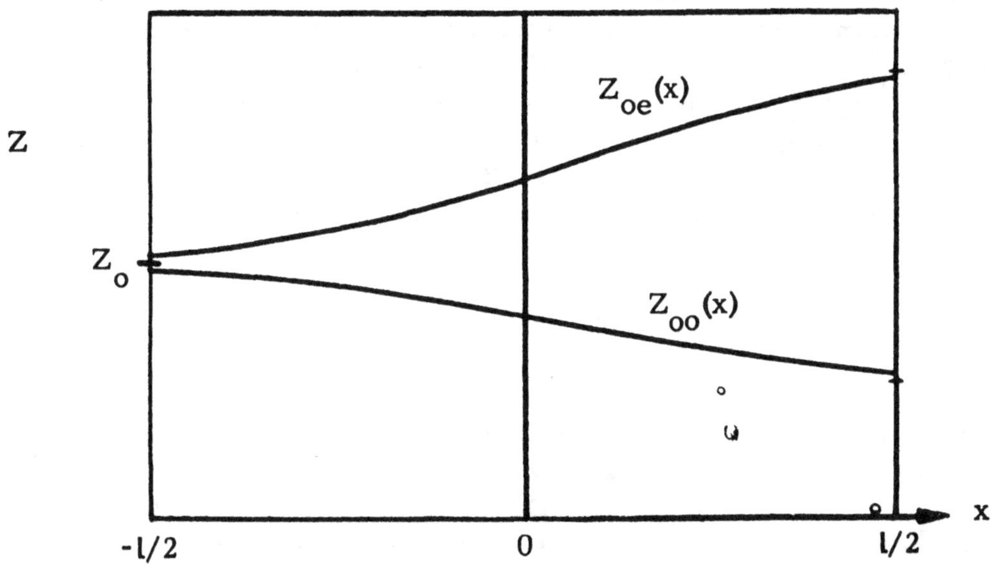

Figure 5. Impedance for Tapered-Line Transformer

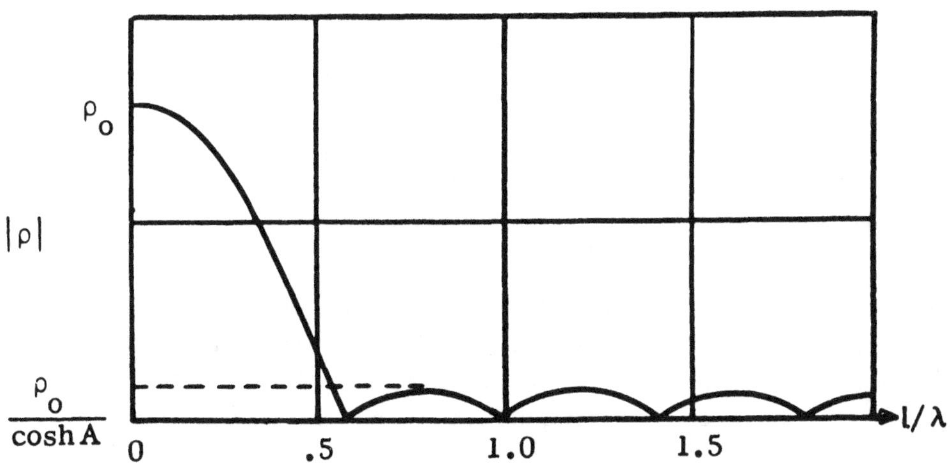

Figure 6. Input Reflection Coefficient

The synthesis of transformers is very similar to the synthesis of radiation patterns for finite line sources. The transformer can be designed to give other types of side lobe response. For example, refer to T.T. Taylor's[12] work. A disadvantage of the equal ripple response is that steps in the impedance function are required (see Figure 5). These steps lower the cutoff frequency but also lead to ripples even at frequencies many times the cutoff frequency.

V EFFECT OF NON-PERFECT TRANSFORMERS

Until this point we have assumed that the tapered lines act as perfect transformers for the even and odd modes of excitation. In the following an approximate analysis will be made to determine the effect of non-perfect transformers. Let us assume that the scattering coefficients for the tapered lines satisfy the following.

$$\Gamma'_{1p} < 0.1 \,, \quad \Gamma'_{2p} < 0.1 \,, \quad T'_1 = e^{-j\theta} \tag{33}$$

Actually the magnitude of T'_1 will be less than one. However, with the restrictions on Γ'_{1p} and Γ'_{2p}, T'_1 should be greater than 0.99.

From (14) the transmission matrix for the tapered lines now becomes

$$A' = A'_p = \begin{bmatrix} e^{j\theta} & -\Gamma'_{2p}\, e^{j\theta} \\ \Gamma'_{1p}\, e^{j\theta} & e^{-j\theta} \end{bmatrix} \tag{34}$$

where a second order term, small with respect to unity has been omitted from A'_{22p}. Performing the matrix multiplication indicated by (17), with A''_p and A''' as shown in (19) and (18) respectively, the transmission matrix for the cascade of the tapered-line and the uncoupled line is given by

$$A_p = \frac{1+k}{2k^{1/2}} \begin{vmatrix} e^{j2\theta} - \Gamma'_{2p}\Gamma_p e^{j2\theta} & \Gamma_p - \Gamma'_{2p} \\ \Gamma'_{1p} e^{j2\theta} + \Gamma_p & \Gamma'_{1p}\Gamma_p + e^{-j2\theta} \end{vmatrix} \tag{35}$$

Then by making use of (13) the scattering coefficients for the cascade are given by

$$\Gamma_{1p} = \frac{\Gamma'_{1p} + \Gamma_p e^{-j2\theta}}{1 - \Gamma'_{2p}\Gamma_p} \approx \Gamma_p e^{-j2\theta} + \Gamma'_{1p} + \Gamma'_{2p}\Gamma_p^2 e^{-j2\theta}$$

$$\Gamma_{2p} = \frac{-(\Gamma_p - \Gamma'_{2p})e^{-j2\theta}}{1 - \Gamma'_{2p}\Gamma_p} \approx -\left[\Gamma_p - \Gamma'_{2p} + \Gamma'_{2p}\Gamma_p^2\right] e^{-j2\theta} \tag{36}$$

$$T_{1p} = \frac{2k^{1/2} e^{-j2\theta}}{(1+k)(1 - \Gamma'_{2p}\Gamma_p)} \approx \frac{2k^{1/2}}{1+k}(1 + \Gamma'_{2p}\Gamma_p) e^{-j2\theta}$$

where again, higher order terms have been dropped.

First, let us assume that the transformers are accurately designed as described in the previous section. We then have

$$\Gamma'_{1p} = \rho_p = \rho_{op} \frac{\cos\,[\theta^2 - A^2]^{1/2}}{\cosh A} e^{-j\theta} = \rho_{op} e^{-j\theta} \psi \tag{37}$$

where ψ is defined in the above equation. Now as pointed out previously, we have

$$\rho_{oe} = -\rho_{oo} = -\Gamma_e \tag{38}$$

$$\Gamma'_{2p} = -\Gamma'_{1p} \tag{39}$$

From these conditions it follows that

$$\Gamma'_{1e} = -\Gamma'_{1o} \quad , \quad \Gamma'_{2e} = -\Gamma'_{2o} \quad , \quad \Gamma_e = -\Gamma_o \quad . \tag{40}$$

From these it follows that

$$\Gamma_{1e} = -\Gamma_{1o} \quad , \Gamma_{2e} = -\Gamma_{2o} \quad , \quad T_{1e} = T_{1o} \quad . \tag{41}$$

The scattering coefficients for the four port magic-T may now be computed from (1) and the above equations. The results are

$$S_{11} = \frac{1}{2} (\Gamma_{1e} + \Gamma_{1o}) = 0$$

$$S_{12} = \frac{k-1}{k+1} e^{-j2\theta} \left[1 - \psi \left(e^{j\theta} - \Gamma_e^2 e^{-j\theta}\right)\right] \tag{42}$$

$$S_{13} = 0$$

$$S_{14} = \frac{2k^{1/2}}{1+k} \left(1 + \psi \Gamma_e^2 e^{-j\theta}\right) e^{-j2\theta}$$

224

$$S_{44} = 0$$

$$S_{43} = \frac{1-k}{1+k} e^{-j2\theta} \left[1 - \psi e^{-j\theta} (1 - \Gamma_e^2)\right] \quad .$$

The ratio of the coupled outputs when the magic-T is excited at port one is given by

$$\frac{S_{12}}{S_{14}} = - \frac{(1 - k)}{2k^{1/2}} \frac{1 - \psi \left(e^{j\theta} - \Gamma_e^2 e^{-j\theta}\right)}{1 + \psi \Gamma_e^2 e^{-j\theta}} \tag{43}$$

or

$$\frac{S_{12}}{S_{14}} \approx - \frac{1 - k}{2k^{1/2}} (1 - \psi e^{j\theta}) \quad . \tag{44}$$

Similarly

$$\frac{S_{43}}{S_{41}} \approx \frac{1 - k}{2k^{1/2}} (1 - \psi e^{-j\theta}) \quad . \tag{45}$$

It is noted that the ratios deviate from the desired value by the factor in parentheses. The unbalance, U, or deviation from this desired value is given by

$$U = 20 \log_{10} (1 + \psi_M) \approx 8.68 \psi_M \tag{46}$$

where

$$\psi_M = \frac{1}{\cosh A} \tag{47}$$

The phase difference between the coupled outputs deviates from the desired value of zero or 180 degrees by

$$\epsilon = \tan^{-1} \psi_M \approx \psi_M \tag{48}$$

The first magic-T was designed so that cosh A was equal to 31.6 (30 dB). As can be seen from the above, this leads to an unbalance of 0.275 dB and a phase error of 1.82 degrees. Later magic-T's were designed for a 40 dB side lobe ratio, that is, cosh A = 100. This would reduce the unbalance to .087 dB and the phase error to .57 degrees. For this design the first null in Fig. 6 would occur at $\ell/\lambda = 0.84$.

For the above formulas we have assumed a prefectly designed tapered-line transformer. In practice this will not quite be the case. It is likely that

$$\Gamma'_{2p} \neq -\Gamma'_{1p} \text{ and } \Gamma'_{1e} \neq -\Gamma'_{1o} \tag{49}$$

In this case the results of (41) will not hold and the magic-T will not be matched nor will the isolation be perfect.

VI EXPERIMENTAL RESULTS

The tapered-line magic-T consists of two 8.36 dB hybrids in cascade* to achieve a 3 dB magic-T. As mentioned earlier use of this technique results in coupling coefficients which are more readily achieved in practice. The coupled hybrids utilize the latest techniques in strip transmission line construction.

A photograph of a 3 dB magic-T designed for the frequency range 2.0 to 12.4 GHz is shown in Figure 7. Both external and internal views are illustrated to indicate the form factor and the circuit proper. The ground planes are spaced 0.142 inches and the dielectric material is polyolefin. The overlapping strip lines on the two boards are separated (when assembled) by a central dielectric sheet with a thickness of 0.17 inch. In the tightest coupled region the two strips are completely overlapped and in the weakly coupled region the two strips do not overlap at all. The width and spacing of the lines were determined from theoretical solutions provided by J.W. Duncan of Hughes Aircraft Company. The circuit dimensions are approximately four inches by five inches.

226 The isolation and coupled output data was obtained by feeding the circuit with sweep oscillators and recording the data on an x-y strip chart recorder. The data illustrated in Figure 8 was reduced from those

---

*The idea of cascading directional couplers is not new. See Monteath, G.D., "Coupled Transmission Lines as Symmetrical Directional Couplers," Proc. IEE, Part B, Radio and Electronic Eng., Vol. 102, No. 3, May 1955, pp. 383-392.

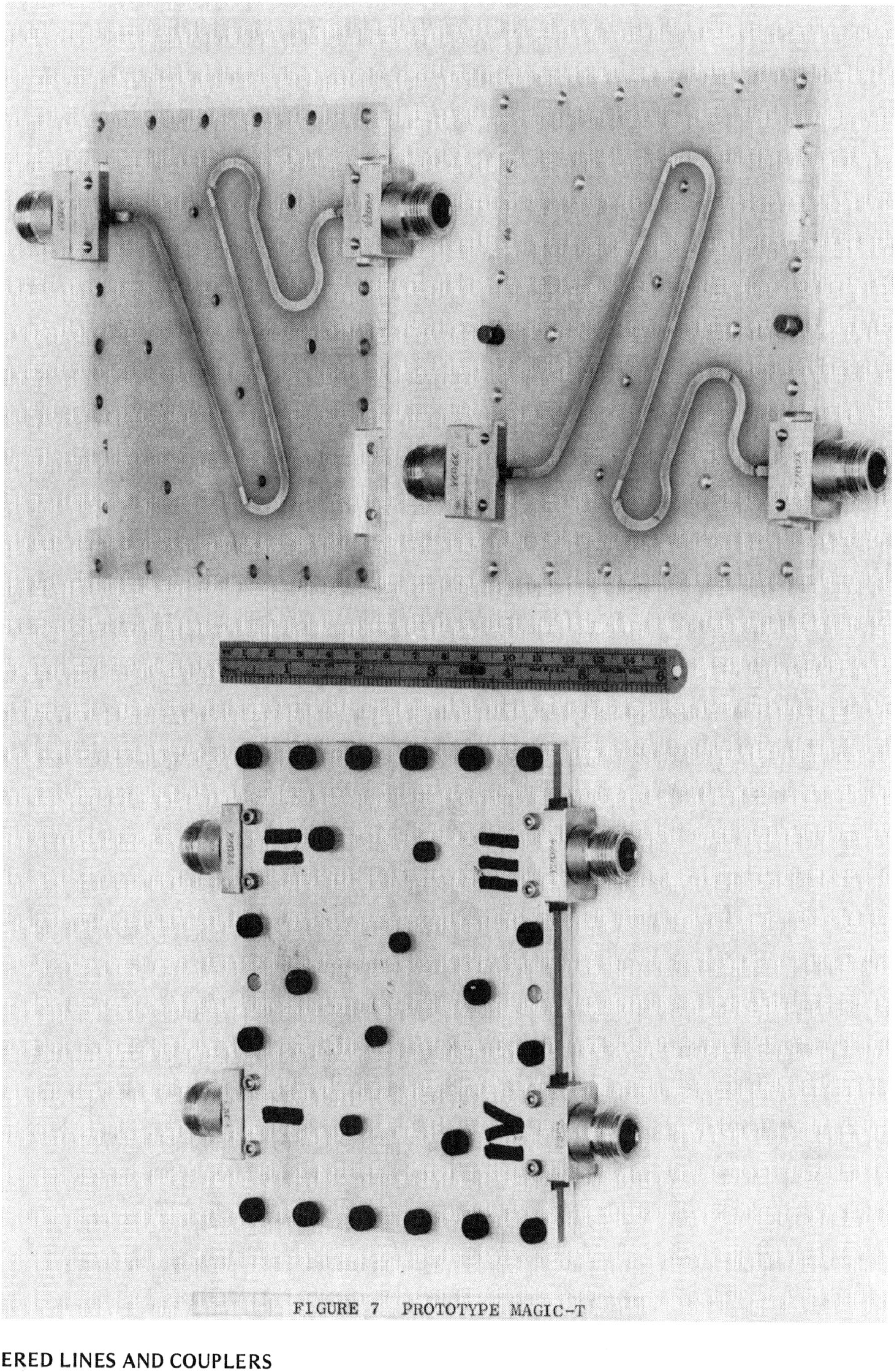

FIGURE 7 PROTOTYPE MAGIC-T

strip chart recordings. It will be observed that the coupled outputs deviate by at most 0.4 dB, and that the isolation is always greater than 24 dB. Furthermore, the coupled outputs deviate from the nominal in phase or out of phase condition by less than $\pm 3^{o}$ over the frequency range. The input VSWR is less than 1.4:1 over the frequency range. A portion of the VSWR is attributable to the coax to strip line transitions and the circuit terminations. Copper and dielectric losses amount to approximately 0.3 dB at the low end to 0.7 dB at the high end of the frequency range.

An important feature of the tapered-line magic-T is that it is based on a high pass design rather than a band pass design. The significance of this is that the circuit has no theoretical upper frequency limit. Its practical upper frequency limit is therefore determined by constructional tolerances and the size of the transition region from coupled to uncoupled lines. The lower frequency limit is determined by the length of the tapered-line coupler. Magic-T's have been constructed with lower cutoff frequencies of 350 MHz and 800 MHz which yield greater bandwidth capabilities than that shown in Figure 7. However, because of the increased length of strip-transmission line these circuits have greater copper and dielectric losses. Some of this loss may, of course, be eliminated by increasing the ground plane spacing for the structure, but then the upper frequency limit will be destroyed due to excitation and propagation of higher order modes. Where application demands, the level of the coupled outputs may be changed simply by changing the coupling coefficient for the tapered-line couplers. For example a magic-T with power split of 1.8 dB and 4.8 dB has been constructed by cascading two 13.25 dB couplers. Experimental measurements on these lower cutoff frequency magic-T's is comparable over their frequency range to the data shown in Figure 8.

VII MATCHING TECHNIQUES

The good isolation and low VSWR in the X-band region were obtained with no small amount of experimentation and analysis. Ideally the transition from tightly coupled to uncoupled lines at the center of the magic-T should occur in an infinitesimal distance. At X-band the transition occurs over a distance which is on the order of a quarter-wavelength. In this transition region the conductors separate at an acute angle with respect to each other. This has the effect of reducing the magnetic coupling more than the electric coupling. From Oliver's[9] fourth equation on p. 1687 this causes the odd mode velocity of propagation to be less than that of the even mode in the transition region. Measurements of even and odd mode impedances from both sides

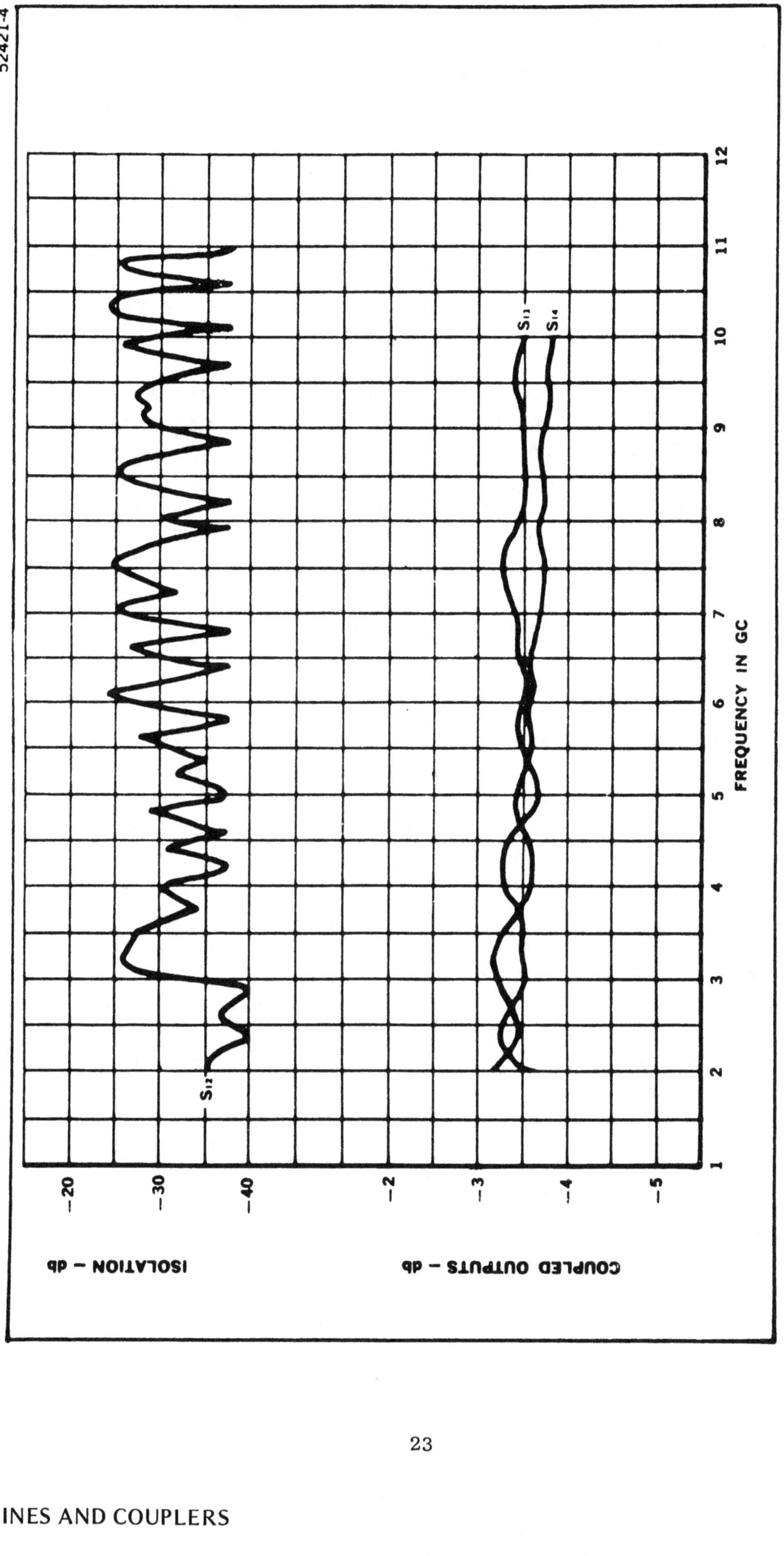

Figure 8. Isolation and Coupled Output Data

of the asymmetrical junction (or transition region) confirmed this. This effect was compensated for by introducing an air gap in the 0.017 inch separator sheet between the tightly coupled lines near the junction. The gap was approximately 0.25 inch long and the width was the same as the strip width. The gap increases the velocity of propagation as well as the impedance of the odd mode but has little effect upon the even mode. The characteristic impedance of the odd mode in the gap region was lowered to the correct value by adding approximately 0.002 inch of metal to each strip in the gap region.

With this technique it was possible to increase the isolation in the **X**-band region by 15 dB.

Small capacitive stubs were also added at the junction region to equalize the magnitudes of the reflection coefficients of the even and odd modes in the **X**-band region.

These same matching techniques have been used on multi-section quadrature hybrids to increase the isolation.

VIII CONCLUSIONS

The tapered-line magic-T represents a very significant advance in the state-of-the-art, both in terms of bandwidth and performance. Better performance over a 17:1 bandwidth[13] than previous broadband devices which covered bandwidths on the order of 2:1 has been obtained. Magic-T's making use of ferrite cores have demonstrated excellent performance over extreme bandwidths, but their upper frequency limit is on the order of 2 GHz. For the microwave band, the best previous broadband device is probably the Asymmetric Coupled-Transmission-Line Magic-T described by Kraker.[14] However, the performance over its 2:1 bandwidth is not as good as that of the tapered-line magic-T. A 3:1 bandwidth log-periodic magic-T was recently described[15] but is a very complicated structure and practical considerations limit it to frequencies below 3 Gc. Quadrature hybrids have recently been built[6] for a 17:1 bandwidth but with isolation as low as 14 dB at the high end of the band.

A bandwidth of 50:1 for the frequency range of 0.25 to 12.5 Gc could easily be achieved with the tapered-line magic-T in a moderately sized package. With good connectors, test equipment and care, the upper frequency limit could be extended to 15 or 20 GHz.

Although the VSWR of the magic-T is quite low (less than 1.35) it could be reduced significantly if better transitions from coax to strip line were available. If reflectionless transitions were available, the VSWR of the tapered-line magic-T would be less than 1.08:1.

The difference between the coupled outputs (magnitude and phase) is much larger than predicted theoretically. This is attributed to constructional errors and imperfect matching of the transition from tightly coupled to uncoupled lines.

An important application of the tapered-line magic-T's is in antenna monopulse and homing feed networks where the magic-T may replace a quadrature hybrid and phase difference network.

REFERENCES

1. V.H. Rumsey, "Frequency Independent Antennas," 1957 IRE National Convention Record, Part I, pp. 114-118.

2. R.H. DuHamel and D.E. Isbell, "Broadband Logarithmically Periodic Antenna Structures," 1957 IRE National Convention Record, Part I, pp. 119-128.

3. E.G. Cristal and L. Young, "Theory and Tables of Optimum Symmetrical TEM-Mode Coupled-Transmission-Line Directional Couplers," IEEE Trans. on Microwave Theory and Techniques, Vol. MTT-13, pp. 544-558, September 1965.

4. J.P. Shelton and J.A. Mosko, "Synthesis and Design of Wide-Band Equal Ripple TEM Directional Couplers and Fixed Phase Shifters," IEEE Trans. on Microwave Theory and Techniques, Vol. MTT-14, pp. 462-473, October 1966.

5. B.M. Schiffman, "A New Class of Broadband Microwave 90-Degree Phase Shifters," IRE Trans. on Microwave Theory and Techniques, Vol. MTT-6, pp. 232-237, April 1958,

6. J.P. Shelton, Jr., J.J. Wolfe, and R.C. Van Wagoner, "Tandem Couplers and Phase Shifters for Multi-Octave Bandwidth," Microwaves, pp. 14-19, April 1965.

7. R. Levy, "General Synthesis of Asymmetric Multi-Element Coupled-Transmission-Line Directional Couplers," IRE Trans. on Microwave Theory and Techniques, Vol. MTT-11, pp. 226-237, July 1963.

8. L. Young, "Microwave Filters--1965," IEEE Trans. on Microwave Theory and Techniques, Vol. MTT-13, pp. 489-508, September 1965. (See page 502.)

9. B.M. Oliver, "Directional Electromagnetic Couplers," Proc. IRE, Vol. 42, pp. 1686-1692, November 1954.

10. J. Reed and G.J. Wheeler, "A Method of Analysis of Symmetrical Four-Port Networks," IRE Trans. on Microwave Theory and Techniques, Vol. MTT-4, pp. 246-252, October 1956.

11. R.W. Klopfenstein, "A Transmission Line Taper of Improved Design," Proc. IRE, Vol. 44, pp. 31-35, January 1956.

12. T.T. Taylor, "One Parameter Family of Line Sources Producing Modified Sin $\pi u/\pi u$ Patterns," Hughes Aircraft Company, Technical Memorandum No. 324, September 4, 1953.

13. M.E. Armstrong, R.H. DuHamel, J.W. Duncan, and H.A. Uyeda, "Design and Development of a Tapered-Line Magic-T," Final Engineering Report, Contract No. N123(62738)-50145A(X), July 1, 1965, Hughes Aircraft Co., Fullerton, California, FR 65-14-133, prepared for U.S. Naval Ordnance Laboratory, Corona, California. See also Allerton Park paper referenced on p. 1 of this paper.

14. D.I. Kraker, "Asymmetric Coupled-Transmission-Line Magic-T," IEEE Trans. on Microwave Theory and Techniques, Vol. MTT-12, pp. 595-599, November 1964.

15. R.H. DuHamel, M.E. Armstrong, and M.A. Meyer, "Log-Periodic Octaline Hybrid Junctions," 1964 PGMTT International Symposium Digest, May 19-21, 1964, Long Island, New York.

Reprinted from IEEE TRANSACTIONS
ON *MICROWAVE THEORY AND TECHNIQUES*
Volume MTT-14, Number 12, December, 1966
pp. 647-656

The Design and Construction of Broadband, High-Directivity, 90-Degree Couplers Using Nonuniform Line Techniques

C. P. TRESSELT, MEMBER, IEEE

***Abstract*—It is possible, at present, to obtain multioctave bandwidth in symmetrical couplers that employ cascaded quarter-wavelength sections of uniformly coupled line. However, the physical junctions between the various sections contribute reactive discontinuities, which significantly degrade coupler directivity. This paper describes a coupler design employing a continuously tapered coupling coefficient that helps to circumvent the directivity problem.**

Two classes of couplers have been investigated, including one which provides optimum equal-ripple performance. Synthesis has been performed with the aid of both digital and Fourier integral computers. Somewhat tighter center coupling is required in the tapered design to produce bandwidth-mean-coupling-level performance comparable to stepped-coupling design.

Experimental data is presented on several models constructed in three-layer polyolefin stripline.

Manuscript received June 29, 1966; revised August 8, 1966.
The author is with The Bendix Corporation, Research Laboratories Division, Southfield, Mich.

I. INTRODUCTION

IT IS POSSIBLE to obtain multioctave bandwidth from a single coupler that employs several cascaded quarter-wavelength sections of uniformly coupled line. The use of a symmetric structure guarantees 90° relative phase lead of the coupled port with respect to the transmitted port at all frequencies. Cristal and Young [1] present tables of coupling coefficients required to produce equal-ripple response for the third- through ninth-order designs, and include a comprehensive bibliography covering the historic development of this form of coupler. Tight coupling normally is required in at least one of the sections of a broadband design. This condition has been alleviated to some extent by the application of tandem interconnection [2].

Unfortunately, the spread in coupling values required in adjacent quarter-wave sections, with or without tandem interconnection, is large enough to produce substantial differences in physical line dimensions. The abrupt transition regions between adjacent sections exhibit reactive discontinuities, which degrade coupler directivity, particularly at the higher microwave frequencies. Some compensation for these effects can be achieved by using tuning screws in the junction regions, or, if a dielectric stripline is involved, by using suitably placed capacitive tuning tabs. In general, however, these techniques have not proved very useful above C-band.

This paper describes a design that eliminates the abrupt interconnections between coupled sections. The coupling coefficient is gradually tapered from loose to tight coupling (in the center) and then back to loose in a symmetrical manner to preserve the 90° character of the design. The familiar condition that $\sqrt{Z_{oe} \cdot Z_{oo}} = 1$ must be observed throughout the coupler length to insure that the port adjacent to the transmitted port remains isolated.

Stepped-impedance couplers are, in a sense, a subclass of tapered-line designs. In stepped-impedance devices, coupling is assumed to change abruptly between quarter-wave sections, whereas, in practice, a finite length is required to effect the desired change. The coupling vs. distance characteristic present in a stepped coupler is thus, in effect, tapered. This distinction may be largely neglected for couplers covering UHF and the lower microwave frequencies. However, in multioctave designs covering frequencies through X-band, transition regions take up a significant fraction of coupler length due to the shortness of the quarter-wave sections involved. Consider, for example, the stepped-impedance tandem coupler covering the 1 to 8 GHz band described by Shelton [2]. The transition regions associated with the center quarter-wavelength section of this design take up at least one-third of the center section. The distributed nature of such transitions usually dictates the use of a cut-and-try procedure to determine the best transition shape and location along the axis of the coupler. Misadjustments are readily apparent in the frequency domain since these two transitions provide the largest amplitude Fourier contribution to the overall coupling vs. frequency characteristic [2].

The nonuniform line couplers treated in this paper are tapered gradually enough to permit usage of physical coupling data based on infinitely long, uniformly-coupled line calculations without modification. This has the practical advantage of allowing the geometry of the coupler to be completely specified without the uncertainty associated with the transition regions of stepped coupler designs.

The primary motivation for investigation of tapered-line designs, however, has been the promise of higher directivity. The reactive discontinuities experienced in stepped designs result from nonpropagating higher order modes present in the vicinity of the transitions. While it is true that such evanescent modes will, to some extent, also be excited along the length of the tapered-line geometry, the amplitude of such modes decreases as the taper employed becomes more gradual with respect to wavelength. The equal-ripple tapered-line coupler described in Section IV utilizes essentially a full quarter wavelength to taper between coupling levels for which an abrupt transition is required in the comparable stepped-coupler model. It is unfortunate that directivity is not more amenable to theoretical calculation in either the stepped or tapered cases, the heart of the problem being the lack of a mathematical description for the modes involved in practical line geometries of interest. Experimental comparison is thus indicated and preliminary data on experimental tapered-line couplers is included for this purpose.

II. Analysis of Nonuniform Line Coupler Performance

Generalized coupler theory can be developed with the aid of the transmission line analogy [1]. A nonuniform transmission line is postulated whose characteristic impedance curve equals the even-mode impedance curve of the coupler to be analyzed. Under these conditions, the reflection coefficient of the transmission line is equal in magnitude and phase to the coupled-arm response of the coupler.

The analysis and synthesis of nonuniform line devices is complicated by the fact that exact closed-form solutions have not been developed [3], [4]. Analysis and synthesis can, however, be performed directly by using a loose-coupling approximation. As will be shown later, such results can be modified to cover tighter coupling without resorting to open-form (series) solutions.

Under the assumption of loose coupling, one can show with the aid of the transmission line analogy that coupling C at frequency ω is given by [3]

$$C(\omega) = \int_0^d e^{-2j\omega x/v} p(x) dx \tag{1}$$

where d is the overall coupler length, v is the velocity of propagation in the medium of interest, and

$$p(x) = \frac{1}{2} \frac{d}{dx} \ln Z_{oe}(x). \tag{2}$$

The quantity $p(x)$ is the reflection coefficient distribution which is related by (2) to the even-mode impedance-vs.-distance characteristic of the coupler. The coupling, as given by (1), is the Fourier transform of $p(x)$. Synthesis can thus be performed directly in the first-order approximation by application of an inverse Fourier transformation.

Second-order coupling theory can be obtained with the aid of the transmission line analogy from Youla's equation (66) [4]. The coupler is matched at its input and output so that $\Gamma_o = \Gamma_l = 0$ for the analogous line. Furthermore, the generator and load impedances are equal, so that Youla's equation (66) becomes, in the terminology of this paper,

$$| C(\omega) | = \frac{\left| \int_0^d e^{-j(2\omega x/v)} p(x) dx \right|}{\sqrt{1 + \left| \int_0^d e^{-j(2\omega x)/v} p(x) dx \right|^2}} . \tag{3}$$

A similarity exists between first- and second-order theory, since the same integral expression appears in both solutions. For small values of coupling, the second-order expression reduces to first-order theory. For tighter values of coupling, second-order theory indicates that larger values of $p(x)$ are required for a given coupling level than are indicated by first-order theory. Flat coupling over broad frequency bands is of primary concern. If a $p(x)$ characteristic produces a flat first-order response, the coupling will remain flat in the second-order theory at a lower mean level, as indicated by (3).

Youla [4] presents an exact description of nonuniform line performance in terms of a Volterra integral equation of a type which may always be solved by successive iterations, the resulting series being absolutely and uniformly convergent under broad conditions. The number of iterations required depends on the mean level of coupling and the bandwidth. Unfortunately, a compact synthesis formula corresponding to higher-order theory has not been found.

Further insight into the design problem can be obtained from first-order considerations. Because the even-mode impedance of the coupler is symmetrical about the coupler center, the derivative of the logarithm of this curve will be odd, as shown in Fig. 1(a). It is convenient to shift axis in the first-order integral to take advantage of this fact, obtaining thereby,

$$C(\omega) = -je^{-j(\omega/v)d} \int_{-d/2}^{d/2} \sin 2 \frac{\omega}{v} u \cdot p(u) du \tag{4}$$

where

$$p(u) = \frac{1}{2} \frac{d}{du} \ln Z_{oe}(u). \tag{5}$$

The magnitude and phase of $C(\omega)$, corresponding to $p(u)$ in Fig. 1(a), are shown in Fig. 1(b). By assuming a desired magnitude-vs.-frequency characteristic, and using phase as shown, one can, by inverse Fourier transformation, obtain the required function of coupling vs. distance. A Scientific-Atlanta CF 1 Fourier integral computer has been used to perform this transformation.

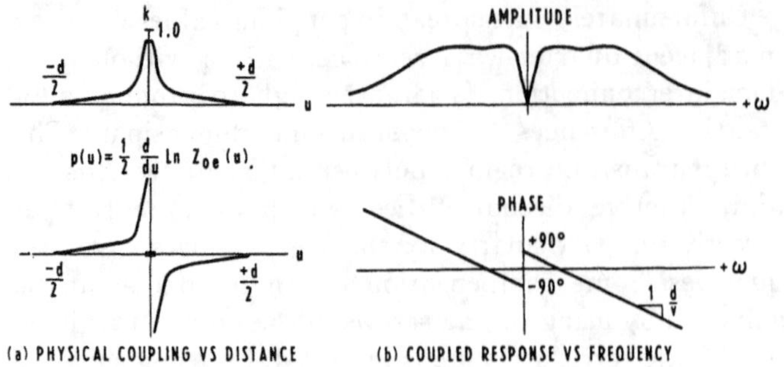

Fig. 1. First-order coupling characteristics.

III. Monotonic $p(u)$ Couplers

The first class of coupler treated possessed a monotonically increasing $p(u)$ characteristic along the length of the coupler, as illustrated in Fig. 1. Pertinent information about the approximation problem for this type of coupler can be obtained by consideration of an impedance-vs.-distance curve that has been found integrable in closed form. Here, it is convenient to use the first-order theory of (4). Let the coupler length d be two units long, and let $p(u)$ be given bv

$$\begin{aligned} p(u) &= \frac{-M(1+u)}{2u}, \qquad -1 \leq u < 0 \\ p(u) &= \frac{-M(1-u)}{2u}, \qquad 0 < u \leq 1. \end{aligned} \tag{6}$$

This characteristic is shown in Fig. 2 for $M = 0.201$. The magnitude of $p(u)$ goes to $\pm\infty$ as u approaches zero from either side. Under these conditions, direct substitution of (6) into (4) leads to the result that

$$C(\theta) = jMe^{-j\theta} \left[Si(\theta) + \frac{1}{\theta} (\cos\theta - 1) \right] \tag{7}$$

where

$$\theta = \frac{2\omega}{V} .$$

Response of this coupler for $M = 0.201$ is plotted in Fig. 3(a). The choice of $M = 0.201$ provides a maximum coupling level $C(\theta)_{max} = 0.3161$, which is -10 dB. Note that coupling rises monotonically with increasing frequency and that coupling is relatively flat for $\theta > 10$.

Equations (5) and (6) require that $Z_{oe} = \infty$ at the coupler center. This is not realizable practically, since the coupling coefficient

$$k(u) = \frac{Z_{oe}^2(u) - 1}{Z_{oe}^2(u) + 1} \tag{8}$$

implies that $k = 1$ at coupler center. Simple truncation of the $p(u)$ characteristic defined by (6) results in a coupling characteristic which droops rather severely with frequency. An iterative technique can be employed

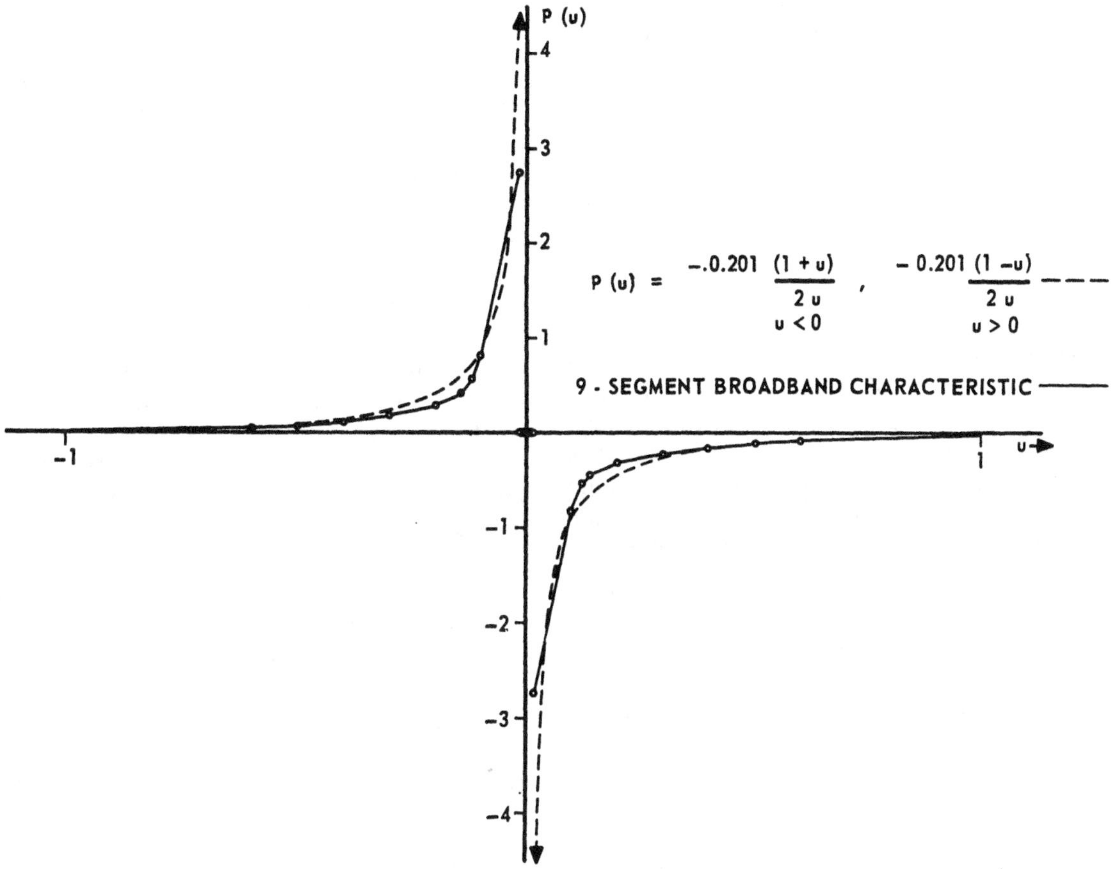

Fig. 2. $p(u)$ vs. u for several monotonic $p(u)$ couplers.

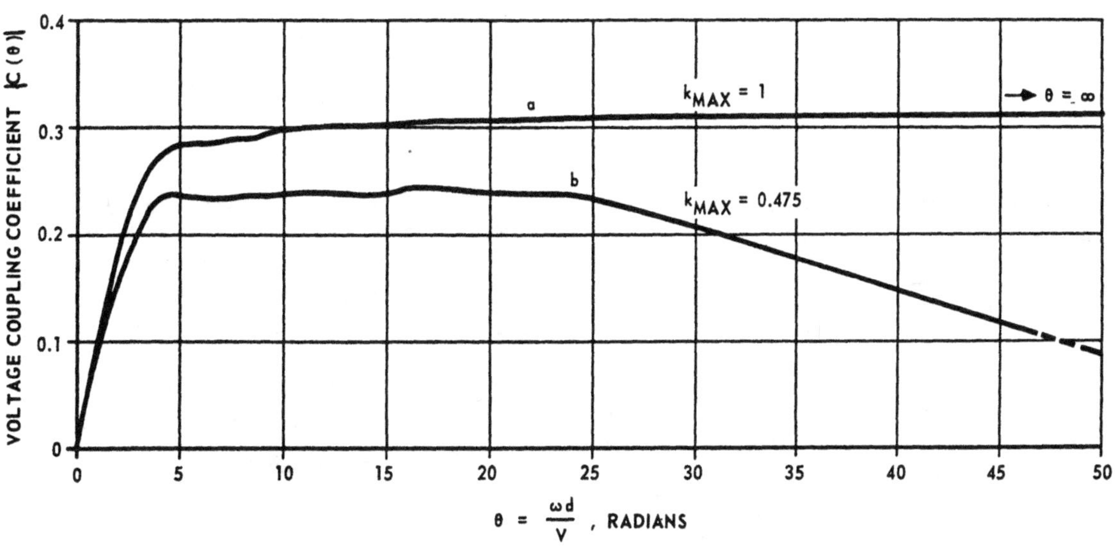

Fig. 3. First-order response of the couplers described in Fig. 2.

to improve such an undesirable response. A curve expressing the difference between a flat coupling level and the given coupler response is formed over the frequency interval in which flat response is desired. Fourier analysis of this error curve will give a set of $p(u)$ values which can be added to the original $p(u)$ characteristic to correct the discrepancy between responses. The response achieved by one particular set of iterations is shown in Fig. 3(b). This characteristic corresponds to the $p(u)$ curve made up of nine straight-line segments shown in Fig. 2. Straight-line segments were used throughout the iterative procedure to minimize the effort required in integrating $p(u)$ to obtain actual coupling coefficients. (A maximum coupling value of $k=0.475$ is required in the center of the coupler.)

One characteristic inherent in the monotonic $p(u)$ type of coupler is the relatively slow dropoff in response at high frequencies. If this response is forced to fall off more rapidly in the correction procedure, a nonmonotonic $p(u)$ function will be produced. The monotonically increasing $p(u)$ coupler was the first tapered design to be investigated because of its relatively smooth coupling characteristic, $k(u)$. This type of coupler is not the most optimum from a mean-coupling bandwidth standpoint as discussed below; however, in moderate bandwidth form, it should prove quite valuable as an instrumentation coupler.

IV. Couplers with Periodic Zeros in $p(u)$

A coupling-vs.-frequency characteristic which would remain flat out to some cutoff frequency, beyond which coupling would rapidly drop to zero, would be highly desirable. All available coupling-bandwidth product would then be employed usefully. One can show, by using first-order coupling theory described earlier in an inverse Fourier transformation sense, that a $p(u)$ distribution given by

$$p(u) = \frac{-1}{\pi} \frac{\sin^2 u/2}{u/2} \tag{9}$$

will provide a first-order mean-coupling level of 1.0 over the frequency range from 0 to 1 radian per second. A 3-dB coupler, then, corresponds to

$$p(u) = \frac{-1}{\pi\sqrt{2}} \frac{\sin^2 u/2}{u/2}.$$

238

A perfectly flat coupling level would require inclusion of all sidelobes of $p(u)$ from $u=-\infty$ to $+\infty$, and hence would be infinitely long. By truncating $p(u)$ at some point in u, an approximation to the desired coupling can be obtained. The general classification of couplers having periodic zeros in $p(u)$ is based on truncation of (9) which exhibits periodic zeros in the sine function.

The Fourier integral computer was used to investigate the approximation problem. A polar recorder was used on the output to permit plotting several solutions on one sheet for purposes of comparison. A series of five runs was made. The first run used the first positive and first negative lobe of

$$\frac{\sin^2 u/2}{u/2}$$

as a $p(u)$ function. (Positive and negative lobes must be paired together to preserve the symmetric nature of the design.) The next run consisted of the first two positive and first two negative lobes of the same function; more lobes were included, a pair at a time, in subsequent runs. The results are plotted in Fig. 4. As is evident from the figure, the addition of a sidelobe pair increases the number of ripples in the output response by one unit. Also evident from the set of responses is the fact that Gibbs' overshoot is present and does not decrease in amplitude with increasing coupler order (number of sidelobes used).

Ripple level may be equalized by using weighting techniques often employed in antenna array synthesis. The ordinates of symmetrically disposed sidelobes can be weighted with Dolph-Chebyshev coefficients to produce approximate equal-ripple performance. Better results can be obtained by appropriate use of a weighting function derived from the equal-ripple trigonometric polynomials used in the synthesis of stepped-coupler designs. As an example, consider the following digital generated ninth-order polynomial which approximates the constant, unity, in an exact equal-ripple manner (± 5 percent voltage) over the interval $0<\theta\leq\pi$:

$$\begin{aligned} f(\theta) &= 1.2626 \sin\theta + 0.3931 \sin 3\theta + 0.2049 \sin 5\theta \\ &\quad + 0.1170 \sin 7\theta + 0.0927 \sin 9\theta. \end{aligned} \tag{10}$$

The conventional Fourier representation in the same interval is given by

$$\begin{aligned} g(\theta) &= \frac{4}{\pi}\left(\sin\theta + \frac{1}{3}\sin 3\theta + \frac{1}{5}\sin 5\theta \right. \\ &\quad \left. + \frac{1}{7}\sin 7\theta + \frac{1}{9}\sin 9\theta\right). \end{aligned} \tag{11}$$

Numerical evaluation of $g(\theta)$ will exhibit the usual Gibbs' overshoot phenomenon. The desired weighting function is obtained from that set of coefficients which, when multiplied by the Fourier coefficients in $g(\theta)$, gives the coefficients of the true equal-ripple function $f(\theta)$. Thus

$$\begin{aligned} w_1 &= 1.2626/(4/\pi) = 0.992 \\ w_3 &= 0.3931/(4/3\pi) = 0.929 \\ w_5 &= 0.2049/(4/5\pi) = 0.805 \\ w_7 &= 0.1170/(4/7\pi) = 0.643 \\ w_9 &= 0.0927/(4/9\pi) = 0.656. \end{aligned} \tag{12}$$

The above weighting coefficients, derived from a discrete Fourier series representation of constant coupling, can be applied to the distributed coupling case in the following manner.

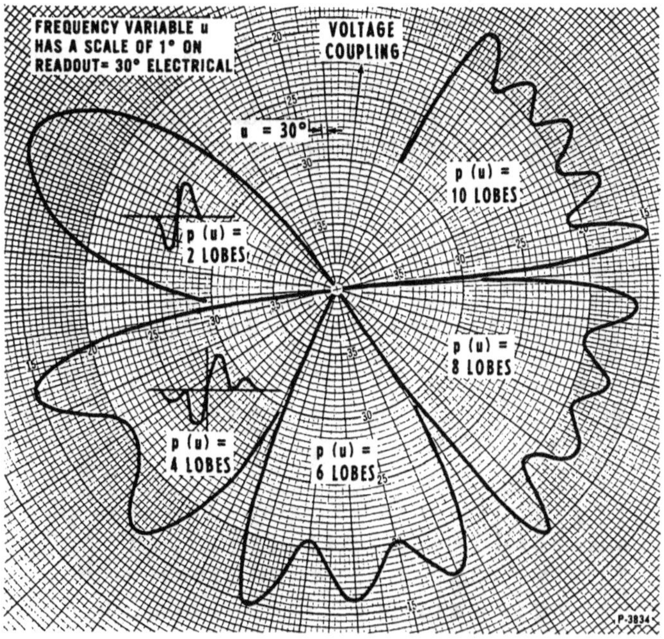

Fig. 4. Frequency response of various truncations of

$$p(u) = \frac{\sin^2 u/2}{u/2}.$$

The first weighting coefficient, w_1, multiplies the ordinates of the first positive and negative lobes of

$$\frac{\sin^2 u/2}{u/2},$$

which extend to $u = \pm 2\pi$. The next coefficient, w_3, multiplies the ordinates of the second set of lobes between $\pm 2\pi$ and $\pm 4\pi$, etc. All five weighting coefficients given in (12) were applied in this manner to the basic $p(u)$ function of (9). The Fourier integral computer was then used to obtain the coupling response.

The resultant coupling is very close to true equal ripple. Because of computer noise, the exact amount of ripple is hard to determine precisely but is about ± 4 percent in voltage. This may be compared with the ± 5 percent voltage ripple associated with the original equal-ripple polynomial from which the weighting function was derived. A more careful analysis using a digital computer to obtain more precision would be desirable. The foregoing weighted $p(u)$ function may be used directly to synthesize various approximate equal-ripple, loose-coupling (first-order) designs.

Couplers designed with the aid of first-order theory, even with exact equal-ripple $p(u)$ data, will not exhibit flat coupling for multioctave bandwidth designs of mean-coupling level greater than about -10 dB. Third-order theory is required as a minimum to describe adequately the effects involved. Fortunately, a closed-form technique, described below, has been found which permits weighting of the $p(u)$ solution to both eliminate Gibbs' overshoot and to include the compensation for tight coupling as required by higher order theory. This technique uses weighting functions derived from exact stepped-coupler designs [1]. Figure 5 indicates approximately a weighted fifth-order $p(u)$ function and the

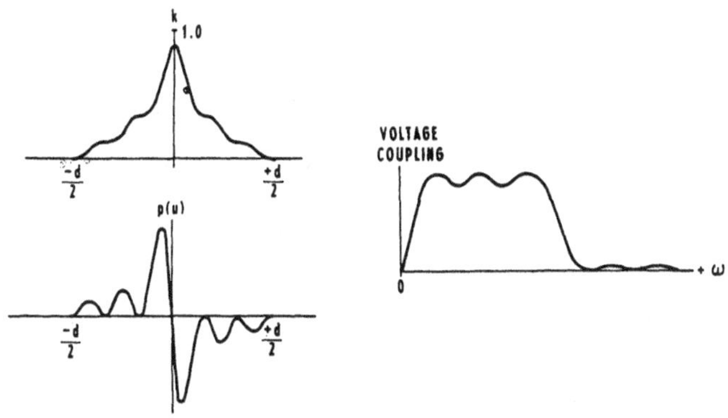

Fig. 5. Response of a typical periodic-zero $p(u)$ coupler.

associated coupling functions in both frequency and distance (k vs. u).

A description of stepped couplers in terms of the

$$p(u) = \frac{1}{2}\frac{d}{du}\ln Z_{oe}(u)$$

domain is required in the process of obtaining the desired weighting function. The coupling coefficients of stepped couplers change discontinuously, so that $p(u)$ consists of a set of Dirac delta functions located periodically with distance, at the discontinuities. The area under each delta function is given by $\frac{1}{2}\ln Z_{i+1}/Z_i$, where Z_i and Z_{i+1} are the even-mode impedances on either side of the discontinuity. Under these conditions, one can show by using (4) that

$$C(\omega) = je^{-j(\omega d/v)}\sum_{n=1}^{(N+1)/2}\ln\frac{Z_{((N+3)/2)-n}}{Z_{((N+1)/2)-n}}\sin(2n-1)\frac{\omega d}{vN} \tag{13}$$

where N is the order of design. (The equal-ripple polynomial listed previously, (10), is of order $N=9$, with $\theta=\omega d/Nv$.)

Cristal and Young's tables [1] give the various even-mode impedances required to build an equal-ripple stepped-coupler for $N=3$, 5, 7, and 9, and for a variety of mean-coupling levels and ripple tolerances. Consider, for example, the case of a stepped ninth-order, 8.34-dB coupler with ripple tolerance of ± 0.30 dB. According to Cristal and Young's tables, such a design exhibits a 10.96:1 bandwidth and consists of the following impedance levels: $Z_1=1.031\,34$, $Z_2=1.076\,97$, $Z_3=1.164\,69$, $Z_4=1.357\,71$, $Z_5=2.253\,15$. One can form the $p(u)$ plane representation of this coupler by calculating the area under the various delta functions as described previously. Equation (13) thus becomes

$$C(\omega) = je^{-j(\omega d/v)}\left[0.5071\sin\frac{\omega d}{9v} + 0.1532\sin\frac{3\omega d}{9v} + 0.0785\sin\frac{5\omega d}{9v} + 0.0433\sin\frac{7\omega d}{9v} + 0.0309\sin\frac{9\omega d}{9v}\right]. \tag{14}$$

The first-order Fourier approximation to 8.34-dB coupling is, from (11),

$$|C(\theta)| = 0.3827 \times \frac{4}{\pi}\left|\sin\theta + \frac{1}{3}\sin 3\theta + \frac{1}{5}\sin 5\theta + \frac{1}{7}\sin 7\theta + \frac{1}{9}\sin 9\theta\right| \tag{15}$$

where again the Gibbs' overshoot is present. One obtains the weighting coefficients from a term-by-term comparison of (14) and (15). Thus,

$$\begin{aligned} w_1 &= 0.5071/(0.3827 \times 4/\pi) = 1.040 \\ w_3 &= 0.1532/(0.3827 \times 4/3\pi) = 0.943 \\ w_5 &= 0.0785/(0.3827 \times 4.5\pi) = 0.805 \\ w_7 &= 0.0433/(0.3827 \times 4/7\pi) = 0.623 \\ w_9 &= 0.0309/(0.3827 \times 4/9\pi) = 0.569. \end{aligned} \tag{16}$$

This set of weighting coefficients can be applied to the respective amplitudes of the first five lobe pairs of

$$\frac{\sin^2 u/2}{u/2}$$

to produce equal-ripple results similar to those obtained in weighting the discrete jump solution. It is instructive to compare the weighting coefficients associated with the 8.34-dB coupler, (16), with the weighting coefficients associated with a loose-coupling design, such as given in (12). Both sets of weighting coefficients tend to de-emphasize the higher order terms in order to eliminate Gibbs' overshoot. In the 8.34-dB coupler, however, the lowest term is accentuated because of the tightness of coupling involved. This effect becomes more prominent when 3-dB coupling is desired, in which case, both w_1 and w_2 are typically greater than unity, emphasizing the effect of the higher order theory.

The even-mode characteristic impedance of the distributed coupler can be obtained by a point-by-point integration of the $p(u)$ distribution, as indicated by (5). Fortunately, the integration can be found in tabular form by noting that

$$2\int_0^a \frac{\sin^2 u/2}{u}\,du = \int_0^a \frac{(1-\cos u)}{u}\,du = \mathrm{Cin}(a) \tag{17}$$

where Cin is one definition of the cosine integral. The Cin is tabulated in King [5] for arguments through 25. Alternately, one might note that $\mathrm{Cin}(a) = C + \ln a - \mathrm{Ci}(a)$, where $C = 0.5772$ is Euler's constant, and $\mathrm{Ci}(a)$ is an alternate form of cosine integral tabulated in a number of references [6], [7]. The integrand function,

$$2\,\frac{\sin^2 u/2}{u},$$

has minima every $2\pi n$ radians and has maxima where $\tan u = 2u$. Solutions to the latter equation are tabulated in [8]. For high order sidelobes, the maxima tend to fall halfway between the minima.

The ninth-order solution previously analyzed covers five positive and five negative lobes of the $p(u)$ function, so that the overall coupler length is $d = 20\pi$ units. One can obtain from the Cin tables the area under various positive lobes of the function

$$\frac{\sin^2 u/2}{u/2}.$$

$$\begin{aligned} &\text{1st lobe:} \quad A_1 = 2.438 \\ &\text{2nd lobe:} \quad A_2 = 0.676 \\ &\text{3rd lobe:} \quad A_3 = 0.402 \\ &\text{4th lobe:} \quad A_4 = 0.287 \\ &\text{5th lobe:} \quad A_5 = 0.223. \end{aligned} \tag{18}$$

The weighted area under positive lobes of the ninth-order, 8.34-dB coupler can be determined from (9) by adjusting the mean first-order coupling level to 0.3827 and by applying the appropriate weighting coefficients:

$$\begin{aligned} &\text{1st lobe:} \quad \frac{0.3827}{\pi} \times 2.438 \times 1.040 = 0.3085 \\ &\text{2nd lobe:} \quad \frac{0.3827}{\pi} \times 0.676 \times 0.943 = 0.0776 \\ &\text{3rd lobe:} \quad \frac{0.3827}{\pi} \times 0.402 \times 0.805 = 0.0394 \\ &\text{4th lobe:} \quad \frac{0.3827}{\pi} \times 0.287 \times 0.623 = 0.0218 \\ &\text{5th lobe:} \quad \frac{0.3827}{\pi} \times 0.223 \times 0.569 = \underline{0.0155} \\ &\text{Total Positive Lobe Area} \qquad 0.4628. \end{aligned} \tag{19}$$

The tapered-line coupler can be fabricated from a plot of coupling coefficient vs. distance. This is obtained by first integrating the weighted

$$\frac{\sin^2 u/2}{u/2}$$

function (using the Cin tables [5]) and equating to the integral of

$$p(u) = \frac{1}{2}\frac{d}{du}\ln Z_{oe}(u);$$

that is, $\frac{1}{2}\ln Z_{oe}(u)$. The coupling coefficient, $k(u) = (Z_{oe}^2 - 1)/(Z_{oe}^2 + 1)$, may then be calculated. For example, the maximum value of coupling results from integrating the total positive lobe area so that, from the above data,

$$\tfrac{1}{2}\ln Z_{oe\max} = 0.4628, \text{ from which } Z_{oe\max} = 2.512 \text{ and } k_{\max} = 0.727.$$

The design produced by the procedure outlined above is shown in Fig. 6.

The actual physical length of the coupled region d can be determined once the desired center frequency f_c

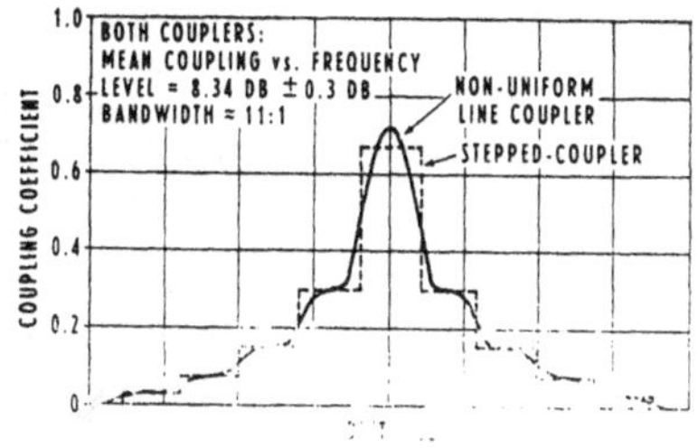

Fig. 6. Comparison of coupling vs. distance characteristics for two classes of coupler design.

of the equal-ripple band has been chosen. It can be shown with the aid of the Fourier integral computer that

$$d = \frac{(N+1)v}{4f_c}. \tag{20}$$

The bandwidth of the tapered-line coupler will be essentially identical to that of the prototype stepped coupler, which, for the example above, is approximately 11:1. If a coupler using this particular taper is to be built to cover 1 to 11 GHz, $f_c = 6 \times 10^9$ Hz. Assuming polyolefin stripline is to be employed, $v = 77.5 \times 10^8$ in/s. Thus,

$$d = \frac{(9+1) \times 77.5 \times 10^8}{4 \times 6 \times 10^9} = 3.23 \text{ inches.}$$

It is interesting to note that (20) states that the equal-ripple tapered-line coupler is exactly one-quarter wavelength longer than the equivalent stepped-impedance coupler.

Figure 6 compares the coupling coefficient for the ninth-order, 8.34 dB, tapered-line design with that of the stepped coupler from which the weighting coefficients were derived. It is evident that the stepped coupler and the tapered-line coupler are related. The tapered-line coupler is about 11 percent longer than the stepped design and requires $k_{max} = 0.727$ in comparison with $k_{max} = 0.672$ for the stepped design. This result is characteristic of the difference between tapered and stepped designs.

The rate of change of coupling with distance is more gradual in the tapered-line design, as is apparent from Fig. 6. Approximately one-quarter wavelength is utilized in tapering between coupling plateaus in the periodic-zero tapered-line model. The abrupt changes in coupling in the corresponding stepped model occur at the center of the tapered regions at the points of maximum rate of change of coupling vs. distance. As mentioned in the Introduction, practical stepped couplers cannot exhibit perfectly abrupt changes in coupling. The present example shows that rather gradual transitions can, in fact, be used while preserving essentially similar response. The chief difference relates to repetitious coupling response at frequencies above the first coupling band: the mathematical model for the ideal stepped coupler repeats at center frequencies which are odd harmonics of the fundamental out to infinite frequency, whereas the periodic-zero tapered-line coupler exhibits essentially zero coupling beyond the first pass band. Practical stepped couplers will exhibit fairly good third harmonic and possibly higher coupling bands if designed for lower frequencies where the transition regions are a small fraction of a quarter wavelength; for higher frequency models, repetition of the second coupling band is typically of poor shape and mean-coupling is lower than the design level showing the presence of appreciable tapering in the physical model.

V. Experimental Models

In reducing to practice the design of couplers having periodic zeros in $p(u)$, a 3-dB coupler was constructed to determine the adequacy of mathematical treatment in predicting mean-coupling level under tight-coupling conditions. Preliminary work on very tight physical coupling indicates that a coupling coefficient of about 0.9 can be achieved in three-layer polyolefin stripline using 0.0015-inch centerboard and 0.062-inch ground plane boards. In this geometry, it is possible to synthesize a tapered-line, 3-dB coupler having more than two-octave bandwidth. The measured response of the coupler is given in Fig. 7. Coupling variation exceeded the design tolerance to ±0.25 dB, primarily because of uncertainty regarding impedances in the stripline geometry. The width of the stripline conductor is approximately one-sixth of that corresponding to $Z_0 = 50$ ohms in the tightly coupled center region (where the thickness of the conductor foil approximately equals the thickness of the dielectric board separating the two lines). The measured isolation on the first trial was, however, a minimum of 21 dB from dc to beyond 7 GHz. The mean coupling of the model was found to be adequately prescribed by the synthesis technique outlined. Broader bandwidth couplers can be obtained by using tandem interconnections of several couplers without resorting to extremely tight coupling levels.

To assure maximum directivity, it is important that the coupler design use fully overlapped coupling in the center to permit crossing of the conductors for purposes of tandem interconnection. A jog in the center caused by overly tight coupling of the lines would defeat the purpose of the tapered-line geometry.

To assure uniform ripple and high isolation, very accurate data on three-layer stripline coupling parameters is necessary. The data employed to date has been obtained from various zero-conductor-thickness, asymptotic solutions. (See, for example, Reference [9].) A 0.010-inch centerboard dielectric thickness, which might typically be used, is small enough in comparison with 1-ounce copper foil (0.0014 inches thick) to cause noticeable deviation from the zero-thickness approximations. In view of the complexity of obtaining useful conformal transformation solutions to the finite conductor-offset stripline problem, an extensive measurement program was initiated to obtain data of the accuracy required. Consequently, a three-layer stripline test fixture

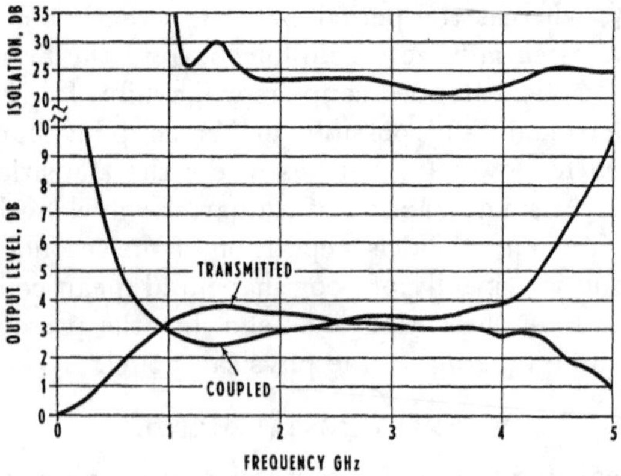

Fig. 7. Measured performance of tapered-line 3-dB coupler having periodic zeros in $p(u)$.

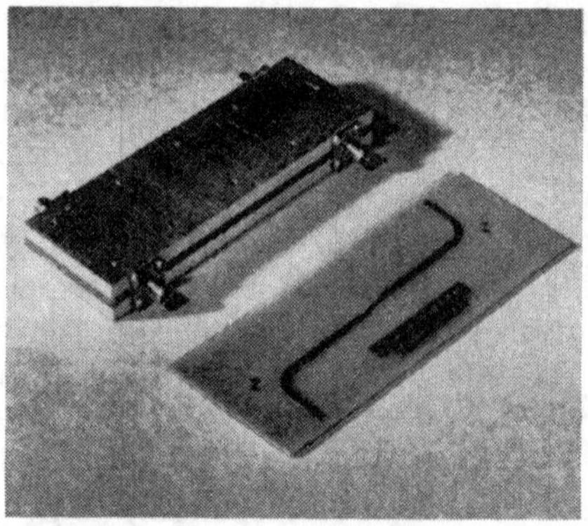

Fig. 8. Ninth-order, 8.34±0.30-dB tapered-line coupler.

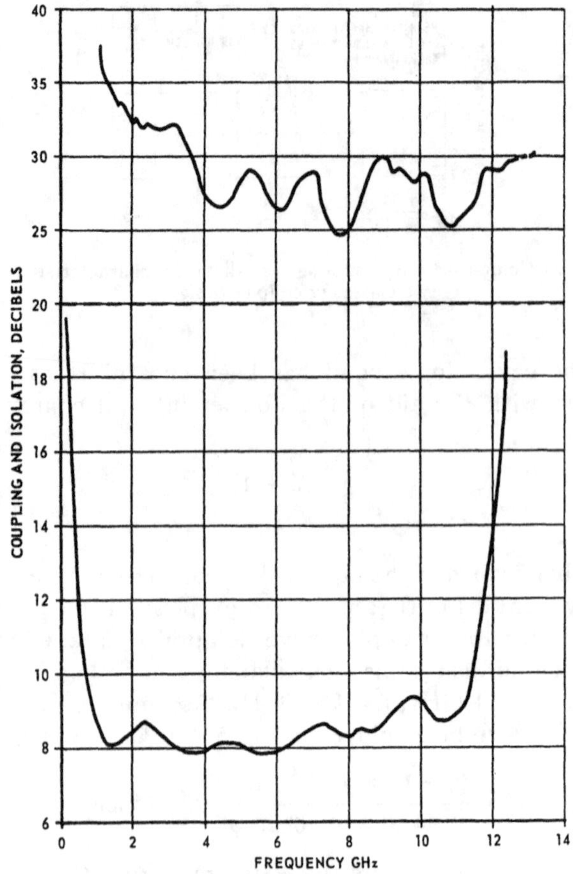

Fig. 9. Measured performance of ninth-order 8.34±0.30-dB tapered-line coupler.

has been constructed recently in which a 6-inch length of uniformly coupled line is measured. A broadband time-domain reflectometer is used to ascertain that overlap has been properly adjusted for a given conductor width to assure that $\sqrt{Z_{oe} \cdot Z_{oo}} = 50$ ohms. Coupling is then measured under matched conditions by both time- and frequency-domain measurements to assure a crosscheck of the values obtained. Such a crosscheck has indicated that preliminary measurements made with this fixture are accurate.

Having obtained somewhat more accurate stripline data, the ninth-order, 8.34±0.30-dB coupler described in the preceding section was constructed, as a general verification of the weighting function theory. The maximum required physical coupling coefficient, $k = 0.727$, can be realized in three-layer polyolefin stripline consisting of 0.010-inch shim stock used with 0.062-inch ground-plane boards when the conductors are fully overlapped. The band from 1 to 11 GHz was chosen for realization, requiring an overall coupled line length of 3.24 inches in polyolefin. Figure 8 shows an exterior view of the coupler and an unmounted ground-plane board of the type contained within the case. With careful inspection, the periodic plateaus in the line geometry can be identified. The measured performance of the coupler is shown in Fig. 9. Mean coupling is reasonably close to 8.34 dB, ripple amplitude is of the right order of magnitude, and the number of major ripples, five, corresponds to the order of the design. The deviation of coupling from 8.34 dB toward the high end of the band is attributable partially to an insertion loss of several tenths of a decibel present at the higher frequencies. The foregoing measurements were made every 200 MHz, the curves shown being drawn through each point measured. Low VSWR BRM miniature loads were used in the measurement of isolation; both the generator and the temperature-compensated bolometer detector were well padded with attenuators. The latter padding was found quite desirable in reducing higher order inaccuracy due to multiple reflections. Low VSWR stripline-to-BRM transitions are an integral part of the coupler.

The above coupler showed sensitivity to pressure exerted on the case. This was ultimately traced to the ability of the thin 0.010-inch shim to assume different depths in the region next to the 0.0014-inch thick foil stripline. The isolation shown in Fig. 9 was improved by several decibels with the addition of 1-mil teflon shim stock to the regions around the foil conductors, thereby virtually eliminating the air gap. For best performance, the coupler should be reconstructed using stripline coupling data measured with the extra shim material present (the shim being polyolefin instead of teflon).

The most significant feature of the data given in Figs. 7 and 9 (both tapered couplers) is that isolation is relatively constant over the band and is actually higher at the high-frequency end of the band. The various strip-line stepped couplers constructed by the author and those reported on in the literature [2], [10] all exhibit a trend towards decreasing isolation throughout the coupling band, often approaching rather low values at the high frequencies. For example, a 1 to 8 GHz stepped-impedance coupler of 8.34-dB mean level was constructed by the author from data given in Reference [2]. It exhibited isolation within the 1 to 8 operating band which in several narrow bands was at least 10 dB poorer than isolation at any point through 12.5 GHz in the tapered-line model of Fig. 9. The isolation of the tandem interconnection of two such stepped couplers is reported to be 10 dB near 8 GHz [10], which result has been duplicated by the author. Reworking transition regions in these and other stepped couplers to include compensation in the form of capacitive tabs or tuning screws can provide some improvement in isolation over most of the coupling band (excluding the high-frequency region in all cases tried by the author). The tapered-line couplers measured to date provide isolation which remains high through the high-frequency band limit, while completely avoiding the rather laborious trial-and-error process of transition location, shaping, and discontinuity compensation present in stepped-coupler design.

A number of monotonic $p(u)$ couplers have been constructed using approximate stripline impedance data for a 0.010-inch polyolefin shim used between 0.062-inch ground-plane boards. One of the ground-plane boards used in a 7-dB coupler of this variety is shown in Fig. 10. The measured response of this coupler is shown in Fig. 11, along with the normalized first-order description of coupling. The minimum directivity of this coupler is 20 dB up to 7.5 GHz. A tandem coupler composed of two sections similar to the above design has been constructed for use in a phase-to-amplitude antenna processing system which operated successfully through 11 GHz.

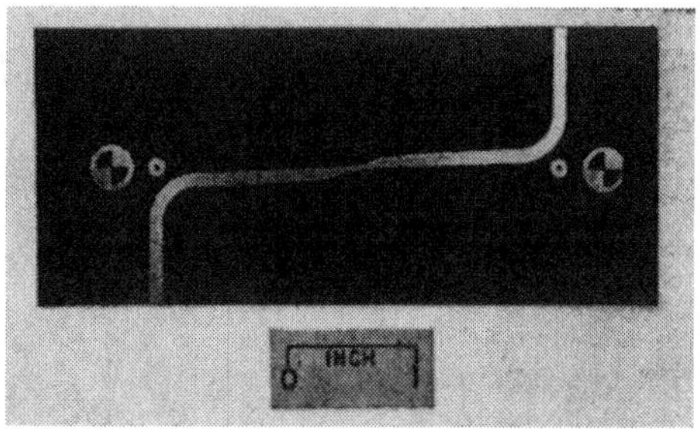

Fig. 10. Stripline board used in 7:1 bandwidth monotonic $p(u)$ coupler.

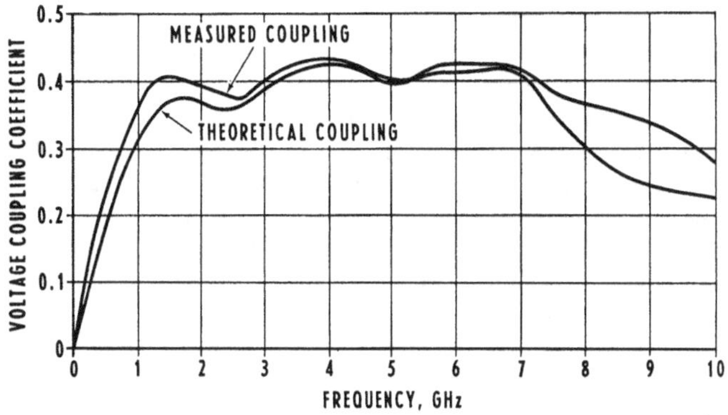

Fig. 11. 7:1 bandwidth monotonic $p(u)$ coupler response.

VI. Conclusions

Of the two classes of tapered coupler described, monotonic $p(u)$ and periodic-zero $p(u)$, the latter class can be designed to produce approximate equal-ripple coupling performance with the aid of published design tables for stepped-coupling designs. In both types, physical coupling vs. distance varies smoothly throughout the coupled region with promise for realization of high directivity at the higher microwave frequencies. Tapered-line coupler geometry can be completely calculated for the length of the coupler eliminating the trial-and-error procedures often associated with locating, shaping, and compensating transitions between sections in stepped-coupler design. Tapered equal-ripple couplers require an extra quarter wave in overall length and a slightly higher center coupling coefficient to achieve mean coupling-bandwidth performance comparable to that provided by stepped-coupler designs.

The limitation on directivity of tapered designs to date has been inaccuracy in knowledge of impedance and coupling of the three-layer stripline configuration. The ultimate limitation on directivity will probably arise from evanescent modes generated in the center of the coupler where the rate of change of line cross section is greatest; the magnitude of this ultimate limit remains to be determined.

References

[1] E. G. Cristal and L. Young, "Theory and tables of optimum symmetrical TEM-mode coupled-transmission line directional couplers," *IEEE Trans. on Microwave Theory and Techniques*, vol. MTT-13, pp. 545–558, September 1965.

[2] J. P. Shelton, R. Van Wagoner, and J. J. Wolfe, "Tandem couplers and phase shifters; a new class of unlimited bandwidth components," presented at the 1964 14th Ann. Symp., USAF Antenna R&D Program, AF Avionics Lab., in cooperation with the University of Illinois, Monticello, Ill., or see *Microwaves*, pp. 14–19, April 1965.

[3] C. B. Sharpe, "An alternative derivation of Orlov's synthesis formula for nonuniform lines," *Proc. IEE (London)*, vol. 109, Monograph 483E, 1962.

[4] D. C. Youla, "Analysis and synthesis of arbitrarily terminated lossless nonuniform lines," *IEEE Trans. on Circuit Theory*, vol. CT-11, pp. 363–372, September 1964.

[5] R. W. P. King, *The Theory of Linear Antennas*. Cambridge, Mass.: Harvard University Press, 1956, pp. 857–864.

[6] *Tables of Sine, Cosine, and Exponential Integrals*, Table III. Federal Works Admin., Works Progress Admin., N. Y., New York, 1940.

[7] *Handbook of Mathematical Functions*, U. S. Dept. of Commerce, NBS, Appl. Math. Ser. 55. Washington, D. C.: U. S. Govt. Printing Office, June 1964, ch. 5.

[8] *Ibid.*, Table 4.19.

[9] J. P. Shelton, "Impedances of offset parallel-doupled strip transmission lines," *IEEE Trans. on Microwave Theory and Techniques*, vol. MTT-14, pp. 7–15, January 1966.

[10] J. P. Shelton, "Synthesis and design of wideband TEM directional couplers," presented at the 1964 Internat'l Conf. of Microwaves, Circuit Theory, and Information Theory, Tokyo, Japan.

Design and Computed Theoretical Performance of Three Classes of Equal-Ripple Nonuniform Line Couplers

C. P. TRESSELT, MEMBER, IEEE

Abstract—The purpose of this paper is to provide an accurate theoretical picture of three classes of tapered-line couplers. The coupler taper is approximated with a cascade of short equal-length coupled line segments whose coupling values match those of the taper at one end of each segment. Exact analysis of the even-mode transmission-line analogy is then performed by evaluating the *ABCD* matrix of the cascade as a function of frequency with the aid of a digital computer. The computed response converges to that of the smooth tapered coupler as the number of line segments increases, total length being held constant.

An exact asymptotic high-frequency model for asymmetric couplers is presented which provides improved prediction of coupler performance, as verified by computer analysis. The associated design procedures are given explicitly.

The insertion phase dispersion of the 90° and $\Sigma-\Delta$ magic tee devices is treated. Comparisons with the insertion phase dispersion of various stepped coupling 90° designs are provided.

I. Introduction

SEVERAL RECENT papers have presented approximate equal-ripple designs for tapered-line couplers including a $\Sigma-\Delta$ magic tee utilizing asymmetrical coupling [1], a symmetrical 90° coupler [2], and a physically short asymmetric design with unspecified phase performance [3]. None of the papers provide an exact theoretical description of expected response due to the mathematical difficulties in evaluating nonuniform line devices. It is the purpose of the present paper to present an accurate theoretical picture of each device by means of analysis on a digital computer. The results can be compared with experimental data; as will become apparent, the theoretical response of the devices is closer to ideal equal-ripple performance than has been achieved experimentally, indicating that further refinements in performance may be achievable.

It is assumed that the four ports of the coupler are terminated in a characteristic impedance of one ohm, and that for match, the relationship $\sqrt{Z_{oe}Z_{o}}=1$ is preserved throughout the design. Under these conditions, the four-port analysis of a coupler can be simplified to the two-port analysis of an analogous nonuniform line whose characteristic impedance is equal to the even mode impedance of the coupled line structure [4].

The analysis of the two-port problem can be treated by approximating the nonuniform line with a large number of short equal-length constant impedance segments whose values of impedance follow the taper of the nonuniform line.

Manuscript received September 19, 1968; revised December 9, 1968.
The author is with Bendix Research Laboratories, Southfield, Mich. 48075.

The impedance values of the N segments in the model are chosen by subdividing the original line into N equal-length pieces. The value of impedance present at the right (or alternately, left) hand ends of the subdivided pieces are assigned to the steps of the model. If the steps are small, the particular choice of right or left end will not materially affect the calculated response, since all devices to be treated possess a continuous impedance characteristic within the tapered region. Exact analysis is performed by evaluating the cascade *ABCD* matrix of the lines as a function of frequency with the aid of a digital computer. The computed response converges to that of the smooth tapered coupler as the number of line segments N increases, total length l being held constant.

II. Detailed Formulation of the Analysis

Let the impedance of the first segment of the model be denoted by Z_1, the second segment be denoted by $Z_2, \cdots$, the final segment corresponding to an impedance Z_N. The cascade matrix for any segment Z_i is given by

$$\begin{bmatrix} A_i & B_i \\ C_i & D_i \end{bmatrix} = \begin{bmatrix} \cos\theta & jZ_i \sin\theta \\ \dfrac{j}{Z_i}\sin\theta & \cos\theta \end{bmatrix} \tag{1}$$

where θ is the electrical length of each segment in degrees. The response of the network is characterized from the overall cascade matrix

$$\begin{bmatrix} A & B \\ C & D \end{bmatrix} \quad \prod_{i=1}^{N} \begin{bmatrix} A_i & B_i \\ C_i & D_i \end{bmatrix}. \tag{2}$$

Scattering parameters can be determined from the cascade parameters by the following relationships which hold for a normalized one-ohm impedance level:

$$\begin{bmatrix} S_{11} & S_{12} \\ S_{21} & S_{22} \end{bmatrix} = \frac{1}{(B+C)+(A+D)} \begin{bmatrix} (B-C)+(A-D) & 2 \\ 2 & (B-C)-(A-D) \end{bmatrix}. \tag{3}$$

In the analogy, coupling corresponds to reflection from the model line and hence is equal in amplitude and phase to S_{11} or S_{22} depending upon which end of the coupler is fed. Transmission in the coupler is identical to transmission in the model and is given by S_{12} or S_{21} which, as shown by (3), is independent of which end is fed since $S_{12}=S_{21}$.

The frequency response of the coupler can most conveniently be scaled in terms of the parameter l/λ where λ is the

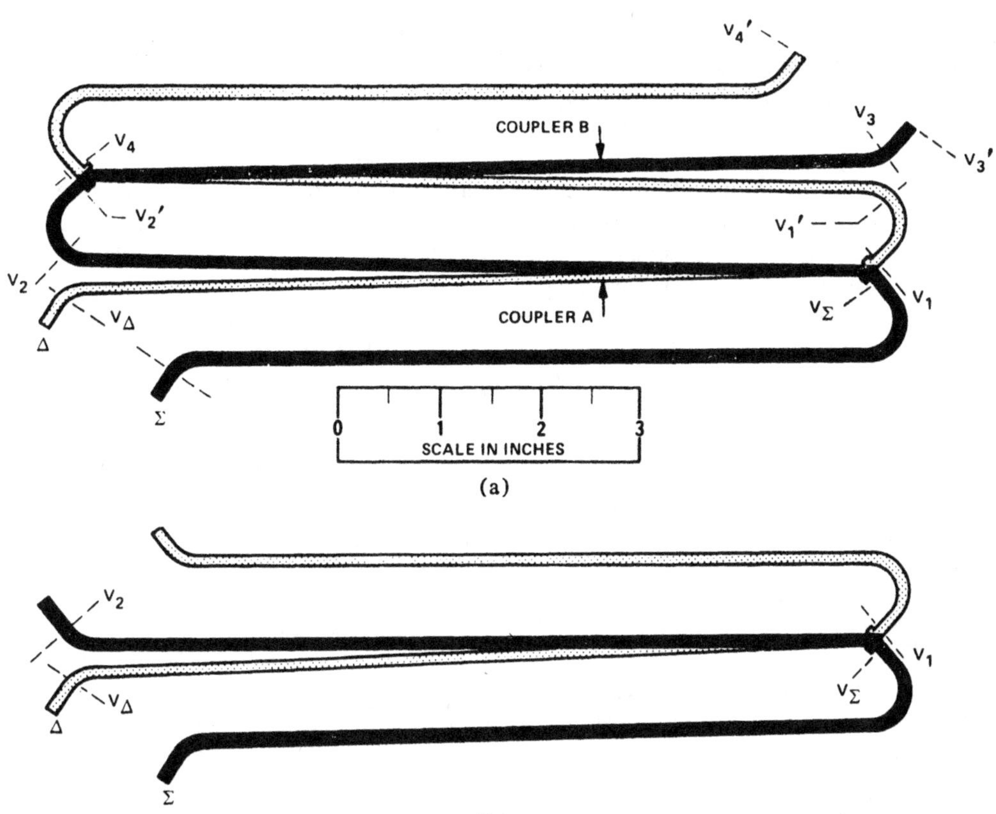

(b)

Fig. 1. X-ray views of asymmetric couplers constructed in three-layer stripline. (a) Tandem interconnection of two asymmetric couplers for use as a 3-dB magic tee. (b) Individual asymmetric coupler including phase reference lines.

wavelength at which the coupler is being evaluated. The approximate design procedures for the various couplers indicate the range of l/λ over which equal-ripple response should be achieved. The corresponding frequency variable θ to be used in the cascade matrix evaluation is given in electrical degrees by

$$\theta = \frac{360l}{N\lambda}. \qquad (4)$$

The choice of a value of N must be such that the frequency parameter θ will be at most several degrees in magnitude at the upper end of the band of interest. As is the case with all commensurate-length unit element cascades [5], computed response is symmetric about $\theta = 90°$ which, in the present application, is unrelated to the actual response of the tapered device. Restated, the restriction is that the length of the steps in the model must be significantly shorter than $\lambda/4$ at the highest frequency to be analyzed to avoid the repetitious behavior of the unit element model.

Several quantities are computed at each frequency. The magnitude of the voltage coupling coefficient $|S_{11}| = |S_{22}|$ is given by

$$|K| = \sqrt{\frac{|B - C|^2 + (A - D)^2}{|B + C|^2 + (A + D)^2}} \qquad (5)$$

For those cases in which two identical couplers are used in tandem, the magnitude of the overall coupling coefficient $|K_T|$ of the tandem combination is given by

$$|K_T| = 2|K|\sqrt{1 - |K|^2}. \qquad (6)$$

The geometry employed in the tandem interconnection of two asymmetric couplers is shown in Fig. 1(a). It will be noted that one of the output lines on each side of the device has been made long enough so as to equalize the distance from the tightly coupled end of the individual couplers to the external terminals. Although this has no effect on the amplitude response of the device as given by (6), it does provide desirable phase performance as will be shown later in Sections IV and V. Similar phase reference lines will be included on all individual asymmetric couplers treated in this paper, as shown in Fig. 1(b).

The phase difference between the coupled and transmitted signals in the case of an individual coupler of length l is given by

$$\phi_0 = \tan^{-1}\left[\frac{-j(B - C)}{\pm(A - D)}\right] - 360l/\lambda. \qquad (7)$$

The first term in (7) represents the angle between the coupled and transmitted signal given by (3), the $\pm$ sign being determined by which end of the coupler is fed, $+$ for S_{11} and $-$ for S_{22}. The second term represents the electrical length of the line used to equalize the distance from the tightly coupled end of the coupler. The phase difference between the output ports of a tandem combination of two such couplers is equal to that of a single coupler as given by (7). This useful relationship can be demonstrated by vectorial addition of the signals present at the output of the coupler shown in Fig. 1(a) as constrained by the scattering relations given in (3). Reference planes pertinent to this exercise have been included in Fig. 1.

The final parameter of interest is the insertion phase dispersion ϕ_d of the coupler, which will be defined as the difference between the phase of the transmitted signal and the phase of a uniform $Z_0=1$-ohm line whose physical length has been chosen to be equal to the length of the coupler. In certain applications, such as antenna feed networks [6]–[8] and swept reflection coefficient measurement systems, a Z_0 reference line is employed in another branch of the system parallel to the coupler to remove as much as possible the insertion phase of the coupler, leaving only a 90° or $\Sigma-\Delta$ phase relation. Coupler insertion phase, however, is dispersive as indicated by the relation

$$\phi_d = \tan^{-1}\left[\frac{-j(B+C)}{A+D}\right] - 360l/\lambda \tag{8}$$

which can be large enough to be of concern in some systems. The first term of (8) comes directly from the transmission coefficient S_{12} or S_{21} in (3), whereas the second term is equal to the electrical length of the coupled section without dispersion. The insertion phase dispersion of a tandem interconnection of two couplers is equal to $2\phi_d$, since the total insertion phase of the transmitted signal can be shown to be twice that of a single coupler (plus nondispersive interconnecting line contributions).

III. Convergence of the Analysis

In the preceding section, it was stated that each segment of the analysis model should be a small fraction of $\lambda/4$ at the highest operating frequency to be analyzed. Further insight into the choice of the total number of segments N to be used in the analysis of a tapered device can be obtained by considering the convergence of a specific example as the number N is increased. The particular coupler used for this study was chosen because of its simple even-mode impedance versus distance characteristics given by $Z=\exp\left[0.024192+0.78220\ (x/l)^2\right]$. The origin of this particular asymmetric -8.34-dB mean coupling design is treated in more detail in Section VI. The computed coupling and phase performance of a tandem combination of two such couplers is shown in Fig. 2 for the choice of $N=300$.

Two of the points shown in the figure have been treated for a wide variey of N values. The first point at $l/\lambda=0.4074$ is at the center of the first coupling maximum, and the second at $l/\lambda=8.00$ is located at a coupling minimum well out in the equal-ripple region. Fig. 3 shows the coupling computed at the two points. It is evident from the figure that a value of N on the order of several thousand is required for four-place accuracy. Although analysis of response at a single frequency entails the matrix multiplication of N 2×2 individual segment matrices, the problem can be handled with ease on many available computers. A check of Fig. 3 shows that the $N=300$ solution plotted in Fig. 2 is within about 0.02 dB of being equal to the response for large N.

Convergence should be checked for each new class of device analyzed. The 90° symmetric coupler analyzed in Section VII possesses a periodically varying even-mode characteristic impedance and a bandpass coupling characteristic. It requires $N=1800$ to achieve accuracy similar to the $N=300$ solution for the short asymmetric design of Fig. 2. The value of θ at the high end of the 90° coupler bandpass (at $l/\lambda=4.58$) is 0.915° or a factor of 10 smaller than that permissible in the asymmetric coupler analysis for similar accuracy. The analysis of the response of the 90° design at a given frequency does not converge monotonically with N as in Fig. 3, but ripples through values both above and below the large N solution. The $\Sigma-\Delta$ magic tee required the smallest value of N of any of the three coupler designs treated in this paper. It is felt that this is due in large measure to the early establishment of both equal-ripple phase and amplitude performance at the low l/λ end of the band.

IV. Asymptotic High-Frequency Solution for Asymmetric Couplers

Asymmetric tapered-line couplers designed to produce high-pass coupling behavior have been known for many years [9], [10]. Construction of such devices has been reported more recently [1], [3]. The most commonly described designs employ exponentially tapered-coupled lines [4], [9]–[11]. In terms of the even-mode impedance analogy, the design consists of an exponentially tapered impedance transformer which transforms the $Z_0=1$-ohm level present at one end of the coupler to a higher impedance level followed by an abrupt return to the one-ohm level. The reflection of energy from the latter discontinuity provides the requisite coupling in the high-pass manner of an exponential impedance taper. Ripple level is greatest at the low end of the band and decreases monotonically with increasing frequency. Any such continuously tapered impedance transformer will exhibit vanishing ripple amplitude at sufficiently high frequencies. A classic example of this effect is provided by the continuous "equal-ripple" tapered transformer of Collin [12]. True equal-ripple transformation, which continues to ripple with undiminished amplitude at high frequencies, requires that a discontinuity be present in the impedance characteristic of the transformer. A first-order treatment of the optimum equal-ripple tapered-line impedance transformer, given by Klopfenstein [13], contains two such discontinuities, one at either end of the structure.

The Klopfenstein taper has been employed in the construction of asymmetric couplers [1], although details of applying the taper to couplers are not included in the reference. Equations relating amplitude ripple and phase error of an individual coupler to the sidelobe level of the Klopfenstein taper are given, although the equation for amplitude ripple appears to be in error by a factor of approximately 2. This is believed to be related to the fact that one of the two discontinuities which normally contribute to ripple amplitude in the prototype transformer is effectively replaced by the large step present in the coupler. An exact high-frequency asymptotic solution is derived below.

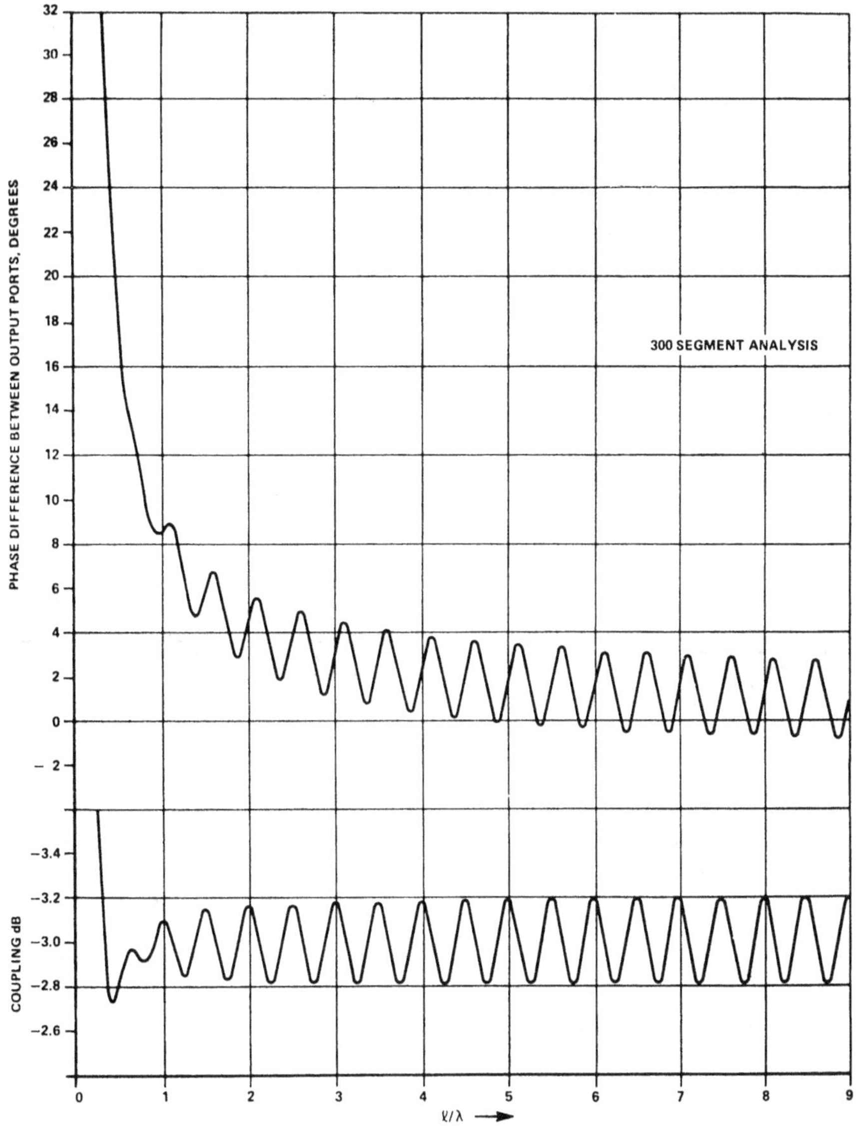

Fig. 2. Phase and amplitude performance of a short asymmetric tapered tandem coupler.

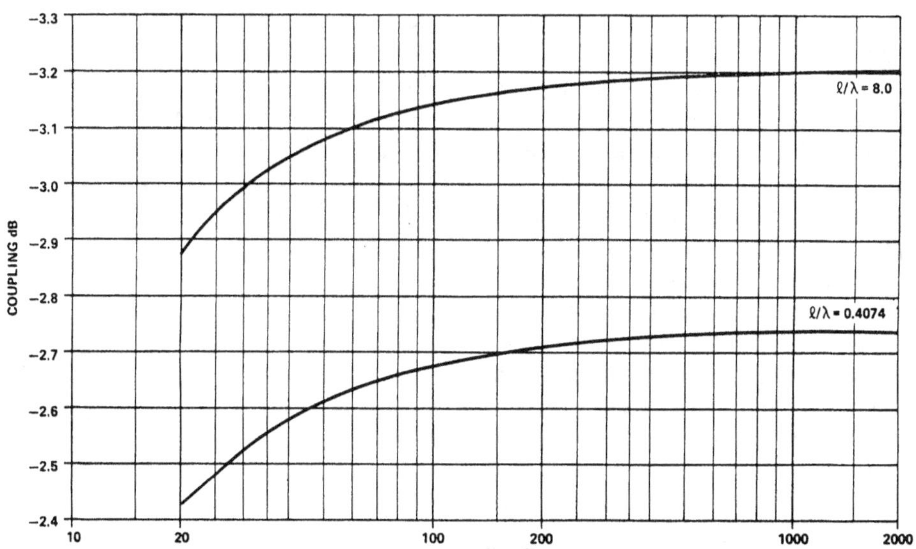

Fig. 3. Computed coupling values at two specific frequencies in Fig. 2 versus N.

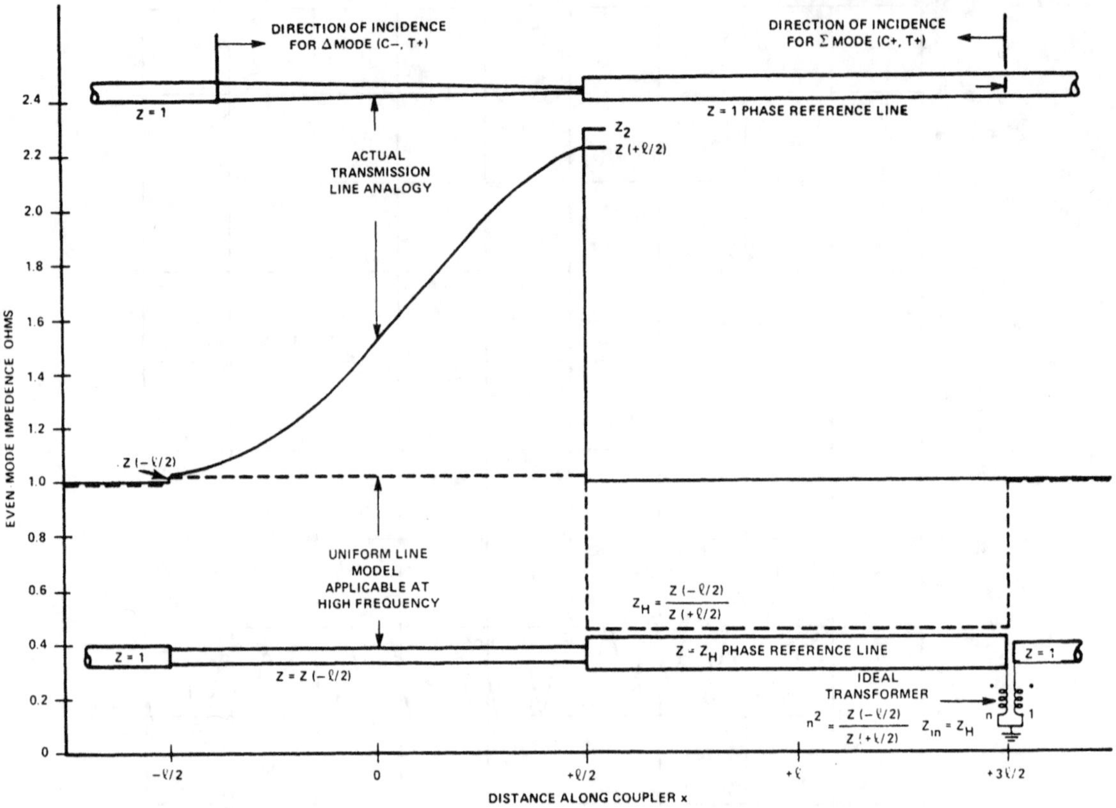

Fig. 4. Transmission-line analogy, Σ−Δ Tee (25-dB sidelobe, −8.34-dB coupling).

The transmission-line analogy of the asymmetric coupler shown in Fig. 1(b) is given in the top of Fig. 4, the specific taper shown being a 25-dB sidelobe Klopfenstein design normalized to produce −8.34-dB mean coupling. In the treatment of the problem to come, S_{11} will be associated with feeding the low impedance end of the taper located at $x=-l/2$. The desired phase response reference is provided in the analogy by using a one-ohm line equal in length to that of the transformer and connected at the impedance step which produces the coupling, as indicated in Fig. 4. As frequency approaches infinity, the impedance transformation within the smooth taper between the level $Z(-l/2)$ and that of $Z(+l/2)$ becomes perfect, with no reflections arising internal to the taper [12]. Only the small initial step at $-l/2$ and the step at $+l/2$ influence the response. This circumstance can be represented by the uniform-line model shown at the bottom of Fig. 4 which was derived by providing the same phase and magnitude of reflection coefficient at the steps as is present in the actual device. An ideal transformer is used in the uniform-line model to represent the effect of the taper; the transformer could be placed in any one of several places in this model. The value of Z_H shown results from equating the reflection coefficient present at $+l/2$ in the uniform-line model with that present in the tapered analogy.

The uniform line model will be first analyzed to determine the maximum and minimum coupling. Consider the coupling provided in the difference mode when the $-l/2$ side of the structure is excited. The corresponding reflection coefficient of the model can be found by using loss-free transmission-line equations, with the result that

$$\Gamma = \frac{[Z(-l/2) - Z(+l/2)] + j[Z(+l/2)Z(-l/2) - 1] \tan \alpha}{[Z(-l/2) + Z(+l/2)] + j[Z(+l/2)Z(-l/2) + 1] \tan \alpha} \tag{9}$$

where α is the electrical length of the uniform $Z(-l/2)$ line in degrees. The maxima and minima of $|\Gamma|$ correspond to maxima and minima of $|\Gamma|^2$ which may be found by differentiation. With the constraint that $1 \le Z(-l/2) < Z(+l/2)$, the zeros of $d|\Gamma|^2/d\alpha$ which occur at $\alpha = 0°, 180°, 360°, \cdots$, correspond to minima in coupling, whereas those zeros at $\alpha = 90°, 270°, 450°, \cdots$, correspond to voltage maxima. Accordingly,

$$|\Gamma|_{\min} \equiv K_{\min} = \frac{Z(+l/2) - Z(-l/2)}{Z(+l/2) + Z(-l/2)} \tag{10}$$

$$|\Gamma|_{\max} \equiv K_{\max} = \frac{Z(-l/2)Z(+l/2) - 1}{Z(-l/2)Z(+l/2) + 1}. \tag{11}$$

The mean coupling level corresponds to the magnitude of the reflection coefficient present at $+l/2$ due to the jump from $Z_j = Z(+l/2)$ down to one ohm in the even-mode analogy. Thus

$$|K|_{\text{mean}} = \frac{Z_j - 1}{Z_j + 1}, \tag{12a}$$

which can be rewritten in the form of a design equation

$$Z_j = \frac{1 + |K|_{\text{mean}}}{1 - |K|_{\text{mean}}}. \tag{12b}$$

For a -8.34-dB mean level, $Z_j = 2.2398$ ohms. The foregoing equation can be used in the design of any asymmetric coupler at frequencies high enough for the taper to provide transformation between two real impedance levels.

The relative phase between the coupled and transmitted signals may be determined from scattering parameters for the asymptotic model. One must use a more general scattering treatment than (3), since the load impedance, as far as the $Z=Z(-l/2)$ unit element is concerned, is Z_H instead of unity. The formation of the scattering parameters from the *ABCD* parameters under these conditions follows from [14].[1] The relative angle between S_{11} and S_{21}, which is associated with exciting the $-l/2$ side of the coupler (Δ mode) can accordingly be shown to be

$$\phi_1 = \tan^{-1} \frac{\sin\alpha\left[Z(-l/2) - \dfrac{Z_H}{Z(-l/2)}\right]}{\cos\alpha(Z_H - 1)} \tag{13}$$

$$\equiv \tan^{-1}(-M\tan\alpha),$$

where

$$M \equiv -\left[\frac{Z(-l/2) - \dfrac{Z_H}{Z(-l/2)}}{Z_H - 1}\right] \tag{14}$$

$$= \frac{Z(-l/2)Z(+l/2) - 1}{Z(+l/2) - Z(-l/2)} \geq 0.$$

In a similar manner, the relative phase between S_{22} and S_{21} can be determined to be

$$\phi_2 = \tan^{-1}(+M\tan\alpha) \tag{15}$$

for the case of exciting the Σ mode with the signal incident on the $+l/2$ side of the coupler.

Phase error can be determined most simply for the Σ-mode excitation. The angle ϕ_2 represents the angle by which the coupled signal leads the transmitted signal with the former being measured at the $+l/2$ plane and the latter at the $-l/2$ plane. The coupled signal must additionally pass through the phase reference line of length α before it actually reaches the external terminal plane, so that the difference in phase ϕ_0 between the coupled and transmitted signals at the output terminals is given by

$$\phi_0 = \phi_2 - \alpha = \tan^{-1}(m\tan\alpha) - \alpha \tag{16}$$

The maximum phase error can be found by differentiation. The lowest order zero of $d\phi_0/d\alpha$ occurs at $\tan\alpha = 1/\sqrt{M}$. The corresponding phase error is

$$\phi_{\max} = \tan^{-1}\left(\frac{M-1}{2\sqrt{M}}\right). \tag{17}$$

The coupled and transmitted ports alternately lead and lag one another, the maximum difference in phase being given by (16). A tandem interconnection of two such couplers will exhibit identical maximum phase error.

Although coupling maxima and minima were determined for Δ-mode excitation, Σ-mode excitation will produce the same maximum and minimum coupling values. Similarly, the maximum phase error predicted for the Σ-mode excitation as shown will be identical with that produced by the Δ mode, the mean phase difference in the latter case being 180 degrees instead of zero degrees.

V. The Tapered $\Sigma-\Delta$ Magic Tee

Sum and difference performance will be provided by asymmetric couplers at high frequencies if the proper phase reference lines are incorporated. In general, however, the phase performance of asymmetric designs will depart from $\Sigma-\Delta$ behavior at lower frequencies unless the taper provides essentially real impedance transformation between one ohm and Z_j. The use of the Klopfenstein equal-ripple impedance taper [13] has been suggested in [1]. This taper provides transformation between two impedance levels Z_1 and Z_2 above a given low-frequency cutoff. As mentioned in Section IV, there are two small discontinuities in Z located at the ends of the tapered transformer; one of the discontinuities loses its identity in the coupler because of the abrupt return to one ohm. The desired prototype Klopfenstein taper corresponds to $Z_1 = 1$ ohm, with Z_2 chosen so that the impedance level on the tapered side of the step at $+l/2$ equals the desired value of Z_j. The design procedure for the coupler, omitted in [1], is given below.

The desired operating parameters of the prototype transformer can be described in terms of a sidelobe parameter

$$S = 20\log_{10}(\cosh A), \tag{18}$$

where S is in decibels. The largest reflection coefficient Γ_0 exhibited by the tapered impedance transformer occurs at dc where the transformer has no effect. This mismatch will be reduced to a maximum (sidelobe) level of

$$\Gamma_{\max} = \frac{\Gamma_0}{\cosh A} \tag{19}$$

by the action of the transformer in an equal-ripple sense for frequencies above a low-frequency cutoff given by

$$l/\lambda_c = \frac{A}{2\pi} \tag{20}$$

[1] Note that in [14], in the expression for $\{S\}$ in terms of *ABCD* parameters, a plus sign belongs between the two expressions in parentheses in the denominator of the common factor:

$$\frac{1}{(B + CZ_{01}Z_{02}) + (AZ_{02} + DZ_{01})}$$

The choice of a larger numerical sidelobe parameter S will reduce the maximum in-band reflection coefficient at the expense of a longer structure.

The even-mode impedance of the Klopfenstein taper, for the choice of $Z_1 = 1$ ohm, is given by

$$\ln Z(x) = \frac{1}{2} \ln Z_2 + \frac{A^2 \ln Z_2}{2 \cosh A} \Phi(z = 2x/l, A), \quad -l/2 \leq x \leq +l/2 \tag{21a}$$

$$\ln Z(x) = 0, \quad x < -l/2 \tag{21b}$$

$$\ln Z(x) = \ln Z_2, \quad x > + /2. \tag{21c}$$

The above equations are corrected forms of [13, eq. (12)]. The substitution suggested in [13, eq. (17)], has been employed above to modify the end points on the taper, which would otherwise be in errer due to the loose coupling approximations involved. The function $\Phi(z, A)$ is given in tabular form in [13] for values of S from 0 to 40 dB in 5-dB steps. Φ is an odd function of Z, so that $\Phi(-z, A) = -\Phi(z, A)$.

The value of $\Phi(1, A)$ present at the tapered side of the discontinuity at $x = +l/2$ is

$$\Phi(1, A) = \frac{\cosh A - 1}{A^2}, \tag{22}$$

according to [13, eq. (15)]. The corresponding impedance of the taper at $+l/2$ may be found by substituting (22) into (21a). This impedance level must be equal to Z_j for the coupler application, so that

$$\ln Z(+l/2) = \ln Z_2 - \frac{\ln Z_2}{2 \cosh A} \equiv \ln Z_j. \tag{23}$$

The value of Z_2, the upper impedance level of the prototype, accordingly is found to be

$$Z_2 = Z_j^{[(2 \cosh A)/(2 \cosh A - 1)]}. \tag{24}$$

As is apparent in Fig. 4, this impedance level has physical significance only in the prototype transformer, since the coupler abruptly returns to the one-ohm level instead of to Z_2. The design equation for the even-mode impedance along the coupler follows from (21a) and (24):

$$Z(x) = \exp\left[\left\{\frac{A^2\Phi\left(z = \frac{2x}{l}, A\right) + \cosh A}{2 \cosh A - 1}\right\} \cdot \ln Z_j\right], \quad -l/2 \leq x \leq +l/2. \tag{25}$$

The performance to be expected from the above design can be obtained by using the asymptotic model given in the last section. The parameters required for evaluating the asymptotic model are $Z(+l/2) = Z_j$ and $Z(-l/2)$. The latter impedance can be found by evaluating (21) at $x = -l/2$, where $\Phi(-1, A) = -\Phi(1, A) = (1 - \cosh A)/A^2$, and where Z_2 is given by (24). The result is

$$Z(-l/2) = Z_j^{[(1)/(2 \cosh A - 1)]}. \tag{26}$$

TABLE I

PREDICTED BEHAVIOR OF A TANDEM INTERCONNECTION OF TWO −8.34 dB Σ-Δ HYBRIDS

S, Sidelobes (dB)	Maximum Coupling (dB)	Minimum Coupling (dB)	ϕ Error (degrees)	Length (l/λ_c)
5.00	−1.1513	−6.6101	24.1256	0.1875
10.00	−1.9651	−4.4365	11.3909	0.2894
15.00	−2.4269	−3.6956	5.8987	0.3839
20.00	−2.6841	−3.3662	3.1784	0.4764
25.00	−2.8276	−3.2021	1.7467	0.5683
30.00	−2.9078	−3.1158	0.9699	0.66
35.00	−2.9528	−3.0689	0.5416	0.7516
40.00	−2.978	−3.0431	0.3034	0.8432

Equations (10), (11), (17), and (20) have been evaluated for $Z_j = 2.2398\Omega$, corresponding to a −8.34-dB design, for various sidelobe parameters, S. The results are listed in Table I in terms of the tandem interconnection of two identical units for −3-dB overall coupling.

A computer analysis of an $S = 25$-dB sidelobe design, including phase error, is shown in Fig. 5. Although the asymptotic model is strictly correct only for high frequencies, it will be noted that the predictions of the model given in Table I hold down to low frequencies despite the approximate nature of the equal-ripple taper design. The amplitude response rather than phase response sets the lowest operating frequency of the device—for maximum 3-degree phase error, coupling exhibits a peak of about −1.5 dB at $l/\lambda = 0.45$. A more reasonable choice of lower cutoff would be $l/\lambda = 0.60$ above which coupling is within 0.31 dB of 3.01 dB, and phase is within 2.2 degrees of ideal. It might be noted that Table I predicts a cutoff of $l/\lambda = 0.5683$, above which the maximum deviation from 3.01 should be 0.19 dB. This deviation from ideal behavior within the coupling band is due to the nature of the first-order design theory.

The insertion phase dispersion of the 25-dB tandem design, which is equal to $2\phi_d$, is shown in Fig. 6. Except for the low-frequency region, the amount of dispersion present is quite small. The choice of a lower frequency cutoff of $l/\lambda = 0.60$, discussed above, results in a maximum of 3.5-degree insertion phase dispersion which rapidly diminishes with increasing frequency.

The model used for the 25-dB sidelobe computer analysis was the even-mode transmission-line impedance given by (25). The function $\Phi(z, A)$ to be employed is tabulated in [13] covering the interval $0 \leq z \leq 1$ with 21 equally spaced points. Since the coupler covers both positive and negative z values, the tabular data of [13] can be used directly for an $N = 40$ segment analysis. Although four-point Lagrangian interpolation was used to obtain a 160-step analysis, the use of a recently published Fortran subroutine for $\Phi(z, A)$ is to be preferred [15].

Both 25-dB and 40-dB sidelobe models have been constructed. Compensation of the junctions at which the abrupt coupling change occurs is quite critical; modifications which tend to improve VSWR and isolation often deteriorate the abruptness of the junction with a resultant reduction in cou-

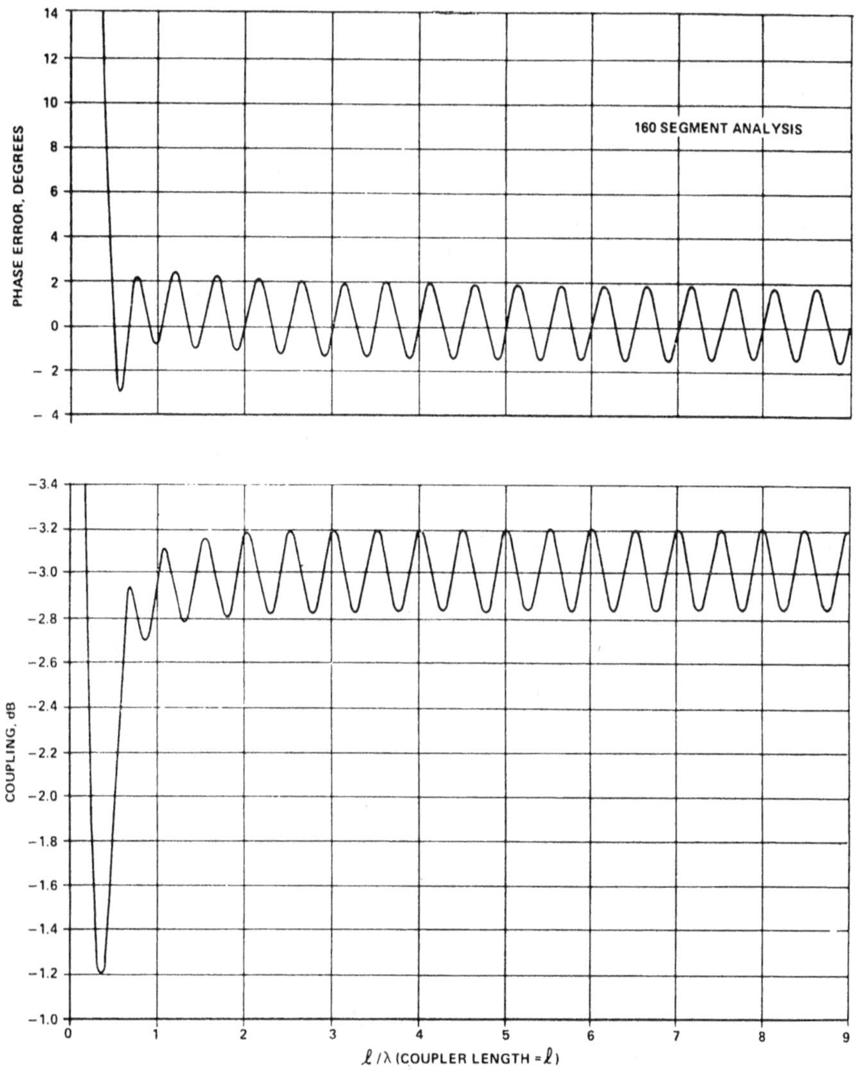

Fig. 5. Theoretical response of two 25-dB sidelobe, −8.34-dB Σ−Δ couplers in tandem.

pling at higher frequencies (see [1], Fig. 6). The tuning tabs at the junctions shown in Fig. 1(a) are symbolic, the final junction compensation being achieved by removing copper in the overlap region at the very end of the junction.

Reference [1] reports that a variable air gap in the vicinity of the center shim was required for maximum directivity. The need for this unwieldy gap has been eliminated by the use of a composite center shim of the basic stripline dielectric surrounded symmetrically by FEP teflon which has a lower dielectric constant. The presence of air or of teflon in the center shim region increases odd-mode propagation velocity with respect to even-mode velocity. It is believed that the location of the reference planes associated with termination of odd- and even-mode propagation tend to occur at slightly different points near the end of the junction, due to the various parasitic reactances present. By using the proper proportion of teflon, the relative mode velocities can be adjusted to effectively equalize the total electrical length encountered in both modes.

Fig. 7 shows the experimental performance achieved in a 40-dB sidelobe model constructed using polyolefin stripline consisting of two 0.062-inch-thick outer boards and a center shim of 14.5 mils of polyolefin surrounded by 0.5-mil FEP teflon on each side. Design cutoff is 800 MHz. Maximum unbalance between coupled and transmitted ports is 1 dB, which corresponds to a 3 ± 0.5-dB coupling specification. Isolation is greater than 26 dB over 86 percent of the band, and greater than 23 dB over 96 percent of the band (*L*, *S*, *C*, and *X*). Deviation in phase from 0° in the Σ mode or from 180° in the Δ mode is less than about 4° from 1 through 7 GHz. These phase measurements were obtained on a swept basis with the aid of a specially calibrated Hewlett–Packard 8410A network analyzer. The tendency toward negative relative phase at higher frequencies is due to a nonoptimum choice of reference planes. If terminal V_4' is shortened by 0.0107 inch relative to terminal V_3' (Fig. 1), phase error is less than 6° from 1 through 12 GHz, as measured from the dashed zero phase error line in Fig. 7.

Upon correcting for the effects of insertion loss, 1.5-dB at 12.5 GHz, the device still exhibits considerably greater coupling ripple than indicated by theory. It is expected that improved performance could be achieved primarily through

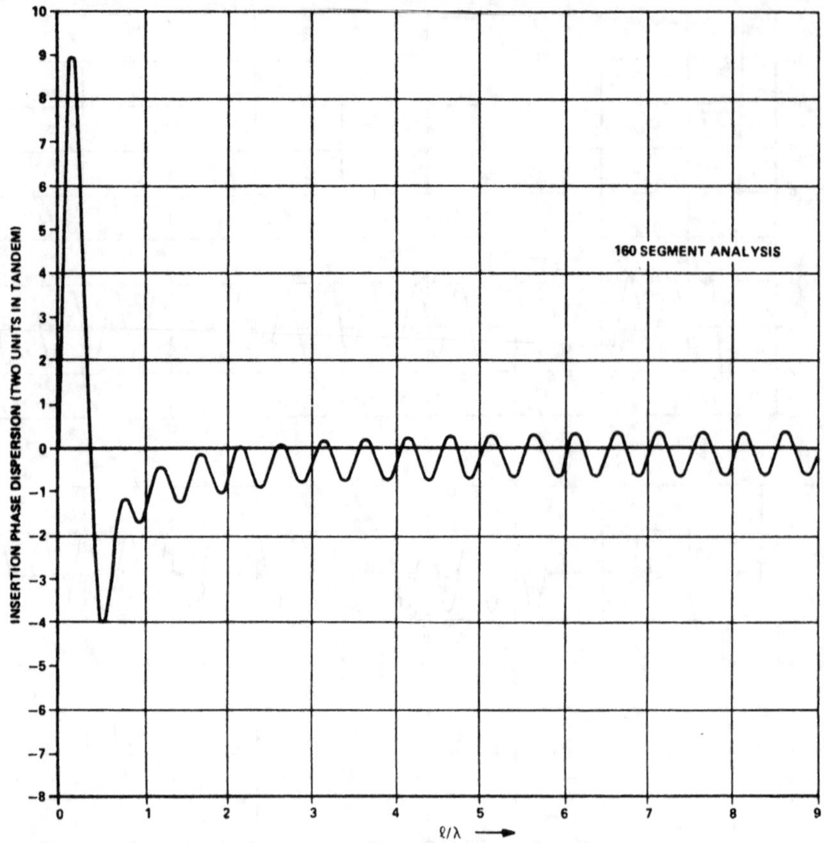

Fig. 6. Insertion phase dispersion of two 25-dB sidelobe, −8.34-dB couplers in tandem.

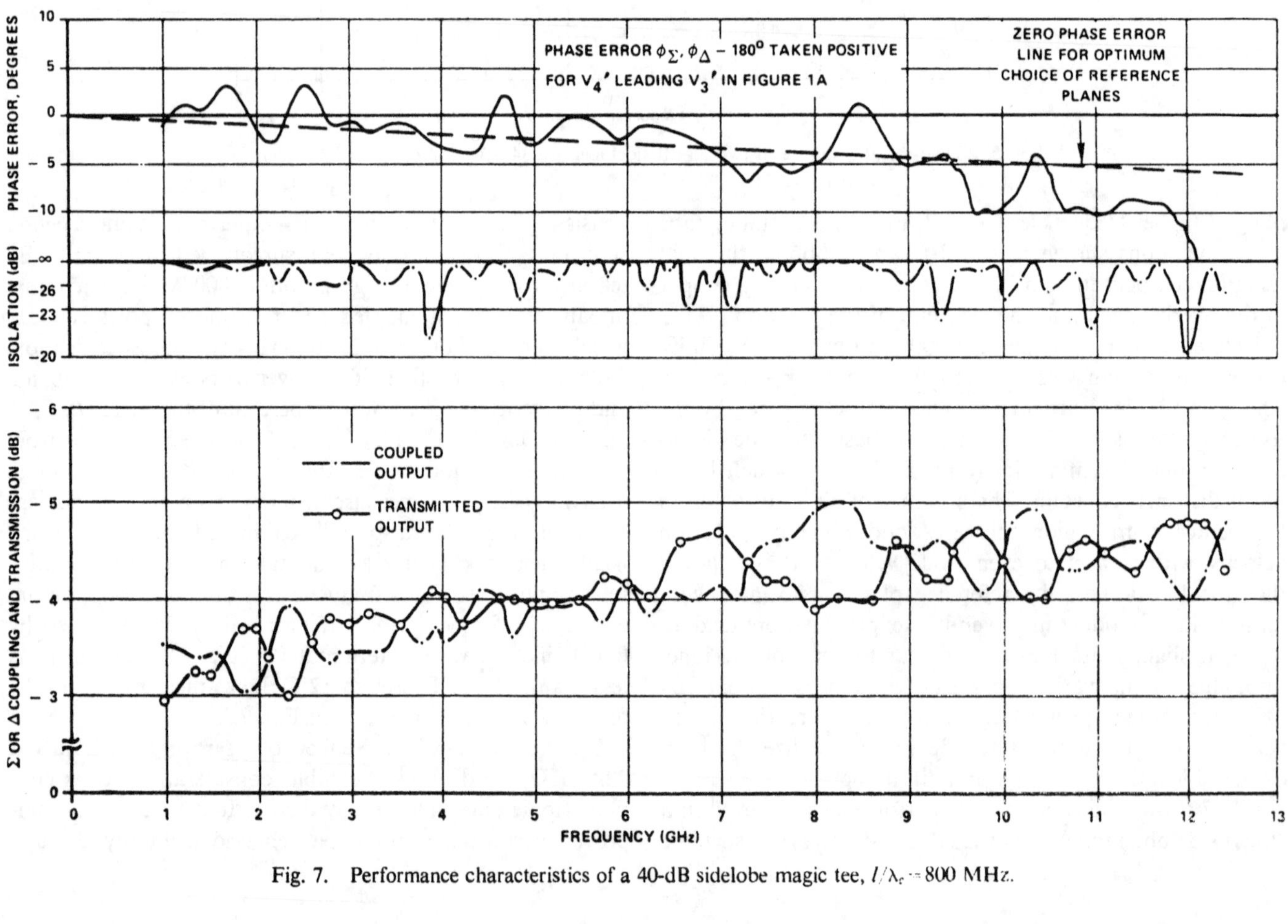

Fig. 7. Performance characteristics of a 40-dB sidelobe magic tee, $l/\lambda_c = 800$ MHz.

252

the use of better camera optics in the photoreduction process. The large size of the X1 negative required aligning the coupler along the diagonal of the sensitized plate, using portions of the field of view known to have distortion. It is doubtful, however, that extremely close correlation with the very low theoretical ripple tolerance indicated could be achieved in practice. It would appear more practical to utilize a sidelobe parameter which provides somewhat larger coupling ripple, thereby reducing the length (and hence camera distortion) and the overall insertion loss of the structure.

Accordingly, a 25-dB sidelobe, 1 GHz lower cutoff design has been constructed in Rexolite 1422 stripline. The unit is about one-half as large as the 800 MHz, 40-dB unit which is shown in scale in Fig. 1(a). Due to the higher dielectric losses of Rexolite, insertion loss remains at 1.5 dB at 12.5 GHz. The temperature stability and cold flow properties of Rexolite are, however, superior to those of polyolefin. Isolation is about 30 dB in L band, decreasing to about 22 dB worst case in X band. Although there are two peaks of 1-dB maximum unbalance between outputs around S band, unbalance above 5 GHz is 0.5 dB maximum, which compares well with the approximate 0.4 dB maximum predicted by the computer.

Although advances in the theory of nonuniform lines may some day provide a more exact equal-ripple design procedure, the performance indicated for the procedure at hand is quite adequate for the majority of engineering applications. Several new and rather interesting mixer and harmonic generator applications can be devised based on the availability of such multi-octave magic tees.

VI. The Tapered Asymmetric Equal-Ripple Coupler

The $\Sigma-\Delta$ magic tee described in the preceding section utilizes a taper whose prototype provides a real impedance transformation between two impedance levels. It is not necessary to employ such a restrictive taper if only coupling amplitude is of concern. Recently Arndt [3] has described an equal-ripple tapered asymmetric design based on amplitude considerations alone. His earliest synthesis approach consisted of adding together cosine-squared, exponential, linear, and constant coupling versus distance characteristics, the relative amplitudes of each term being adjusted until an approximate equal-ripple coupling amplitude has been achieved [16]. Amplitude equalization has been treated more compactly by using a simple power series representation of the coupling versus distance characteristic [3], [16].

The design results given in [3] are quite adequate for loose-coupled designs of 20-dB mean level and below, as demonstrated by the experimental results presented. The design has the advantage of short length for a given low-frequency cutoff. One cannot, however, directly employ the results of Arndt for a -8.34-dB mean level design due to the failure of the loose coupling approximation used in his design procedure, as illustrated below. A -8.34-dB mean level corresponds to 0.3827 mean voltage coupling. According to [3, eq. (5)], the coupling coefficient k_0 present at the tightly coupled end of the design should be twice the desired mean voltage coupling level, or $k_0=0.7654$. The actual mean coupling level that would be provided by such a design is given by (12), where

$$Z_j = \sqrt{\frac{1+0.7654}{1-0.7654}} = 2.7432.$$

Accordingly, $K_{\text{mean}}=0.46569$ or -6.64 dB, which is greater than desired.

One possible correction, which should prove to be superior under tightly coupled conditions, employs a change in variable to a form associated with the differential reflection coefficient [2], [4] $p(u)=\frac{1}{2}d/du \ln Z_{oe}(u)$. The basic substitution, $k=\ln Z_{oe}$, is similar in effect to a substitution used to improve the approximate synthesis of stepped impedance transformers [17]. A numerical example will be employed to illustrate the technique. According to [3], the coupling coefficient distribution for an asymmetric coupler possessing ± 3 percent voltage coupling ripple is

$$\frac{k}{k_0}\left(\frac{z}{l}\right) = 1 - 1.940\left(\frac{z}{l}\right) + 0.970\left(\frac{z}{l}\right)^2. \tag{27}$$

The distribution has maximum coupling at $z=0$. In order to be consistent with the previous treatment in which a minimum coupling occurs at $x=0$, one makes the substitution $z=l-x$ with the result that

$$k\left(\frac{x}{l}\right) = k_0\left[0.03 + 0.970\left(\frac{x}{l}\right)^2\right]. \tag{28}$$

The cancellation of the linear term is a coincidence associated with the particular ripple level. If the even-mode impedance present at $x/l=1$ is chosen to be Z_j for a given mean coupling level, (25), the change of variables results in the expression

$$\ln Z_{oe}\left(\frac{x}{l}\right) = \ln Z_j\left[0.03 + 0.970\left(\frac{x}{l}\right)^2\right]. \tag{29}$$

The response of this design for a -8.34-dB mean level is shown in Fig. 2 for two such units in tandem. The convergence of the analysis of this particular design was treated in Section III because of its simple even-mode versus distance characteristic.

It will be noted in Fig. 2 that the low-frequency overshoot present is within 0.08 dB of the desired equal-ripple level. The low-frequency cutoff is approximately $l/\lambda_c=0.29$, at which point the $\Sigma-\Delta$ phase error is approximately 37°. As is evident from the figure, the convergence of phase error to a zero mean occurs quite slowly. The device could be used as a magic tee above $l/\lambda=6.6$ with a maximum phase error of less than 3°. The structure would have to be $6.6/0.6=11$ times as long as the $\Sigma-\Delta$ tee of the last section for roughly similar magic tee performance. Neglecting phase performance, however, the modified Arndt design is less than half the length of the preceding 25-dB $\Sigma-\Delta$ tee for the same low-frequency amplitude cutoff. The size advantage can be enhanced by removing the phase reference lines for those applications in which $\Sigma-\Delta$ performance at high frequencies

TABLE II
PREDICTED BEHAVIOR OF A TANDEM INTERCONNECTION OF TWO −8.34-dB MODIFIED ARNDT DESIGNS

Voltage Ripple ΔA (percent)	Maximum Coupling (dB)	Minimum Coupling (dB)	ϕ Error (degrees)	Length (l/λ_c)
10	−2.4137	−3.714	6.0447	0.185
9	−2.469	−3.6383	5.4388	0.19
8	−2.5253	−3.5638	4.8334	0.20
7	−2.5825	−3.4905	4.2284	0.21
6	−2.6407	−3.4185	3.6238	0.23
5	−2.6998	−3.3477	3.0194	0.25
4	−2.7599	−3.278	2.4152	0.27
3	−2.821	−3.2094	1.8112	0.30
2	−2.8831	−3.142	1.2074	0.33
1	−2.9462	−3.0756	0.6037	0.41

is not required, thereby returning the design to the basic physical form presented by Arndt.

Reference [3] contains power series representations of k/k_0 for voltage ripple levels ΔA covering from 1 to 10 percent of the mean coupling level. A table of the performance expected from two −8.34-dB couplers in tandem has been prepared, based on the substitution of $\ln Z/\ln Z_j = \ln Z/\ln 2.2398$ for k/k_0 in Arndt's design equations. Coupling amplitude and high-frequency phase error predictions are based on the asymptotic model of Section IV [(10), (11), (17)]. The low-frequency cutoff values listed in Table II correspond to the design values of [3].

The $N=300$ computer analysis of the 3 percent ripple case shown in Fig. 2 substantiates the tabular predictions closely. Note that, although phase error has not settled down to its asymptotic value by $l/\lambda=9$, the peak-to-peak variation at high frequencies correspond well to twice the peak value listed in Table II.

VII. THE 90° TAPERED-LINE COUPLER

A design procedure for tapered-line 90° couplers which employs weighting functions from equal-ripple stepped designs is given in [2]. Fourier integral computer analysis and the experimental results for a ninth-order tapered design presented in [2, Fig. 6] indicate that the design procedure is at least quasi-equal ripple. The analysis techniques described earlier permit a more accurate determination of the response characteristic. An $N=1800$ step analysis of this design is shown in Fig. 8. The ninth-order stepped prototype used exhibits a −8.34-dB±0.30-dB coupling which is equivalent to a maximum voltage coupling of 0.396 and a minimum of 0.370. The low-frequency ripples of the tapered design, Fig. 8, are within 0.001 of these extremes, gradually exhibiting smaller ripples at the higher frequencies. Bandwidth of the tapered design, figured at the 0.370 level is 11.1:1, whereas that of the prototype weighting function design is 10.96:1. Similar results have been obtained for a number of 90-degree tapered designs. It appears that the design procedure produces a tapered coupler which is within the ripple specification of the stepped prototype. The bandwidth is slightly greater than that of the prototype due to the more gradual slopeoff of coupling on the high-frequency end of the band, a feature inherent in the nonrepetitive nature of the tapered coupler. It is conceivable that an iterative synthesis technique could be employed to produce equal coupling ripple, although the benefit to be obtained over the decreasing ripple of the present procedure might not justify the effort required.

The experimental data of [2, Fig. 9] exhibit a dropoff in coupling at high frequencies that is not entirely attributable to insertion loss. Empirical correction procedures for this reduced coupling are discussed in [18] (which describes the construction of a broader bandwidth eleventh-order tapered design). A tentative explanation for part of the effect has been found by a computer analysis of the ninth-order tapered design truncated at the length of the stepped prototype. Response is reasonably similar to the normal tapered design, except for diminished coupling of the highest frequency ripple, the effect noted in actual tapered models. The tapered coupler is normally one-quarter wavelength longer than the stepped prototype, the difference being in the very loosely coupled tails at each end of the coupler, as shown in [2, Fig. 6]. These regions correspond to circular bends in the stripline as is visible in [2, Fig. 8]. Magnetic coupling tends to be excessively diminished by the large angle between the lines in these regions. Although overall coupling is low, there will be an adverse contribution to directivity [7]. The spacing between the lines can exceed $\lambda/4$ at high frequencies, which is believed to result in relative time phasing which does not permit the coupling in these regions to add correctly to the overall coupled response. The use of miters in this region has provided the only specialized circumstance found to date in which the use of stepped coupling has resulted in improved directivity.

The theoretical insertion-phase dispersion of the unmodified −8.34-dB ninth-order tapered design, used in tandem for 3-dB overall coupling, is shown in Fig. 9 along with the insertion phase dispersion calculated for a stepped ninth-order −3.01±0.20-dB symmetrical design [19]. It will be noted that dispersion of the tandem tapered design is on the same order of magnitude as the nontandem stepped model at high frequencies. The insertion phase dispersion of two stepped ninth-order couplers connected in tandem has been calculated for the case of −8.34-dB±0.30-dB coupling which served as the prototype for the tapered design. Phase dispersion of the tapered coupler is nearly equal to that of the prototype stepped design at low frequencies, but exceeds the dispersion of the stepped design by a factor of 2 at the high end of the band.

A plot of insertion phase dispersion for equal-ripple symmetric stepped couplers of various mean levels and ripple tolerances measured at the amplitude bandwidth edge is shown in Fig. 10. Although the insertion phase dispersion of a tapered 90° coupler should ideally be computed for each case of interest, the lower band edge dispersion will be quite close to that predicted by Fig. 10 for the stepped prototype. Dispersion at the upper band edge will be higher, a factor of 2 being suggested based on the limited analysis obtained to date. This increased high-frequency dispersion, which may be attributed to the nonrepeating nature of the tapered cou-

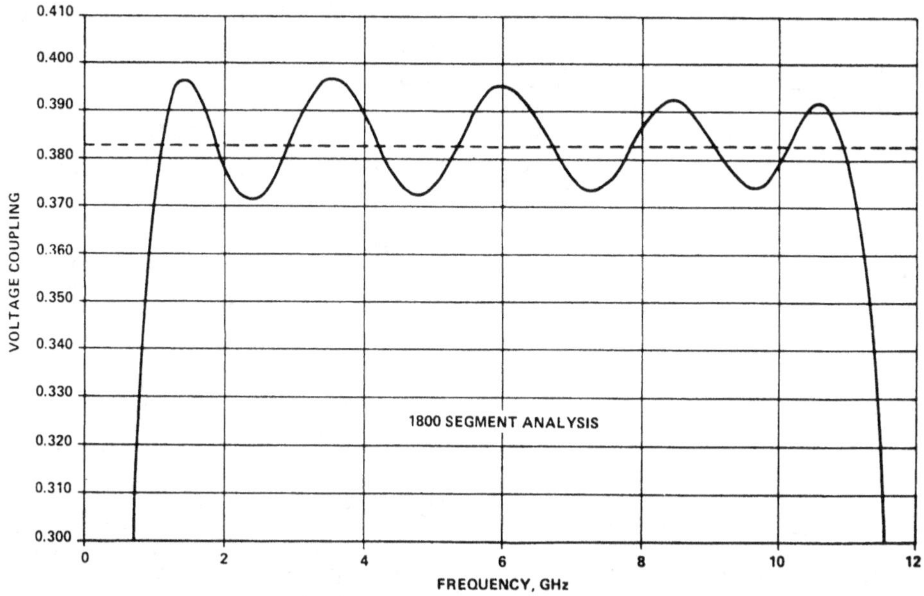

Fig. 8. −8.34-dB tapered 90° coupler theoretical response.

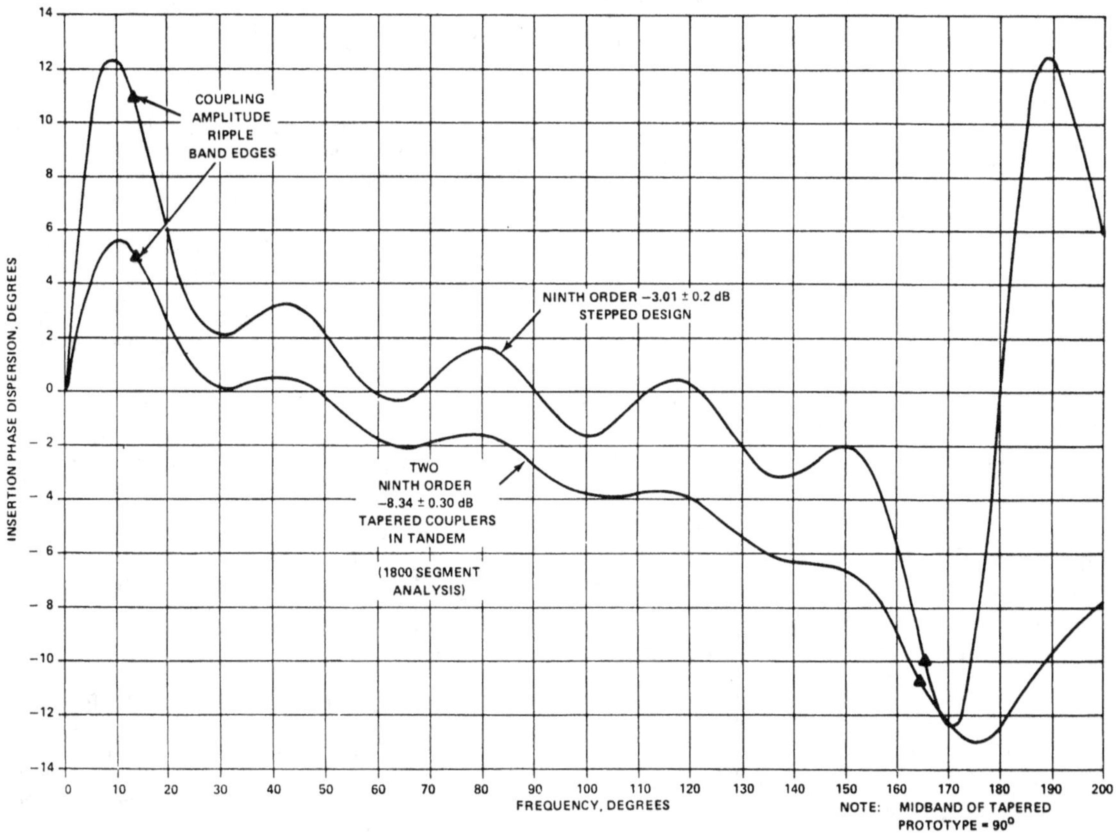

Fig. 9. Comparison of the insertion phase dispersion of stepped and tapered 90° couplers.

pling response, can be of serious consequences in some applications [6], [8]. It is evident from Fig. 9 that dispersion is considerably lower in the midband region of both the tapered and stepped couplers, so that a more restrictive operating band can be defined in which dispersion will be considerably lower than at the amplitude ripple band edges. At least one class of antenna feed network exists which is independent of the phase dispersion of the 90° hybrids used in the design [7].

VIII. Conclusions

An analytical evaluation of tapered-line couplers has been presented which employs cascade analysis of a multiple-step model of the tapered device. The procedure provides good accuracy within a reasonable computation time on a digital computer of 6-μs memory cycle.

Three different classes of equal-ripple tapered couplers, each of practical system value, have been investigated by

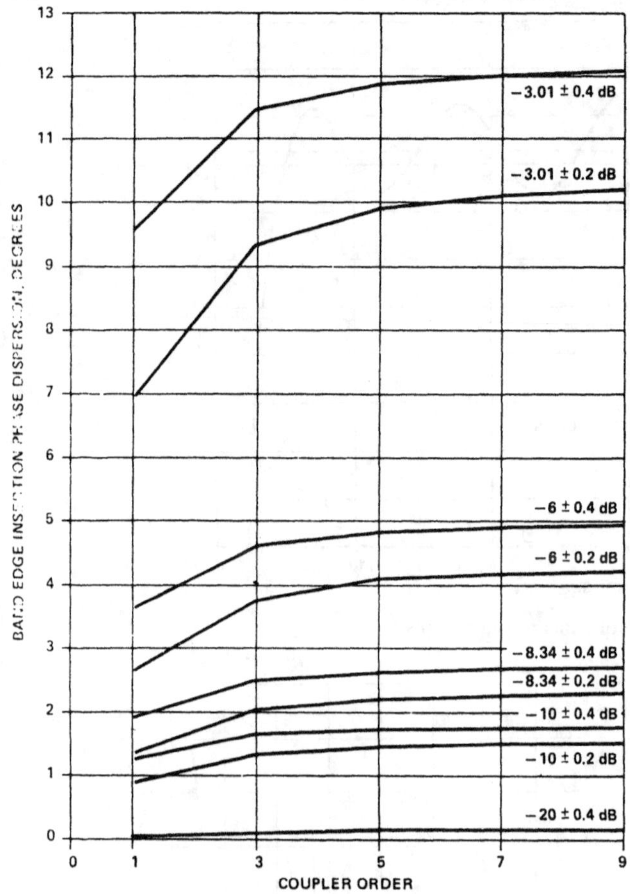

Fig. 10. Band edge insertion phase dispersion of various 90° stepped designs.

using the analysis technique: the $\Sigma-\Delta$ magic tee, a physically short asymmetric design, and the 90° symmetric coupler. A high-frequency asymptotic model has been presented which permits accurate specification of the performance of the first two designs; the 90° design is found to be adequately described by performance of its stepped prototype. Although all three design procedures are approximate, the computed responses are shown to be close enough to equal-ripple to satisfy most engineering requirements. Comparison with experiment indicates that further performance refinements must be achieved through improved experimental technique rather than through the removal of flaws in the design procedures. Experimental L to X band 3-dB tapered magic tees and 90° hybrids designed by these procedures typically exhibit coupling ripple of ± 0.5 dB and a minimum directivity of 20 dB. Phase error of the tapered 3-dB magic tee is 6° maximum, whereas that of the 90° 3-dB hybrid is 3° maximum throughout the same multi-octave region.

A seldom discussed parameter of couplers, insertion phase dispersion, has been treated for the magic tee and the 90° designs. The magnitude of dispersion is shown to be large enough at the high-frequency edge of the 90° design to be of concern in certain antenna and reflection coefficient measurement systems.

References

[1] R. H. DuHamel and M. E. Armstrong, "A wide-band monopulse antenna utilizing the tapered-line magic-T," presented at the 15th Annual USAF Antenna Symposium, University of Illinois, Monticello, October 12–14, 1965.
[2] C. P. Tresselt, "The design and construction of high-directivity, 90-degree couplers using nonuniform line techniques," *IEEE Trans. Microwave Theory and Techniques*, vol. MTT-14, pp. 647–656, December 1966.
[3] F. Arndt, "High-pass transmission-line directional coupler," *IEEE Trans. Microwave Theory and Techniques* (Correspondence), vol. MTT-16, pp. 310–311, May 1968.
[4] C. B. Sharpe, "An equivalence principle for nonuniform transmission-line directional couplers," *IEEE Trans. Microwave Theory and Techniques*, vol. MTT-15, pp. 398–405, July 1967.
[5] R. J. Wenzel, "Exact design of TEM microwave networks using quarter-wave lines," *IEEE Trans. Microwave Theory and Techniques*, vol. MTT-12, pp. 94–111, January 1964.
[6] R. H. DuHamel, M. E. Armstrong, and W. F. Pedler, "Dual polarization monopulse antennas using conical arrays of log-periodic antennas," *Digest, 1966 IEEE G-AP Internatl. Antenna and Propagation Symp.* (Palo Alto, Calif.).
[7] P. C. Goodman and C. P. Tresselt, "Development of stripline antenna feed network," USNWC, China Lake, Calif., Contract N60530-67-C-1193, Final Rept., April 4, 1968. (Applications for this report should be addressed to the Defense Documentation Center.)
[8] S. Walker, "Broadband stripline balun using quadrature couplers," *IEEE Trans. Microwave Theory and Techniques* (Correspondence), pp. 132–133, February 1968.
[9] B. M. Oliver, "Directional electromagnetic couplers," *Proc. IRE*, vol. 42, pp. 1686–1692, November 1954.
[10] A. L. Feldshtein, "Synthesis of stepped directional couplers," *Radiotekhn. i Electron.*, vol. 6, pp. 234–240, February 1961. (English transl.: *Radio Engrg. Electronics*, vol. 6, pp. 74–85, 1961.)
[11] S. Yamamoto, T. Azakami, and K. Itakura, "Coupled nonuniform transmission line and its applications," *IEEE Trans. Microwave Theory and Techniques*, vol. MTT-15, pp. 220–231, April 1967.
[12] R. E. Collin, "The optimum tapered transmission line matching section," *Proc. IRE*, vol. 44, pp. 539–548, April 1956.
[13] R. W. Klopfenstein, "A transmission line taper of improved design," *Proc. IRE*, vol. 44, pp. 31–35, January 1956 (see *Proc. IRE* (*Correspondence*), vol. 44, pp. 1055–1056), August 1956.
[14] R. W. Beatty and D. M. Kerns, "Relationships between different network parameters, not assuming reciprocity or equality of the waveguide or transmission line characteristic impedances," *Proc. IRE*, vol. 52, p. 84, January 1964.
[15] M. A. Grossberg, "Extremely rapid computation of the Klopfenstein impedance taper," *Proc. IEEE* (Letters), vol. 56, pp. 1629–1630, September 1968.
[16] F. Arndt, "High-pass directional coupler for TEM waves with space-dependent coupling factor," (in German), *Arch. Elek. Übertragung*, vol. 21, pp. 139–146, March 1967.
[17] G. L. Matthaei, L. Young, and E. M. T. Jones, *Microwave Filters, Impedance Matching Networks and Coupling Structures.* New York: McGraw-Hill, 1964, pp. 284–300.
[18] C. P. Tresselt, "A tapered-line 3-dB coupler," Research Laboratories, Bendix Corporation, Southfield, Mich., Naval Ordinance Lab., Corona, Calif. Contract N123-(62738)-56114-A, Final Rept. 3772. (Copies of this report are obtainable from Defense Documentation Center.)
[19] E. G. Cristal and L. Young, "Theory and tables of optimum symmetrical TEM-mode coupled-transmission line directional coupler," *IEEE Trans. Microwave Theory and Techniques*, vol. MTT-13, pp. 545–558, September 1965.

Slot line was introduced by Cohn (p. 259) in 1968, and is analogous to microstrip; in a sense it is the dual of microstrip. It can be used in some interesting ways in combination with microstrip, as has been shown by deRonde[27] and Hunton and Takeuchi.[28] It is only a matter of time before parallel-slot-line couplers will come in use. When two slots run parallel, one can again use an even and odd mode analysis. Now for one mode, when the electric fields point in the same direction in both slots, the thin metal strip separating the two slots will have little effect, and this mode corresponds essentially to propagation in a wider slot. However for the other mode,when the electric fields point into or out of the center dividing strip together, the two coupled slots essentially form a coplanar coaxial line, or what Wen calls a coplanar waveguide (p. 270). Therefore taken together, the two reprints by Cohn and Wen contain the ingredients for the design of slot line directional couplers. (As remarked in the companion volume on **MICROWAVE FILTERS USING PARALLEL COUPLED LINES**, these two papers could also be used for the design of parallel coupled slot fulters.)

SLOT LINE IX

If the coplanar coaxial line or waveguide is the prime transmission line, then two such lines can also be run parallel to one another to form a directional coupler. Wen[29] has also investigated this coupler.

Slot Line on a Dielectric Substrate

SEYMOUR B. COHN, FELLOW, IEEE

***Abstract*—Slot line consists of a narrow gap in a conductive coating on one side of a dielectric substrate, the other side of the substrate being bare. If the substrate's permittivity is sufficiently high, such as $\varepsilon_r = 10$ to 30, the slot-mode wavelength will be much smaller than free-space wavelength, and the fields will be closely confined near the slot. Possible applications of slot line to filters, couplers, ferrite devices, and circuits containing semiconductor elements are discussed. Slot line can be used either alone or with microstrip line on the opposite side of the substrate. A "second-order" analysis yields formulas for slot-line wavelength, phase velocity, group velocity, characteristic impedance, and effect of adjacent electric and magnetic walls.**

Manuscript received February 17, 1969; revised May 15, 1969. This work was performed for Stanford Research Institute as a part of their study program for U. S. Army Electronics Command, Contract DAAB07-68-c-0088. This paper covers material presented at the 1968 G-MTT Symposium, Detroit, Mich., May 20–22, 1968.

The author is a consultant to Stanford Research Institute, Menlo Park, Calif.

I. Introduction

SLOT LINE consists of a slot or gap in a conductive coating on a dielectric substrate, as shown in Fig. 1. Both resonant and propagating slots in metal sheets have been used as radiating antenna elements (see [1], [2], and [3], and their bibliographies). For slot line to be practical as a transmission line, radiation must be minimized. This is accomplished through the use of a high permittivity substrate, which causes the slot-mode wavelength λ' to be small compared to free-space wavelength λ, and thereby results in the fields being closely confined to the slot with negligible radiation loss. For example, if $\epsilon_r = 20$, then $\lambda'/\lambda \approx \frac{1}{3}$, and analysis shows the slot-mode fields to be sharply attenuated at a distance $r/\lambda = \frac{1}{8}$, or $r = 0.5$ inch at 3 GHz.

Fig. 2(a) shows the slot-mode fields in a cross-sectional view. A voltage difference exists between the slot edges. The

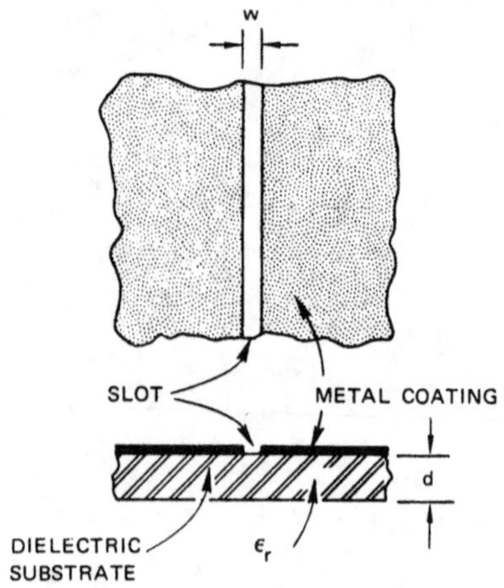

Fig. 1. Slot line on a dielectric substrate.

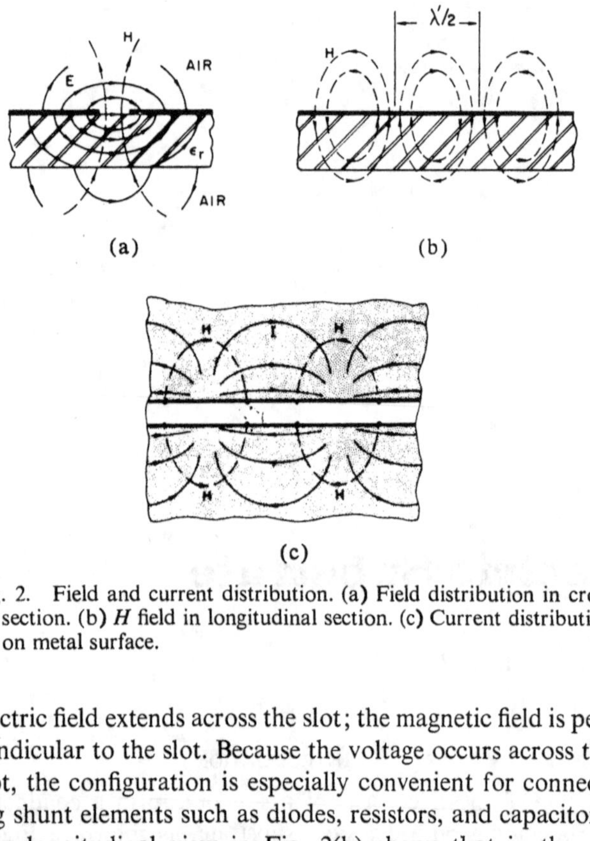

Fig. 2. Field and current distribution. (a) Field distribution in cross section. (b) H field in longitudinal section. (c) Current distribution on metal surface.

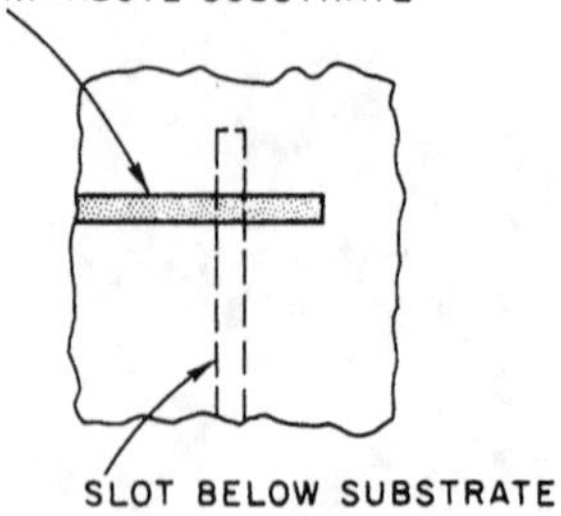

Fig. 3. Simple transition between slot line and microstrip.

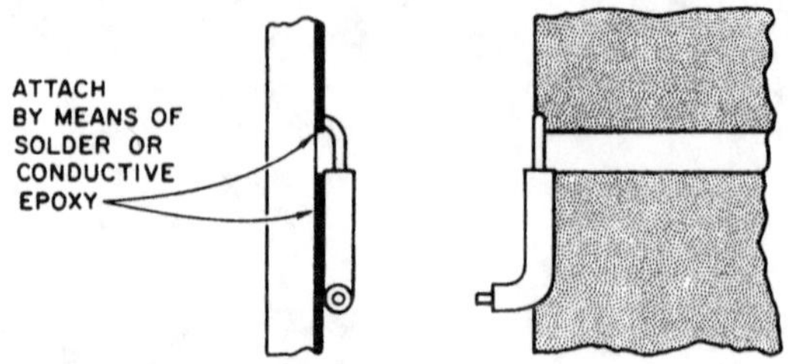

Fig. 4. Broad-band transition between slot line and miniature semirigid coaxial line.

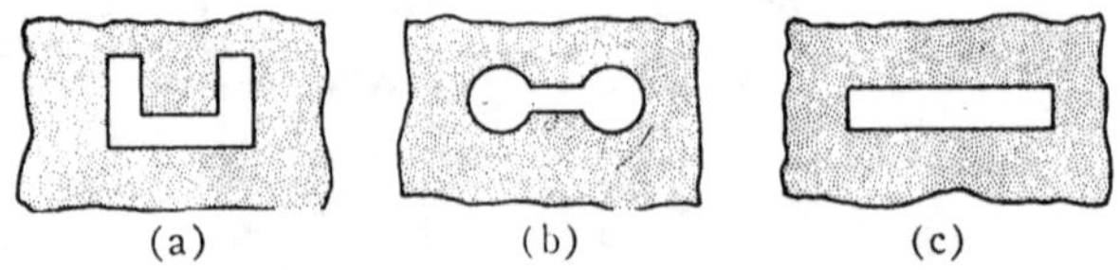

Fig. 5. Resonant slots.

electric field extends across the slot; the magnetic field is perpendicular to the slot. Because the voltage occurs across the slot, the configuration is especially convenient for connecting shunt elements such as diodes, resistors, and capacitors. The longitudinal view in Fig. 2(b) shows that in the air regions the magnetic field lines curve and return to the slot at half-wavelength intervals. Consequently, a propagating wave has elliptically polarized regions that can be usefully applied in creating certain ferrite components. The current paths on the conducting surface are shown in Fig. 2(c). The surface-current density is greatest at the edges of the slot and decreases rapidly with distance from the slot. A propagating wave has regions of elliptically polarized current and magnetic field in this view, also.

An interesting possibility for microwave integrated circuits is the use of slot lines on one side of a substrate and microstrip lines on the other. When close to each other, coupling between the two types of lines will exist, and when sufficiently far apart they will be independent. Coupling between a slot and a strip can be used intentionally in certain components. For example, parallel lengths of slot and strip can be made to act as a directional coupler. If a slot and strip cross each other at right angles, as in Fig. 3, coupling will be especially tight, and a transition covering approximately 30-percent bandwidth can be achieved when the strip and slot widths are optimally related, and when the strip and slot are extended approximately one-quarter wavelength beyond the point of crossing. With matching techniques, a bandwidth of an octave or so should be feasible.

Fig. 4 shows one way that a wide-band transition between miniature-cross-section coaxial line and slot line can be made. Additional structural details may be needed to obtain optimum matching and to prevent radiation loss.

A half-wavelength slot, as shown in Fig. 5(a), can be used as a resonator. If desired, the resonant slot may be made more compact by capacitively loading its center, as in Fig. 5(b), or by bending it, as in Fig. 5(c). Applications to bandstop and bandpass filters are shown in Fig. 6. In the bandstop example, the terminating lines are microstrip, and in the bandpass example they are slot lines. Many other filter configurations are feasible, using slots alone or slots with strips on the opposite side of the substrate.

The basic electrical parameters of slot line are the characteristic impedance Z_0 and the phase velocity v. Because of the non-TEM nature of the slot-line mode these parameters are not constant, but vary with frequency at a rather slow rate per octave. This behavior contrasts with quasi-TEM

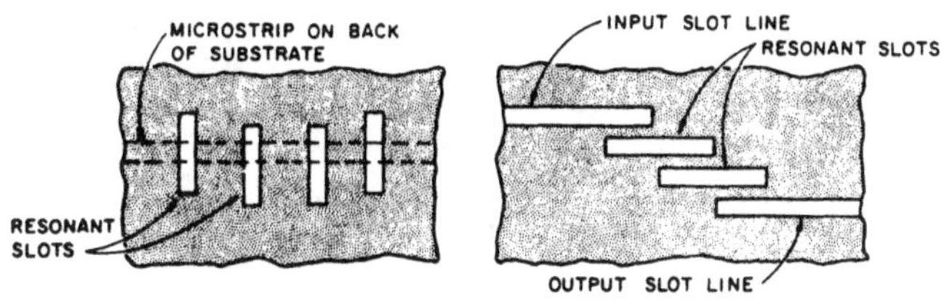

(a) (b)

Fig. 6. Filter applications. (a) Bandstop filter. (b) Bandpass filter.

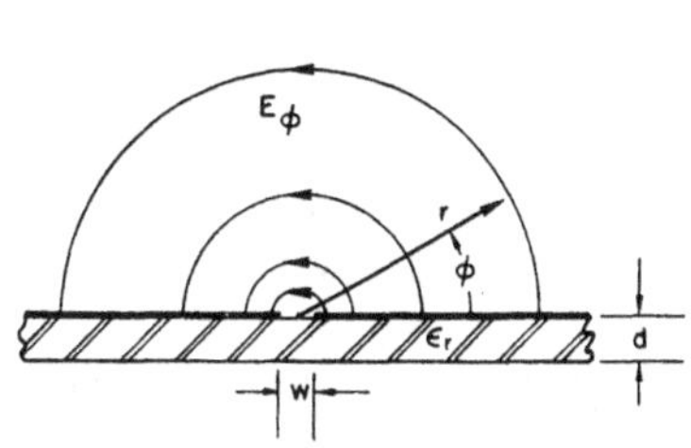

Fig. 7. Cylindrical coordinates with axis on center line of slot.

microstrip line, whose Z_0 and v are first-order independent of frequency. On the other hand, slot line differs from waveguide in that it has no cutoff frequency. Propagation along the slot occurs at all frequencies down to $f=0$, where, if the metal-coated substrate is assumed infinite in length and width, v/c approaches unity and Z_0 approaches zero. Other parameters treated in this paper are the ratio of phase and group velocities v/v_g and the effect of adjacent walls. Attenuation has not yet been treated, but data thus far indicates it to be about the same as for microstrip on the same substrate [20].

II. Approximations for Slot Line

Several references on slot antennas were found in which the presence of a substantial amount of dielectric material in or near the slot was taken into account. Strumwasser, Short, Stegen, and Miller [4] have studied experimentally the effects of filling a slot in a thick metal plate with dielectric material. They give data on resonant-length reduction and radiation resistance coupled into an air-dielectric TEM line. Bailey [5] has measured resonant length and radiation conductance of a slot in a waveguide wall covered by a protective layer of dielectric material. Galejs [6] has analyzed theoretically a slot in a zero-thickness, perfectly conducting sheet separating free space from a lossy dielectric medium of infinite extent. For example, a slot radiator in a wire mesh on the surface of the ground would be simulated by this model.

Galejs utilizes an integral-equation method to obtain complex expressions for radiation efficiency and other parameters of the slot antenna. His *zero-order* solution for the propagation constant along the slot can be easily modified into the following simple formula for relative wavelength:

$$\frac{\lambda'}{\lambda} = \sqrt{\frac{2}{\epsilon_r + 1}}\,. \tag{1}$$

Since wavelength is inversely proportional to the square root of permittivity, an effective permittivity of a uniform medium replacing the two different dielectric half spaces may be defined as

$$\epsilon_r' = \frac{\epsilon_r + 1}{2}\,. \tag{2}$$

The *second-order* solution for slot line derived in this paper shows that (1) is a fair approximation for slot line, yielding values within about 10 percent in typical slot-line cases. The second-order solution shows quantitatively how λ'/λ varies with the parameters d, w, ϵ_r, and λ.

The field components on the air side of the slot can be computed quite easily as a function of λ, λ', and distance r from the slot (Fig. 7). If we assume $w/\lambda \ll 1$, then the electric voltage across the slot may be replaced by an equivalent line source of magnetic current. At a distance r at least several times larger than w the longitudinal component of magnetic field is given by [7]

$$H_z = A H_0^{(1)}(k_c r) \tag{3}$$

where $H_n^{(1)}(x)$ is the Hankel function of first kind, order n, and argument x. The coefficient k_c is

$$k_c = \sqrt{\gamma_z^2 + k^2} = j\frac{2\pi}{\lambda}\sqrt{\left(\frac{\lambda}{\lambda'}\right)^2 - 1} \tag{4}$$

since $\gamma_z = j2\pi/\lambda_z = j2\pi/\lambda'$ and $k = 2\pi/\lambda$.

By (1), a zero-order value of k_c is

$$k_c = j\frac{2\pi}{\lambda}\sqrt{\frac{\epsilon_r - 1}{2}}\,. \tag{5}$$

The other field components are H_r and E_ϕ. They are related to H_z by [7]

$$H_r = -\frac{\gamma_z}{k_c^2}\frac{\partial H_z}{\partial r} = \frac{A}{\sqrt{1 - \left(\frac{\lambda'}{\lambda}\right)^2}} H_1^{(1)}(k_c r) \tag{6}$$

$$E_\phi = \frac{j\omega\mu}{k_c^2}\frac{\partial H_z}{\partial r} = \frac{-\eta(\lambda'/\lambda)A}{\sqrt{1 - \left(\frac{\lambda'}{\lambda}\right)^2}} H_1^{(1)}(k_c r) \tag{7}$$

where the identity $d[H_0^{(1)}(x)]/dx = -H_1^{(1)}(x)$ was used.

The Hankel function of imaginary argument, $H_n^{(1)}(j|x|)$, approaches zero proportional to $e^{-|x|}/\sqrt{|x|}$ for $|x|$ large. Equation (4) shows that the argument $k_c r$ is imaginary for $\lambda'/\lambda < 1$. Hence a relative wavelength ratio less than unity is a sufficient condition to ensure decay of the slot-mode field with radial distance. As λ'/λ is decreased, the decay becomes sharper and the fields become more tightly bound to the slot.

A radius r_{cp} of circular polarization of the magnetic field requires $|H_z/H_r| = 1$. By means of (3) and (6), r_{cp} must satisfy

$$\left|\frac{H_1^{(1)}(k_c r_{cp})}{H_0^{(1)}(k_c r_{cp})}\right| = \sqrt{1 - \left(\frac{\lambda'}{\lambda}\right)^2}\,. \tag{8}$$

However, tables show that $|H_1^{(1)}(j|x|)| > |H_0^{(1)}(j|x|)|$ for all $|x|$ [8]. Since the right-hand side of (8) is less than one, a solution for r_{cp} does not exist. Nevertheless, elliptical polarization occurs for all r, and low axial ratios occur for r sufficiently large.

Also of interest is the ratio of voltage along a semicircular path at constant radius divided by the voltage directly across the slot. This ratio is

$$\frac{V(r)}{V} = \frac{k_c r H_1^{(1)}(k_c r)}{\lim\limits_{|x|\to 0} [|x| H_1^{(1)}(j|x|)]} = \frac{\pi}{2} k_c r \, | H_1^{(1)}(k_c r) | . \tag{9}$$

As an example of field decay, let $\epsilon_r = 16$, $f = 3$ GHz, and $\lambda = 4$ inches. The zero-order value of λ'/λ is 0.343, and of $k_c r$ is 4.30 r, where r is in inches. At $r = 0$, 0.7, 1.0, and 1.3 inches, $V(r)/V = 1$, 0.120, 0.0382, and 0.0118, respectively. If a plane metal wall is positioned perpendicular to the radius vector at distance $r/2$ from the slot, an image of the slot will appear at distance $r/2$ behind the wall. The effect will be that of two equally excited parallel slots spaced by r. Thus if the metal wall is at distance 0.5 inch in the above example, $r = 1$ inch and $V(r)/V$ of one slot is 0.0382, or -28.36 dB, at the other slot. Coupling between slots is even weaker than this, since only part of the total voltage $V(r)$ of one slot affects the other slot. Therefore, a wall or other perturbing object can be as close as $r = 0.5$ inch with little effect on λ' or Z_0, and the fields and stored energy of the slot mode are mainly confined within this same radial distance. These conclusions have been verified both by experiment and by computations using the second-order solution.

III. Basis of Second-Order Solution

An analytical approach for slot line offering high accuracy is described in this section. The solution obtained by this approach will be referred to as *second-order*. A first-order solution offering intermediate complexity and accuracy between zero-order (1) and second-order equations would also be useful, but has not yet been completed.

The key feature of the approach used in the second-order solution is the introduction of boundary walls permitting the slot-line configuration to be treated as a rectangular-waveguide problem rather than as a problem in cylindrical coordinates. Thus the infinite orthogonal sets of relatively simple rectangular-waveguide modes apply rather than sets of cylindrical modes embodying all orders of Hankel functions.

Parameters evaluated in this paper are: relative-wavelength ratio λ'/λ, characteristic impedance Z_0, ratio of phase velocity to group velocity v/v_g, and the effect of nearby electric and magnetic walls.

Future plans are to adapt the second-order solution to yield the even- and odd-mode characteristic impedances and velocities of parallel slots, the effect of metal-coating thickness greater than zero, attenuation per unit length, unloaded

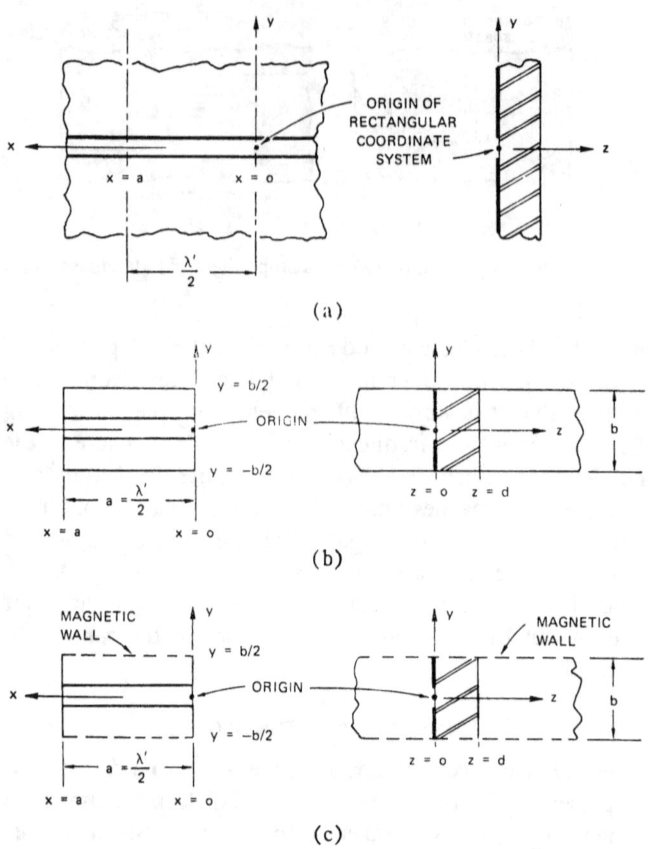

Fig. 8. Development of waveguide models for slot-line solution. (a) Insertion of transverse conducting planes at $x=0$ and a. (b) Insertion of conducting planes at $y = \pm b/2$. (c) Insertion of magnetic walls at $y = \pm b/2$.

Q of a resonant slot, and presence of more than one dielectric substrate on one or both sides of the slotted conductive sheet.

Conversion of the slot-line configuration into a rectangular-waveguide problem is illustrated in Fig. 8. First assume that slot waves of equal amplitude are traveling in the $+x$ and $-x$ directions. Then transverse planes spaced by $\lambda'/2$ exist where the transverse E field and normal H field cancel to zero. Let two such planes occur at $x=0$ and $x=\lambda'/2=a$ in Fig. 8(a). Conducting (or electric) walls of infinite extent may be inserted in these planes without disturbing the field components between the planes, and the semi-infinite regions at $x<0$ and $x>a$ may be eliminated. The section of slot line between the transverse planes supports a resonant slot-wave mode with no loss of energy, if the dielectric substrate and the conducting walls are assumed dissipationless.

Next, conducting walls are inserted in planes parallel to the slot and perpendicular to the substrate at $y = \pm b/2$. The region separated out of the original infinite space has the rectangular-waveguide boundary shown in Fig. 8(b). Since the fields are tightly bound to the vicinity of the slot, the walls at $y = \pm b/2$ will have negligible effect for b sufficiently large (typically one inch at 3 GHz), yet they serve the important function of enabling the use of rectangular-waveguide mode sets, thereby greatly simplifying the analysis.

Magnetic walls may be placed at $y \doteq \pm b/2$ instead of electric walls. The result is the boundary shown in Fig. 8(c), where two magnetic walls spaced by b and two electric walls spaced by a are used.

Images of the slot in the electric or magnetic walls at $y=\pm b/2$ result in an infinite array of parallel slots in the $z=0$ plane having center-to-center spacing of b. Therefore, the effect of adjacent slots may be computed for both electric- and magnetic-wall imaging, allowing the even- and odd-mode characteristic impedances and wavelengths to be evaluated for a slot in an infinite array. For one pair of slots rather than an array, these even- and odd-mode quantities are given approximately by the $b\rightarrow\infty$ values modified by one half the change computed for the infinite array.

Thus, the introduction of walls in Fig. 8 has created the configuration of a capacitive iris in a rectangular waveguide, with air and dielectric regions as indicated. Consider the metal-walled case in Fig. 8(b). All waveguide modes must have the $\lambda'/2$ variation of the slot wave in the x direction. Also, because of the symmetry of the structure, all modes must have an E-field maximum at the center of the slot. Therefore, the full set of modes satisfying the boundary conditions are TE_{10}, TE_{12}, TE_{14}, $\cdots$, and TM_{12}, TM_{14}, $\cdots$; that is, $TE_{1,2n}$ for n an integer ≥ 0 and $TM_{1,2n}$ for $n\geq 1$.

For the slot wave, $\lambda'<\lambda$ and hence $a<\lambda/2$. Therefore, the TE_{10} and all higher modes are cut off, or nonpropagating, in the air regions. In the dielectric region the TE_{10} mode is propagating, and the first few higher modes may propagate or all higher modes may be cut off, depending on the size of b. Since all modes are cut off in the two air regions, the energy of the resonant slot-wave mode is trapped near the slot. The amplitude of each mode in each region must be such that when the full set of modes are superimposed, the boundary conditions in the iris plane at $z=0$ will be met, and all field components on either side of the dielectric-to-air interface at $z=d$ will be matched. An alternative but equivalent condition is that transverse resonance occur; that is, the sum of the susceptances at the iris plane must equal zero. This sum includes the susceptances of the TE_{10} mode looking in the $-z$ and $+z$ directions, and the capacitive-iris susceptance representing higher modes on the $-z$ and $+z$ sides of the iris.

A formula for characteristic impedance is derived in Appendix II by a method utilizing the total-susceptance formula. Because of the non-TEM nature of the slot wave, definition of characteristic impedance is somewhat arbitrary. The reasonable and useful definition chosen here is $Z_0=V^2/2P$, where $V=-\int E_y dy$ is peak voltage amplitude across the slot and P is average power flow of the wave. As in waveguide, this definition does not necessarily yield the best match in a transition to coaxial or microstrip line. In fact, experimental results indicate that a 50 ohm coaxial or microstrip line requires about 75 ohm slot impedance computed by the voltage power definition [19], [20]. Reactive discontinuity effects may also play a part in this discrepancy.

IV. Parameter Formulas—Second-Order Solution

Define as an independent variable $p=\lambda/2a$. At the transverse-resonance frequency, $a=\lambda'/2$ and hence $p=\lambda/\lambda'$ for $B_t=0$ where B_t is the total susceptance at the plane of the slot. Once $p=\lambda/\lambda'$ has been determined as the solution of $B_t=0$ for a given set of parameters ϵ_r, w, d, b, and a, the wavelength and frequency for this solution are simply $\lambda=2a(\lambda/\lambda')$ and $f=c/\lambda$.

The formula for B_t is as follows for the case of electric walls at $y=\pm b/2$:

$$\eta B_t=\frac{a}{2b}\left[-v+u\tan\left(\frac{\pi du}{ap}-\tan^{-1}\frac{v}{u}\right)\right]+\frac{1}{p}\left\{\left(\frac{\epsilon_r+1}{2}-p^2\right)\ln\frac{2}{\pi\delta}+\frac{1}{2}\cdot\sum_{n=1,2,3,\cdots}\left[v^2\left(1-\frac{1}{F_n}\right)+M_n\right]\frac{\sin^2(\pi n\delta)}{n(\pi n\delta)^2}\right\},\tag{10}$$

and as follows for magnetic walls at $y=\pm b/2$:

$$\eta B_t=\frac{1}{p}\left\{\left(\frac{\epsilon_r+1}{2}-p^2\right)\ln\frac{8}{\pi\delta}+\frac{1}{2}\cdot\sum_{n=\frac{1}{2},\frac{3}{2},\frac{5}{2},\cdots}\left[v^2\left(1-\frac{1}{F_n}\right)+M_n\right]\frac{\sin^2(\pi n\delta)}{n(\pi n\delta)^2}\right\}\tag{11}$$

where $\eta=\sqrt{\mu_0/\epsilon_0}=376.7$ ohms, $\delta=w/b$, and

$$u=\sqrt{\epsilon_r-p^2},\qquad v=\sqrt{p^2-1}\tag{12}$$

$$F_n=\sqrt{1+\left(\frac{b}{2an}\cdot\frac{v}{p}\right)^2},\qquad F_{n1}=\sqrt{1-\left(\frac{b}{2an}\cdot\frac{u}{p}\right)^2}.\tag{13}$$

For F_{n1} real, M_n is

$$M_n=\frac{\epsilon_r\tanh r_n-p^2F_{n1}^2\coth q_n}{\left[1+\left(\frac{b}{2an}\right)^2\right]F_{n1}}-u^2\tag{14}$$

where

$$r_n=\frac{2\pi ndF_{n1}}{b}+\tanh^{-1}\left(\frac{F_{n1}}{\epsilon_rF_n}\right)\tag{15}$$

$$q_n=\frac{2\pi ndF_{n1}}{b}+\coth^{-1}\left(\frac{F_n}{F_{n1}}\right).\tag{16}$$

For F_{n1} imaginary, M_n is

$$M_n=\frac{\epsilon_r\tan r_n'-p^2|F_{n1}|^2\cot q_n'}{\left[1+\left(\frac{b}{2an}\right)^2\right]|F_{n1}|}-u^2\tag{17}$$

where

$$r_n'=\frac{2\pi nd|F_{n1}|}{b}+\tan^{-1}\left(\frac{|F_{n1}|}{\epsilon_rF_n}\right)\tag{18}$$

$$q_n'=\frac{2\pi nd|F_{n1}|}{b}+\cot^{-1}\left(\frac{F_n}{|F_{n1}|}\right).\tag{19}$$

Equations (10) and (11) for B_t are valid for $\delta=w/b\le 0.15$, $w<\lambda/(4\sqrt{\epsilon_r})$, and $w\le d$. These ranges are more than adequate for usual slot-line dimensions.

The procedure in using the above equations is to substitute a set of values ϵ_r, w, d, b, and $a=\lambda'/2$, and then to solve (10) or (11) for the value of p at which $\eta B_t=0$. This p is equal to λ/λ'. Wavelength and frequency are then given by $\lambda=2a(\lambda/\lambda')$ and $f=c/\lambda$. Solution of the above equations is most conveniently accomplished with an electronic computer, arriving at $B_t=0$ by an iterative process.

The group velocity is $v_g=d\omega/d\beta$, where $\omega=2\pi f$ and β is the slot-wave phase constant in radians per unit length. From this definition of v_g, one may obtain the following relations for the ratio of phase velocity to group velocity:

$$\frac{v}{v_g}=1+\frac{f}{\lambda'/\lambda}\cdot\frac{-\Delta(\lambda'/\lambda)}{\Delta f}=1+\frac{f}{\lambda/\lambda'}\cdot\frac{\Delta(\lambda/\lambda')}{\Delta f} \qquad (20)$$

where $\Delta(\lambda'/\lambda)$, $\Delta(\lambda/\lambda')$, and Δf are computed from two separate solutions of $\eta B_t=0$ for fixed values of ϵ_r, w, d, and b, and for two slightly different values of $a=\lambda'/2$ incremented plus and minus from the desired a. The frequency f may be assumed to lie midway in the Δf interval.

The slot-wave characteristic impedance is obtained as follows from the ηB_t formulas:

$$Z_0=376.7\,\frac{v}{v_g}\cdot\frac{\pi}{p}\cdot\frac{\Delta p}{-\Delta\eta B_t}\ \text{ohms.} \qquad (21)$$

In this equation $\Delta\eta B_t$ is computed from (10) or (11) with ϵ_r, w, d, b, and a held constant, and with p incremented slightly plus and minus from the value $p=\lambda/\lambda'$ at $\eta B_t=0$; v/v_g is obtained from (20) for the same parameters. Equations (20) and (21) may be used with (10) to yield values for electric walls at $y=\pm b/2$, or with (11) for magnetic walls.

V. Computed Data

The second-order solution equations in Section IV were programmed on an electronic computer and the parameters λ'/λ, v/v_g, and Z_0 were computed for various slot-line cases.[1]

Fig. 9 shows the effect of wall spacing b on the parameters for both electric and magnetic walls. In this computation, $\epsilon_r=20$, $d=0.137$ inch, $w=0.025$ inch, and λ' is constant at 1.360 inches. The corresponding curves for electric and magnetic walls merge together for $b>1.5$ inches and are only slightly separated for $b=1$ inch. As b decreases below 1 inch, the λ'/λ curves diverge more and more rapidly. Since λ' is held constant, λ varies inversely and f proportionally with λ'/λ. The ratio v/v_g decreases with b for magnetic walls and increases for electric walls. The behavior of the characteristic impedance is more complex, the electric-wall curve rising and then falling sharply as b decreases, and the magnetic-wall curve falling and then rising sharply. This reversal of direction is caused by the v/v_g factor in (21). The quantity $Z_0/(v/v_g)$ is virtually constant for the two types of walls for

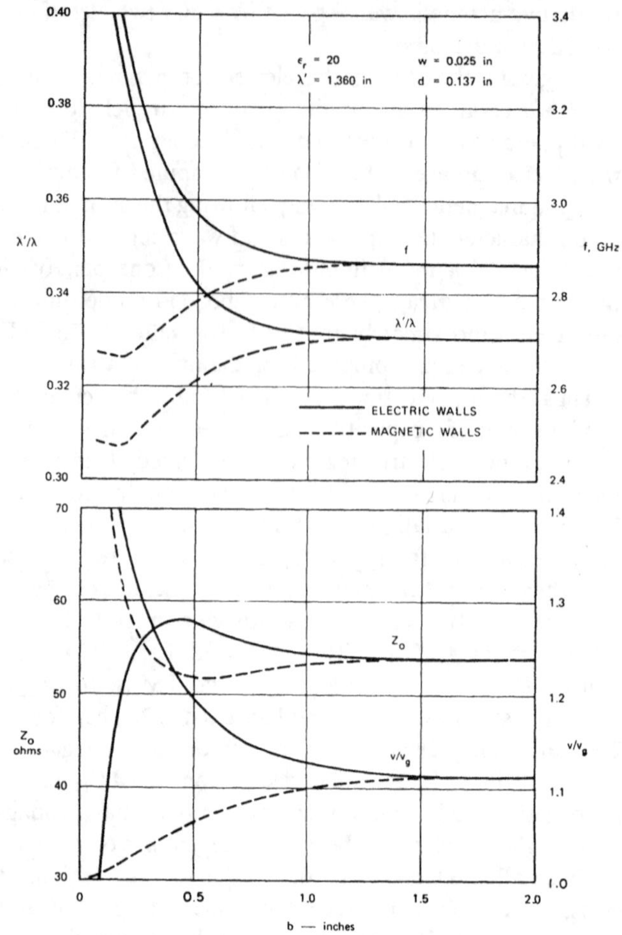

Fig. 9. Slot-line parameters versus b with ϵ_r, w, d and $a=\lambda'/2$ fixed.

b as low as 0.6 inch, while for b smaller, this quantity diverges without changing direction.

The variation of Z_0 and λ'/λ with frequency is shown in Fig. 10 for $\epsilon_r=20$, $d=0.137$ inch, $w=0.025$ inch, and for $\lambda'=2a$ stepped from 0.5 to 150 inches in the computation. Curves are plotted for $b=3$ inches and $b\to\infty$ (that is, b sufficiently large at each λ' value such that Z_0 and λ'/λ are essentially independent of b). The ratio λ'/λ decreases monotonically in the plotted range. For increasing f, λ'/λ falls toward $1/\sqrt{\epsilon_r}=0.224$, while for f approaching zero λ'/λ rises to 0.390 for magnetic walls at $b=3$ inches and to 1.0 for the two other curves. The zero-order value of λ'/λ calculated from (1) is indicated for comparison.

The characteristic impedance plotted in Fig. 10 has a broad maximum, varying by only ± 7 percent between 1 and 6 GHz. As f approaches zero, Z_0 approaches 40.3 ohms for magnetic walls at $b=3$ inches, and approaches 0 ohms for the other two cases.

A few experimental λ'/λ points are shown in Fig. 10. The substrate used in the test piece had a specified ϵ_r of 20. Length, width, and thickness were approximately 5 by 1.75 by 0.137 inches. For the circled points one side was covered with Scotch Brand No. 51 aluminum tape, about 0.0007 inch thick, including 0.0001 inch of adhesive. A slot 0.025 inch wide and 3 inches long was left bare. A probe terminating a

[1] Normalized graphs of λ'/λ and Z_0 versus d/λ for $w/d=0.02$, 0.1, 0.2, 0.4, and 0.6, and for $\epsilon_r=13$, 16, and 20 have been prepared and are included in another paper [19].

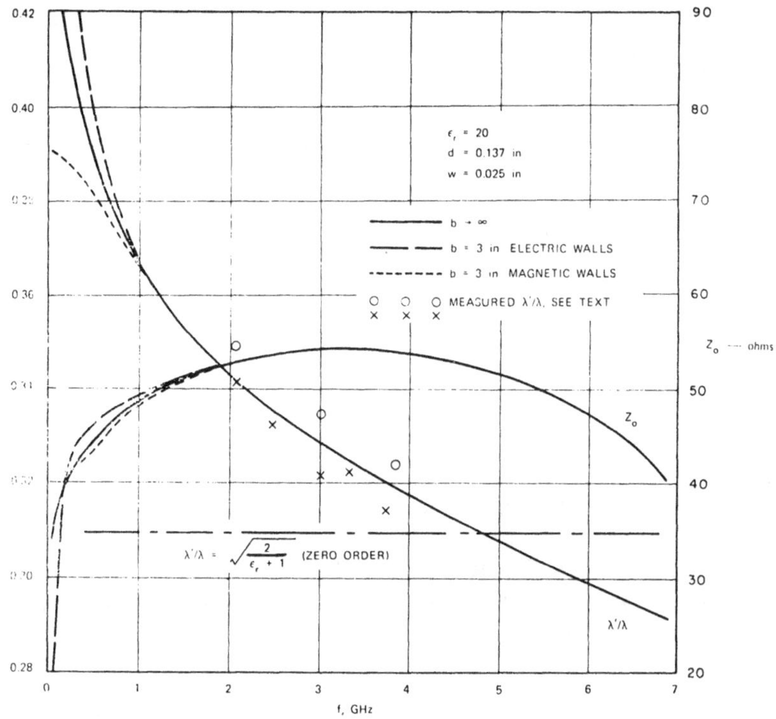

Fig. 10. Graph of λ'/λ and Z_0 versus frequency.

reflectometer coupler was coupled to the slot near one end. Resonance was indicated by a dip in the reflected wave. The slot length was varied by sliding a thin steel scale over the slot from the end opposite the probe. At each of the test frequencies the overlapping length of steel scale was varied until resonance was found. Several lengths for resonance were obtained differing in $\lambda'/2$ steps, thus yielding λ' and λ'/λ. The circled points lie parallel to the theoretical λ'/λ curve, but about 2.3 percent above it. For the crossed points, the same substrate sample was tested with copper plating directly applied. The thickness was about 0.0008 inch and slot width was 0.0235 inch. These points lie about 1.1 percent below the theoretical curve. The difference between the two sets of points may be attributed to the thin layer of adhesive on the aluminum tape. Additional tests with other permittivities and other w, d, and λ yield similar agreement with λ'/λ computations [19].

The effect of varying slot width w on λ'/λ and Z_0 is shown in Fig. 11 for two different cases of ϵ_r and d. In each case λ' is held constant. The ratio λ'/λ increases with w, but only slightly. The characteristic impedance Z_0 increases substantially, although far less than in proportion to w.

Measurements of Z_0 have not yet been made.[2] However, the computed Z_0 values for the magnetic-wall case can be checked when $b/\lambda' \ll 1$, since in the static limit this slot-line cross section becomes equivalent to a TEM-mode transmission line containing an equivalent dielectric medium $\epsilon_r' = (\lambda/\lambda')^2$. The Z_0 formula for this TEM line may be obtained by a straightforward modification of a case treated in [18]:

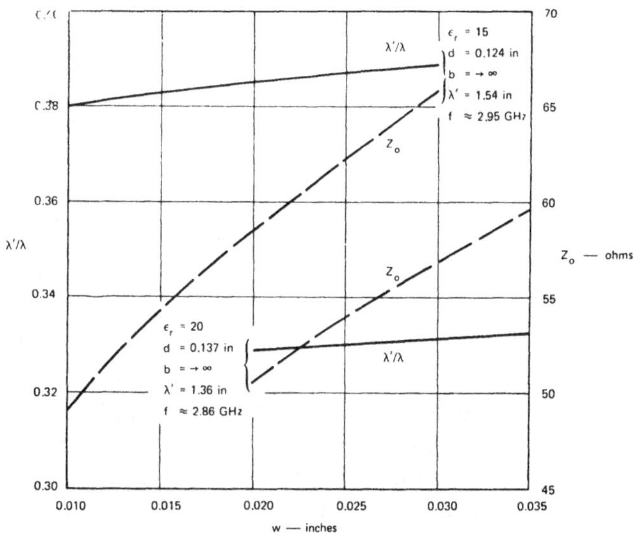

Fig. 11. Graph of λ'/λ and Z_0 versus slot width.

$$Z_0 = \frac{591.7}{\sqrt{\epsilon_r'}\,\ln\left(\frac{8b}{\pi w}\right)} = \frac{591.7(\lambda'/\lambda)}{\ln\left(\frac{8b}{\pi w}\right)}\ \text{ohms} \qquad \begin{array}{l} b/w > 3 \\ b/\lambda' \to 0. \end{array} \tag{22}$$

[2] Good transitions have been achieved at C and X bands between slot line and 50 ohm coaxial and microstrip lines with $\epsilon_r = 16$, $d = 0.055$ inch, and $w = 0.021$ inch [20]. Another good transition at S band had $\epsilon_r = 16$, $w = 0.031$ inch, and $d = 0.062$ inch [19]. See the discussion at the end of Section III on the arbitrary nature of slot-line Z_0 definition.

TABLE I

COMPARISON OF SECOND-ORDER AND STATIC SOLUTIONS FOR Z_0

	Second-Order Solution		Eq. (22)
b (inches)	λ'/λ	Z_0 (Ω)	Z_0 (Ω)
0.10	0.30752	78.38	78.40
0.14	0.30707	68.35	68.37
0.20	0.30763	60.45	60.39
0.30	0.31230	54.60	54.04
0.40	0.31640	52.42	50.50
0.60	0.32384	51.77	46.59

In Table I, (22) is tested against data computed for the curves of Fig. 9 ($\epsilon_r=20$, $d=0.137$ inch, $w=0.025$ inch, $\lambda'=1.360$ inches, $f\approx 2.670$ GHz). Agreement is excellent for $b\leq 0.20$ inch or $b/\lambda'\leq 0.15$. A second test is afforded by a computed point used in plotting the graph in Fig. 10. For $\epsilon_r=20$, $d=0.137$ inch, $w=0.025$ inch, $b=3.00$ inches, and $\lambda'=150.0$ inches, the second-order solution gives $\lambda'/\lambda=0.38987$, $f=30.67$ MHz, and $Z_0=40.33$ ohms. Equation (22) yields 40.32 ohms.

APPENDIX I

DERIVATION OF B_t FORMULAS

Fig. 12 shows a longitudinal yz-plane view through the waveguide model for the slot-line analysis. The waveguide boundaries in this case are conducting, or electric, walls. The TE_{10}-mode susceptance at the iris plane ($z=0$) is B_d looking to the right into the dielectric slab and B_a looking to the left into the air region. The total susceptance in the iris plane is $B_t=B_d+B_a$.

A formula for B_d will be derived first and then modified to yield B_a. The approach is based on a previously published analysis of a waveguide filter consisting of an alternating series of steps between two cross-section heights [9], [10], and on earlier analyses applied to other discontinuities [11], [12]. Fig. 13 shows the basic structure treated in [9] and [10], with notation modified for slot line. A waveguide of height b is driven by a waveguide of height w, with an abrupt step at $z=0$. A reactive plane is placed transverse to the waveguide at $z=l$. This terminating plane can be either an electric or magnetic wall.

Since all walls and the dielectric material are assumed lossless, the admittance looking to the right at $z=0$ is a pure susceptance, B_d. Because of symmetry, only $TE_{1,2n}$ (with $n\geq 0$) and $TM_{1,2n}$ (with $n\geq 1$) modes are present. In a transverse plane the E_y and H_x components of each mode are proportional to $\sin \pi x/a \cos 2\pi ny/b$. Thus the total E_y and H_x fields at the $z=0$ plane and $x=a/2$ are functions of y as follows:

$$E_y = R_0 + \sum_{n>0} R_n \cos\frac{2\pi ny}{b} \tag{23}$$

$$H_x = -y_{i0}R_0 - \sum_{n>0} y_{in}R_n \cos\frac{2\pi ny}{b} \tag{24}$$

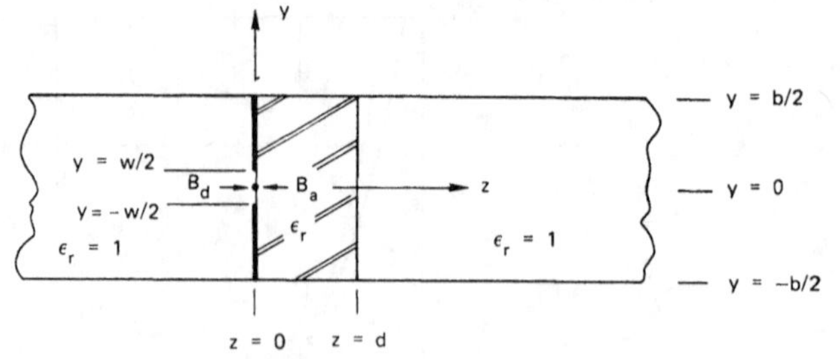

Fig. 12. Waveguide model containing capacitive iris and dielectric slab.

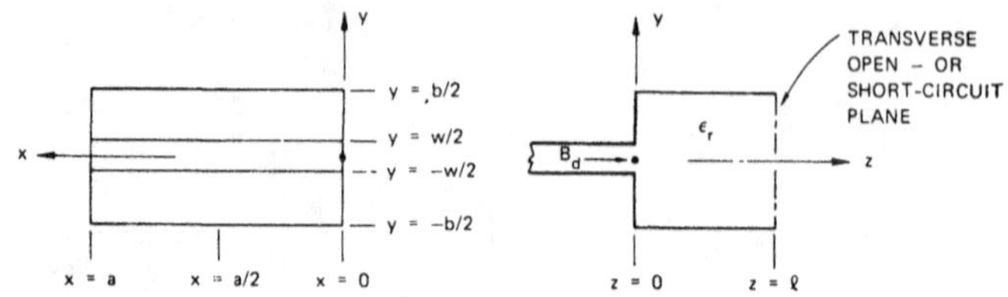

Fig. 13. Waveguide of height b driven by waveguide of height w.

where $n=1, 2, 3, \cdots$, R_0 and R_n are constants, and input wave admittances y_{i0} and y_{in} are defined by

$$y_{i0} = -\left(\frac{H_x}{E_y}\right)_{TE_{10}} \tag{25}$$

$$y_{in} = -\left(\frac{(H_x)_{TE_{1,2n}} + (H_x)_{TM_{1,2n}}}{(E_y)_{TE_{1,2n}} + (E_y)_{TM_{1,2n}}}\right). \tag{26}$$

The analysis will be simplified by assuming that only the TE_{10} mode is present in the waveguide of height w at $z=0$. Therefore, E_y and H_x are constants as follows:

$$E_y = \begin{cases} C_0 & \text{for } |y| \leq w/2 \\ 0 & \text{for } w/2 < |y| \leq b/2 \end{cases} \tag{27}$$

$$H_x = -C_0 y_i' \qquad \text{for } |y| \leq w/2 \tag{28}$$

where y_i' is input wave admittance in the waveguide of height w at the plane $z=0$. The error in the analysis due to neglecting higher modes in the region $z<0$ is very small for $\delta=w/b$ small, as would be usual for slot line.

Equation (23) has the form of a Fourier series. Let this series equal E_y as defined by (27). Then R_0 and R_n are determined as follows:

$$R_0 = C_0\delta, \qquad R_n = 2C_0\delta\frac{\sin \pi n\delta}{\pi n\delta}. \tag{29}$$

Next, set the right-hand sides of (24) and (28) equal, and integrate with respect to y over the opening of height w to obtain

$$C_0 y_i' = y_{i0}R_0 + \sum_{n>0} y_{in}R_n\frac{\sin \pi n\delta}{\pi n\delta} \tag{30}$$

and, by (29),

$$(b/w)y_i' = y_{i0} + 2\sum_{n>0} y_{in}\left(\frac{\sin \pi n\delta}{\pi n\delta}\right)^2. \tag{31}$$

At this point replace the wave admittances by guide admittances defined on the TE_{10}-mode voltage power basis in the complete waveguide cross section [13]; that is, replace y_{in} by $Y_{in}=(a/2b)y_{in}$ and y_i' by $Y_i=(a/2w)y_i'$. Then

$$Y_i = jB_d = Y_{i0} + 2\sum_{n>0} Y_{in}\left(\frac{\sin \pi n\delta}{\pi n\delta}\right)^2. \qquad (32)$$

With reference to Fig. 12, Y_{i0} is the admittance seen by a TE_{10} wave directed into a dielectric-filled waveguide region of length d terminated by an infinite air-filled region:

$$Y_{i0} = jY_{o1}\tan\left(\beta_1 d + \tan^{-1}\frac{Y_0}{jY_{o1}}\right). \qquad (33)$$

In the notation used in (33), symbols with the subscript 1 apply to the dielectric-filled region between $z=0$ and $z=d$; without this subscript they apply to the air regions. Y_{o1} and Y_o are TE_{10}-mode characteristic admittances and $\beta_1=\gamma_1/j=2\pi/\lambda_{g1}$ is the TE_{10}-mode phase constant.

$$Y_0 = \frac{a\gamma}{j2b\eta k}, \quad Y_{o1} = -j\frac{a\gamma_1}{2b\eta_1 k_1} = \frac{a}{2b\eta}\cdot\frac{\lambda}{\lambda_{g1}} \qquad (34)$$

where $\eta=\sqrt{\mu_0/\epsilon_0}=376.7$ ohms and $\eta_1=\eta/\sqrt{\epsilon_r}$; γ and γ_1 are the TE_{10}-mode z-directed propagation constants; $k=2\pi/\lambda$ and $k_1=2\pi\sqrt{\epsilon_r}/\lambda$ are the plane-wave constants; and λ_{g1} is the TE_{10}-mode guide wavelength in the ϵ_r region.

For $n>0$, both $TE_{1,2n}$ and $TM_{1,2n}$ modes are present. For each n, the corresponding TE and TM amplitudes must be chosen so that E_x exactly cancels at $z=0$. In this way the total E_x field will be zero at $z=0$, as is required by the boundary conditions in that transverse plane. From (26) we obtain for the nth mode

$$Y_{in} = \frac{Y_{iTMn} + Y_{iTEn}D_n}{1 + D_n} \qquad (35)$$

where

$$D_n = \frac{(E_y)_{TE1,2n}}{(E_y)_{TM1,2n}} \text{ when } (E_x)_{TE1,2n} + (E_x)_{TM1,2n} = 0. \qquad (36)$$

From the field-component formulas for the TE and TM modes [7], [9] we obtain

$$D_n = \left(\frac{b}{2an}\right)^2. \qquad (37)$$

The input admittances Y_{iTMn} and Y_{iTEn} for each n are[3]

$$Y_{iTMn} = Y_{o1TMn}\tanh\left(\gamma_{n1}d + \tanh^{-1}\frac{Y_{oTMn}}{Y_{o1TMn}}\right) \qquad (38)$$

$$Y_{iTEn} = Y_{o1TEn}\coth\left(\gamma_{n1}d + \coth^{-1}\frac{Y_{oTEn}}{Y_{o1TEn}}\right) \qquad (39)$$

where the characteristic admittances and propagation constants are

[3] In (38) and (39), the pair of functions tanh and $\tanh^{-1}$ is interchangeable with coth and $\coth^{-1}$. The selection here is appropriate since $Y_{oTMn}/Y_{o1TMn}<1$ and $Y_{oTEn}/Y_{o1TEn}\geq 1$.

$$Y_{o1TMn} = \frac{jak_1}{2b\eta_1\gamma_{n1}}, \quad Y_{oTMn} = \frac{jak}{2b\eta\gamma_n} \qquad (40)$$

$$Y_{o1TEn} = \frac{a\gamma_{n1}}{j2b\eta_1 k_1}, \quad Y_{oTEn} = \frac{a\gamma_n}{j2b\eta k} \qquad (41)$$

$$\gamma_{n1} = \sqrt{\left(\frac{2\pi n}{b}\right)^2 + \left(\frac{\pi}{a}\right)^2 - \left(\frac{2\pi}{\lambda}\right)^2\epsilon_r} = \frac{2\pi n}{b}\sqrt{1 - \left(\frac{b}{n\lambda_{g1}}\right)^2} \qquad (42)$$

$$\gamma_n = \frac{2\pi n}{b}\sqrt{1 + \left(\frac{b\gamma}{2\pi n}\right)^2}. \qquad (43)$$

Then (35) and (37) through (41) yield

$$\eta Y_{in} = j\frac{ak}{2b\gamma_{n1}}\cdot\frac{\epsilon_r\tanh r_n - \left(\frac{b\gamma_{n1}}{2ank}\right)^2\coth q_n}{1 + \left(\frac{b}{2an}\right)^2} \qquad (44)$$

$$r_n = \gamma_{n1}d + \tanh^{-1}\left(\frac{\gamma_{n1}}{\epsilon_r\gamma_n}\right) \qquad (45)$$

$$q_n = \gamma_{n1}d + \coth^{-1}\left(\frac{\gamma_n}{\gamma_{n1}}\right). \qquad (46)$$

Now let $p=\lambda/2a$ and

$$u = \frac{\lambda}{\lambda_{g1}} = \frac{\gamma_1}{jk} = \sqrt{\epsilon_r - p^2} \qquad (47)$$

$$v = \frac{\gamma}{k} = \sqrt{p^2 - 1} \qquad (48)$$

$$F_{n1} = \frac{b\gamma_{n1}}{2\pi n} = \sqrt{1 - \left(\frac{b}{n\lambda_{g1}}\right)^2} = \sqrt{1 - \left(\frac{bu}{2anp}\right)^2} \qquad (49)$$

$$F_n = \frac{b\gamma_n}{2\pi n} = \sqrt{1 + \left(\frac{bv}{2anp}\right)^2}. \qquad (50)$$

Equations (32) through (34) and (44) through (50) yield

$$\eta B_d = \frac{au}{2b}\tan\left(\frac{\pi ud}{ap} - \tan^{-1}\frac{v}{u}\right) + \frac{1}{2p} \sum_{n=1,2,\cdots}\left[\frac{\epsilon_r\tanh r_n - p^2F_{n1}^2\coth q_n}{\left[1 + \left(\frac{b}{2an}\right)^2\right]F_{n1}}\right]\cdot\frac{\sin^2 \pi n\delta}{n(\pi n\delta)^2}. \qquad (51)$$

The first term is the TE_{10}-mode susceptance; the second term is the discontinuity susceptance representing the effect of higher modes. The rate of convergence of the series is very slow but will be improved by the following procedure. If we let $d\to\infty$ and $f\to 0$, then $\lambda'=2a\to\infty$ and each term S_n of the series becomes

$$S_n' = \frac{u^2\sin^2 \pi n\delta}{n(\pi n\delta)^2}. \qquad (52)$$

The original series may be replaced by

$$\sum_{n=1,2,\cdots} S_n = \sum_{n=1,2,\cdots} (S_n - S_n') + \sum_{n=1,2,\cdots} S_n'. \quad (53)$$

The first series on the right of the equal sign converges rapidly compared to the series in (51), while the second may be summed in closed form by the following identity [11], [15]:

$$\lim_{\delta\to 0} \sum_{n=1,2,3,\cdots} \frac{\sin^2 \pi n\delta}{n(\pi n\delta)^2} = \ln\frac{1}{2\pi\delta} + \frac{3}{2} = \ln\frac{1}{\delta} - 0.3379. \quad (54)$$

The accuracy of this is excellent for $\delta \leq 0.15$. At $\delta = 0.15$ the error is only 0.4 percent. Therefore, the susceptance of both sides of a capacitive iris is approximately

$$\eta B_C \approx \frac{u^2}{p}\left(\ln\frac{1}{\delta} - 0.3379\right). \quad (55)$$

However, the exact formula in the limit $\lambda_{g1}\to\infty$ is [14]

$$B_C = \frac{4bY_{01}}{\lambda_{g1}} \ln\csc\frac{\pi\delta}{2} = \frac{u^2}{p}\ln\csc\frac{\pi\delta}{2}, \quad \delta \leq 1$$
$$\approx \frac{u^2}{p}\ln\frac{2}{\pi\delta} = \frac{u^2}{p}\left(\ln\frac{1}{\delta} - 0.452\right), \quad \delta \ll 1. \quad (56)$$

The small difference between the constants in (55) and (56) is the result of assuming in the analysis that E_y and H_x are constant across the slot instead of being the correct functions of y. At this point (53) may be written

$$\sum_{n=1,2,\cdots} S_n = \sum_{n=1,2,\cdots} (S_n - S_n') + u^2 \ln\frac{2}{\pi\delta} \quad (57)$$

where (56) is used instead of (55) for better accuracy. Equation (51) for the susceptance looking into the dielectric section is now

$$\eta B_d = \frac{au}{2b}\tan\left(\frac{\pi ud}{ap} - \tan^{-1}\frac{v}{u}\right) + \frac{u^2}{2p}\ln\frac{2}{\pi\delta} + \frac{1}{2p} \sum_{n=1,2,\cdots}\left[\frac{\epsilon_r \tanh r_n - p^2F_{n1}^2 \coth q_n}{\left[1+\left(\frac{b}{2an}\right)^2\right]F_{n1}} - u^2\right]\cdot\frac{\sin^2 \pi n\delta}{n(\pi n\delta)^2}. \quad (58)$$

The susceptance B_a looking toward the left from the iris in Fig. 12 may be obtained from (58) by letting $\epsilon_r = 1$ or $d = 0$. With either substitution, the following equation results:

$$\eta B_a = -\frac{av}{2b} - \frac{v^2}{2p}\ln\frac{2}{\pi\delta} + \frac{1}{2p}\sum_{n=1,2,\cdots} v^2\left(1 - \frac{1}{F_n}\right)\frac{\sin^2 \pi n\delta}{n(\pi n\delta)^2}. \quad (59)$$

When (58) and (59) are added, the total susceptance B_t given by (10) is obtained for electric walls at $y = \pm b/2$.

The susceptance B_t for the case of magnetic walls at $y = \pm b/2$ may be obtained easily from the electric-wall solution. Let the cross section be as shown in Fig. 8(c). Modes excited in the waveguide are TE_{11}, TE_{13}, TE_{15}, $\cdots$ and TM_{11}, TM_{13}, TM_{15}, $\cdots$; that is, $TE_{1,2n}$ and $TM_{1,2n}$ where $n = \frac{1}{2}, \frac{3}{2}, \frac{5}{2}$, etc. The TE_{10} mode cannot exist within this boundary. Careful examination of each step of the above analysis for electric walls shows that (51) applies when the first term representing the TE_{10} contribution is dropped, and when the summation is performed for $n = \frac{1}{2}, \frac{3}{2}, \frac{5}{2}, \cdots$ rather than $n = 1, 2, 3, \cdots$. Evaluation of (54) with $n = \frac{1}{2}, \frac{3}{2}, \frac{5}{2}, \cdots$, yields [15]

$$\lim_{\delta\to 0}\sum_{n=\frac{1}{2},\frac{3}{2},\cdots}\frac{\sin^2 \pi n\delta}{n(\pi n\delta)^2} = \ln\frac{4}{\delta} - 0.3379. \quad (60)$$

Therefore, in (58) for ηB_d and (59) for ηB_a drop the first term, replace $\ln 2/\pi\delta$ by $\ln 8/\pi\delta$, and change the summation index from $n = 1, 2, 3, \cdots$ to $n = \frac{1}{2}, \frac{3}{2}, \frac{5}{2}, \cdots$. The resulting expression giving ηB_t for magnetic walls at $y = \pm b/2$ is (11).

Appendix II

Derivation of Formulas for Z_0 and v/v_g

Define the slot-line characteristic impedance Z_0 by

$$Z_0 = \frac{V_+^2}{2P_+} \quad (61)$$

where P_+ is the average power flow of a slot wave traveling in the $+x$ direction (Fig. 8) and V_+ is the peak amplitude of the voltage across the slot. Now assume a resonant length $\lambda'/2$ of slot line having waves of equal power P^+ and P^- traveling in the $+x$ and $-x$ directions. The total stored energy in this length is $W_t = (P^+ + P^-)(\lambda'/2v_g) = P^+\lambda'/v_g = (2\pi P^+/\omega')(v/v_g)$, where v_g is group velocity, or velocity of energy transport along the slot line [16]. Let V_0 be the maximum voltage at the center of the resonant length of slot. Since the waves in the $+x$ and $-x$ directions have equal voltages $V_+ = V_-$, the maximum voltage is $V_0 = 2V_+$, and $Z_0 = (V_0/2)^2/2P^+ = \pi V_0^2 v/4\omega W_t v_g$.

The following general relation holds at a port of a cavity at resonance [16]: $W_t = (V^2/4)(dB/d\omega)$. In the case of the $\lambda'/2$ resonant slot, we shall assume the port to be at the iris plane $z = 0$ in Fig. 8. We shall set B equal to the total waveguide susceptance B_t at that plane and V equal to the slot voltage V_0 at $x = \lambda'/4$. These choices of B and V are consistent with the waveguide impedance and slot impedance definitions, both of which are on a voltage power basis. Therefore,

$$Z_0 = \frac{\pi}{\omega(dB_t/d\omega)}\cdot\frac{v}{v_g} \quad (62)$$

where $dB_t/d\omega$ is evaluated at the resonant frequency; that is, at $B_t = 0$. Let $\omega = 2\pi c/\lambda$ and $p = \lambda/2a$. Then $\omega = \pi c/ap$ and $d\omega = -(\pi c/ap^2)dp = -\omega(dp/p)$. Substitution of these relations in (62) yields (21).

The ratio v/v_g will now be evaluated. Phase and group velocity are given by $v = \omega/\beta_x = f\lambda'$ and $v_g = d\omega/d\beta_x$ [16], [17] where β_x is the slot-wave phase constant. Since $\beta_x = 2\pi/\lambda'$, we obtain $v_g = -\lambda'^2/(d\lambda'/df)$. Differentiate λ'/λ as follows:

$$\frac{d(\lambda'/\lambda)}{df} = \frac{1}{\lambda}\frac{d\lambda'}{df} - \frac{\lambda'}{\lambda^2}\frac{d\lambda}{df}.$$

Solve this for $d\lambda'/df$ and substitute $d\lambda/df = d(c/f)/df = -\lambda/f$. The resulting relations yield the first part of (20). The second part is obtained in a similar manner.

Acknowledgment

The author wishes to thank E. G. Cristal of Stanford Research Institute for preparing the computer program of the second-order solution, and J. P. Agrios, C. Heinzman and E. A. Mariani of U. S. Army Electronics Command for their experimental studies of slot line. The above people, and also L. A. Robinson and L. Young of Stanford Research Institute, participated in numerous discussions with the author.

References

[1] H. Jasik, Ed., *Antenna Engineering Handbook*. New York: McGraw-Hill, 1961, chs. 8, 9.
[2] A. F. Harvey, *Microwave Engineering*. New York: Academic Press, 1963, pp. 633–638.
[3] W. H. Watson, *Waveguide Transmission and Antenna Systems*. New York: Oxford University Press, 1947.
[4] E. Strumwasser, J. A. Short, R. J. Stegen, and J. R. Miller, "Slot study in rectangular TEM transmission line," Hughes Aircraft Company, Tech. Memo 265, Air Force Contract AF 19(122)-454, January 1952.
[5] M. C. Bailey, "Design of dielectric-covered resonant slots in a rectangular waveguide," *IEEE Trans. Antennas and Propagation*, vol. AP-15, pp. 594–598, September 1967.
[6] J. Galejs, "Excitation of slots in a conducting screen above a lossy dielectric half space," *IRE Trans. Antennas and Propagation*, vol. AP-10, pp. 436–443, July 1962.
[7] S. Ramo and J. R. Whinnery, *Fields and Waves in Modern Radio*, 2nd ed. New York: Wiley, 1953, pp. 357–358.
[8] E. Jahnke and F. Emde, *Tables of Functions with Formulae and Curves*. New York: Dover, 1943, pp. 236–243.
[9] S. B. Cohn, "A theoretical and experimental study of a waveguide filter structure," Cruft Lab., Harvard University, Cambridge, Mass., Tech. Rept. 39, Contract N5 ORI-76, Task Order 1, April 25, 1948.
[10] ——, "Analysis of a wide-band waveguide filter," *Proc. IRE*, vol. 37, pp. 651–656, June 1949.
[11] W. C. Hahn, "A new method for the calculation of cavity resonators," *J. Appl. Phys.*, vol. 12, p. 62, 1941.
[12] J. R. Whinnery and H. W. Jamieson, "Equivalent circuits for discontinuities in transmission lines," *Proc. IRE*, vol. 32, pp. 98–115, February 1944.
[13] S. A. Schelkunoff, *Electromagnetic Waves*. Princeton, N. J.: Van Nostrand, 1943.
[14] N. Marcuvitz, *Waveguide Handbook*, M.I.T. Rad. Lab. Ser., vol. 10. New York: McGraw-Hill, 1951, pp. 218–219.
[15] R. E. Collin, *Field Theory of Guided Waves*. New York: McGraw-Hill, 1960. Use $\Sigma_{1,2,3,\cdots} e^{jnx}/n^3$ on p. 579 and $\Sigma_{1,3,5,\cdots} e^{jnx}/n^3$ on p. 580.
[16] C. G. Montgomery, R. H. Dicke, and E. M. Purcell, *Principles of Microwave Circuits*, M.I.T. Rad. Lab. Ser., vol. 8. New York: McGraw-Hill, 1948. See p. 230, eq. (40) for relation between W_t, V, and $dB_t/d\omega$; see p. 53, eqs. (103) and (104) for v and v_g.
[17] L. Brillouin, *Wave Propagation in Periodic Structures*. New York: Dover, 1953, pp. 72–76.
[18] F. Oberhettinger and W. Magnus, *Anwendung der Elliptischen Funktionen in Physik und Technik*. Berlin: Springer, 1949, pp. 63, 114–116.
[19] E. Mariani, C. Heinzman, J. Agrios, and S. B. Cohn, "Measurement of slot-line characteristics," presented at the 1969 IEEE G-MTT Internatl. Symp., Dallas, Tex., May 5–7, 1969, to be published in *IEEE Trans. Microwave Theory and Techniques.*
[20] G. H. Robinson and J. L. Allen, "Applications of slot line to miniature ferrite devices," presented at the 1969 IEEE G-MTT Internatl. Symp., Dallas, Tex., May 5–7, 1969.

Coplanar Waveguide: A Surface Strip Transmission Line Suitable for Nonreciprocal Gyromagnetic Device Applications

CHENG P. WEN, MEMBER, IEEE

***Abstract*—A coplanar waveguide consists of a strip of thin metallic film on the surface of a dielectric slab with two ground electrodes running adjacent and parallel to the strip. This novel transmission line readily lends itself to nonreciprocal magnetic device applications because of the built-in circularly polarized magnetic vector at the air–dielectric boundary between the conductors. Practical applications of the coplanar waveguide have been experimentally demonstrated by measurements on resonant isolators and differential phase shifters fabricated on low-loss dielectric substrates with high dielectric constants. Calculations have been made for the characteristic impedance, phase velocity, and upper bound of attenuation of a transmission line whose electrodes are all on one side of a dielectric substrate. These calculations are in good agreement with preliminary experimental results. The coplanar configuration of the transmission system not only permits easy shunt connection of external elements in hybrid integrated circuits, but also adapts well to the fabrication of monolithic integrated systems. Low-loss dielectric substrates with high dielectric constants may be employed to reduce the longitudinal dimension of the integrated circuits because the characteristic impedance of the coplanar waveguide is relatively independent of the substrate thickness; this may be of vital importance for low-frequency integrated microwave systems.**

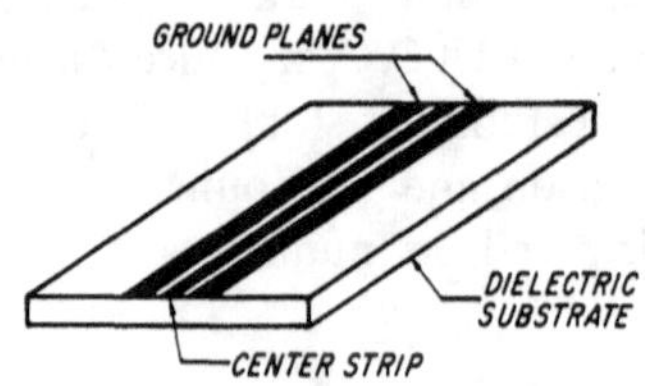

Fig. 1. Coplanar waveguide (CPW), a surface strip transmission line.

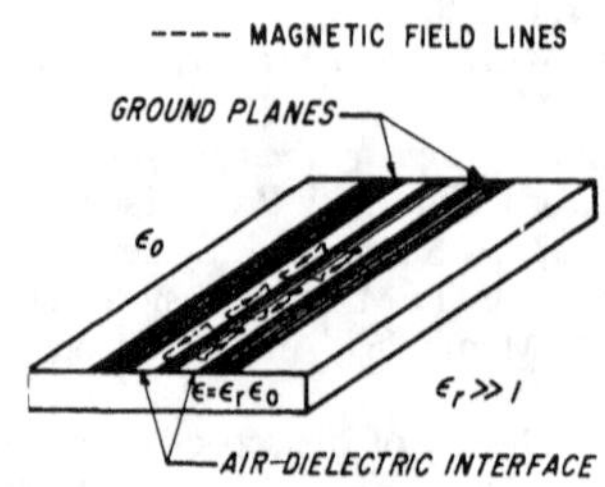

Fig. 2. RF magnetic field configuration in a CPW.

MICROWAVE integrated circuits offer system engineers prospects of small, batch processed modules for radar and communication systems. In the past, radar and communication systems included a variety of nonreciprocal magnetic devices. These devices require circularly polarized RF magnetic fields for their operation, and present microstrip and strip lines do not provide such fields. In addition, the ground plane of these lines, which is located on the opposite side of a dielectric substrate, is not easily accessible for shunt connections necessary for many active microwave devices. Direct dependence of the characteristic impedance on the thickness of the substrate makes it difficult to use a low-loss, high-dielectric-constant material, such as a temperature-compensated ceramic. This is a definite drawback for low-frequency operation where size consideration dominates. All these disadvantages may either be overcome or alleviated by a novel integrated-circuit transmission-line configuration in which all conducting elements, including the ground planes, are on the same side of a dielectric substrate. This is called the coplanar waveguide, a surface-strip transmission line.

A coplanar waveguide (CPW) consists of a strip of thin metallic film deposited on the surface of a dielectric slab with two ground electrodes running adjacent and parallel to the strip on the same surface, as shown in Fig. 1. There is no low-frequency cutoff because of the quasi-TEM mode of propagation. However, the RF electric field between the center conducting strip and the ground electrodes tangential to the air–dielectric boundary produces a discontinuity in displacement current density at the interface, giving rise to an axial, as well as transverse, component of RF magnetic field. These components provide the elliptical polarization needed for nonreciprocal gyromagnetic microwave device applications [1]. If the relative dielectric constant ϵ_r of the substrate is very large compared to unity, the magnetic field at the interface is nearly circularly polarized with the plane of polarization perpendicular to the surface of the substrate, as shown in Fig. 2. Such a transmission line readily lends itself to integrated-circuit fabrication techniques and nonreciprocal gyromagnetic device applications because of the built-in circularly polarized magnetic vector which is easily accessible at the surface of the substrate. The coplanar configuration of all conducting elements permits easy connection of external shunt elements such as active devices as well as the fabrication of series and shunt capacitances. It is also ideal for connecting various elements in monolithic microwave integrated circuits built on semiconducting substrates or ferromagnetic semiconductors. All ground planes may be connected together through a metallic capsule as shown in

Manuscript received March 28, 1969; revised July 9, 1969. This paper was presented at the International Microwave Symposium, Dallas, Tex., May 5–7, 1969.

The author is with RCA Laboratories, David Sarnoff Research Center, Princeton, N. J. 08540.

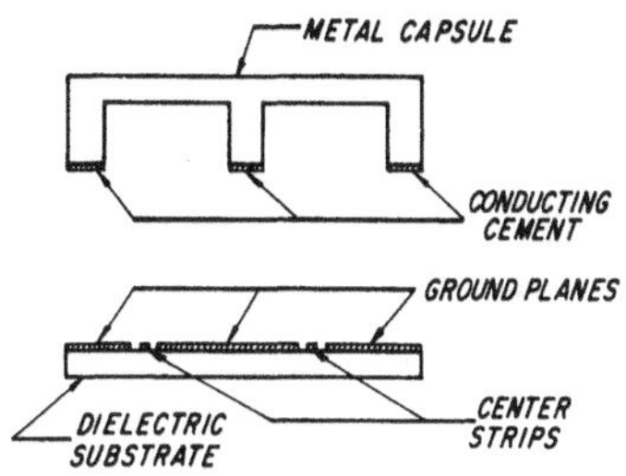

Fig. 3. Metallic capsule for CPW ground connections.

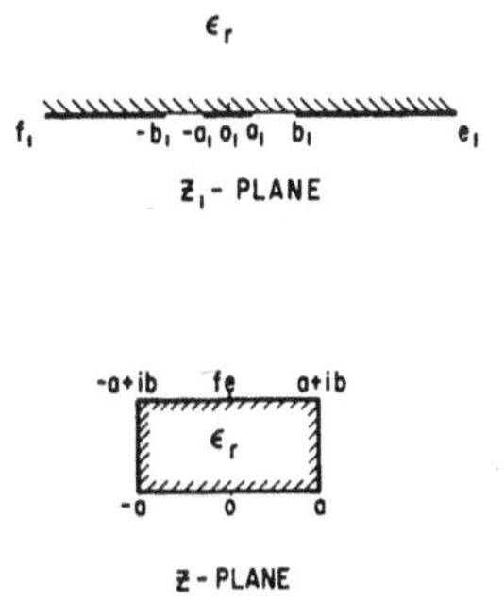

Fig. 4. Conformal mapping transformation of the upper half-plane of a CPW into the interior of a rectangle.

Fig. 3, serving both as a common ground and a protective cover. Because of the high dielectric constant of the substrate, most of the RF energy is stored in the dielectric and the loading effect of the grounded cover is negligible if it is more than two slot widths away from the surface.

Characteristic impedances Z_0 of CPWs fabricated on dielectric half-planes with relative dielectric constants ϵ_r have been calculated as a function of the ratio a_1/b_1, where $2a_1$ is the width of the center strip and $2b_1$ is the distance between two ground electrodes. A zeroth-order quasi-static approximation is employed. The dielectric half-plane Z_1 in Fig. 4 may be transformed to the interior of a rectangle in the Z-plane by conformal mapping [2]:

$$\frac{dZ}{dZ_1} = \frac{A}{(Z_1^2 - a_1^2)^{1/2}(Z_1^2 - b_1^2)^{1/2}} \tag{1}$$

where A is a constant. The ratio a/b of the rectangle in the Z plane may be evaluated by multiplying both sides of (1) by dZ_1 and carrying out the integration

$$a + jb = \int_0^{b_1} \frac{A\,dZ_1}{(Z_1^2 - a_1^2)^{1/2}(Z_1^2 - b_1^2)^{1/2}}\,\cdot \tag{2}$$

As a result

$$\frac{a}{b} = \frac{K(k)}{K'(k)} \tag{3}$$

where

$k = a_1/b_1$,
$K(k)$ = complete elliptical integral of the first kind [3],
$K'(k) = K(k')$,
$k' = (1-k^2)^{1/2}$.

If the relative dielectric constant of the material filling the rectangle in the Z plane of Fig. 4 is ϵ_r, a uniform electric field E is set up in the capacitor with the top and bottom plates charged up to opposite polarities and the capacitance per unit length of the line, including the empty space half-plane, is

$$C = (\epsilon_r + 1)\epsilon_0 \frac{2a}{b}\,\cdot \tag{4}$$

A zeroth-order quasi-static approximation has been employed to estimate the phase velocities and the characteristic impedances of CPWs. The approximation simply treats the CPW as a transmission line totally immersed in a dielectric with effective dielectric constant $(\epsilon_r+1)/2$. The resulting phase velocity is

$$v_{ph} = \left(\frac{2}{\epsilon_r + 1}\right)^{1/2} c \tag{5}$$

where c is the velocity of light in free space and the characteristic impedance is

$$Z_0 = \frac{1}{C v_{ph}}\,\cdot \tag{6}$$

The ratio v_{ph}/c is shown in Fig. 5 as a function of ϵ_r. In Fig. 6 the characteristic impedance Z_0 is shown as a function of a_1/b_1 with the relative dielectric constant ϵ_r as a parameter ranging from unity to 250. Experimental confirmation has been obtained at three points shown on the same figure with transmission lines fabricated on substrates of relative dielectric constant $\epsilon_r = 9.5$, 16, and 130, respectively. The calculated characteristic impedance of a parallel-strip coplanar line, the dual of the CPW, is shown in Fig. 7 as a function of a_1/b_1 with the relative dielectric constant ϵ_r of the substrate as a parameter [4]. The configuration of this line can be found on the same figure.

The thickness of the dielectric substrate becomes less critical with higher relative dielectric constants. In the CPW configuration the characteristic impedance increases by less than 10 percent when the thickness of the substrate is reduced from infinity to $b_1 - a_1$, the width of the slots, for infinitely large ϵ_r. In practice, the thickness t of the substrate should be one or two times the width of the slots. It is obvious that the finite thickness of the substrate will influence the dispersion characteristics of the transmission line but no estimate has been made on the extent of the effect. The experimentally measured dispersion characteristic of a coplanar waveguide fabricated on a single-crystal rutile substrate is shown on the frequency versus βL plot of Fig. 8. It is not known what portion of the dispersion is caused by the crystal anisotropy of the substrate and what portion is attributed to the inherent characteristics of the CPW mode of propagation on a dielectric half-space.

At microwave frequencies up to X band the attenuation of a CPW is due mainly to the copper loss of the conductors if the loss tangent of the dielectric is 0.001 or smaller. Measurements made on a 16.6-ohm CPW fabricated on a rutile substrate ($\epsilon_r \cong 130$, $k = \frac{1}{3}$, center conductor width is 0.025

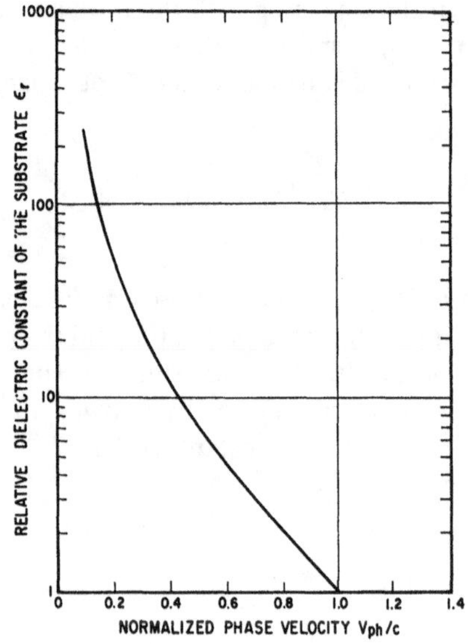

Fig. 5. Relationship between the relative dielectric constant ϵ_r of the substrate and the normalized phase velocity v_{ph}/c in a CPW.

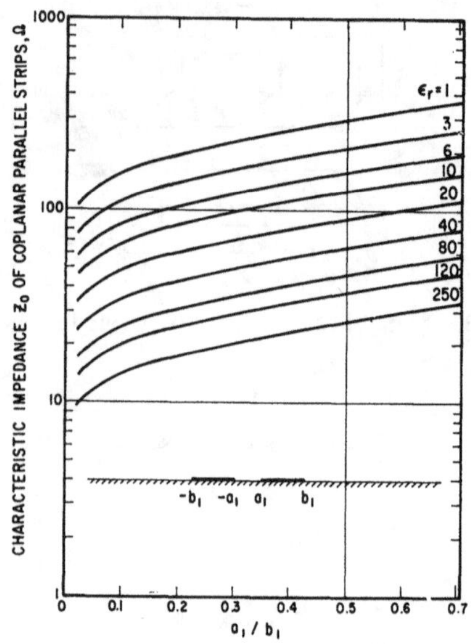

Fig. 7. Characteristic impedance Z_0 of coplanar parallel strips as a function of the ratio a_1/b_1 with the relative dielectric constant ϵ_r as a parameter.

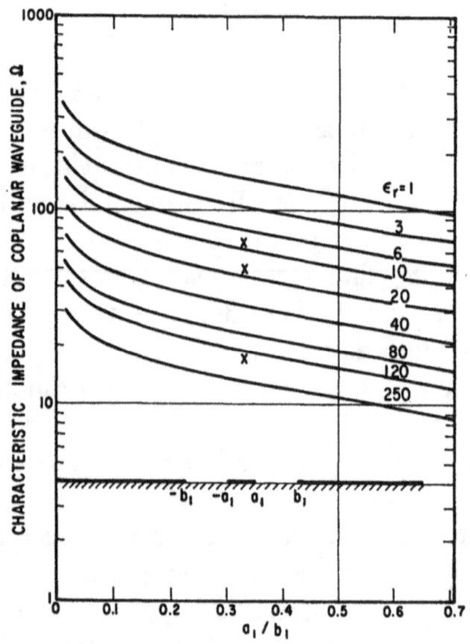

Fig. 6. Characteristic impedance Z_0 of CPW as a function of the ratio a_1/b_1 with the relative dielectric constant ϵ_r as a parameter.

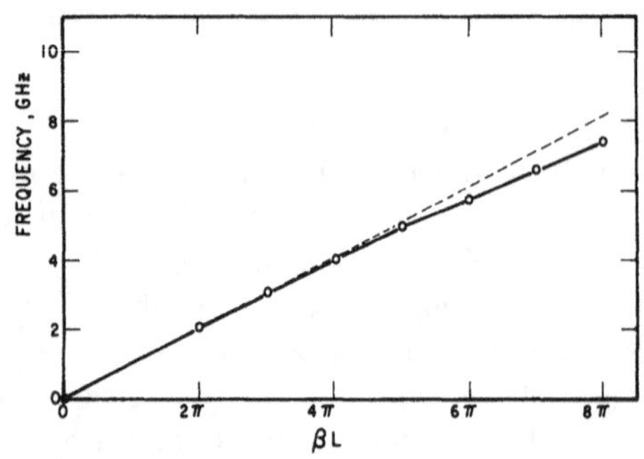

Fig. 8. Dispersion characteristics of a CPW on a TiO_2 substrate.

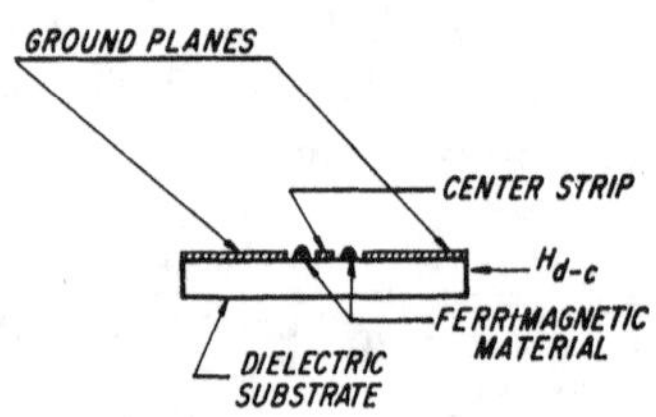

Fig. 9. Resonant isolator or differential phase shifter in CPW configurations.

inch, thickness of gold film is 2 microns, thickness of the dielectric substrate is 0.025 inch), yielded a Q of 173 at 4 GHz, corresponding to an attenuation of only 0.158 dB per wavelength.

Other design considerations include radiation problems when the total distance from one ground plane to another approaches $\lambda/2$. This will pose a limit on the dielectric constant of the substrate employed at a given frequency and the transverse dimension of the transmission line. CPWs can be easily adapted to coaxial systems using OSM connectors. Broad-band (2–12 GHz) matching has been achieved with a 50-ohm CPW deposited on a magnesium oxide substrate with $\epsilon_r = 16$.

Nonreciprocal gyromagnetic devices such as resonant isolators and differential phase shifters have been fabricated by attaching ferrimagnetic slabs at the air–dielectric interface between the conductors, as shown in Fig. 9. A differential phase shifter fabricated on an all-magnetic garnet substrate has also been built. A transverse dc magnetic field parallel to the surface of the substrate is required to provide appropriate bias conditions. As shown in Fig. 10, a CPW ferrimagnetic-resonance isolator built on the surface perpendicular to the c axis of a single rutile crystal provides 37-dB isolation at the center frequency of 6 GHz while the forward attenuation is below 2 dB. Overall length of the line is 0.8 inch including a quarter-wave transformer at each end. The center conductor width is 0.030 inch, $k = 0.33$, and the substrate is 0.025 inch thick. Strips of Trans-Tech G-1000 YIG

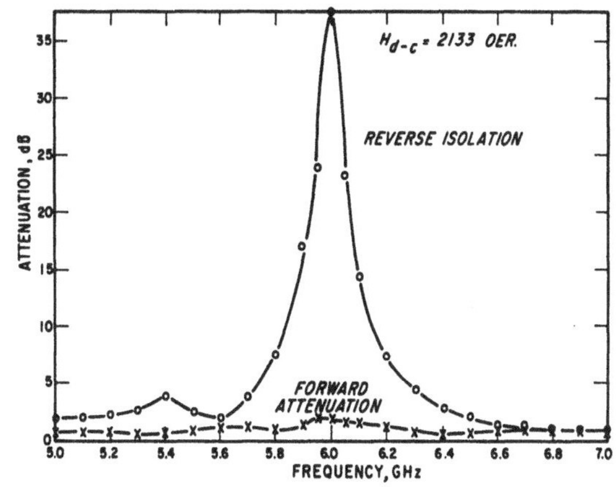

Fig. 10. Performance of a CPW ferrimagnetic-resonant isolator on a single-crystal rutile substrate.

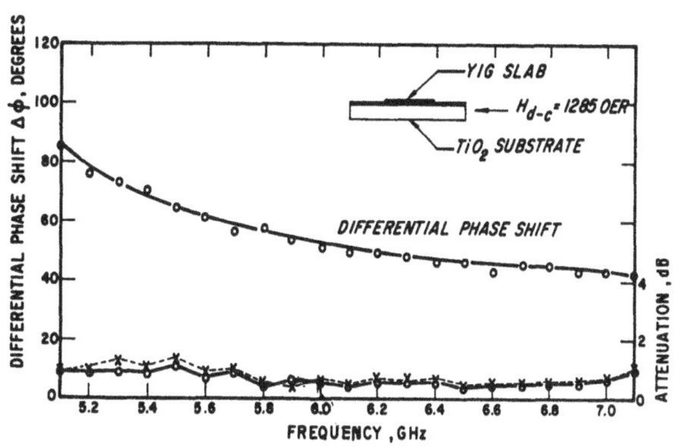

Fig. 11. Performance of a CPW ferrimagnetic differential phase shifter on a single-crystal rutile substrate.

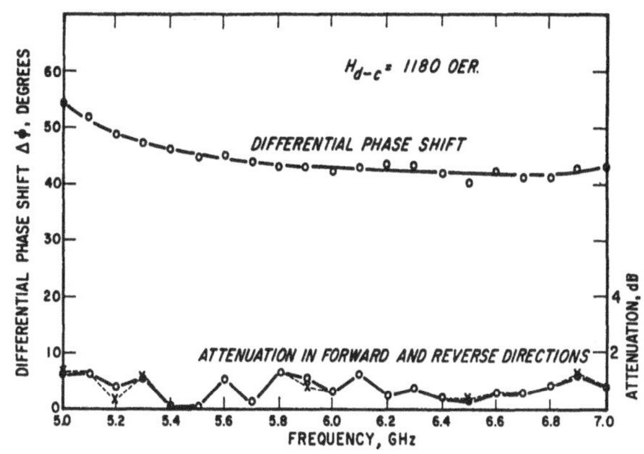

Fig. 12. Performance of a CPW ferrimagnetic differential phase shifter on a YIG substrate.

(0.010 inch×0.005 inch×0.60 inch) are attached by low-loss cement to the rutile surface with the center line of the YIG 0.010 inch from the ground planes. The performance of a phase shifter whose configurations are similar to the previously discussed isolator, are shown in Fig. 11. No attempt has been made to equalize the amount of differential phase shift across the band of frequencies. Average differential phase shift is over 45° while the insertion loss in either direction is less than 1 dB between 5.6 GHz and 7.1 GHz. Higher loss is observed at the lower frequencies which are near the ferrimagnetic resonance. Differential phase shift and loss data for a CPW fabricated on a YIG substrate is shown in Fig. 12. A tapered ceramic piece (0.800 inch long, 10° tapering on both ends, $\epsilon_r = 83$) is placed on top to provide the circularly polarized magnetic vector in the ferrite. The amount of differential phase shift varies little with frequency beyond 5.5 GHz. Insertion loss of 1.3 dB or less has been measured from 5.0 GHz to 7.0 GHz without special effort to match the device to the 50-ohm test system. These preliminary results demonstrate the gyromagnetic nonreciprocal device capabilities of coplanar waveguides.

In summary, the practicality of a novel microwave integrated circuit transmission line suitable for nonreciprocal gyromagnetic device applications has been demonstrated. Some preliminary calculations for characteristic impedance, phase velocity, and attenuation characteristics are presented and compared with experimental results. The coplanar configuration of the transmission system not only permits easy shunt connections of external elements, it also adapts well to the fabrication of monolithic integrated circuits.

Acknowledgment

The author wishes to thank H. Davis for his able assistance in preparation and during experimentation and R. Goodrich for the deposition of thin film circuits. Constant encouragement and valuable suggestions from B. Hershenov and L. S. Napoli are gratefully acknowledged.

References

[1] B. J. Duncan, L. Swern, K. Tomiyasu, and J. Hannwacker, "Design considerations for broad-band ferrite coaxial line isolators," *Proc. IRE*, vol. 45, pp. 483–490, April 1957.

[2] W. R. Smythe, *Static and Dynamic Electricity*. New York: McGraw-Hill, 1950.

[3] E. Jahnke and F. Emde, *Tables of Functions with Formulae and Curves*, 4th ed. New York: Dover, 1945.

[4] R. F. Frazita, "Transmission line properties of coplanar parallel strips on a dielectric sheet," M.S. thesis, Polytechnic Institute of Brooklyn, Brooklyn, N. Y., 1965.

www.ingramcontent.com/pod-product-compliance
Lightning Source LLC
LaVergne TN
LVHW080035210926
848481LV00011B/3200